AF823473

EAI/Springer Innovations in Communication and Computing

Series editor
Imrich Chlamtac, European Alliance for Innovation, Ghent, Belgium

Editor's Note

The impact of information technologies is creating a new world yet not fully understood. The extent and speed of economic, life style and social changes already perceived in everyday life is hard to estimate without understanding the technological driving forces behind it. This series presents contributed volumes featuring the latest research and development in the various information engineering technologies that play a key role in this process.

The range of topics, focusing primarily on communications and computing engineering include, but are not limited to, wireless networks; mobile communication; design and learning; gaming; interaction; e-health and pervasive healthcare; energy management; smart grids; internet of things; cognitive radio networks; computation; cloud computing; ubiquitous connectivity, and in mode general smart living, smart cities, Internet of Things and more. The series publishes a combination of expanded papers selected from hosted and sponsored European Alliance for Innovation (EAI) conferences that present cutting edge, global research as well as provide new perspectives on traditional related engineering fields. This content, complemented with open calls for contribution of book titles and individual chapters, together maintain Springer's and EAI's high standards of academic excellence. The audience for the books consists of researchers, industry professionals, advanced level students as well as practitioners in related fields of activity include information and communication specialists, security experts, economists, urban planners, doctors, and in general representatives in all those walks of life affected ad contributing to the information revolution.

About EAI

EAI is a grassroots member organization initiated through cooperation between businesses, public, private and government organizations to address the global challenges of Europe's future competitiveness and link the European Research community with its counterparts around the globe. EAI reaches out to hundreds of thousands of individual subscribers on all continents and collaborates with an institutional member base including Fortune 500 companies, government organizations, and educational institutions, provide a free research and innovation platform.

Through its open free membership model EAI promotes a new research and innovation culture based on collaboration, connectivity and recognition of excellence by community.

More information about this series at http://www.springer.com/series/15427

Dagmar Cagáňová • Natália Horňáková
Editors

Mobility Internet of Things 2018

Mobility IoT

Springer

EAI
RESEARCH MEETS INNOVATION

Editors
Dagmar Cagáňová
Institute of Industrial Engineering
and Management
Faculty of Materials Science
and Technology in Trnava
Slovak University of Technology
in Bratislava
Slovakia

Natália Horňáková
Institute of Industrial Engineering
and Management
Faculty of Materials Science
and Technology in Trnava
Slovak University of Technology
in Bratislava
Slovakia

ISSN 2522-8595 ISSN 2522-8609 (electronic)
EAI/Springer Innovations in Communication and Computing
ISBN 978-3-030-30910-7 ISBN 978-3-030-30911-4 (eBook)
https://doi.org/10.1007/978-3-030-30911-4

© Springer Nature Switzerland AG 2020
This work is subject to copyright. All rights are reserved by the Publisher, whether the whole or part of the material is concerned, specifically the rights of translation, reprinting, reuse of illustrations, recitation, broadcasting, reproduction on microfilms or in any other physical way, and transmission or information storage and retrieval, electronic adaptation, computer software, or by similar or dissimilar methodology now known or hereafter developed.
The use of general descriptive names, registered names, trademarks, service marks, etc. in this publication does not imply, even in the absence of a specific statement, that such names are exempt from the relevant protective laws and regulations and therefore free for general use.
The publisher, the authors, and the editors are safe to assume that the advice and information in this book are believed to be true and accurate at the date of publication. Neither the publisher nor the authors or the editors give a warranty, expressed or implied, with respect to the material contained herein or for any errors or omissions that may have been made. The publisher remains neutral with regard to jurisdictional claims in published maps and institutional affiliations.

This Springer imprint is published by the registered company Springer Nature Switzerland AG.
The registered company address is: Gewerbestrasse 11, 6330 Cham, Switzerland

Preface

This scientific monograph consists of 38 cutting-edge and insightful research chapters dealing with the topic of innovation in various points of view.

The authors are individuals as researchers, developers, and practitioners from all around the world, all of whom share a common interest in the area of innovation with focus on mobility in IoT.

The editors would like to express their sincere thanks to the authors who have contributed with their knowledge and latest research to the creation of the scientific monograph.

Trnava, Slovakia

Dagmar Cagáňová
Natália Horňáková

Contents

FMWare: IoT-Based Fleet Management System

Patrick Flinner, Jungsoo Lim, Russell Abbott, Sammy Urbina, and Mohammed Al Rawi

1 Introduction

In this paper, we discuss an Internet of Things (IoT)-based fleet management system, named FMWare. Because FMWare is built with open-source software and hardware, its cost is quite modest. Current industry solutions come with monthly subscription fees or pay-as-you-go charges; FMWare has neither. In addition, FMWare is highly customizable and extendable.

The Los Angeles County Department of Parks and Recreation oversees over 70,000 acres and manages 182 parks across all of Los Angeles County [2]. To cover this vast area, the County utilizes over 600 vehicles. Currently, these vehicles are managed by a paper-based system. Each time a County employee uses a County vehicle, the employee must fill in a paper form. This form consists of numerous fields such as purpose of trip, mileage of the car, gas level, etc.

This paper-based process is time consuming for both drivers and service personnel. The current system also makes monitoring the usages of vehicles and scheduling maintenance service extremely difficult. The inability to perform maintenance in a timely manner leads to safety issues for drivers, as well as misallocation of County resources. In addition, the paper-based system provides no way to detect unauthorized use of vehicles.

P. Flinner · J. Lim (✉) · R. Abbott
Department of Computer Science, California State University, Los Angeles, Los Angeles, CA, USA
e-mail: pflinner@calstatela.edu; jlim34@calstatela.edu; rabbott@calstatela.edu

S. Urbina · M. Al Rawi
Los Angeles County Department of Parks and Recreation, Alhambra, CA, USA
e-mail: SUrbina@parks.lacounty.gov; MAlRawi@parks.lacounty.gov

© Springer Nature Switzerland AG 2020
D. Cagáňová, N. Horňáková (eds.), *Mobility Internet of Things 2018*,
EAI/Springer Innovations in Communication and Computing,
https://doi.org/10.1007/978-3-030-30911-4_1

All current industry solutions for fleet management require hefty monthly subscription fees ranging from $50 to $150, as well as an initial device charge [4, 7, 13]. These devices are built to be company specific and cannot be used outside their own company's services. Were the County to opt to change to a different service provider, it would have to purchase a new set of devices. Also, the cost of tailoring the service to suit County needs is prohibitively expensive.

As an alternative, we developed FMWare, an IoT-based Fleet Management System. FMWare eliminates the current paper form process and allows County employees expedited use of County vehicles. Driver information is collected through the use of the County's employee badge system.

To use a vehicle, drivers must scan their badge on a Radio Frequency Identification (RFID) reader, which is attached to the FMWare Edge device installed in each vehicle. The badge reader collects and stores the employee ID.

While the vehicle is in motion, FMWare collects Global Positioning System (GPS) data and vehicle diagnostics data, including mileage, gas level, and diagnostics trouble codes (DTCs).

When a vehicle returns to base, the data stored on the FMWare device is uploaded to County servers through secure wireless gateways which are accessible in County parking lots. Data collected from the FMWare Edge Device is uploaded via Apache's open-source data flow control software, NiFi and MiNiFi.

FMWare monitors vehicle operation in near real-time, which facilitates identification of unauthorized vehicle use in a timely manner. FMWare can assist with effective scheduling of preventive maintenance and regular service for vehicles. Hence, staff safety is enhanced while reducing operating costs.

The FMWare Edge device is composed of a single-board computer, a Global Positioning System (GPS) module, an RFID reader, an on-board diagnostics (OBD) adapter, a Liquid-Crystal Display (LCD) screen, and our power supply unit (PSU). Our PSU is composed of a buck converter and two 2 A fuses. The device records the diagnostic information retrieved from the vehicle's OBD system through IEEE 802.15.4 WPAN for communication. The GPS module is used to collect route data. Upon the vehicle's return to a County parking lot, the FMWare Edge device transmits the stored data through secure wireless access points available within County parking lots. The data that is uploaded to County servers can be subsequently viewed on our web application by managers.

In summary, we show the feasibility of using IoT technology by implementing an affordable and easily customizable fleet management system, design and implement a fleet management system to remotely collect vehicle diagnostic data and vehicle operating data, and present an affordable IoT-based fleet management system for monitoring and managing a large number of vehicles.

2 Related Work

Penna et al. developed an IoT-based fleet monitoring system mainly to monitor fuel consumption and to optimize fuel usages [8]. They installed a stainless steel (SS) pin fuel sensor to the vehicle to collect the level of fuel. The traveled distance was computed by using GPS based odometer. The collected fuel level data was uploaded to the server through a cellular network. Since the authors were only relying on fuel level sensors and GPS based odometer, detecting issues with a vehicle is not feasible.

A fleet management system using vehicle tracker, named Rad100 was designed and implemented by Saghaei [10]. Rad100 is a self-contained portable unit which consists of a micro-controller, GPS module, motion detection sensor, and a cellular communication module. The main functionality of Rad100 is to find a vehicle's location. When a vehicle is moving, the GPS data from Rad100 is sent to the back-end server through a cellular network and the location of the vehicle can be monitored. Although Rad100 can assist to recover a stolen vehicle, it is not capable of detecting any issues with the vehicles.

A solution for fleet management using pre-fabricated modules was presented by Istrefi et al. [6]. The author used various pre-built factory modules including GPS modules, collision prevention assist modules, and cellular network communication modules. The author computed fuel consumptions based on GPS data instead of fuel flow meter due to safety and accuracy issues. They noticed that a small amount of fuel consumption could not be measured accurately by a fuel flow meter. The system supports tracking vehicle's location in real-time, estimating fuel consumption, and assisting collision prevention. However, as the modules are not integrated, inter-operating the modules is a challenging task.

Dhall et al. presented a solution how to effectively schedule preventive vehicle maintenance by connecting vehicles [3]. The author proposed to connect vehicles by implementing Message Queue Telemetry Transport (MQTT) messaging protocol. However, the author did not address the underlying network connectivity and actual vehicle data collection.

The proof of concept of an automated fleet management system based on cloud computing was proposed by Sodeyama et al. [11]. The author analyzed the operation and maintenance (O&M) cost associated with client/server approach and cloud-based approach, and then proposed cloud based fleet management. However, the author did not address the underlying vehicle data collection.

In addition, the following commercial products are available:

Fleet Genius by Prova Systems and Technologies, Inc. [4] provides packaged vehicle management service. Through the use of a proprietary plug-in device along with fleet management software, Fleet Genius provides a solution for small to large scale fleets wanting to keep track of and maintain their vehicles. Their software supports a range of functionalities, some of which include driver performance, vehicle trip tracking, vehicle diagnostics and maintenance. Data collection from their plug-in device to their cloud server is done via their proprietary wireless access points or available cellular network via Android mobile device. All of these are available at significant costs. The Annual plan is $159 per device with a free plug-in device. The cost for the pay-as-you-go plan is priced at $14.95 per month per device with $69.95 for the Device.

US Fleet Tracking [13] utilizes a proprietary plug-in OBD-II device that contains a GPS tracker inside to keep track of the vehicle. The system keeps track of the vehicles' location, diagnostics, and maintenance. However, the plug-in OBD-II device comes with a price tag of **$249.00 excluding tax**. In addition, in order to use the plug-in device, users are encouraged to pay a subscription fee to the services which start at **$29.95 per month per device**. With offers of new additional services or improvements to the network connectivity, the monthly subscription fee goes up to **$79.95 per month per device**.

MasTrack [7] by Mobile Asset Solutions utilizes a proprietary plug-in OBD-II device that contains a built-in GPS tracker. The device sends out data in predefined intervals that are dependent on the service plan a user has purchased from MasTrack. The plug-in device comes with a price tag of $78.00 excluding applicable tax, with services for data reporting ranging from $109 to $171 annually per device. They also offer monthly plans ranging from an average of $15.83–$17.91 per month per device.

However, all the commercial vehicle management systems use proprietary devices and software. In addition, they offer only packaged services. Thus, users are made to choose from the predefined service packages. Customizing service to match a company's needs is either very costly or not possible.

3 FMWare System Requirements

Our goal was to build an IoT-based fleet management system that satisfies the following requirements:

1. The system should automate the collection of vehicle diagnostics data while the vehicles are moving. To aid in preventing safety hazards, continuous collection of vehicle diagnostics data is critical.
2. The system should identify the driver of each vehicle, record the route and distance driven. The system should flag unauthorized drivers or vehicle usage by unauthenticated drivers.
3. The system should be scalable so that it can manage a large number of vehicles effectively. The department maintains over 600 vehicles to cover 70,000 acres.
4. The system should use open-source software and hardware. The system should be affordable, customizable, and extendable. At present, the monthly subscription fee for over 600 vehicles is cost-prohibitive.
5. The system should not change any of the vehicle's factory wiring. To keep the manufacturer warranty in force, modification or cutting of factory-installed wiring is prohibited.

4 FMWare System Design

To satisfy these requirements, we developed an IoT-based fleet management system named FMWare. The overall FMWare architecture is shown in Fig. 1. It consists of the following components:

(a) **Raspberry Pi B3+**. We use a Raspberry Pi 3B+ as our IoT edge device and sensor board. The collected data is stored on on-board flash memory until a secure wireless gateway is available.
(b) **Global Positioning System (GPS) module**. We use a GPS module to collect route data and compute the distance of a trip.
(c) **On-board diagnostics II (OBD-II) adapter**. We use an OBD adapter to collect a vehicle's diagnostics data.
(d) **Radio Frequency ID (RFID) reader**. We use an RFID scanner to collect a driver's badge id and identify unauthorized use of vehicles.
(e) **Liquid-Crystal Display (LCD) screen**. We use an LCD screen to send messages to a driver.
(f) **Power supply unit (PSU)**. We use a customized PSU to reduce a vehicle's battery output to the Raspberry Pi from 12 to 5 V.
(g) **Apache's NiFi and MiNiFi**. We use NiFi and MiNiFi to control and secure data transmission.
(h) **Back-end servers**. An SQL Database server and a web server.
(i) **Web-Apps**. We implement web-apps to allow users to interact with the collected data.

Upon a vehicle's return to the base station, the data collected on the Raspberry Pi is transmitted to County servers through secure Wi-Fi access points available in the County parking lot. To encrypt data and to control data transmission flow we use NiFi and MiNiFi, an open-source data flow control software.

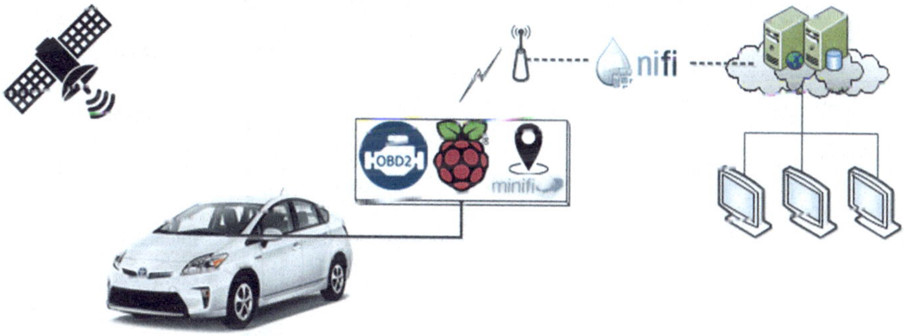

Fig. 1 FMWare architecture overview

4.1 Edge IoT Sensor Board

We use a Raspberry Pi 3B+, a single-board micro-controller. Raspberry Pi 3B+ provides 1.4 GHz 64-bit quad-core ARM Cortex-A53 processor [9]. Also, Raspberry Pi 3B+ is equipped with 1 GB of RAM as well as on-board flash storage. In addition, Raspberry Pi 3B+ features dual-band IEEE 802.11b/g/n/ac Wi-Fi, IEEE 802.15 Bluetooth 4.2, and 10/100/1000 Mbit/s Ethernet. Moreover, Raspberry Pi 3B+ offers 40 pin general purpose input–output (GPIO) connectors. Figure 2 illustrates the GPIO layout of the Raspberry Pi 3B+. Also, a Raspberry Pi 3B+ weights 45 g and its physical dimension is shown in Fig. 3.

The Raspberry Pi is a $35 light-weight portable fully equipped computer. It can be comfortably installed in most vehicle makers and models.

4.2 Global Positioning System (GPS) Module

To collect trip route data, a GPS module is used to reconstruct the path traveled and to compute the distance traveled. In this project, we use a Neo6m to track where a vehicle travels as shown in Fig. 4. The module comes equipped with antennae and a five pin set: VCC, Ground, TX, RX, PPS. The VCC, Ground, TX, and RX pins are connected to the Raspberry Pi via wires. The Neo6m GPS unit costs approximately $13.50.

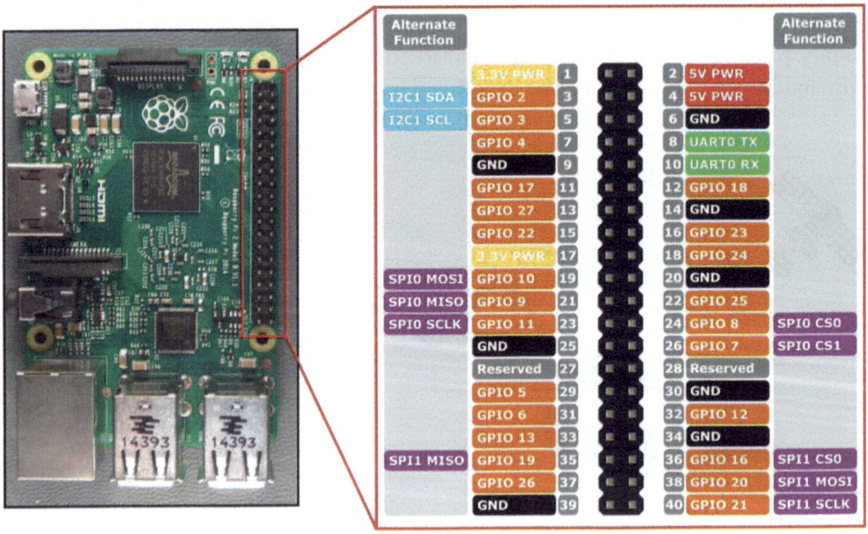

Fig. 2 Raspberry Pi 3

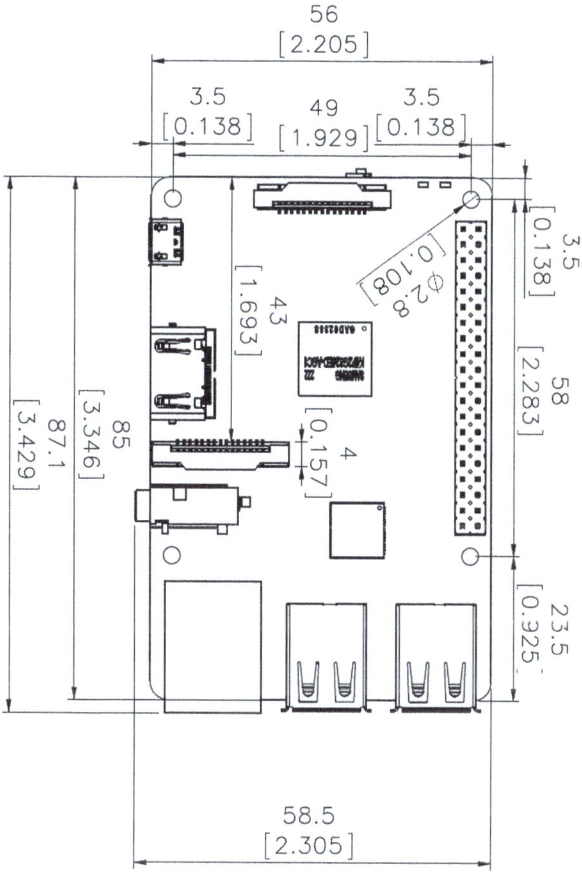

Fig. 3 Raspberry Pi 3 dimension

Fig. 4 GPS module

Reconstructing Traveled Route

We poll GPS data every second. The polled data is compared with the previous data. If the difference between the two is less than a preset delta, we do not store the new data assuming there is a traffic jam. The delta threshold can be reset using a command-line parameter. Currently, we set the delta to 10 ft or 3.048 m. The collected GPS data is stored in on-board flash memory of the Raspberry Pi until the coordinates are uploaded to the server. On the back-end server, we perform reverse geocoding on the collected GPS data and reconstruct the traveled route of the trip.

Computing Distance Traveled

The vehicle's odometer data is protected and cannot be read through standard PID request calls. Thus, to compute the distance traveled, we utilize GPS data.

We use the haversine formula [5] to determine the distance between two points on the surface of the earth. (See Listing 1.)

Based on our experiments, the distance traveled computed by the haversine formula provides excellent matches with odometer readings.

Listing 1 Haversine formula

The haversine formula is defined as

$$\text{hav}(\theta) = \sin^2\frac{\theta}{2} = \frac{1 - \cos\theta}{2}. \tag{1}$$

For adjacency pair of longitude and latitude of the collected GPS data, we set

- λ_1, λ_2: longitude of point 1 and longitude of point 2,
- φ_1, φ_2: latitude of point 1 and latitude of point 2,
- r is the radius of the sphere,
- d is the distance between the two points.

Then, based on the haversine formula,

$$\text{hav}\frac{d}{r} = \text{hav}(\varphi_2 - \varphi_1) + \cos\varphi_1 \cos\varphi_2 \text{hav}(\lambda_2 - \lambda_1), \tag{2}$$

and we let

- $\theta_1 = \varphi_2 - \varphi_1$,
- $\theta_2 = \lambda_2 - \lambda_1$.

Solve for d by applying the inverse haversine,

$$d = 2r \arcsin\left(\sqrt{\text{hav}(\theta_1) + \cos\varphi_1 \cos\varphi_2 \text{hav}(\theta_2)}\right) \tag{3}$$

$$= 2r \arcsin\left(\sqrt{\sin^2\left(\frac{\theta_1}{2}\right) + \cos\varphi_1 \cos\varphi_2 \sin^2\left(\frac{\theta_2}{2}\right)}\right). \tag{4}$$

4.3 On-Board Diagnostics II Adapter

In order to collect a vehicle's diagnostics data, we use an on-board diagnostics (OBD-II) adapter. We utilize a scanning tool, which interfaces with a vehicle's control subsystems to retrieve data. We use an ELM 327 scanner as shown in Fig. 5. ELM 327 OBD scanner costs approximately $10.

For data requested through the OBD, parameter identification (PID) codes are used. Every vehicle built after 1996 has the On-Board Diagnostics Version II built in. Since 1996 the pin set and PIDs for every vehicle have been standardized as shown in Fig. 6. This allows for the creation of general tools that can be used to diagnose vehicle problems for different car manufacturers.

First, we developed software to request data through the OBD. We utilized an open-source library called Python-OBD to translate the information that is being sent from the ELM 327. Second, we connected an OBD to a data link connector (DLC) usually located under the panel on the driver's side. Once the OBD-II tool is plugged into the DLC, the tool can communicate with the engine control unit (ECU), a computer that controls all engine activities within every car. The software installed on the Raspberry Pi sends PID requests to the OBD-II tool through a Bluetooth connection and receives diagnostic data through the same connection.[1]

The amount of diagnostics data collected can be increased significantly depending on the number of diagnostic trouble codes (DTCs) and the parameter ids (PIDs) collected. OBD-II provides 192 standardized PIDs [12]. Each PID is represented by 4 bytes. The size of data returned by each PID varies from PID to PID. The smallest is 1 byte; the largest is 21 bytes. OBD-II diagnostic trouble codes (DTCs) are used

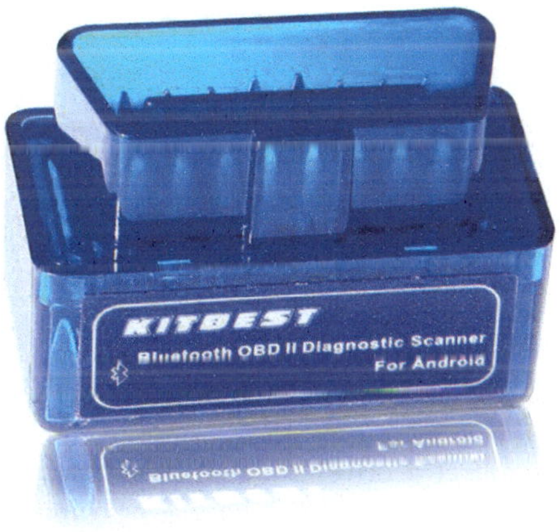

Fig. 5 OBD adapter

[1]Due to space limitation in an instrument panel, we opted to install the Raspberry Pi inside the passenger side dashboard. For a vehicle manufactured after 1996, an OBD-II DLC is usually located under the instrument panel on the driver side.

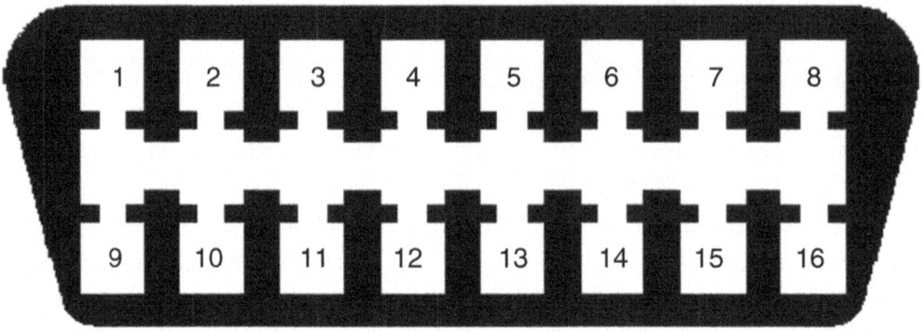

PIN	DESCRIPTION	PIN	DESCRIPTION
1	Vendor Option	9	Vendor Option
2	J1850 Bus +	10	j1850 Bus
3	Vendor Option	11	Vendor Option
4	Chassis Ground	12	Vendor Option
5	Signal Ground	13	Vendor Option
6	CAN (J-2234) High	14	CAN (J-2234) Low
7	ISO 9141-2 K-Line	15	ISO 9141-2 Low
8	Vendor Option	16	Battery Power

OBD-II Connector and Pinout

Fig. 6 OBD-II pin set

to present faults detected in the vehicle. Each DTC is represented by 4 bytes. For our initial implementation, we collect few PIDs as listed in Table 1 and all DTCs returned by ODB-II.

4.4 Radio Frequency ID (RFID) Reader

To collect a driver's identification, we use the County employee badge system. All County employees have a proximity I.D. card containing an employee I.D., which can be read by the Radio Frequency Identification Reader (RFID scanner). A driver's I.D. badge must be scanned with the RFID reader installed in the dashboard before they operate a vehicle. After being scanned, the driver's employee I.D. is

Table 1 List of sample OBD PIDs collected

PID (in decimal)	Description	Data size (in bytes)
3	Fuel system status	2
12	Engine RPM	2
13	Vehicle speed	1
16	MAF air flow rate	2
31	Run time since engine start	2
32	PIDs supported [21–40]	4
33	Distance traveled with malfunction indicator lamp on	2
47	Fuel tank level input	1
51	Absolute barometric pressure	1
64	PIDs supported [41–60]	4

Fig. 7 LCD module

stored on the Raspberry Pi along with the other data. FMWare uses this information to identify unauthorized use of vehicles and to verify that the scanned badge is that of a legitimate County employee.

4.5 *Liquid-Crystal Display (LCD) Screen*

To display the device status to the driver, we use a 16X2 LCD screen as shown in Fig. 7. For example, we display "Ready" to let a driver know that a vehicle is ready for driving. We chose a I2C backpack module to reduce pin out to just 4 pins.

4.6 Power Supply Unit (PSU)

We built a custom power supply unit (PSU) to run the Raspberry Pi off a 12 V vehicle battery. To step down the voltage from 12 to 5 V we use voltage buck converters as shown in Fig. 8. Also, we placed 2 A fuses between the Raspberry Pi, PSU, and vehicle battery for circuit protection.

4.7 Apache's NiFi and MiNiFi

With the amount of data we collect on a Raspberry Pi, we need a tool that allows us to monitor the data transmission flow, to identify any faults occurring during data transmission, to handle the massive concurrent influx of new requests, and to encrypt personal data. Although there are numerous solutions, we wanted to utilize an open-source tool so that we can extend it as needed. To do so, we chose NiFi and MiNiFi, which is an open-source data flow control software system built by the Apache Software Foundation [1]. This software provides data flow management, including data buffering, controlling the flow, prioritized queuing, flow specific quality of service (QoS), data recovery, and data encryption. MiNiFi is a light weight version of NiFi, which allows for simple NiFi integration of edge devices.

Apache NiFi provides a web-based user interface (UI) that allows one to build an automated data flow control system. It also provides visualizing, editing, monitoring, and administration of those data flows. The NiFi UI offers a component tool bar, navigation palette, operations palette, and status bar. Figure 9 shows a small subset of the current project's data flow implementation.

Processors are the most common NiFi/MiNiFi component. They manage data inflow, outflow, routing, and manipulations. Processors provide numerous attributes that enable data source labeling, data streaming, data filtering, data modification, and data provenance tracking. Figure 10 shows an example of adding a processor.

When a processor is added to the program, the next step is to configure its various properties. For example, Fig. 11 shows the properties that can be set up for

Fig. 8 PSU module

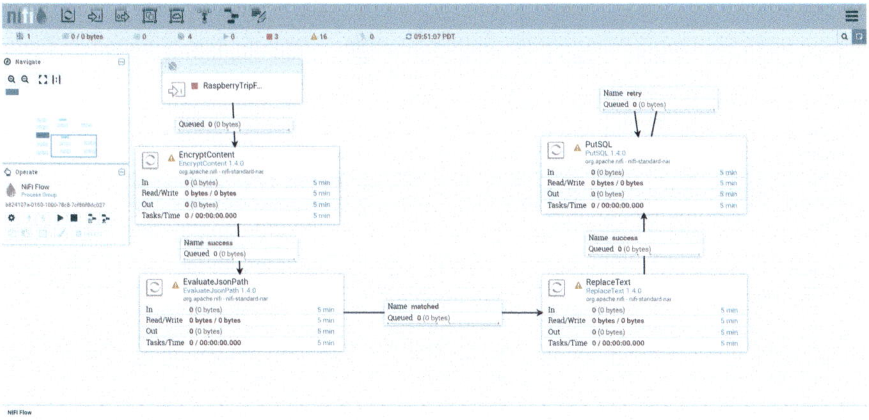

Fig. 9 Example of NiFi interface

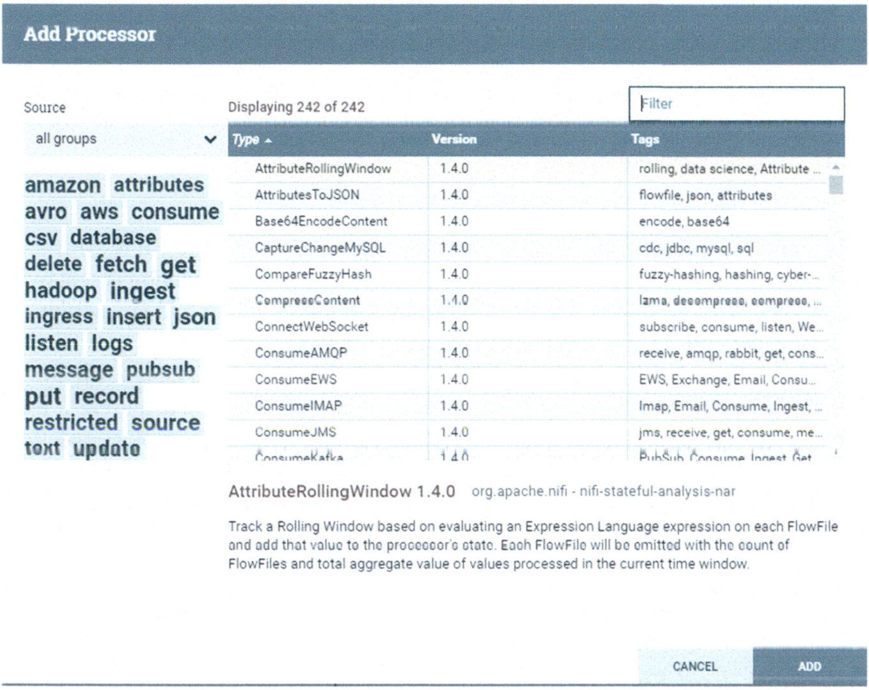

Fig. 10 Example of adding processor

the EvaluateJSONPath processor. This processor receives a JSON formatted text file and extracts specific data points from the file and ingests them into the JSON embedded data flow.

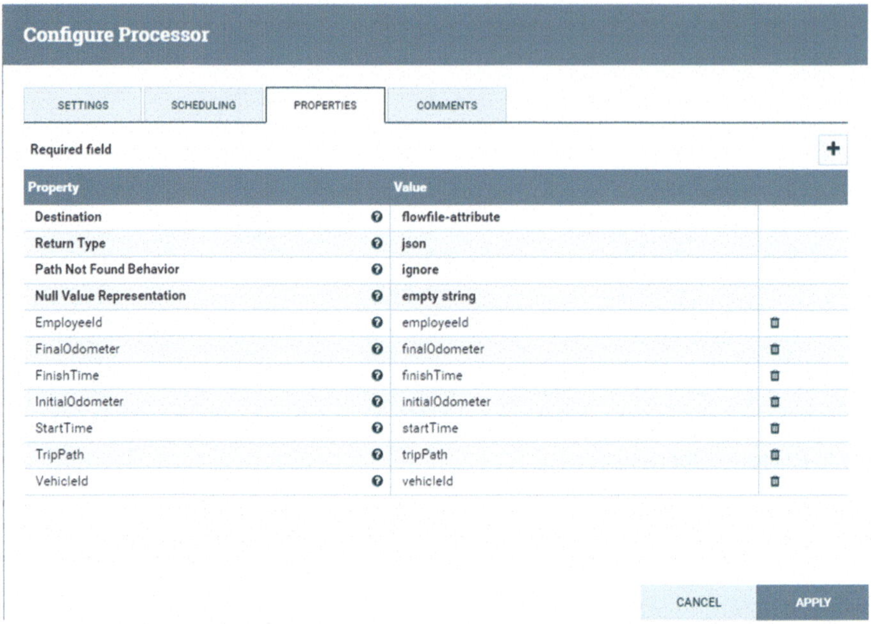

Fig. 11 Example of properties configuration

In addition, NiFi provides built-in data encryption functionalities. It is critical to protect the data acquired by the Raspberry Pi since personal information is sent over the Internet. NiFi is equipped with multiple encryption protocols and an encryption processor that both encrypt outgoing data and decrypt received data.

We built three data flow controls: MiNiFi Flow, Trip Flow, and Diagnostic Flow. The MiNiFi agent on Raspberry Pi retrieves and encrypts the data acquired. It also manages the uploading of the data to FMWare servers when a network connection is available. Upon upload, the data is sent to the Trip Flow or Diagnostic Flow.

When a Trip Flow or Diagnostic Flow processor receives data, it first decrypts the data. Then, it parses the JSON files and constructs SQL insert statements. Finally, the data acquired by the Raspberry Pi is stored in a database.

4.8 Back-End Servers

We configured and implemented an SQL Database server to manage our data. For server back-end and front-end connections, we used the Entity Framework (EF) Core to provide RESTFul data access service.

4.9 *Web-Apps*

We used the Angular Framework to build the user front-end. We implemented two web-Apps, one drivers and the other for managers. The Driver App allows a driver to supplement the trip information with information that cannot be collected automatically. This includes such items as project title, project function, task order, and purpose of trip. The Manager App allows managers to review and approve trips and to add/delete/update drivers and vehicles.

5 Testing and Evaluation

We performed component testing on a simulator. The first complete end-to-end field test was performed in April 2018. During the first field test, we identified a few software glitches. After we fixed all known issues, the system performed successfully in the second field test.

County officials found that the system meets their requirements. They plan to deploy it department-wide.

The FMWare project was presented to the Los Angeles County Board of Supervisors in March 2018. It was received with great acclaim.

6 Conclusion

FMWare successfully integrates a number of hardware and software components. We were able to reduce the amount of time it takes for county employees to document their vehicle usage.

Future work will integrate machine learning to allow the Department of Parks and Recreation to determine when preventative maintenance should be performed.

We have demonstrated that one can construct a complete IoT system using inexpensive hardware components and open-source software. The system is easily extendable to include additional services and components.

Acknowledgements A special thanks to Professor Abbott for inspiring me for this work.

References

1. Apache NiFi Architecture. (2018, May 3). Retrieved from https://nifi.apache.org/docs/nifi-docs
2. County of Los Angeles Annual Report 2009–2010. Public Affairs, Chief Executive Office (p. 23).

3. Dhall, R., & Solanki, V. (2017). An IoT based predictive connected car maintenance approach. *International Journal of Interactive Multimedia and Artificial Intelligence, 4*(3), 16–22.
4. Fleet Genius: Fleet Management Software. (2018, May 3). Retrieved May 3, 2018, from urlhttps://www.fleet-genius.com
5. Goodwin, H. B. (1910). The haversine in nautical astronomy. *Naval Institute Proceedings, 36*(3), 735746
6. Istrefi, D., & Cico, B. (2013). Fleet Management Solution Fuel consumption and collision prevention system modules. *IRACST-International Journal of Computer Science and Information Technology and Security, 3*(3), 2249–9555.
7. Mastrack: GPS Tracking The Way You Need It. (2018, May 3). Retrieved from https://www.mastrack.com
8. Penna, M., Shivashankar, Arjun, B., Goutham, K. R., Madhaw, L. N., & Sanjay, K. G. (2017). Smart fleet monitoring system using internet of things (IoT). In *2nd IEEE International Conference On Recent Trends in Electronics Information and Communication Technology (RTEICT)*. Piscataway, NJ: IEEE.
9. RASPBERRY PI. (2018, May 3). Retrieved from https://www.raspberrypi.org
10. Saghaei, H. (2016). Design and implement of a fleet management system using novel GPS/GLONASS tracker and web-based software. In *1st International Conference on New Research Achievements in Electrical and Computer Engineering*.
11. Sodeyama, K., Takada, M., Hane, S., Furuno, Y., & Watanabe, H. (2015). Cloud-based fleet management system proof of concept. *Featured Articles, Hitachi Review, 64*(1), 29–34
12. State of California Air Resource Board, Workshop Report. Technical Status and Revisions to Malfunction and Diagnostic System Requirements for 2010 and Subsequent Model Year Heavy-Duty Engines (HD OBD). March 2012, CA. (2018, May 3). Retrieved from https://www.epa.gov/state-and-local-transportation/vehicle-emissions-board-diagnostics-obd
13. US Fleet Tracking Co. (2018, May 3). Retrieved from https://www.usfleettracking.com

Communication Management in Multiclient System Using SMS Gateway

Peter Pistek and Martin Hudec

1 Introduction

Internet of Things (IoT) is an indivisible part of every city. As penetration of informatics into other areas of life rises, the need for specialized devices also rises. IoT devices usually monitor several parameters and send them to further processing on servers and personal devices. With this trend there is a demand for communication and the most spread networks are the mobile (or cellular) networks. In this chapter we focus mainly on SMS messages as an easy-to-use and reliable service for communication.

Usually, it is interesting to use P2P (person to person) transmission; for IoT it is also very interesting to use A2P (application to person) and A2A (application to application) that are necessary for deployment in IoT.

SMS has a built-in form of compression that allows to send 160 characters in 140 bytes. This only applies to standard SMS which uses ASCII characters. Text encoding may also be 8 or 16 bits, but it decreases the message length.

For SMS gateways (also known as ESME—external short message entity) we can use the SMPP protocol in a standard TCP/IP network that uses direct communication with short message service center (SMSC).

Security in SMS is mostly provided by security in GSM network itself because SMS has a lack of security [1]. Data algorithms from the A5 family are used to secure data transmission over the wireless GSM interface. These algorithms, however, work on the principle of stream ciphers and all of them have been managed to break through [2]. In addition to the weak cryptographic algorithms (with

P. Pistek (✉) · M. Hudec
Faculty of Informatics and Information Technologies, Slovak University of Technology in Bratislava, Bratislava, Slovakia
e-mail: peter.pistek@stuba.sk

© Springer Nature Switzerland AG 2020

D. Cagáňová, N. Horňáková (eds.), *Mobility Internet of Things 2018*,
EAI/Springer Innovations in Communication and Computing,
https://doi.org/10.1007/978-3-030-30911-4_2

proven vulnerabilities), there are several other vulnerabilities in SMS messages. The most significant transmission vulnerability is the time of active encryption. The content is encrypted only during wireless transmission. Subsequently, the message is decrypted and transmitted without data protection over the operator's network. It is often the case that, for example, SMS messages sent from abroad are transmitted via different types of networks, often via the Internet. This leaves them exposed to vulnerabilities and attacks [3]. Of course, in newer generations of mobile networks, there are several new security features, but even 4G networks are still susceptible to man-in-the-middle attack [4].

Lower layers in RM OSI (bellow application layer) have problem with inclusion of new security standards due to existing GSM network entities. There are several approaches in this area. SeeSMS [5] provides a robust solution but has the disadvantage of using RSA algorithms that may be too demanding for use in IoT devices [6]. Larger keys (such as 1024 in RSA vs. 160 bits in ECC) must be used compared to equivalent ECC. Implementation of public keys guarantees significantly greater security, but again, it brings a lot of complexity into the system. Asymmetric encryption is usually used only for key exchange. This key is subsequently used in symmetric encryption which is used for data communication. The state-of-the-art encryption algorithm is AES which is still not breached and it is effective even on embedded systems and mobile devices [7].

SMSSec [8] proposed a protocol that can be used to secure an SMS communication. It uses two-phase protocol (asymmetric cryptography for handshake and symmetric cryptography for communication). Process of asymmetric handshake takes 50 s. At work [9], 50% less bandwidth was achieved during the authentication process, due to the more complex security architecture that is used predominantly outside of end clients.

For use in the IoT area, it is worth mentioning that the encryption of SMS messages themselves may not be of great importance. At present, usually at least on the one side, the content of messages can be sent to TCP/IP networks, by mail, or by the Web service [10], until it reaches its goal (on a special server, to a user, etc.). Therefore, it is also necessary to remember the security standards of these transmission channels.

The organization of this chapter is as follows. Next section contains information about compression algorithms and our experimental results. Section SMS gateway is focused on our approach as the whole system with introduction of our protocol. This is followed by secure communication including description of key exchange with experiments focused on time needed for this exchange and overhead caused by use of compression and encryption in the communication.

2 Compress Algorithms

Although SMS uses (for ASCII encoding) an 8- to 7-bit conversion, it may still be necessary to transmit more characters (in particular a parameter form different sensors). Using standard gzip algorithms (Lempel-Ziv coding, LZ77), lzma

(Lempel-Ziv-Markov chain algorithm) or zip is not appropriate because they need to generate a header for the compressed file. With short texts (like SMS), however, an increase in the text itself can be achieved.

There are special algorithms for use on short texts such as xz,[1] b64pack [11], shoco,[2] smaz[3], and deflater.[4] For their comparison, we performed a short three-sentence test (text in alphabet using 16-bit encoding with numbers, English with numbers, and randomly generated text with predominant representation of characters other than letters).

2.1 *Experiments: Compression Algorithms*

In experiments, we found shoco and smash algorithms to be comparable, and their effectiveness also depends on the type of compressed text. For the purposes of compressing text messages (SMS), the algorithm smaz has proven to be the most effective (Table 1).

In Table 2 we can see the best results of each algorithm for specific input texts. This suggests the suitability of testing multiple algorithms before sending the message. It leads to use of the best compression algorithm for the shortest possible output. This approach is interesting in cases where the amount of data transmitted is prioritized and we have spare computational time on IoT devices.

3 SMS Gateway

For the best use of SMS gateways in IoT field and user interactions, we designed the use of these entities:

- Email communication—redirects secure SMS communication to email.
- XML Web Service Interface—secure Web service for other systems.
- Modems for communication via GSM interface.
- Managing users and their content.

For easy deployment on end devices, we designed a program library (in Java as a proof of concept) which ensures security with the proposed SMS gateway (Fig. 1).

[1]https://linux.die.net/man/1/unxz

[2]https://github.com/Ed-von-Schleck/shoco

[3]https://github.com/antirez/smaz

[4]https://docs.oracle.com/javase/8/docs/api/java/util/zip/Deflater.html

Table 1 Results of compression algorithms

	Sentence no. 1			Sentence no. 2			Sentence no. 3		
Algorithm	Compression	Decompression	Effectiveness	Compression	Decompression	Effectiveness	Compression	Decompression	Effectiveness
xz	70 ms	13 ms	−25%	74 ms	9 ms	−23%	74 ms	9 ms	−26%
b64pack	–	–	–	12 ms	12 ms	29%	–	–	–
shoco	8 ms	8 ms	5%	8 ms	8 ms	36%	8 ms	8 ms	0%
smaz	101 ms	28 ms	16%	72 ms	24 ms	34%	70 ms	13 ms	−5%
deflater	11 ms	3 ms	6%	3 ms	2 ms	21%	3 ms	2 ms	13%

Table 2 Verify the effectiveness of using compression algorithms

Algorithm	Type of data	Input size	Size after compression	Effectiveness of compression	Compression time	Decompression time
Shoco and deflater	JSON file	327 B	165 B	49.54%	9 ms	11 ms
	Random characters	141 B	139 B	1.42%	1 ms	2 ms
Smaz and deflater	ASCII text	201 B	110 B	45.27%	1 ms	1 ms
	Long JSON file	51863 B	494 B	99.05%	159 ms	172 ms
Deflater	Text in 16-bit alphabet	166 B	132 B	20.48%	1 ms	1 ms
	Long text	1062 B	586 B	44.82%	5 ms	7 ms

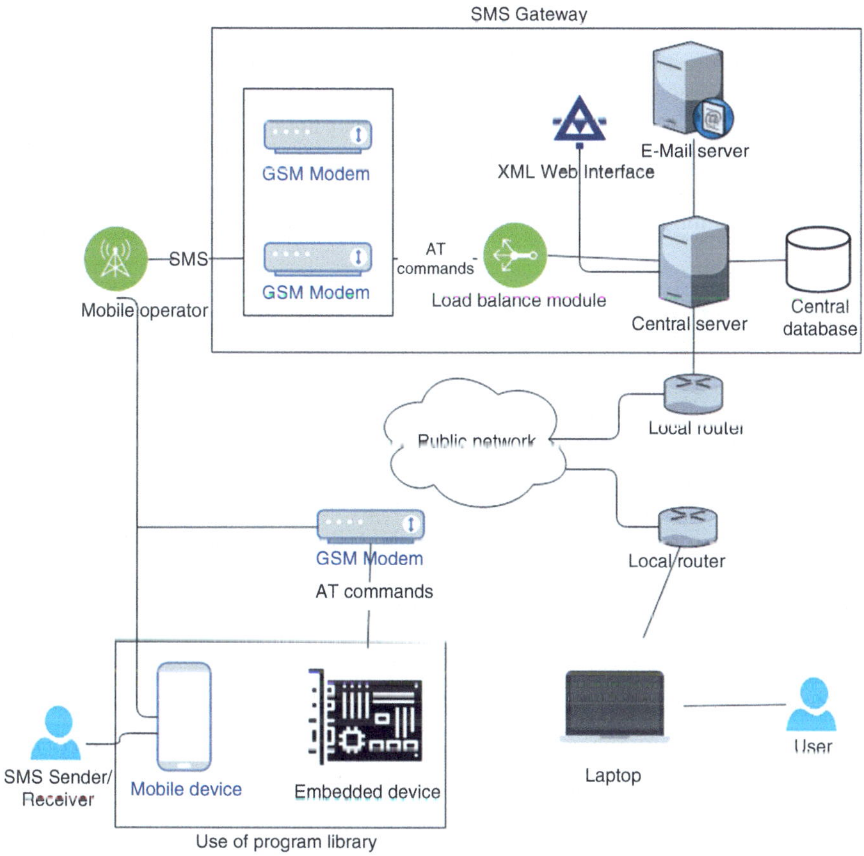

Fig. 1 System architecture using the software SMS gateway on the application server

3.1 Header and Specific Frames

We designed the protocol to take up as little space as possible in the body of SMS, as the size of the message is limited to 140 bytes. The necessary header parameters were specified. Also, messages have been broken down by type to key exchange, data transmission, and technical messages.

All frames have the same first three starting fields and one final field in the header. The designed version works with a custom 4-byte header, leaving 136 bytes of data usage.

Common Header Fields

In order to determine the type of message, we suggested the following parameters to be placed in each header:

- *Sequence Number*—The automatically incremented number (after sending the message).
- *Content Type*—The type of message content, which determines what message type is transmitted, can acquire three values: DATA, KEY, and TECHNICAL.
- *Cypher Active*—Active encryption, says whether the content behind the header is encrypted or not.
- *Checksum*—Checksum of the header contains the mathematical sum of all the header fields modulated by the maximum size of the checksum field.

Key Exchange Frame

The key exchange frame includes, in addition to the base fields, one extra field that defines the current key exchange status. All fields in the header are shown in Table 3. For the full process of key exchange see Chapter "Performance Comparison of Zoning Techniques in Geographic Forwarding Rules for Ad Hoc Networks".

Key Exchange Stage—serves for business logic to determine the correct process of handover key management. It can acquire the following values:

Table 3 Header format for key exchange

Name	Sequence number	Content type	Cypher active	Key exchange stage	Padding	Checksum
Bits in the header	0–7	8–9	10–10	11–15	16–16	17–31
Description	Message sequence number	Message content type	Is encryption active?	State of key exchange	Empty content	Header checksum

Table 4 Header format for data transfer

Name	Sequence number	Content type	Cypher active	Destination type	Compression type	Checksum
Bits in the header	0–7	8–9	10–10	11–13	14–16	17–31
Description	Message sequence number	Message content type	Is encryption active?	Final message form	Compression type	Header checksum

- REQUEST—request to start the key exchange.
- RESPONSE—response to key exchange requests.
- ACK—confirmation of correct key exchange.
- API_KEY_REQUEST—request to exchange the API key.
- API_KEY_ACK_SUCCESS—confirmation of successful exchange of API key.
- API_KEY_ACK_ERROR—unsuccessful API key replacement.

API key is a unique key generated by central server for each device.

Data Transfer Frame

In addition to the base fields, the data transfer frame is enriched by two fields that define the type of compression of the transmitted text and the target format to which the message should be translated (see Table 4).

- *Destination type*—the target form of the message that serves for business logic to determine the correct process of processing the message with the transmitted data. It can acquire the following values:
  - WS—Web service.
  - EMAIL—email.
  - STORE—save in server SMS gateway without forwarding.
  - DEFAULT—leaving the decision on the configuration of the server (SMS gateway).
- *Compression type*—the selected compression type of message that serves for business logic to determine the correct process for decompressing the message with the transmitted data. It can acquire the following values:
  - SHOCO—shoco algorithm.
  - SMAZ—smaz algorithm.
  - CUSTOM_ALPHABET—compression using a specific device alphabet.
  - DEFLATER—deflater algorithm.
  - NONE—inactive compression.

Table 5 Header format of technical message

Name	Sequence number	Content type	Cypher active	Technical message type	Compression type	Checksum
Bits in the header	0–7	8–9	10–10	11–13	14–16	17–31
Description	Message sequence number	Message content type	Is encryption active?	Final message form	Compression type	Header checksum

Frame for Technical Message Transfer

In addition to the basic fields, the technical messaging frame includes two fields that define the type of data compression of the transmitted text and the type of transmitted technical message (see Table 5).

- *The technical message type*—serves the business logic to determine the correct process of processing the technical message. It can acquire the following values:
  - PING—communication test.
  - REQUEST_KEY_EXCHANGE—request for reexchange of keys.
- *Compression type*—the type of compression selected is the same as the compression type field at the header for data transfer.

4 Secure Communication

The basic pillars of the SMS gateway include message encryption. The Diffie-Hellman key exchange method (variant for ECC and ECDH method) was used to ensure secure transmission of the content. The exchange itself consists of four steps; the first exchange is initiated by the client device.

The key exchange on the server side begins by accepting the SecureSMSKey message, with the REQUEST parameter, which the client device defines for the key exchange request. The message contains a public key of the client device that will be loaded by the SMS gateway and use it together with its own private key to generate a shared secret. The shared secret created together with the API key is used to generate a symmetric AES key.

Since the API key should be transmitted by a different form of transfer (recommended: via the cable when initializing the device), it is not necessary to generate a new key during encryption nor to put it in front of encrypted text. Amount of encrypted content being transferred is saved this way.

The key exchange continues by sending a response to the client device in the SecureSMSKey message, with the RESPONSE parameter, in which the SMS gateway sends the device its own public key, so the device can generate a shared secret.

After this exchange, both devices have generated a shared secret and a symmetric key. Verification messages are designed to verify that the shared secret is created correctly. Symmetric encryption is implemented in the AES/CTR/NoPadding mode, that is, as a stream cipher, ensuring that the size of the content is not increased.

After receiving the SecureSMSKey message with the ACK and active encryption indicator, the server (SMS gateway) decrypts the message content that contains the control sum of the generated client's symmetric key. The SMS gateway generates the same checksum from its own symmetric key and compares both sums.

If both checksums are equal, it will again generate SecureSMSKey with the ACK and active encryption indicator. SMS gateway inserts the encrypted checksum, which it generates from its own symmetric key, and sends the message to a specific client device.

The server waits for confirmation of the delivery of the message and then sets the status device in the database to say that the symmetric keys are successfully generated. During the entire exchange of keys, the active states (in which both sides of the key exchange occur) are stored that can also be displayed to the user via the Web interface.

To improve the security of transmission, a lightweight of exchanging keys is designed using periodic API key renewal, which is used as an initialization vector for symmetric encryption. This process is again initialized by the client device. The SMS gateway begins to exchange the API key after receiving the SecureSMSKey message with the API_KEY_REQUEST parameter and the active encryption indicator. A new generated API key is transmitted in the message body that generated the client device along with the checksum of the newly generated symmetric key. The server SMS gateway, like in a normal exchange, calculates the checksum and checks it with the values received in the message body.

4.1 Experiments: Key Exchange

Based on experiments, we found that the library initialization for the ability to secure the communication lasts for 19 s (Table 6). Periodic key exchange along with time of its generation takes 14 s (Table 7). These times are, however, dependent on the quality and overload of the GSM network. During this time, four SMS messages with the custom protocol are transmitted in the case of the initial key exchange; only two SMS messages are exchanged for the key API.

Table 6 Time view on key exchange

Process on a client device	Time from beginning	SMS gateway process	Time from beginning
Request for key exchange	0 s	Receive key exchange request	3 s
Respond receive	9 s	Send reply for request	4 s
Send ACK message	9 s	Receive ACK message	14 s
Receive ACK message	17 s	Send ACK message	14 s
Successful key exchange	17 s	Wait for delivery message	19 s
		Successful key exchange	19 s

Table 7 Time view on API key exchange

Process on a client device	Time from beginning	SMS gateway process	Time from beginning
Request for API key exchange	0 s	Receive API key exchange request	8 s
Respond receive	12 s	Send reply for request	9 s
Successful API key exchange	12 s	Wait for delivery message	14 s
		Successful API key exchange	14 s

Comparison to other solution:

- SMSSec [8]: 50 s.
- EasySMS [9]: 25.5 s.
- Our solution: 19 s.

Compared to SMSSec, our solution is more than 62% faster and even 25% faster than the EasySMS (Tables 6 and 7).

4.2 Experiments: Overhead

In experiments, we found that delivering the SMS itself over the operator's network takes an average of 3.14 s. The increase of the time to send the message in the case of additional encryption and compression is approximately 14 ms in the worst test time, representing 4.3% of the time of sending the message. This time is negligible in terms of transmission of the message over the operator's network.

With our solution there is a possibility to more than double the possible message length and with the secure encrypted communication with just small overhead. Based on the final use of end IoT devices the secure unencrypted message can be achieved with only 0.8% time overhead or only compressed messages with just 3.18% time overhead (Table 8).

Table 8 Comparison of the impact for each function to the time of message creation

	Message length	Compression Encryption	Compression Encryption	Compression Encryption	Compression Encryption
Duration of message creation	360 B	21.73 ms	24.55 ms	32.95 ms	35.32 ms
Duration of message creation	180 B	17.5 ms	20.69 ms	26.67 ms	30.52 ms
Duration of message creation	140 B	16.24 ms	18.51 ms	27.49 ms	26.92 ms
Duration of message creation	90 B	17.03 ms	18.64 ms	25.96 ms	27.80 ms
Average		18.12 ms	20.60 ms	28.27 ms	30.14 ms

5 Conclusion

We have proposed a new solution[5] for IoT communication. Its advantage is that it uses SMS to transmit data over mobile networks, ensuring the largest coverage. Although the use of the GSM module is energy intensive, we have experience with the solution where the modules are connected once a day or in the event of a critical event, which significantly improves its usability. The SMS communication feature shows that the device got its SMS messages even after being off-line (of course, this time cannot be too long).

Due to the wide range of use of IoT devices, this has to be considered in our proposal. Messages can be sent via SMS in encrypted and unencrypted form, depending on security demands. The time required for key exchange was reduced up to 62% compared to similar solutions.

As an effort is made to connect the devices to an energy-intensive GSM module as least as possible, the goal can be to compress the data sent over the mobile network using a minimum number of SMS messages. Our solution looks for the best result from several compression algorithms to be used. In one extreme case, we have succeeded in placing a long JSON file in 4 SMS messages instead of the original 370 SMS messages. Common text that can be placed in one SMS message has 304 bytes (+ our 4-byte header), which is an increase in transmission efficiency by 117%.

Each case of using IoT devices can choose their own communication method with/without encryption with/without compression to best suit the intended use. Even when using secure and compressed communications, SMS messaging is

[5] https://gitlab.com/m.hudec92

prolonged from 18 to 30 ms (+66%), but this represents only a 4.3% increase in the process from sending to receiving messages via GSM network.

The impact of our approach in praxis to IoT community could differ depending on the final usage (and configuration); for example the amount of transferred data is not always important in small sensors. If our library is used on IoT devices (or the library is reimplemented to other languages) it will improve two crucial parameters (one of them or both) in IoT communication—security and amount of transferred data. This will happen with only a small amount of work from end users. During initialization phase of IoT device one step has to be added: import of API key to the device.

Other extensions of our approach may include direct connection to SMSC of mobile operators through the SMPP protocol, and the use of other types of mobile networks (such as LoRa, SigFox) that are specialized in data transmissions.

Acknowledgement This work was supported by the Cultural and Educational Grant Agency of the Slovak Republic (KEGA 011STU-4/2017) and ITMS 26240220084.

References

1. Thomas, M., & Panchami, V. (2015). An encryption protocol for end-to-end secure transmission of SMS. In *2015 International conference on circuits, power and computing technologies [ICCPCT-2015]* (pp. 1–6). Nagercoil, India: IEEE. https://doi.org/10.1109/ICCPCT.2015.7159471.
2. Upadhyay, D., Shah, A., & Sharma, P. (2014). Design, implementation and analysis of GSM stream cipher: Software simulators vs. real test bed—FPGA. In *2014 Sixth International Conference on Computational Intelligence and Communication Networks* (pp. 930–934). IEEE. doi: https://doi.org/10.1109/CICN.2014.196.
3. Toorani, M., & Beheshti, A. (2009). Solutions to the GSM security weaknesses. In *International Conference on Next Generation Mobile Applications, Services and Technologies, 2008. NGMAST '08. The Second International Conference*. Cardiff, UK: IEEE. https://doi.org/10.1109/NGMAST.2008.88.
4. Kamal, A. A., et al. (2014). An efficient authentication and key agreement protocol for 4G (LTE) networks. In *2014 IEEE Region 10 Symposium*. Kuala Lumpur, Malaysia: IEEE. https://doi.org/10.1109/TENCONSpring.2014.6863085.
5. De Santis, A., et al. (2010). An extensible framework for efficient secure SMS. In *2010 International Conference on Complex, Intelligent and Software Intensive Systems* (pp. 843–850). Krakow, Poland: IEEE. https://doi.org/10.1109/CISIS.2010.81.
6. Zamfir, S., Balan, T., Iliescu, I., & Sandu, F. (2016). A security analysis on standard IoT protocols. In *2016 International Conference on Applied and Theoretical Electricity (ICATE)* (pp. 1–6). Krakow, Poland: IEEE. https://doi.org/10.1109/ICATE.2016.7754665.
7. Moschos, A., Papadimitriou, G., & Nicopolitidis, P. (2018). Proactive encryption of personal area networks and small office-home office networks under advanced encryption standard application. *Security and Privacy, 1*(1), e10, John Wiley & Sons. https://doi.org/10.1001/spy2.10.
8. Li-Chang Lo, J., Bishop, J., & Eloff, J. H. P. (2008). SMSSec: An end-to-end protocol for secure SMS. *Computers and Security, 27*(5–6), 154–167. https://doi.org/10.1016/j.cose.2008.05.003.

9. Saxena, N., & Chaudhari, N. S. (2014). EasySMS: A protocol for end-to-end secure transmission of SMS. *IEEE Transactions on Information Forensics and Security, 9*(7), 1157–1168). IEEE. https://doi.org/10.1109/TIFS.2014.2320579.
10. Songtao, L., & Ming, Q. (2015). The sender controlled security model for message service. In *2015 IEEE Twelfth International Symposium on Autonomous Decentralized Systems* (pp. 187–191). Taichung, Taiwan: IEEE. https://doi.org/10.1109/ISADS.2015.46.
11. Kalajdzic, K., Patel, A., & Ali, S. H. (2014). Rapid lossless compression of short text messages. *Computer Standards & Interfaces, 37*, 53–59). LCNS. https://doi.org/10.1016/j.csi.2014.05.005.

Software-Defined Networking for Smart Buildings: Advances and Challenges

Zoubir Mammeri and Muhammad U. Younus

1 Introduction

Due to advances in wireless and mobile networks and in microelectronics—including miniaturization of sensors and cameras—cities are becoming smart. As elements of smart cities, buildings also are becoming smart.

Smart devices, such as sensors and actuators, are used to automate processes and improve efficiency and comfort of building occupants. Being composed of smart devices, smart buildings (SBs) are typical domain of application of Internet-of-Things (IoT) technologies [1, 2]. The killer application in SB field is energy saving, which should help the deployment of green cities and drastic reduction of pollution [3–7].

SBs are interdisciplinary and involve not only information and communication technologies (ICT), but also huge skills, methods, tools, and products from mechanics, solar energy harvesting, electricity, electronics, chemistry, sociology, economy, etc. In this chapter, SBs are only considered from a networking point of view.

The core of smart buildings is smart building management systems (BMS). All SBs are not expected to provide the same services. Thus, BMSs differ from each other depending on the nature and extent of services offered in buildings. In literature, one may find *intelligent buildings* (IBs) and *intelligent building systems* (IBSs) [8] instead of SBs and BMSs. IB and IBS terms are no longer (or rarely) used in recent literature.

SBs use multiple types of networks to collect data (i.e., measurements) from sensors and cameras and transmit commands to actuators. The first objective of this chapter is to provide a brief state of the art regarding networking aspects of SBs.

Z. Mammeri · M. U. Younus (✉)
IRIT—Paul Sabatier University, Toulouse, France
e-mail: zoubir.mammeri@irit.fr; Muhammad.younus@irit.fr

© Springer Nature Switzerland AG 2020
D. Cagáňová, N. Horňáková (eds.), *Mobility Internet of Things 2018*,
EAI/Springer Innovations in Communication and Computing,
https://doi.org/10.1007/978-3-030-30911-4_3

Software-defined networking (SDN) emerged recently as a solution to facilitate the management of complex networks [9]. Status information about network devices—including routers, switches, and links—is centralized on an SDN controller, which has a coherent view of the network, and consequently it is able to make the best decisions for managing the network devices to provide the required levels of QoS and security to SB applications. The second objective of this chapter is to discuss the contributions of SDN approach to SB networks and highlight some challenges to address before SDN-based smart building networks become a reality.

The remainder of the chapter is structured as follows. Section 2 overviews services provided to smart building users and presents networking technologies to support SBs. Section 3 briefly presents SDN architecture and reviews literature, which focused on SDN networking for SBs. Section 4 highlights some challenging issues to address for making the networks SDN compliant to support future SBs. Section 5 concludes the chapter.

2 Smart Buildings

2.1 Concepts and Services

Before emergence of smart building, *building automation* and *intelligent building* were used. The objectives of all three terms are the same; only the approaches and technologies differ. Recently, SBs extended previous concepts with additional subsystems for managing and controlling energy sources, house appliances, and energy consumption using most often wireless networks and smart devices. SB objectives are to provide new levels of comfort to the occupants with minimum possible energy consumption. Notice that *smart homes* are specific cases of smart buildings.

A huge variety of applications (services) are provided in SB context, including [1, 2, 10] the following:

Energy consumption. In almost all countries, buildings account for more than 1/3 of total energy consumption and it is continuously increasing [3, 11]. Commercial, administration, and home/habitation buildings consume a huge amount of energy for lightening. Thus, reducing energy consumption (due to lightening, air conditioning, home electric and electronic devices) is one of the primary objectives of smart buildings. Building materials and ICT may contribute to reduce energy consumption. In addition to SB services, many countries are encouraging their citizens to isolate their homes to reduce energy consumption.

Energy storage control. Electrical energy (such as electricity generated by solar and wind powers) in smart building can convert the electric energy directly into water heat energy, and then the heat stored in water is released when needed.

Quality of life and comfort. Users can easily control huge devices and machines installed at home in such a way that their comfort is improved.

Healthcare. Advances in medicine have significantly increased life expectancy of populations around the world. Smart buildings may provide assistance to elderly and avoid their placement at medical centers. In addition, by using wireless technologies, any building occupant may be rescued in case of emergency.

Localization of people and devices. It is an SB service to quickly find objects at home, locate people, etc.

Security and surveillance. Cameras and infraction detectors provide powerful means to enhance security of buildings, and consequently increase the quietude of individuals living there. SBs may also provide intelligent interconnections with medical centers, fire department, and police department to send emergency notifications.

Remote control. Monitoring homes and commercial buildings is made easy thanks to smart devices to which users can connect at anytime and from anywhere to see and monitor their homes, devices, and belongings.

Prediction of building operations. The objective is to detect abnormal consumption of energy, wears of installations, and so on, and then to recommend appropriate actions to maintenance staff.

Specific services. In case of commercial and company buildings, services such as meeting room management, music broadcast, and oven control may be provided.

Overall, management system of SB is composed of three layers:

Sensor layer. It is composed of a set of sensor nodes deployed inside a building that periodically, or on demand, measure specific phenomena (such as temperature, pressure, electricity consumption, space occupancy) and send measurements to the control layer. Huge types of sensor are available for measuring electricity consumption, luminance, temperature, humidity, air quality, carbon dioxide, carbon oxide, sound, motion, image, and so on.

Control layer. It receives measurements from sensors and instructions from management staff to take, in real time, decisions regarding the control of physical devices. For example, reduce energy consumption by controlling electrical appliances such as air-conditioning level and lights. A variety of building automation and control systems have been proposed. Their intelligence and complexity depend on the SB services to provide. Recent control systems are based on machine learning algorithms. SB control system proactively learns from occupants' behavior and adapts their operations based on the indoor and outdoor conditions.

Actuation layer. It includes a huge variety of actuators such as comb drives, motors, servomechanisms, hydraulic cylinders, piezoelectric actuators, pneumatic actuators, and screw jacks. Actuators are deployed in SBs to lock/unlock doors, switch on/off lights, raise/lower shutters, open/close valves, and start/stop air-conditioning.

2.2 Networking Technologies for SBs

SB should not be regarded as composed of islands, each associated with a specific service. Rather, it should be considered as composed of interconnected smart devices.

Thus, networking plays a vital role. Wireless sensor networks play a primary role for continuously sensing and monitoring the building energy, environment, and users' behavior and interactions.

Communication requirements significantly vary from one application to another. For example, temperature sensing requires much less bitrate than building hall monitoring through cameras deployed at each corner. Thus, bitrates span a wide spectrum from some kb/s to hundreds of Mb/s. Also required transmission ranges vary from very short range (e.g., in centimeters to connect RFID readers) to long range (e.g., in kilometers to connect to cloud through gateways and cellular base stations).

Today's dominant technologies in SB include Zigbee, Bluetooth, WiFi, UWB, 6LoPAN, LoraWAN, IPV6, cellular technologies (3G, 4G, and forthcoming 5G in few years), Power line communication, and Ethernet. These technologies are ready to support internal (indoor) and external (outdoor) SB communications [12–14].

Data packet size, packet transmission frequency, latency, distance, environmental conditions (including walls), reliability, and cost are the main factors to select communication technologies to fulfil specific SB communication requirements.

In the last years, a number of research works have been undertaken to deploy prototypes of experimental SBs over WSNs. In particular, the following SB services have been considered: electricity metering and energy saving [4, 15–17]; light and fan speed control [18]; ventilation system control [19, 20]; temperature sensing [21]; occupancy measurement, localization, and motion tracking [5, 6, 22, 23]; building surveillance [24]; and inhabitant wellness [25]. It should be noticed that Zigbee is the dominant technology used in the experimentations previously mentioned.

As the references below show, WSNs have been intensively addressed in smart building context. However, as far as we know, WSNs have been addressed on per-use case basis. In other words, published work showed deployment and performance analysis of WSNs to meet specific application requirements (e.g., monitoring indoor air quality in a specific building). Generic paradigms to address WSNs in SBs are still missing.

2.3 *IoT and SB*

IoT is a paradigm to enable interconnection of huge types of devices through sensors and actuators to the Internet. IoT enables a wide range of applications including energy saving, health, transportation, security, and leisure [26–28]. IoT application domains mainly include smart cities, smart grids, smart streets, smart buildings, and smart homes.

According to Ericsson's forecast [29], there will be around 30 billion IoT devices in 2022. Among that huge device number, a significant amount of smart devices would be used in smart buildings.

Some work addressed smart buildings from IoT point of view and suggested to consider SBs as applications of IoT. Carrillo et al. [30] proposed an IoT framework to enable smart building to communicate with each other (for example, in order to warn the locals around in case of fire or critical situations). Bellangente et al. [7] proposed an IoT application for energy management in smart buildings on a university campus.

3 SDN for Smart Buildings

3.1 Overview of SDN Architecture

Networks and services they provide become more and more complex, which results in huge difficulties in managing them. Configuration of traditional networks to fulfil new requirements and demands takes time and comes at high cost. Recently, software-defined networking (SDN) emerged as a solution to ease network management and to reduce costs [9].

SDN is based on three pillars: (1) softwarization of network device management, (2) centralization of decisions, and (3) separation of control plane from data plane of forwarding device. Softwarization provides a high level of flexibility and reduces manual efforts and human errors. Centralization provides a holistic view of the network, which enables to make optimal management decisions. Decoupling control plane from data plane enables virtualization of networks (i.e., decoupling applications and services from physical network specificities). Broadly, SDN consists of three planes (also called layers): application, control, and data planes as shown in Fig. 1 and it's detail is given below:

Application plane. It includes network management applications (also called *network applications*) such as routing, traffic engineering, load balancing, mobility management, security and intrusion detection, and fault tolerance. Application layer is still an open area.

Control plane. It centralizes the control tasks (i.e., configuring, monitoring, collecting forwarding statistics ...) regarding how data packets are handled at network devices (i.e., routers and switches). Implementation of control plane tasks results in a *networking operating system* (NOS). NOS is a centralized system, but it may be supported by one or several interconnected machines. Today, huge commercial and open-source SDN controllers are available.

Data plane (also called *infrastructure layer*). It is composed of forwarding devices (i.e., routers and switches), which forward data packets. At this plane, network devices are in charge of data packet-related actions—including receiving, forwarding, and dropping data packets—depending on the instructions from control plan.

Southbound interface—or application programming interface (API)—defines the protocol of communication between control plane and forwarding devices. Northbound interface is an API, which provides facilities for developing network applications.

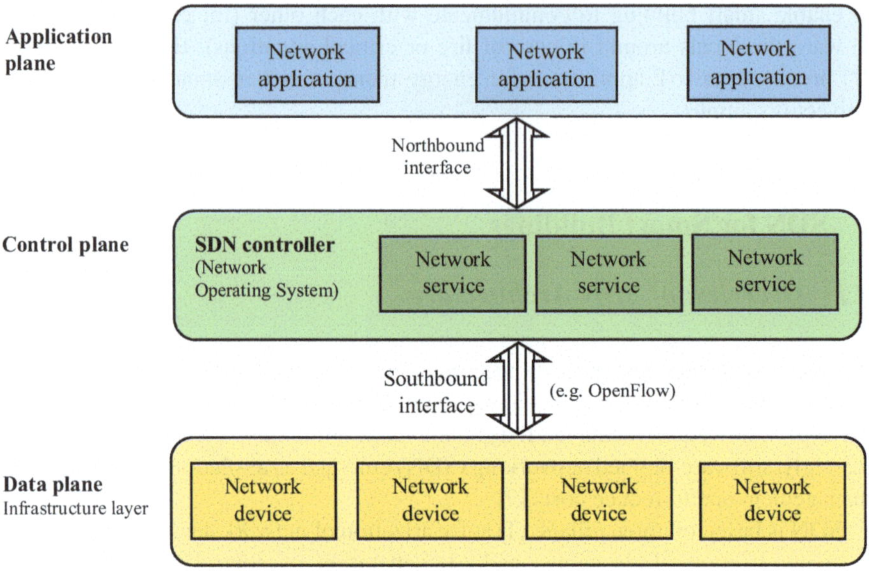

Fig. 1 Software-defined networking architecture

OpenFlow is the standard specification of southbound interface provided by open networking foundation (ONF) [31]. It is a protocol to allow SDN controllers deploy forwarding rules on switches and routers. OpenFlow was designed to work with existing products, without the need of specialized hardware. Today, a number of OpenFlow-compliant forwarding devices are available.

OpenFlow-compliant forwarding devices maintain tables (called OpenFlow tables). Each flow table in a forwarding device contains a set of flow entries; each flow entry consists of three components: match fields (including MAC and IP source and destination addresses, protocol type, port numbers, and other fields), instructions to apply to matching packets (including forward, drop, and update packet ...), and counters (to keep track of how many packets matched the rule). Flow entries are configured by SDN controllers through rules included in messages sent to forwarding devices. Controllers can add, update, and delete flow entries in flow tables, reactively or proactively, depending on network application requirements.

3.2 *SDN for Smart Buildings*

In standard Internet, nodes are often autonomous and may connect to each other at unpredictable times to receive data with very different characteristics. Thus, each node is in charge of almost networking functions, which results in complex and expensive nodes. In SBs, the context is different: (1) SBs are composed of objects (or smart devices) with limited resources, (2) data to/from objects is (often) known a priori in terms of characteristics and frequency of exchange, and (3) smart objects are deployed to participate to deliver services to building occupants. Thus, SDN approach is a good candidate, which offers a centralized view, to manage those objects and facilitate their deployment at low cost. Through SDN, it is easier to adapt (adding, moving, and removing nodes) the network topology to address new users' requirements.

As shown in Fig. 2, basic application of SDN paradigm to SB networking results in an architecture composed of SB services, an SDN controller, and switches to connect sensors and actuators. Management—including routing table configuration—of switches is made through OpenFlow, which is a *de facto* standard.

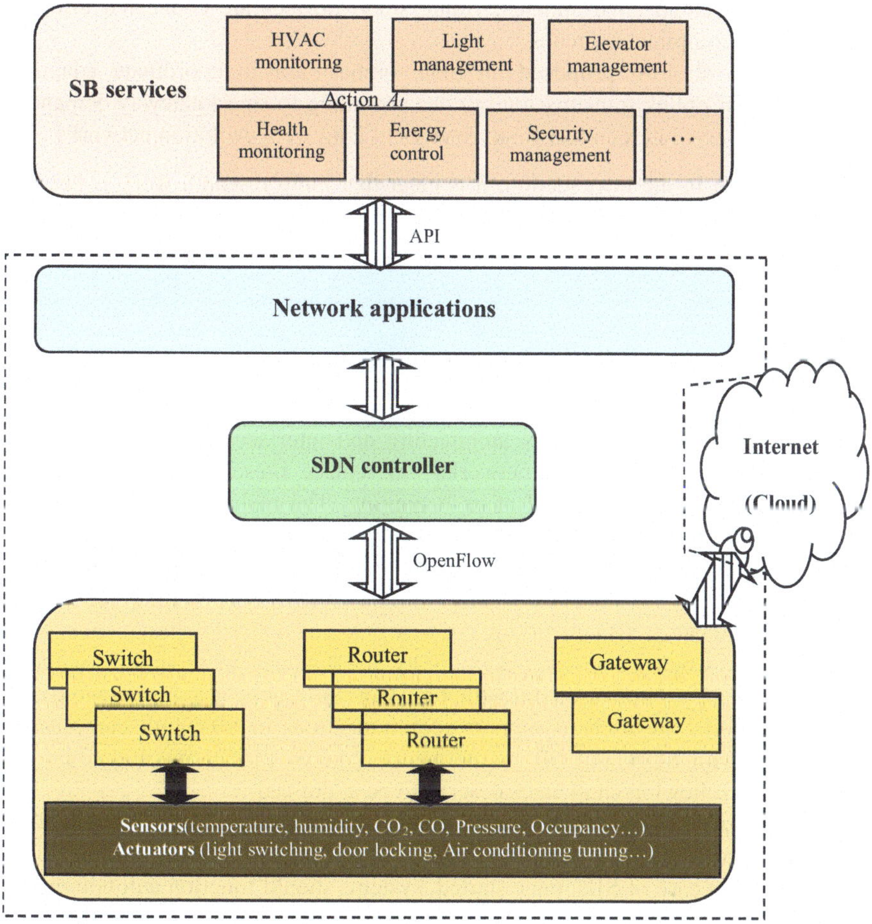

Fig. 2 SDN-based smart building architecture

While SDN has been intensively addressed in huge types of networks to facilitate their design, management, and maintenance [32], there are few works, which considered SDN for SB networks. Below is a summary of state of the art regarding SDN-based networks for SBs.

Tosic et al. [33] proposed a framework to manage forwarding tables of switches to collect data useful for context-aware service provisioning while balancing load in building networks to receive a high level of QoS.

Xue et al. [34] and Xu et al. [35] provided an OpenFlow-based framework to address authentication between central controller and smart devices. OpenFlow is regarded as a powerful tool to implement security rules on smart devices.

Kathiravelu et al. [36] proposed a context-aware middleware to seamlessly connect smart devices to the SDN controller.

Lee et al. [37] proposed a SDN-based method for auto-configuration of home networks in cloud environments. Home smart devices are connected through OpenFlow-compliant switches.

Thyagaturu et al. [38] proposed an SDN orchestrator to coordinate adaptive allocation of uplink transmission bitrates according to smart gateway demands. Smart gateways interconnect smart devices to long-term evolution networks.

As discussed in the next section, it is clear that more research work is needed before SDN-based networks would be capable of meeting communication requirements in SBs.

4 Challenges in SDN-Based Networking for SB

Roadmap of application of SDN architecture and interfaces for the design and deployment of SDN-compliant networks—to replace present networks used in different activity sectors—is still at its beginning. The real world of operational networks is not easy to evolve, because of huge hardware, protocols, and applications that are already deployed. Coexistence of current networks and future SDN-based networks would take a long time before SDN-based networks would dominate. Fortunately, SB networking is a new field and real SB deployment is very scarce. Thus, it would be easier to adopt SDN in SB management systems without jeopardizing legacy automation infrastructure. Currently, general-purpose wired and wireless networks are used to support communications in SBs. Full compliance of networks with SDN and SB requirements requires further investigations. In particular, the following challenges deserve to be addressed.

Autonomous management. BS management systems should not be designed as "traditional" computerized systems, which rely on frequent human administrator interventions. In case of SBs, once started, systems should function autonomously. In particular, networking aspects managed by SDN controllers should be autonomous. Adapting, reconfiguring, adding, and removing smart devices—without human intervention—to face any change in buildings are challenging.

Intelligent radiation control. As almost all devices used in SBs communicate wirelessly, radiations occur, which impacts not only network throughput, but—more important—health of building occupants. It is of paramount importance that radiations are to be kept at a minimum level to preserve health of people. Signal strength may be collected by SDN controller and analyzed to either reduce transmission range of some devices or stop some of them while limiting the overall QoS and not shutting down critical devices.

Robustness and availability of SB management system. Whenever some networking elements are unavailable, some SB services may become partially or totally unavailable. No users would agree on the installation of SB management system, which frequently enforces power shutdown or prevents them from accessing their homes, offices, or stores, because of network failures. Thus, SDN networks for SBs should be redundant based. Design and deployment of communication links and SDN controllers providing a given level of availability, at a reasonable cost, are challenging. SDN approach is based on centralization of decisions. How redundant SDN controllers may collaborate to make the whole system robust is challenging.

Quality of service guarantees. Some SB services (such as comfort service) do not have tight QoS requirements. However, other services (such as electricity management, health emergency assistance, fire emergency, and intrusion detection) are critical and have QoS requirements in terms of delay and reliability. To provide real-time and reliability guarantees for SBs—where various types of traffic coexist—new protocols and algorithms, in particular dynamic and adaptive routing protocols and resource reservation, are required. Very few works addressed QoS guarantees in SDN-oriented networks in SB context.

Security and privacy. Security and privacy are aspects of paramount importance. In cyber context, attacks may affect any element of SBs. Today's physical keys to access buildings will be replaced by smart devices to lock and unlock doors. To avoid cyberattacks and prevent unauthorized people from entering inside buildings or to intercept private data through cameras or other smart devices, efficient security mechanisms should be integrated. In case a single SDN controller is used to manage security, SBs become (very) vulnerable to attacks. Specification, verification, and distribution of security rules in SDN controllers and smart devices become a critical challenge. SBs could not be deployed in real world unless users trust them.

Collaboration of SB SDN controllers. SBs, urban traffic monitoring systems, public lightening systems, water supply systems, and many other systems are all elements of smart cities. Thus, collaboration between SBs is required to coordinate actions in cities (such as electricity provisioning from smart grids and emergency in case of fire or water flooding). SBs' collaboration at application level results in coordination at networking level. Design of SB SDN controllers to allocate network resources and provide priority to urgent traffic, at city level, is a challenge to move from SB islands to collaborating SBs.

Hierarchical SDN control and evolvability. On the one hand SBs provide services different in nature even if they share computing and networking resources. Networks used in SBs also are different from each other according to several dimensions (range, bitrate, protocols ...) on the other hand. Combining together multiple networking technologies to support various services would result in complex management and SBs hard to evolve. One way to overcome complexity and facilitate evolvability, in case of technology evolution, is the design of hierarchy of SDN controllers.

Intelligent maintenance. In conventional buildings, some elements (such as light bulbs and locks) are replaced when they do not work or are likely to not work in the near future. SBs are composed of software and hardware elements. Software require updates and hardware may fail or become obsolete. Thus, preventive maintenance of smart devices is useful for SBs. From the networking point of view, SDN design should consider smooth software updating and preventive maintenance using appropriate mechanisms to detect abnormal functioning of devices. In case some communication links are broken, SDN controllers should be designed to limit inconvenience to building users.

Machine learning-based SDN for SBs. High-quality SB services (e.g., keeping energy consumption at its minimum or privacy protection at its maximum) may be provided when SB management systems apply powerful learning models on collected data to derive behaviors that impact some indicators (e.g., high consumption of energy or vulnerability of some surveillance cameras). When networking is of concern, machine learning techniques may be integrated in SDN devices and controllers to optimize network resources including sensor node energy, predict nodes (objects) mobility, provide better QoS, enhance security, and anticipate sensor battery replenishment. Machine learning-based SDN networks for smart building are challenging.

5 Conclusions

To improve energy consumption and to provide comfort and assistance to building occupants—in case of emergency—buildings are becoming smart. In smart buildings, wireless technologies play a paramount role in smart device interconnection to enable the delivery of various services at low cost. This chapter discusses some issues regarding networking technologies used as underlying communication infrastructures to support smart buildings.

SDN recently emerged as a promising solution to facilitate network management. Smart buildings are new applications of ICT and they are more likely to use SDN in their network management without the replacement of the existing hardware and software components being a barrier. Existing work on SDN-based networks for SBs is very scarce and focused on integration of some existing technologies, like Zigbee, in SBs. However, full compliance of networks with SDN and SB requirements requires to address much more challenges regarding methodology, protocols, optimization techniques, and standardization. This chapter highlights some challenges regarding network management, radiation control, network robustness and availability, QoS, security and privacy protection, network maintenance, network control hierarchization, and integration of machine learning in SB network management to put SDN at the core of SB networks.

References

1. Bajer, M. (2018). IoT for smart buildings—long awaited revolution or lean evolution. In *IEEE 6th International Conference on Future Internet of Things and Cloud* (pp. 149–154). Barcelona: IEEE.
2. Minoli, D., Sohraby, K., & Occhiogrosso, B. (2017). IoT considerations, requirements, and architectures for smart buildings—energy optimization and next-generation building management systems. *IEEE Internet of Things Journal, 4*(1), 269–283.
3. Perez-Lombard, L., Ortiz, J., & Pout, C. (2008). A review on buildings energy consumption information. *Energy and Buildings, 40*(3), 394–398.
4. Hannan, M. A., et al. (2018). A review of internet of energy based building energy management systems: Issues and recommendations. *IEEE Access, 6*, 38997–39014.
5. Papatsimpa, C., & Linnartz, J. P. (2018). Energy efficient communication in smart building WSN running distributed hidden Markov chain presence detection algorithm. In *IEEE 4th World Forum on Internet of Things* (pp. 112–117). Singapore: IEEE.
6. Putra, G. D., Pratama, A. R., Lazovik, A., & Aiello, M. (2017). Comparison of energy consumption in Wi-Fi and Bluetooth communication in a smart building. In *IEEE 7th Annual Computing and Communication Workshop and Conference* (pp. 1–6). Las Vegas: IEEE.
7. Bellangente, P., Ferrari, P., Flammini, A., & Rinaldi, S. (2015). Adopting IoT framework for energy management of smart building: A real test-case. In *IEEE 1st International Forum on Research and Technologies for Society and Industry Leveraging* (pp. 1–6). Torino: IEEE.
8. Wong, J. K. W., Li, H., & Wang, W. (2005). Intelligent building research: A review. *Automation in Construction, 4*, 143–159.
9. Kruetz, D., et al. (2015). Software-defined networking: A comprehensive survey. *Proceedings of the IEEE, 103*(1), 14–76.
10. Alam, M. R., Bin Ibne Reaz, M., & Mohd Ali, M. A. (2012). A review of smart homes—past, present and future. *IEEE Transactions on Systems, Man, and Cybernetics, 42*(6), 1190–1203.
11. Lior, N. (2012). Sustainable energy development: The present (2011) situation and possible paths to the future. *Energy, 43*(1), 174–191.
12. Gungor, V., et al. (2011). Smart grid technologies: Communication technologies and standards. *IEEE Transactions on Industrial Informatics, 7*(4), 529–539.
13. Kuzlu, M., Pipattanasomporn, M., & Rahman, S. (2015). Review of communication technologies for smart homes/building applications. In *IEEE Innovative Smart Grid Technologies Conference—Asia* (pp. 1–6). Bangkok: IEEE.
14. Lobaccaro, G., Carlucci, S., & Lofstrom, E. (2016). A review of systems and technologies for smart, homes and smart grids. *Energies, 9*(348), 1–33.
15. Zhou, Z., Zhao, X., & Chong, P. H. (2013). Optimal relay node placement in wireless sensor network for smart buildings metering and control. In *15th IEEE International Conference on Communication Technology* (pp. 456–461). Guilin: IEEE.
16. Suryadevara, N. K., Mukhopadhyay, S. C., Kelly, S. D., & Singh Gill, S. P. (2015). WSN-based smart sensors and actuator for power management in intelligent buildings. *IEEE/ASME Transactions on Mechatronics, 20*(2), 564–571.
17. Byun, J., et al. (2012). An intelligent self-adjusting sensor for smart home services based on Zigbee communications. *IEEE Transactions on Consumer Electronics, 58*(3), 794–802.
18. Kaiwen, C., Kumar, A., Xavier, N., & Kumar Panda, S. (2016). An intelligent home appliance control-based on WSN for smart buildings. In *IEEE International Conference on Sustainable Energy Technologies* (pp. 282–287). Hanoi: IEEE.
19. Brito, J., et al. (2014). An intelligent home automation control system based on a novel heat pump and wireless sensor networks. In *IEEE 23rd International Symposium on Industrial Electronics* (pp. 1448–1453). Istanbul: IEEE.
20. Taneja, J., Krioukov, A., Dawson-Haggerty, S., & Culler, D. (2013). Enabling advanced environmental conditioning with a building application stack. In *International Green Computing Conference* (pp. 1–10). Arlington: IEEE.

21. Chen, X., et al. (2016). Energy-efficient wireless temperature sensoring for smart building application. In *IEEE International Conference on Solid-State and Integrated Circuit Technology* (pp. 680–683). Hangzhou: IEEE.
22. Brennan, C., Taylor, G. W., & Spachos, P. (2018). Distributed sensor network for indirect occupancy measurement in smart buildings. In *14th International Wireless Communications & Mobile Computing Conference* (pp. 1290–1295). Limassol: IEEE.
23. Chen, P., Ou, K.-S., & Chen, K.-S. (2010). IR indoor localization and wireless transmission for motion control in smart building applications based on Wiimote technology. In *SICE Annual Conference* (pp. 1781–1785). Taipei: IEEE.
24. Saha, S., & Neogy, S. (2014). A case study on smart surveillance application system using WSN and IP webcam. In *Applications and Innovations in Mobile Computing Applications and Innovations in Mobile Computing Conference* (pp. 36–41). Kolkata: IEEE.
25. Ghayvat, H., Liu, J., Mukhopadhyay, S., & Gui, X. (2015). Wellness sensor networks: A proposal and implementation for smart home for assisted living. *IEEE Sensors Journal, 15*(12), 7341–7348.
26. Uckelmann, D., Harrison, M., & Michahelles, F. (2011). *Architecting the Internet of Things*. Berlin: Springer.
27. Ng, I., & Wakenshaw, S. (2017). The Internet-of-Things: Review and research directions. *International Journal of Research in Marketing, 34*(1), 3–21.
28. Al-Fuqaha, A., et al. (2015). Internet of things: A survey on enabling technologies, protocols, and applications. *IEEE Communication Surveys & Tutorials, 17*(4), 2347–2376.
29. Ericsson. The connected future. Retrieved October 30, 2018, from https://www.ericsson.com/en/mobility-report/internet-of-things-forecast
30. Carrillo, C., Benitez, V., Mendoza, C., & Pacheco, J. (2015). IoT framework for smart buildings with cloud computing. In *IEEE 1st International Smart Cities Conference* (pp. 1–6). Guadalajara: IEEE.
31. ONF. (2015). Open Networking Foundation—OpenFlow Switch Specification. Version 1.5.1.
32. Haque, I., & Abu-Ghazaleh, N. (2016). Wireless software defined networking: A survey and taxonomy. *IEEE Communications Surveys & Tutorials, 18*(4), 2731–2737.
33. Tosic, M., Ikovic, O., & Boskovic, D. (2016). SDN based service provisioning management in smart buildings. In *39th International Convention on Information and Communication Technology, Electronics and Microelectronics* (pp. 754–759). Opatija: IEEE.
34. Xue, N., Huang, X., & Zhang, J. (2016). S^2Net: A security framework for software defined intelligent building networks. In *15th IEEE International Conference on Trust, Security and Privacy in Computing and Communications* (pp. 654–661). Tianjin: IEEE.
35. Xu, R. Y., et al. (2015). Software defined intelligent building. *International Journal of Information Security and Privacy, 9*(3), 84–99.
36. Kathiravelu, P., Sharrifi, L., & Veiga, L. (2015). Cassowary: Middleware platform for context-aware smart buildings with software-defined sensor networks. In *2nd Workshop on Middleware for Context-Aware Applications in the IoT* (pp. 1–6). New York: ACM.
37. Lee, M., Kim, Y., & Lee, Y. (2015). A home cloud-based home network auto-configuration using SDN. In *12th International Conference on Networking, Sensing and Control* (pp. 444–449). Taipei: IEEE.
38. Thyagaturu, A. S., Dashti, Y., & Reisslein, M. (2016). SDN-based smart gateways (Sm-GWs) for multi-operator small cell network management. *IEEE Transactions on Network and Service Management, 13*(4), 740–753.

Performance Comparison of Zoning Techniques in Geographic Forwarding Rules for Ad Hoc Networks

Mohammed Souidi and Ahmed Habbani

1 Introduction

The broadcasting is widely used in the MANETs. Nodes broadcast specific messages to share and to gather link state information in order to update the routing tables on large scale. In OLSR [5, 12], we distinguish two types of broadcast messages: HELLO messages are destined to sense links between nodes and to build local topology. The HELLO messages are never forwarded, and TC messages help to build the global topology. They are forwarded into the entire network. They transport the local information on large scale. To decrease the number of TC messages, the OLSR has adopted the Multipoint Relay (MPR) concept [9]. Only the MPR nodes can generate and forward TC messages, and only links between the MPRs and their selectors are shared. The concept of MPR minimizes the number of nodes participating into the flooding of TC messages. However, the Default Forwarding Rules (DFR) of OLSR only take into account the source of the message, which leads to redundant retransmissions and causes more collisions and congestion in the radio channel [15, 16]. The GFR technique [13] uses the location information of nodes to make selective forwarding taking into consideration the positions of destination nodes. It divides the network into virtual partitions based on the originator's position and avoids forwarding TC messages between them. The algorithm considers a transmission as redundant when a MPR forwards a message when all nodes in its neighborhood have received it or can receive it from other MPRs. In [13], the authors have presented several ways to compute the propagation zones. Nevertheless, they studied only vertical division with two zones. In this paper,

M. Souidi (✉) · A. Habbani
SSL Lab, ENSIAS, Mohammed V University in Rabat, Rabat, Morocco
e-mail: mohammed_souidi@um5.ac.ma; ahmed.habbani@um5.ac.ma; http://ensias.um5.ac.ma

© Springer Nature Switzerland AG 2020
D. Cagáňová, N. Horňáková (eds.), *Mobility Internet of Things 2018*,
EAI/Springer Innovations in Communication and Computing,
https://doi.org/10.1007/978-3-030-30911-4_4

we study other zoning techniques with 2, 4, and 8 zones. We draw analysis on how the multiplication of the propagation zones can affect the overhead and the performance of the network. We compare their impacts on the exchange of TC messages and we measure the reliability of the network by analyzing the packets exchanged at the physical layer of nodes.

Our paper is organized as follows: Sect. 2 presents some techniques used to reduce the overhead in MANETs. Section 3 presents the mathematical model of OLSR. Section 4 describes the integration of the GFR to the OLSR protocol. Section 5 presents the propagation zones studied in this paper. Section 6 describes the simulations' parameters and results. Finally, Sect. 7 concludes the paper.

2 Related Work

Many researches have used the location information of nodes to reduce the cost of overhead in MANETs. They aim to decrease the number of retransmissions or the number of nodes responsible for flooding messages. The Location Aided Routing (LAR) [7, 10] forwards the Route Request (RR) toward the destination node. It delimits the nodes participating to the flooding of the RR inside a forwarding zone such as only the farthest node rebroadcasts the message. However, LAR is a centralized algorithm. It needs to know the positions of all nodes to work properly. In [6, 16, 17], the authors have proposed to organize nodes in hierarchical structure called Cluster. Only the central unit, the Cluster-head (CH), broadcasts messages, while the Gateways (GW) are responsible of inter-cluster communications. The network does not start until clusters are constructed. This technique uses an initial phase to form the groups of nodes. It is computed every time the hierarchical structure falls, which may increase the lost packets. Moreover, the collision rises at the level of the CH causing more collisions [2]. Greedy forwarding in [11] divides the network into several zones based on the positions of the most distant two nodes. Nodes freely advertise broadcast messages inside each grid but only one node forwards them to the adjacent partition. Like LAR, all nodes have to know the positions of the other nodes to compute the grids. This information is not always available and takes much time to be updated via the network. The GFR technique removes all these drawbacks. It functions without any initial phase in a decentralized manner. The algorithm starts from the start of the network and requires no latency. Only the position of the originator of TC message is required to form the zones. The GFRs can reduce the number of retransmissions even with partial network topology information.

3 OLSR

The MPR concept has been well treated in the literature. It decreases the number of nodes participating into the flooding of broadcast messages. Every node u in OLSR exchanges HELLO messages to discover the local information and to elect its MPRs. Each time u receives an OLSR packets, it immediately updates its topology information denoted G'. Let $N1(u)$, $N2(u)$, $M(u)$, and $MS(u)$ denote, respectively, the 1-hop neighbor set, the 2-hop neighbor set, the MPR set, and the MPR selector set of node u. R is the signal range of u. For simplicity R is uniform for all nodes. A MPR x generates a TC message every 5 s to inform the entire network about the local topology. x advertises links between itself and nodes in $MS(x)$. If $x \in M(u)$ and u sends a TC_u message, x rebroadcasts it immediately. To avoid loops, each MPR maintains a table of shared TC messages. Figure 1 shows the Default Forwarding Rules of the native OLSR.

4 Geographic Forwarding Rules

The GFRs presented in [13] use the system created by the location service [4], such as GPS [8] in open areas or anchors [3] in the case of indoors, to locate the positions of nodes. Each node u is identified by its coordinates (X_u, Y_u). The GFR technique partitions the network into propagation zones using the coordinates of the originator of the TC message.

When a MPR o generates a TC message at time t, it introduces its coordinates (X_o^t, Y_o^t) to the message header. Figure 2 shows the integration of the originator's position to the message header of an OLSR packet.

Fig. 1 Default forwarding rules of OLSR

```
Procedure M(u)
If x ∈ MS(u)
  If TCx is not already forwarded
  u forwards TCx
End procedure
```

```
 0                   1                   2                   3
 0 1 2 3 4 5 6 7 8 9 0 1 2 3 4 5 6 7 8 9 0 1 2 3 4 5 6 7 8 9 0 1
+-+-+-+-+-+-+-+-+-+-+-+-+-+-+-+-+-+-+-+-+-+-+-+-+-+-+-+-+-+-+-+-+
|  Message Type |     Vtime     |         Message Size          |
+-+-+-+-+-+-+-+-+-+-+-+-+-+-+-+-+-+-+-+-+-+-+-+-+-+-+-+-+-+-+-+-+
|                      Originator Address                       |
+-+-+-+-+-+-+-+-+-+-+-+-+-+-+-+-+-+-+-+-+-+-+-+-+-+-+-+-+-+-+-+-+
|  Time To Live |   Hop Count   |    Message Sequence Number    |
+-+-+-+-+-+-+-+-+-+-+-+-+-+-+-+-+-+-+-+-+-+-+-+-+-+-+-+-+-+-+-+-+
|                         Node Position                         |
+-+-+-+-+-+-+-+-+-+-+-+-+-+-+-+-+-+-+-+-+-+-+-+-+-+-+-+-+-+-+-+-+
```

Fig. 2 The modified Header message format

```
Procedure M(x)
    If s ∈ MS(x)
        If TC_o^t is not already forwarded
        If( s is the originator of TC_o^t
      or ∃ u ∈ N^{-s}(x) | u is located on the boundary
  of Z_o^t(u) or ∀ u ∈ N(x) | Z_o^t(u) = Z_o^t(x))
            x forwards TC_o^t
        Else if ( ∀ v ∈ N^{-s}(x), ∃ P_v ⊂ G'_v(x) from
        which v will receive the TC_o^t)
            x does not forward TC_o^t
            Else x forwards TC_o^t
End procedure
```

Fig. 3 Geographic forwarding rules of OLSR

Let $|Z_o^t|$ and $|B_o^t|$ denote the propagation zone set and their borders, respectively, computed from the position (X_o^t, Y_o^t). $Z_o^t(u)$ is the zone of node u. When a MPR x receives the TC_o^t message from node $s \in MS(x)$, it retransmits it immediately if:

1. s is the originator of the TC_o^t message, $s = o$. The MPRs of the originator always forward TC messages to ensure a good dissemination at the beginning of the flooding process.
2. Or $\exists u \in N(x)$ as u is located between two zones on the border $B_o^t(u)$.
3. Or $\forall u \in N(x)$, the zones of u and x are the same, $Z_o^t(u) = Z_o^t(x)$.

However, if it exists a neighbor u in another zone, $Z_o^t(x) \neq Z_o^t(u)$, x will try to avoid the transmission of TC_o^t. Based on $G'(x)$, x will try to find a path P_v for every neighbor $v \in N^{-s}(x)$ as $N^{-s}(x) = N(x) \backslash \{N(x) \cap N(s), s\}$. P_v is a reverse route from MPR to its MPR selector until reaching the originator o as $\forall q \in P_v$, $Z_o^t(q) = Z_o^t(v)$. Figure 3 details the GFRs algorithm used with OLSR.

5 Propagation Zones

The propagation zones are computed from the originator's position. They separate the network into groups and define their borders. In [13], several types of propagation zones have been presented. Nevertheless, the authors had studied only vertical division with two zones. Figure 4 presents other types of propagation zones we study in this paper.

In Fig. 4a the network is divided vertically into two propagation zones. The border line is $X = X_o^t$. We mention this partitioning method as 2ZV-GFR:

- u is located in Z_1 if $X_u > X_o^t$
- u is located in Z_2 if $X_u < X_o^t$
- u is on the boundary line if $X_u = X_o^t$

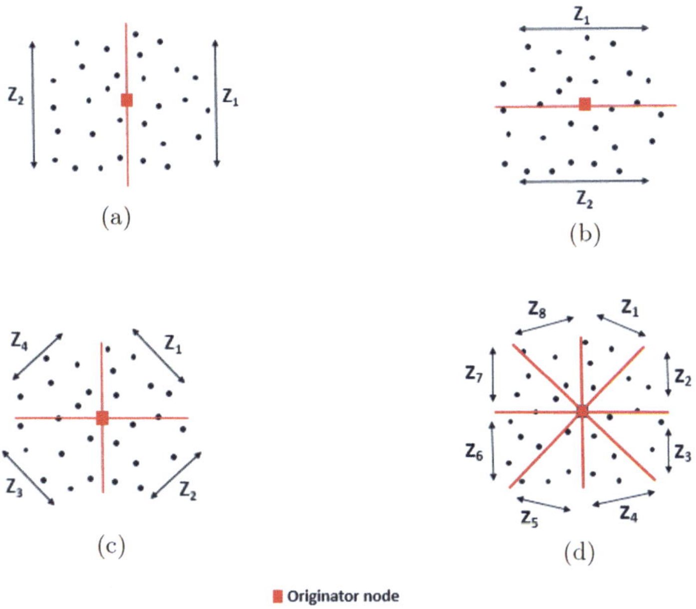

Fig. 4 Studied propagation zones. (**a**) 2ZV-GFR. (**b**) 2ZH-GFR. (**c**) 4Z-GFR. (**d**) 8Z-GFR

In Fig. 4b the network is divided horizontally into two propagation zones. The border line is $Y = Y_o^t$. We mention this partitioning method as 2ZH-GFR:

- u is located in Z_1 if $Y_u > Y_o^t$
- u is located in Z_2 if $Y_u < Y_o^t$
- u is on the boundary line if $Y_u = Y_o^t$

In Fig. 4c, nodes are separated by four propagation zones. The boundaries are the axis $X = X_o^t$ and the axis $Y = Y_o^t$, they form a system of origin (X_o^t, Y_o^t). We mention this partitioning method as 4Z-GFR:

- u is located in Z_1 if $X_u > X_o^t$ and $Y_u > Y_o^t$
- u is located in Z_2 if $X_u > X_o^t$ and $Y_u < Y_o^t$
- u is located in Z_3 if $X_u < X_o^t$ and $Y_u < Y_o^t$
- u is located in Z_4 if $X_u < X_o^t$ and $Y_u > Y_o^t$
- u is on the boundary line if $X_u = X_o^t$ or $Y_u = Y_o^t$

In Fig. 4d, the network is divided into eight propagation zones. We mention this partitioning method as 8Z-GFR. Each partition forms a triangle adjacent to the Y-axis or the X-axis shaped with an angle equal to 45°. To determine the zone of any node u, we infer from its position (X_u, Y_u) the coordinates of the originator (X_o^t, Y_o^t). Then the new coordinates of the node u are $X_u^o = X_u - X_o^t$ and $Y_u^o = Y_u - Y_o^t$. The boundary lines are $X = 0$, $Y = 0$, $Y = X$, and $Y = -X$. The origin of the system is $(X_o^t = 0, Y_o^t = 0)$. Four cases arise:

1. $X_u^o > 0$ and $Y_u^o > 0$
   - if $0 < \frac{X_u^o}{Y_u^o} < 1$, u belongs to Z_1
   - if $\frac{X_u^o}{Y_u^o} > 1$, u belongs to Z_2
   - if $\frac{X_u^o}{Y_u^o} = 1$, u is on the border line between the two zones Z_1 and Z_2
2. $X_u^o > 0$ and $Y_u^o < 0$
   - if $\frac{X_u^o}{Y_u^o} < -1$, u belongs toZ_3
   - if $-1 < \frac{X_u^o}{Y_u^o} < 0$, u belongs to Z_4
   - if $\frac{X_u^o}{Y_u^o} = -1$, u is on the boundary between the two zones Z_3 and Z_4
3. $X_u^o < 0$ and $Y_u^o < 0$
   - if $0 < \frac{X_u^o}{Y_u^o} < 1$, u belongs to Z_5
   - if $\frac{X_u^o}{Y_u^o} > 1$, u belongs to Z_6
   - if $\frac{X_u^o}{Y_u^o} = 1$, u is on the boundary between the two zones Z_5 and Z_6
4. $X_u^o < 0$ and $Y_u^o > 0$
   - if $\frac{X_u^o}{Y_u^o} < -1$, u belongs to Z_7
   - if $-1 < \frac{X_u^o}{Y_u^o} < 0$, u belongs to Z_8
   - if $\frac{X_u^o}{Y_u^o} = -1$, u is on the boundary between the two zones Z_7 and Z_8

The GFRs are not limited only to this kind of partitioning. Other zoning techniques can be used with the GFR algorithm based on the position of the originator. In our paper, we have focused our study on these methods.

6 Simulations and Results

We have used the OLSR module of the Network Simulator NS-3 [14] and BonnMotion [1] to generate the same mobility of nodes for realistic comparison. We have implemented the four types of propagation zones with the GFRs. Then we experimented simulations with different densities of nodes moving randomly in a fixed area of 1000 m $\times$1000 m. We measured the impact of the partitioning method on the exchange of TC messages, the OLSR packets, and the dropped packets. The communication radio of each node was set to 100 m. Table 1 summarizes the simulations' parameters.

Figure 5 shows the quantity of topology control messages generated by all MPRs. The GFRs do not affect the TC message generation function or the number of MPRs. The number of TC messages is the same for all protocols regardless of the broadcasting technique.

Table 1 Simulations' parameters

Simulation environment	Parameters
Zone size	1000 m × 1000 m
Density of nodes	20, 40, 60, 80, 100, 120
Communication radio	$R = 100$ m
Modulation	802.11b peer to peer mode
Data mode	DsssRate11 Mbps
Control mode	DsssRate11 Mbps
Mobility model	Random waypoint
Simulation time	100 s

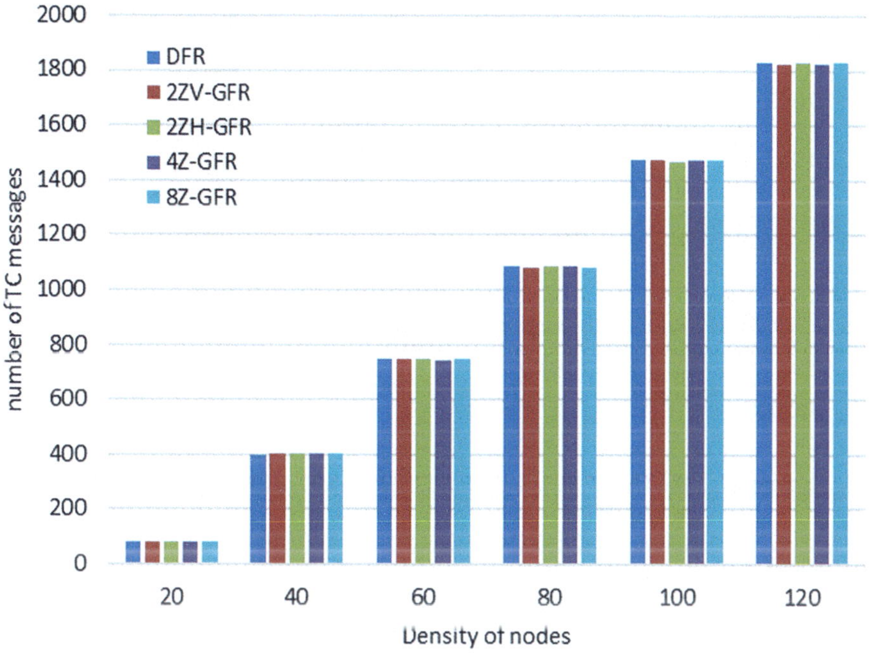

Fig. 5 Generated TC messages by the MPRs

Figure 6 illustrates the retransmissions of TC messages by all MPRs. As expected, the GFR technique reduces the number of retransmissions compared to the native OLSR which decreases the number of receptions as shown in Fig. 7. The gain becomes high as the number of nodes grows, which makes the network more reliable (Fig. 7).

For 2ZV-GFR and 2ZH-GFR, the number of banned retransmissions is approximately the same. The small difference is due to the network topology. The two partitioning techniques create different network topologies in each zone, which affects the search for the alternative path P_v.

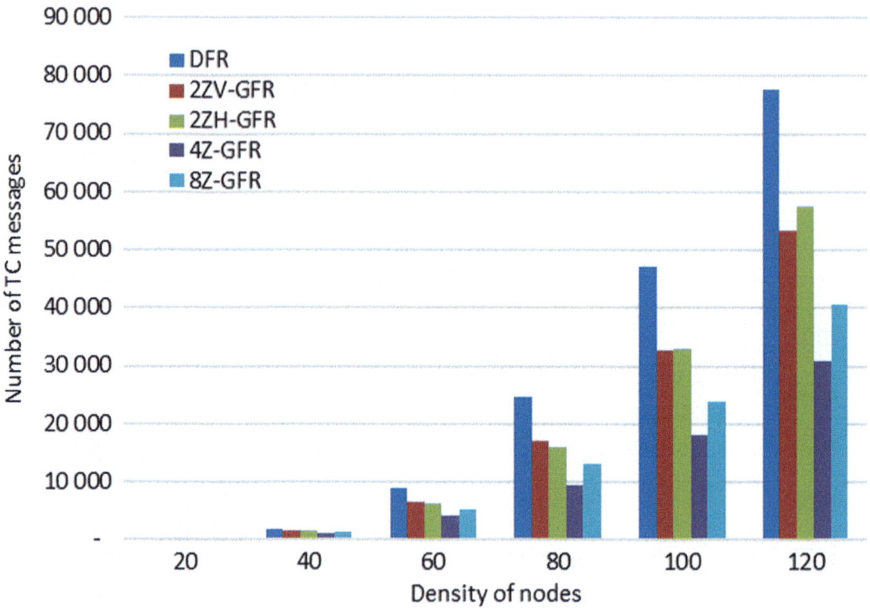

Fig. 6 Retransmissions of TC messages by the MPRs

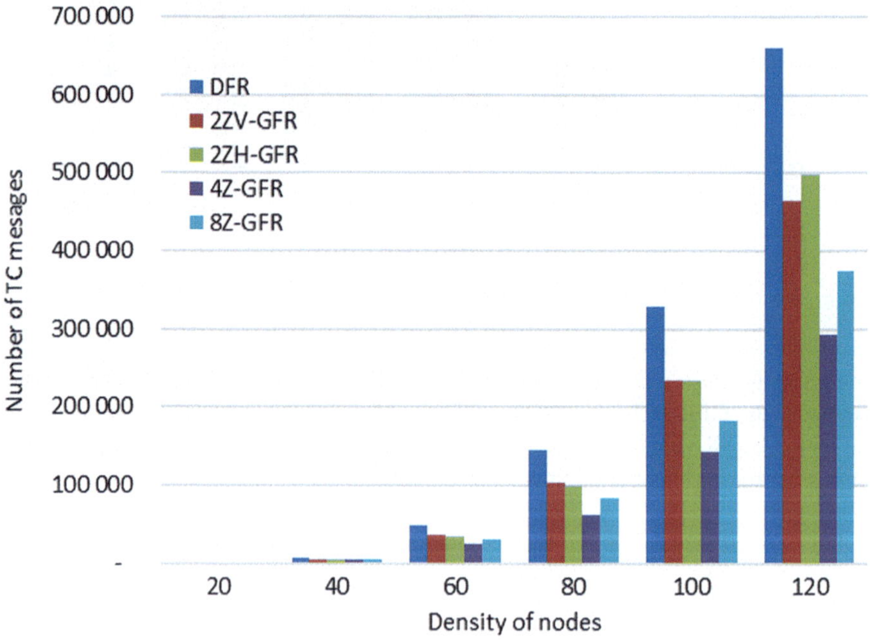

Fig. 7 Number of receptions of TC messages by all nodes

The 4Z-GFR technique has more opportunities to reduce the overhead. The algorithm bans useless retransmissions on two axes. The gain is approaching half of GFRs with two zones results.

Despite the multiplication of partitions in 8Z-GFR, the reduction of useless retransmissions has gone down with eight zones. It is less than that of four zones and better than that of two zones. This comes from the fact that the density of MPRs has decreased in each zone especially around the originator. It is difficult to avoid redundant messages in this area due to the lack of MPRs. However, as the network is going large, the banned retransmissions increase due to multiplication of border lines. These offer more opportunities to reduce the overhead than other methods with few zones.

Figures 8 and 9 present the number of packets sent to the MAC layers of nodes. The GFRs reduce the useless messages, which reduces the number of packets needed to hold them. We notice that 2ZV-GFR and 2ZH-GFR use less packets than the native OLSR. The difference recorded between the two partitioning methods is related to unsent TC messages showed in Fig. 6. The 4Z-GFR technique has reduced more useless retransmissions which permitted to scale down the number of packets. The gain reached 10% for large networks. The 8Z-GFR algorithm has used more packets than 4Z-GFR method because it has transmitted more TC messages.

Nodes in MANETs share the same radio channel. Since TC messages are sent in a broadcast manner, all nodes are affected. Figure 10 displays dropped packets

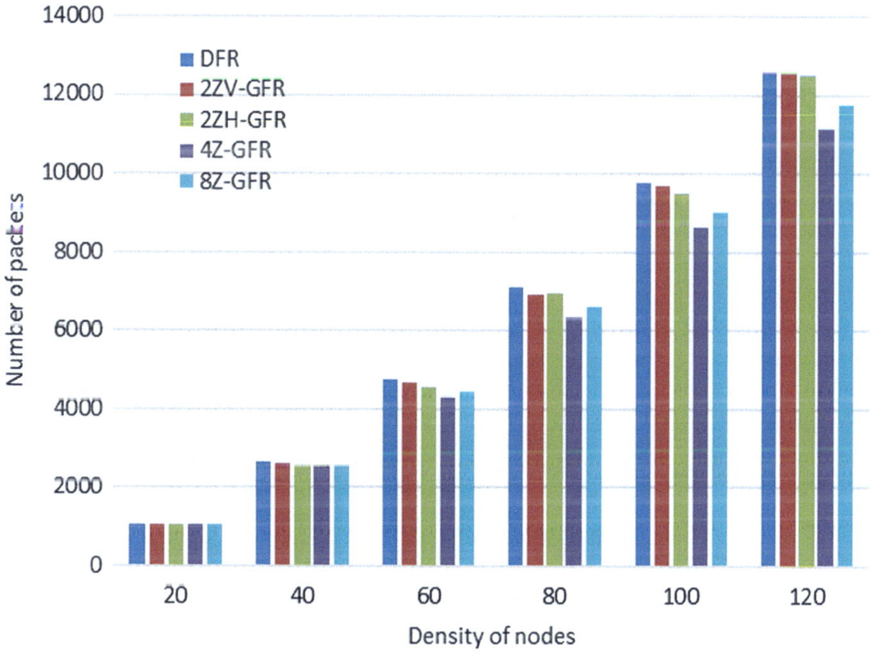

Fig. 8 Transmitted packets to the MAC layers of nodes

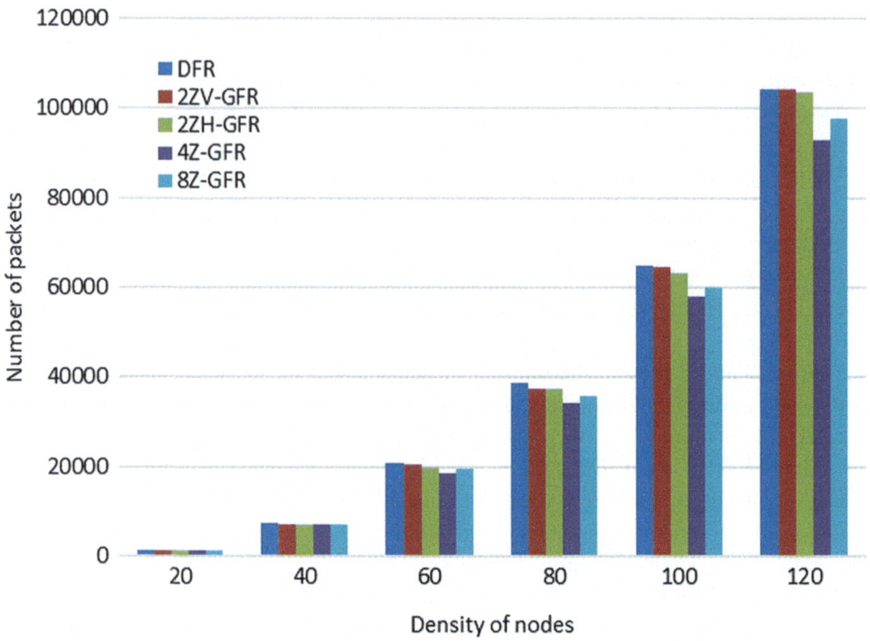

Fig. 9 Received packets at the MAC layers of nodes

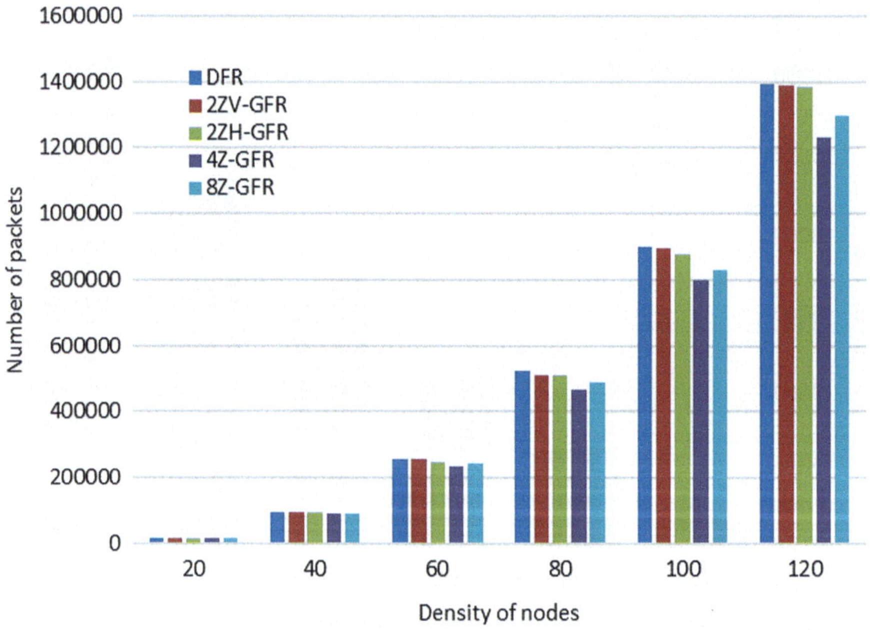

Fig. 10 Dropped packets at the physical layers of nodes

at the physical layers of nodes. The bigger the number of exchanged packets is, the bigger the number of interferences would be. All algorithms use different number of packets, which will be reflected on dropped packets. The dropped packets decrease as the network is getting denser, especially in 4Z-GFR and 8Z-GFR algorithms. They dropped, respectively, 12% and 8% of packets less than the native OLSR. This performance will enhance the reliability of the network.

7 Conclusion

The broadcasting is widely used in MANETs. It helps to disseminate information without infrastructure. However, the broadcasting increases the collision and the energy consumption. The Geographic Forwarding Rules have proved their ability to scale down the cost of flooding by avoiding the useless retransmissions. They use the location information of nodes to divide the network into propagation zones and they ban retransmissions between them. In this paper, we have studied different types of propagation zones with the GFRs implemented in the context of OLSR. Then we have compared the gain of each partitioning technique to the performance of the Default Forwarding Rules of OLSR in terms of traffic overhead including TC messages and exchanged packets. We conclude that the multiplication of propagation zones offers more opportunities to avoid the redundant retransmissions on several border lines. However, the multiplication of zones is proportionally related to the size of the network and the density of MPRs in each partition. In the future, we will try to adapt dynamically the choice of the partitioning technique to the density of nodes.

References

1. Aschenbruck, N., Ernst, R., Gerhards-Padilla, E., & Schwamborn, M. (2010). *BonnMotion: A mobility scenario generation and analysis tool*. Paper presented at the Proceedings of the 3rd International ICST Conference on Simulation Tools and Techniques, Torremolinos, Malaga, Spain.
2. Fujii, S., Takyu, O., Sasamori, F., Handa, S., Fujii, T., & Ohta, M. (2017). Optimal cluster head selection and rotation of cognitive wireless sensor networks for simultaneous data gathering. In *2017 International Conference on Information Networking (ICOIN), 11–13 January, 2017* (pp. 303–308).
3. Han, G., Jiang, J., Zhang, C., Duong, T. Q., Guizani, M., & Karagiannidis, G. K. (2016). A survey on mobile anchor node assisted localization in wireless sensor networks. *IEEE Communications Surveys and Tutorials, 18*(3), 2220–2243.
4. Hightower, J., & Borriello, G. (2001). Location systems for ubiquitous computing. *Computer, 34*(8), 57–66.
5. Jacquet, P., Muhlethaler, P., Clausen, T., Laouiti, A., Qayyum, A., & Viennot, L. (2001). Optimized link state routing protocol for ad hoc networks. In *IEEE INMIC 2001: Proceedings of IEEE International Multi Topic Conference, 2001 Technology for the 21st Century, 2001* (pp. 62–68).

6. Jamaesha, S. S., & Bhavani, S. (2018). A secure and efficient cluster based location aware routing protocol in MANET. *Cluster Computing*, 1–8.
7. Pandey, K., Raina, S. K., Raw, R. S., & Singh, B. (2017). Throughput and delay analysis of directional-location aided routing protocol for vehicular ad hoc networks. *International Journal of Communication Systems, 30*(9), e3192.
8. Parmar, B. M., & Maradia, K. G. (2018). GPS-Aided AODV Routing Protocol for MANET. In *Optical and wireless technologies* (pp. 585–598). Springer: Berlin.
9. Qayyum, A., Viennot, L., & Laouiti, A. (2000). *Multipoint relaying: An efficient technique for flooding in mobile wireless networks*. INRIA: Rocquencourt
10. Rana, K. K., Tripathi, S., & Raw, R. S. (2018). Analytical analysis of improved directional location added routing protocol for VANETS. *Wireless Personal Communications, 98*(2), 2403–2426.
11. Si, W., Tse, Q., Mao, G., & Zomaya, A. Y. (2018). On the performance of greedy forwarding on Yao and Theta graphs. *Journal of Parallel and Distributed Computing, 117*, 87–97. https://doi.org/10.1016/j.jpdc.2018.02.006
12. Souidi, M., Habbani, A., Berradi, H., & Essaid, B. (2017). Node localization to optimize the MPR selection in smart mobile communication. In *Proceedings of the 2017 International Conference on Smart Digital Environment* (pp. 8–13). ACM: New York.
13. Souidi, M., Habbani, A., Berradi, H., & El Mahdi, F. (2018). Geographic forwarding rules to reduce broadcast redundancy in mobile ad hoc wireless networks. *Personal and Ubiquitous Computing*, 1–11.
14. *The network simulator NS-3*. https://www.nsnam.org/
15. Tseng, Y.-C., Ni, S.-Y., Chen, Y.-S., & Sheu, J.-P. (2002). The broadcast storm problem in a mobile ad hoc network. *Wireless Networks, 8*(2–3), 153–167.
16. Vegni, A. M., & Natalizio, E. (2015). Forwarder smart selection protocol for limitation of broadcast storm problem. *Journal of Network and Computer Applications, 47*, 61–71. https://doi.org/10.1016/j.jnca.2014.08.009
17. Wang, Z., Liu, L., Zhou, M., & Ansari, N. (2008). A position-based clustering technique for ad hoc intervehicle communication. *IEEE Transactions on Systems, Man, and Cybernetics, Part C (Applications and Reviews), 38*(2), 201–208. https://doi.org/10.1109/TSMCC.2007.913917

State of the Art and Challenges of Robotics in Slovak Republic

Miroslav Balog, Ľubica Kotorová Slušná, and Jozef Herčko

1 Introduction

Global competition influences the development of all sectors and economic activities. Industry is among the sectors exposed to the highest competition due to its global character. One of the answers to the increasing demands of the competitive environment is to increase productivity and to introduce product and process innovation through a significant technological upgrading, the so-called fourth industrial revolution. So, it is necessary to transform traditional approach production systems to digital production systems [1]. The growing fusion of industrial production and information communication technologies of the fourth industrial revolution is called Industry 4.0 [2]. The concept of Industry 4.0 links information, objects, and people due to the convergence of the physical and virtual (cyberspace) world in the form of cyber-physical systems, enabling companies to transform into the so-called smart environment [3, 4]. The onset of the fourth industrial revolution is a major breakthrough in the development of the industry, putting pressure on the redefinition of companies' business models [5].

Industry 4.0 represents the transition from "centralized" to "decentralized" smart manufacturing and production. Smart production is a system where devices,

M. Balog (✉)
Center of Social and Psychological Sciences of Slovak Academy of Sciences, Bratislava, Slovakia
e-mail: progmiba@savba.sk

Ľ. K. Slušná
Strategos, s.r.o., Bratislava, Slovakia

J. Herčko
University of Žilina, Žilina, Slovakia

© Springer Nature Switzerland AG 2020
D. Cagáňová, N. Horňáková (eds.), *Mobility Internet of Things 2018*,
EAI/Springer Innovations in Communication and Computing,
https://doi.org/10.1007/978-3-030-30911-4_5

systems, and networks are able to independently collaborate with the objective of managing the process of industrial manufacturing.

The fourth industrial revolution will bring in new technologies and concepts such as cyber-physical systems, product life cycle management systems, digital manufacturing, digital twin, exponential technology, big data, cloud computing, Internet of Things (and services), robotics, telematics, and cyber security.

The fourth industrial revolution will bring significant productivity gains and growth of the overall efficiency through the implementation of new technologies.

Industry 4.0 is considered to be the next phase of the manufacturing sector digitization driven by the astonishing rise in data volumes, computational power, and connectivity, especially new low-power wide-area networks; the emergence of analytics and business intelligence capabilities; new forms of human-machine interaction such as touch interfaces and augmented reality systems; and improvements in transferring digital instructions to the physical world, such as advanced robotics and 3D printing [6]. The Industry 4.0 technological field is not new but it is highly heterogeneous (actually it is the aggregation point of more than 30 different fields of technology) [7]. Digital technology compass (Fig. 1) of the Industry 4.0 covers a variety of state-of-the-art technologies [8].

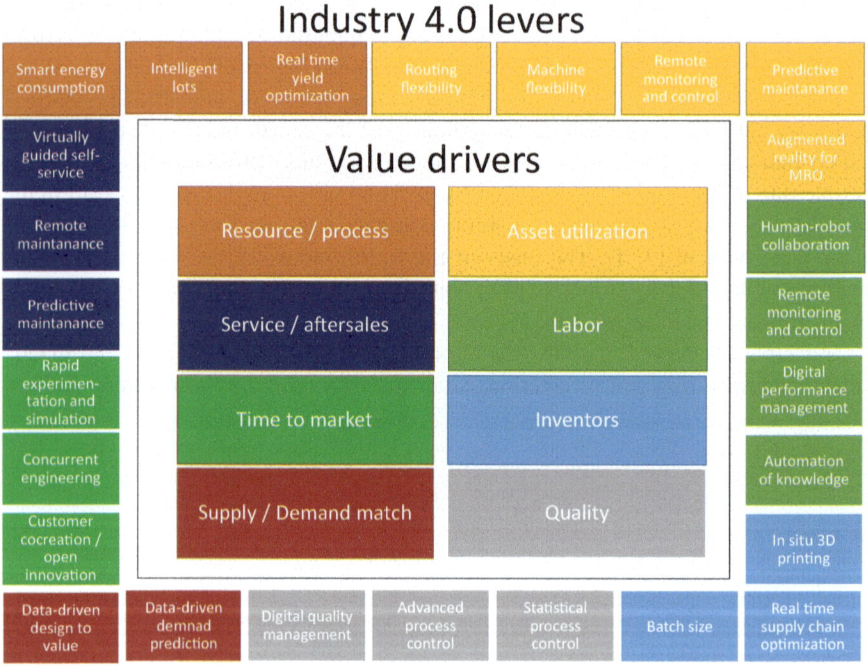

Fig. 1 Digital compass of the Industry 4.0 (own procession based on [8])

Industry 4.0 can be perceived in terms of two different theoretical approaches: long-wave theory or policy-driven discourse. Policy-driven innovation discourse in manufacturing industries aims to institutionalize innovation systems that encompass business, academia, and politics, an innovation system mode known as triple helix mode of innovation. A comparison of these approaches is presented in Table 1 [9].

One of the consequences of increased use of ICT is the pressure put on companies to change from product to service mindset [10]. Germany was the first country to officially introduce this concept as Industry 4.0. Industry 4.0 has highlighted the potential for making policies that create momentum for innovations beyond the frontiers of innovative policies [9].

Despite the benefits of implementing the Industry 4.0 concept, there are barriers to its implementation. Selected barriers to adoption of Industry 4.0 have been extensively studied, for example in India (Table 2) where the relationship between the individual barriers was investigated. The study used structural modelling and fuzzy MICMAC analysis. Contextual relationships between observed barriers have been identified [11].

Government efforts should focus on supporting the implementation of the Industry 4.0 concept, including digital economy, even though there is uncertainty whether government interventions will be able to compensate the "losers": it means that reaching Kaldor-Hicks efficiency is extremely difficult. Therefore, it is necessary to implement an upgrade in governance known as governance 2.0, which will support structural changes through Industry 4.0 and digitalization [12].

The World Economic Forum in Davos has identified advanced robotics as one of the industry's leading technology drivers of the Industry 4.0 concept. According

Table 1 Industry 4.0 from theoretical point of view

	Industry 4.0 as long wave	Industry 4.0 as policy-driven discourse
Core identity	Technological revolution	Policy-driven innovation discourse
Outcome	Changed social and economic life	Institutionalized innovation systems
Time frame	Long term	Medium term
Relevance of actors	Low	High
Possibility for codetermination	Low	High

Table 2 Barriers for Industry 4.0

Employment disruptions	Lack of standards and reference architecture
High implementation cost	Lack of Internet coverage and IT facilities
Organizational and process changes	Security and privacy issues
Lack of clear comprehension about IoT benefits	Seamless integration and compatibility issues
Need for enhanced skills	Regulatory compliance issues
Lack of knowledge management systems	Legal and contractual uncertainty

to key robotic business executives, in the next decades, collaborative robots will become the main robotic technology [13]. Robotics is one of the key technologies that create the image of the technological changes associated with the fourth industrial revolution.

The global growth of the robotics market occurred after the crisis in 2009, when the global demand started to grow. In recent years, the industrial robotics segment has seen truly impressive, dynamic growth. Productivity of production process, capital, and investment returns and consequently the entire profit of companies all have a bearing on it [14]. Global sales have doubled over the last 5 years (see chart below). In 2017, 381,335 industrial robots were sold worldwide. For comparison, between 2005 and 2008, the average number of robots sold was 115,000 units. Between 2011 and 2017, 236,000 robots were sold. In 2017, sales of industrial robots worldwide increased by 30% compared to 2016.

Up to 73% of global industrial robot sales go to five countries, namely China, Japan, South Korea, the United States, and Germany. Since 2013, China has been the world's largest robotic market with continual dynamic growth.

In 2017, the average robot density (number of industrial robots installed) in the worldwide manufacturing industry reached 85 robots per 10,000 employees. Europe is the region with the highest robot density per number of employees; in 2017, the average robot density per 10,000 employees reached 106 robots, 91 robots in the United States, and 75 robots in Asia/Australia. Looking at individual countries, Korea is at the top of the list; the number of robots per 10,000 employees reached 710 in this country followed by Singapore with 658 robots per 10,000 employees. In Slovakia, this indicator reached 151 robots in 2017 [15]. For the future, the robot application is expected to grow further (Fig. 2).

The automotive industry remains the most important customer for industrial robots. Investments into new industrial capacities in developing countries and into the modernization of production capacities by major automotive producers have contributed to growing number of new robot installations. Industry 4.0 is one of the main drivers of the development of industrial robotics.

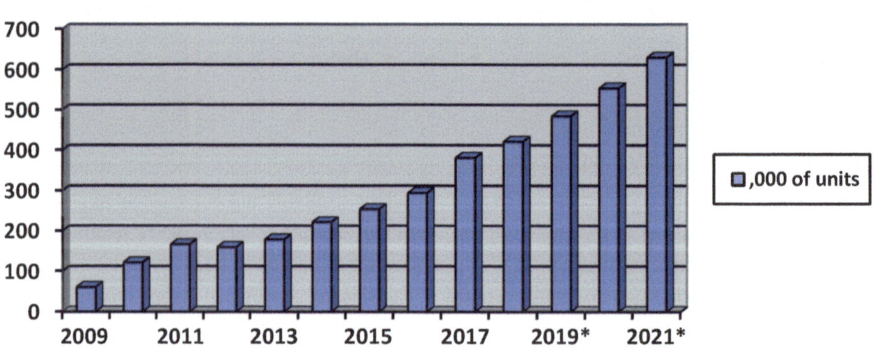

Fig. 2 Estimated annual worldwide supply of industrial robots [15]

2 Robotics in Slovakia

Robotics became a center of interest in the former socialist Czechoslovakia in 1971. Conditions for the development of this area were made possible by existing research and development workplaces, university study branches related to automation and robotics, in the Czechoslovak and Slovak Academy of Sciences.

The first Czechoslovak robots were produced in 1973 under the name QJN in cooperation with VÚSTE Praha and VÚTS Brno. The production of industrial robots and manipulators was concentrated mainly in companies in Slovakia: VUKOV Prešov, ZPA Prešov, ZŤS Detva, Vihorlat Snina, and VÚMA Nové Mesto nad Váhom. In Bohemia, the companies were TOS Kouřim, DESTA, TST Praha, ZSE Praha, and Motocyklové závody Strakonice. By 1982, according to the data of the Statistical Office of the Czechoslovak Socialist Republic, a total of 1009 automated technological sites were operating in Czechoslovakia. The robotics program has had a positive impact on the development of engineering, electrical engineering, electronics, computing, and material engineering. A positive point for Slovakia was that robotics were mostly developed within its territory [16].

One of the factors that had an influence on the relatively dynamic development of robotics within Czechoslovakia was the more than critical concentration of the necessary development resources not only in the area of research and development, including the environment of universities and Slovak Academy of Sciences (SAS), but also in the effective use of the potential of the large so-called production-economic units. The orientation of some production-economic units on robotics has created prerequisites for the development and growth of the sector through the practical application of inventive solutions in industrial practice. Dissolution, or the liquidation of production-economic units during the transformation period in the 1990s, caused for a temporary slowdown and lagging of Slovak robotics compared to the advanced robotic world.

At present, several dozens of highly innovative businesses operate within the Slovak Republic. Since 1989, these companies have undergone dynamic development. There are currently three types of businesses in Slovakia:

- Traditional, with history dating back before 1989 (e.g., Matador, EVPÚ)
- Enterprises risen from the remains of production-economic units (e.g., ZTS VVÚ Košice, Spinea)
- New companies (e.g., Robotec, Automatica, Kybernetika, Spinea—created from the teams of the former VUKOV Prešov)
- Foreign enterprises with affiliated branches in Slovakia (e.g., ABB, SCHUNK, iRobot)

A large part of the businesses that have been established since 1989 have emerged from the environment of scientific research organizations, especially universities (e.g., CEIT, MicroStep, Prvá Zváračská, Kybernetika). They are predominantly small- and medium-sized enterprises (SMEs) located mainly in the Bratislava, Košice, and Prešov regions. Geographical location of the enterprises is largely correlated with the original organizational units managed within the production-economic units, or with the location of public research organizations.

2.1 Analysis of the Robotics Environment

Businesses operate in a complex environment and a specific institutional arrangement that determines their development. Robotics environment in the Slovakia has been mapped through a comprehensive approach using the so-called diamond of competitive advantages (Fig. 3). The competitive advantage does not depend on individual elements of the Porter model or individual companies but on the interconnection of all elements within the diamond [17]; it has also been used to map the robotics sector in the United States [18].

The diamond of competitive advantages puts into context factors that affect the competitiveness of companies and clusters, but also of entire regions or nations. *Business strategy and rivalry*: First, this involves the competition between companies, the way companies operate, how they are managed, how they communicate, and what goals they have. A competitive environment is important because it puts pressure on innovation, but cooperation is beneficial, for example, by helping to reduce some costs, and allowing a critical mass of competences to be reached in key development areas.

Related and supporting industries: Creating inputs that are important for the competitiveness and innovative performance of the industry as such, they represent the ability of local companies to deliver cost-effective inputs and to participate in innovative business processes. A successful and internationally competitive industry can also create a competitive edge for other related industries operating in the region.

Terms of entry factors: We can divide these into five main groups: human resources, raw materials, knowledge, capital, and infrastructure. The existence of a specific resource is often a key competitive advantage of a business, industry, or country.

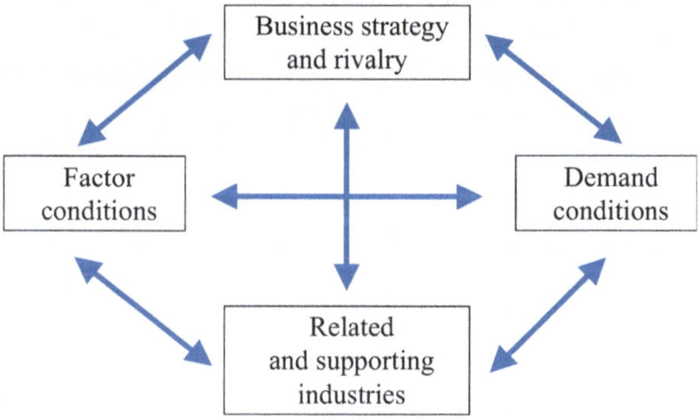

Fig. 3 The determinants of competitive advantage

Demand conditions: Demand is determined primarily by the size of the market, and by its needs and wishes. It is precisely the conditions of demand that create prerequisites for a competitive advantage in the case of pressure from customers to carry out innovative activities.

Detailed mapping of the robotic environment was undertaken as a qualitative research in 2015 and 2016. The methods such as questionnaire and interview were used to understand main trends in the national environment. Within the scope of the survey, various areas of operation of companies were considered in terms of their significance for the development of the enterprise itself, the dynamics of trends in the development of these factors, and the activity of the government in their development. The individual factors were proposed in interviews with the actors and subsequently evaluated in the framework of the questionnaire survey (Table 3). In business strategy and rivalry, ten areas were identified by companies as key areas. In related and supporting areas six areas were identified, in factor conditions eight, and in demand conditions seven key areas. The questionnaire survey involved 25 established companies operating in the field of robotics, which have their headquarters registered in the Slovak Republic. Overall, we have received 15 correctly filled out questionnaires representing 60% return rate. The results obtained were evaluated using descriptive methods and using mathematical data processing. The evaluation was performed using the Likert scale. Significance was assessed on a scale of 1–4: insignificant, not significant, significant, and very significant. Factor development trends were assessed on a scale of 1–5, very low, falling, without change, growing, and very steep, and government activity was rated on a scale of 1–4: negatively acting, not supporting, supporting poorly, and promoting sufficiently. The data for each of the areas listed below represent the average values of the individual parameters for the entire selection group.

Business strategy and rivalry. At present, in the automation, robotics, and digital technologies sector, several dozen companies operate in Slovakia as system integrators. Barriers to entry of new players are high, both because of the need for specific know-how and the high capital investment needed. Businesses create barriers to new players entering the market through their own research and development. However, it is not forbidden for new start-up players to enter the market in these technology domains. In general, the level of cooperation is relatively low among businesses in Slovakia. Some businesses cooperate on a supplier-customer base. Some Slovak businesses offer unique, specific products and services that cannot be duplicated thanks to in-house R&D. They can compete with the price, quality, and long-term know-how.

Robotic firms consider, from the importance of their development point of view, as the most important factors, rated in the category of significant-very significant:

- R&D activities, ownership of technological capacities, and know-how (3,6)
- Unique solution supply (3,8)
- Qualified human resources (3,7)

This shows the realization of own innovative solutions by enterprises, even in limited cooperation with public R&D organizations. For the implementation of the

Table 3 Key aspects of activity

Business strategy and rivalry	Possibility of specialization Domestic competitor activities Activities of foreign competitors Emergence of new companies in the sector (e.g., FDI, start-ups) Collaboration with other companies within the sector (competition) R&D activities, ownership of technological capacities and know-how Unique solution supply Benefits and investment support (taxation) IPR protection of proprietary solutions Qualified human resources
Related and supporting industries	Presence of companies with specific know-how and technologies Presence of foreign/global supply company branches Ability of companies to cooperate on creating innovative solutions Presence of organizations supporting the sector (e.g., business chamber, association for robotic) Defined sector standards Quality human resources with knowledge of the robotics industry
Factor conditions	Availability of qualified human resources Brain drain Wage growth Availability of quality R&D capacities (human resources, infrastructure) Connection of quality public R&D organization to practice/industry needs Functional technology transfer mechanisms from public R&D organizations Available funding (banks, venture capital) Availability of grant support Appropriate support infrastructure (broadband, road infrastructure)
Demand conditions	Presence of strong industries (manufacturing plants) Wider understanding of technology transformation by customers (Industry 4.0) National market size Demand generated by public sector (e.g., defense, safety, healthcare) Costs of entry to foreign markets (access to the single market) Joint export initiatives and consortia Global demand growth forecasts

in-house R&D, quality human resources are necessary considered to be an internal competitive advantage that needs to be developed and maintained [19]. The activity of domestic competitors (2,6) and the activities of foreign competitors (2,6), which are essential for companies, were also marked as important. Other factors were in the category insignificant-not very significant, including specialization. State of the industry does not create prerequisites for business specialization in specific domains. The factor of benefits and investment support (taxation) was considered to be completely insignificant for the development of the companies, since the government's activities are not focused on the development of robotics.

From the point of view of development trends, R&D activities, ownership of technological capacities, and know-how (4) was identified as the growing factor, followed by factors:

- Activities of foreign competitors (3,4)
- Emergence of new companies in the sector (e.g., FDI, start-ups) (3,4)
- Collaboration with other companies within the sector (competition) (3,2)
- Unique solution supply (3,4)
- Qualified human resources (3,13)

Other factors are in terms of trends without change. This shows the increasing need for R&D, but also the need for in-house technological and knowledge capacities that businesses are trying to develop. There is also an increase in the activity of competitors, but also the overall effort of companies and industries to offer innovative solutions.

For the development of companies, the government activities are important; they were evaluated in a different way; the most implemented measures were aimed at:

- R&D activities, ownership of technological capacities, and know-how (2,9)
- Unique solution supply (2,9)

The implemented measures are focused on the R&D activities and partly on the technological upgrading and R&D activities. In other areas, government activity was low. In the case of the emergence of new firms in the industry, the protection of IPR and qualified human resources were seen by the respondents as not specifically supported by the government.

Related and supporting industries. Under the related and supporting industries, we understand mainly but not exclusively sectors with high value added such as engineering, material processing, ICT, or electrical engineering. A certain number of input operations or the products are necessary to ensure that the operation of companies is available within Slovakia. Deliveries of many essential components are mediated by business representatives of multinational companies producing "technology blocks" used to develop complex solutions. In some cases, companies are forced to accommodate the OEM requirements when selecting subcontractors, as components and materials for their production must be standardized. However, there is a fairly large space for expanding the local subcontracting base.

Robotic enterprises identified the presence of the foreign/global supply companies (3,1) as the most important factor, followed by:

- Presence of companies with specific know-how and technologies (2,27)
- Ability of companies to cooperate on creating innovative solutions (2,47)

- Quality human resources with knowledge of the robotics industry (2,73)

This points to the significant position of complementary companies for the development of robotics sector, which should have sufficient technological capacities tailored to some specific needs of robotics. It can be concluded that the potential supply sector is inadequately developed in some areas, i.e., there is a technological gap. Exceptions include the representation of multinational companies offering comprehensive solutions. Other factors were seen as insignificant-little important.

From the point of view of development trends, the most important factors were:

- Activities of foreign/global supply companies (3,4)
- Ability of companies to cooperate on creation of innovative solutions (3,5)
- Quality human resources of suppliers with the robotics sector knowledge (3,1)

Other factors are not subject to change. This indicates the consistency between the perception of companies and the overall trends in the sector.

The government according to the respondents carries out activities only in the area of the presence of companies with specific know-how and technologies (2,3) and partly in the area of increasing the companies' ability to cooperate in the development of innovative solutions (2,3). The measures are mainly focused on the technology transfer, R&D support, and partly the development of cluster cooperation. The other areas, according to the respondents, were not really supported by the government.

Factor conditions. Robotics belongs to the knowledge-intensive sectors and therefore requires highly qualified human resources with specific skills not only in the given field, but also in other technological areas. There is a high demand for qualified designing engineers, hardware or software specialists, designers, technologists, developers, and more. The problem is the human capital flight of qualified professionals. Moreover, the quality of the graduates of the education system is not enough for the needs of practice, despite the growing cooperation between academia and industry. Access to bank capital is good; banks have no problem to provide commercial loans to entities operating in these areas in case of a favorable financial situation. Alternative forms of financing as venture capital are not widely used by companies, but they are considering these options for the future. Research and development are funded from own resources, but, considering the risk nature of some projects, from the public resources as well. Own research innovation infrastructure but also the technological capacities are not sufficiently developed. Public R&D organizations work with businesses to address their problems, but the intensity of collaboration has the potential to grow. About 36 universities and 56 institutes at Slovak Academy of Sciences operate in Slovakia. A number of workplaces with a focus relevant to the automation, robotics, and digital technologies were identified. At eight universities and three SAS institutes there are more than 140 professors, 280+ associate professors, 530+ PhDs, as well as several hundred PhD students in the same or related subject fields. In this context, intercultural management is a challenge that should be addressed in order to support mutually convenient collaboration [20]. The current cooperation is inadequate, both because of the unwillingness to cooperate and because of the misunderstanding of the needs of enterprises, the diversity of "life cycles" in enterprises, and public scientific research organizations. Public research and development organizations have sufficient infrastructure; however, it does not sufficiently reflect the needs of the real economy. The current focus of scientific research projects is distant from practice, and businesses do not consider the research focus of public research organizations to be relevant for their development. Mechanisms of technology transfer from the environment of public scientific research organizations are inadequate.

Respondents considered the most important factor affecting firms to be the increasing wage growth (3,7) along with the availability of skilled human resources (3,7). This is related to other high-ranking factors, namely brain drain (3,4) and availability of grant support (3,4). The sufficiency of human resources in the market is considered to be critically important to ensure the long-term competitiveness of companies. This is also related to the perception of brain drain into other industries and outside of Slovakia. In the category of little significant-significant factors, there were availability (2,5) and an appropriate supporting infrastructure (2,0). Others were in the category insignificant-not very significant. At present, companies have no problem obtaining private resources (especially bank loans) to finance their activities. They consider nonrepayable grant support as a more advantageous form of funding, supporting their technological development.

Most of the observed factors were evaluated in terms of trends in the category without change-increasing, except the availability of qualified human resource factor, which has declining tendencies. On the other hand, the brain drain of experts outside the sector or outside the country is considered to be significantly increasing. Increased expert outflow may affect the competitiveness and growth of the industry.

The activities of government are perceived in many different ways, the most important factors being:

- Availability of qualified human resources (2,5)
- Availability of quality R&D capacities (2,1)
- Functional technology transfer mechanisms from public R&D organizations (2,1)
- Available funding (2,2)
- Availability of grant support (2,9)
- Sufficient supporting infrastructure (2,6)

In critical areas, the brain drain (1,6), wage growth (2), and connection of quality public R&D and educational organizations to the industry needs (1,9) were perceived to be more negative. The government's efforts to improve the activities of companies in robotics should be prioritized especially in these areas.

Demand conditions. Customers include all key industries in Slovakia (automotive, engineering, food, wood processing, etc.). However, the size and capacity of the Slovak market are small and for some specific products the national market virtually does not exist. For this reason, many companies focus primarily on exporting their production. The real availability of foreign markets is significantly limited for Slovak SMEs, mainly because of the cost barriers to entering and establishing in these markets. Nevertheless, some businesses have successfully established themselves in foreign markets. Their products are mainly supplied to the EU market but also to Russia or Asia. Some Slovak companies have foreign representations or branches in several countries of the world. These businesses are known within their technology area virtually worldwide, or they are known to potential customers as well as to competitors. In the area of export, there is no common cooperation, coordinated operation, and promotion. Collaboration with large companies or with globally established enterprises regarding the demand is absent.

In terms of demand-side conditions, the most important areas for business development were the following:

- Presence of strong industries (3,7)
- Wider understanding of technology transformation by customers (Industry 4.0) (3,8)
- Costs of entry to foreign markets (3,9)

The most important sector is the automotive industry, which has a number of OEMs and their sub-supplier network in Slovakia and in the wider area of Central Europe. Mainly the key industry sectors are exposed to the technological transformation associated with the Industry 4.0 concept implementation.

The size of the national market (2,8) and the growth forecast of global demand (2,9) can also be considered significant. The demand generated by the public sector (1) and joint export initiatives and consortia (1,3) were marked as nonsignificant. Thus, the public sector's effort does not create conditions for the development of the industry, which is in contradiction with the export efforts of companies and their perception of the costs associated with establishing themselves on foreign markets.

In terms of trends, the fastest growing factor was the global demand growth forecast (4,5), followed by:

- Presence of strong industries (3,5)
- Wider understanding of technology transformation by customers (Industry 4.0) (3,7)
- Size of the national market (3,3)
- Costs of entry to foreign markets (3,6)

Significance of critical factors thus sees significant growth, which has a positive impact on the development of companies. For other factors, no changes were identified. Respondents consider cumulative government activities to be suboptimal. The highest scores were in the following areas:

- Costs of entry to foreign markets (2,4)
- Joint export initiatives and consortia (2,4)
- Size of national market (2,3)

The government supports the export sector, a support, which, according to respondents, is inappropriately targeted. Support for growth in the domestic market is perceived through the support of technological upgrading of manufacturers that

Table 4 Order of development factors

	Perception	Trends	State
Business strategy and rivalry	3	2	1
Related and supporting industries	4	3	3
Factor conditions	2	4	2
Demand conditions	1	1	4

Fig. 4 Comparison of determinants

are customers of robotic companies. Other areas by respondents are not supported according to the respondents.

From the cumulative point of view (Table 4), it is obvious that the most important factor in terms of companies is demand, both in terms of its perception of the development of the company and in terms of development trends. The state implements activities mainly in areas relevant to business strategy and rivalry.

The second most important factor is factor conditions in terms of both the perception of companies and the activities of the state. Development trends are most prominent in business strategy and rivalry.

The state appears to be partially orienting its measures outside the needs of the robotics sector. As an example, demand conditions have been identified as being relatively essential for the development of robotics with positive development trends, with state activity being relatively the lowest (1,87) compared to other areas (Fig. 4). Overall, it can be said that the implemented support has a traditional broad-spectrum character, which does not sufficiently take into account the particularities of the robotics sector.

The business strategy and rivalry area is perceived as less significant with a lower intensity of change, but according to respondents the state is doing the most activities in this area. It seems that at the time of the survey, there was a certain mismatch between the state's activity and the needs of the robotics businesses. Implementation policies therefore require a higher degree of consideration of the expectations of the needs of the robotics sector.

2.2 Policy Status Quo

The Slovak Republic does not systematically support the development of robotics and does not have an integrated strategy in this area. A certain exception is the strategy Through Knowledge Towards Prosperity - Research and Innovation Strategy for Smart Specialization of the Slovak Republic approved in 2013. The strategy was prepared according to the concept of Innovation Strategies for Smart Specialization (RIS3) prepared for EU member states. The concept assumes that strategies for smart specialization (RIS3) are integrated, place-based economic transformation agendas that do five important things [21]:

- They focus policy support and investments on key national/regional priorities, challenges, and needs for knowledge-based development, including ICT-related measures.
- They build on each country's/region's strengths, competitive advantages, and potential for excellence.
- They support technological as well as practice-based innovation and aim to stimulate private sector investment.
- They get stakeholders fully involved and encourage innovation and experimentation.
- They are evidence based and include sound monitoring and evaluation systems.

Slovak RIS3 has defined the vision "To drive a structural change of the Slovak economy towards growth based on increasing innovation capability and R&I excellence to promote sustainable growth in income, employment and standard of living." Three types of priority areas, namely areas of economic specialization, prospective areas of specialization, and areas of specialization from the point of view of available scientific and research capacities (Table 5), have been identified to achieve this vision (Table 5).

The strategy aims to make the effects take place in areas such as the employment of young people under changing conditions, population aging and quality of life, marginalized groups and social inclusion, reduction of emissions, protection and better use of natural resources (mainly water, land, and forests), or adaptation to climate change. The strategy has identified four strategic objectives, namely:

1. Deepening integration and embeddedness of key major industries increasing local value added through the cooperation of the local supply chains and turning local supply chains into embedded clusters
2. Increased contribution of research to the economic growth via global excellence and local relevance
3. Creating a dynamic, open, and inclusive innovative society as one of the preconditions for the increase in the standard of living
4. Improving the quality of human resources for an innovative Slovakia

On the basis of the national RIS3, the so-called Operational Program for Research and Innovation (OP R&I), which allocated €2,266,776,537 in research,

Table 5 Slovak smart specialization strategy

Areas of economic specialization	Automotive and mechanical engineering industries Consumer electronics and electrical equipment Information and communication products and services Production and processing of iron and steel
Prospective areas of specialization	Automation, robotics, and digital technologies Processing and increasing the value of light metals and their alloys Production and processing of polymers and progressive chemical substances (including smart fertilizations) Creative industry Increasing the value of domestic raw material base Support of smart technologies in the area of processing raw materials and waste in the regions of their occurrence
Available scientific and research capacities	Research of materials and nanotechnologies Information and communication technologies Biomedicine and biotechnology Environment and agriculture including modern environment-friendly chemical technologies Sustainable energy and energetic

development, and innovation, has been prepared. Resources were to be directed into priority areas. Subsequently, a RIS3 Implementation Plan was developed in 2017 to identify the technological priorities in detail. Five domain platforms have been created for this purpose:

- Vehicles for the twenty-first century
- Industry for the twenty-first century
- Digital Slovakia and creative industry
- Population health and medical technologies
- Healthy food and environment

The platforms identified technological priorities for the implementation of RIS3 and the effective targeting of OP R&I resources, with robotics identified as a priority area in several domains in various forms (Table 6).

In addition, various related technology areas have been identified such as simulation; modelling of industrial, transport, and other systems, or machine vision and other methods of detection and object analysis; interactive education-stereoscopy; mobile applications; Web applications; and conversational platforms. Robotics have become part of a wide range of applications.

Nevertheless, it can be concluded that national as well as regional authorities do not implement sufficient measures in favor of the development of robotics in Slovakia. The only exception is the effort of the Slovak Government to support the development of R&D and innovation activities or export. However, this support is not targeted and systemic; it lacks a connection to the real needs of the economy. The support provided so far appears to be ineffective for the development of the perspective disciplines being monitored. The elaboration of the national smart

Table 6 Technological priorities

Vehicles for the twenty-first century	Progressive production and material processing technologies, powder technologies, vacuum metallurgical technologies, precise casting, 3D composite printing, additive industrial production, advanced surface coating technologies, automated and robotized manufacturing technologies Research and development of components and nodes for automatization, robotization, and digitization of processes Research and development of complex robotized systems
Industry for the twenty-first century	Progressive production and material processing technologies, powder technologies, vacuum metallurgical technologies, precise casting, 3D composite printing, additive industrial production, advanced surface coating technologies, automated and robotized manufacturing technologies Complex robotized systems including autonomous
Digital Slovakia and creative industry	Systems for security of shared space between people and robotic systems Innovative and intuitive forms of robotic structure control using visual, haptic, and audio links Creating a transformable and scalable concept of mobile manipulation robotic systems to solve internal logistics of industrial production Automatization and robotization of key enabling technologies
Population health and medical technologies	"Ambient assistant living" products
Heathy food and environment	3D scanning technologies for increased yield and finalization of wood processing, robotization, and automation of wood processing

specialization strategy, which concentrates on a limited number of priorities, which are determined by strengths and international specialization of Slovakia, can be considered a positive step on the part of the government. Within this strategy, robotics is identified as a perspective area. To meet the complex challenges associated with the implementation of the Industry 4.0 concept, a new system for managing and management of changes and processes [12] needs to be introduced. This means that strategies and action plans relevant to Industry 4.0 should support the creation of a flexible economic and social system capable of responding to changes. Additionally, cluster should be the overall goal of any relevant policy interventions. It is widely recognized that cluster policy if truly implemented has a significant potential to stimulate cluster development [22]. Therefore, efficient mechanisms would be launched to boost the sector [23]. It is essential to create conditions for maximizing the effectiveness of initiatives and policies and to create an appropriate infrastructure [24].

3 Conclusions

Industrial companies all over the world are exposed to growing competition due to globalization and specialization that has affected productivity growth. A response to these global trends is a further increase in productivity within the forthcoming fourth industrial revolution, called Industry 4.0. The Industry 4.0 concept incorporates several technology sub-concepts and technology areas. One of the key ones is robotics, which represents an innovative technology domain with a high potential to contribute to the productivity growth of manufacturing (not only) industrial enterprises. In the context of global technological changes and gradual widespread adoption of the Industry 4.0 concept, the global growth of robotics and robotics solutions has been observed. These trends were also reflected in Slovakia. As part of the robotics environment analysis, a number of involved parties, such as innovative businesses, as well as public R&D organizations have been identified. Interviews with business representatives and R&D organizations have revealed several important findings. Robotic companies have many years of experience and know-how thanks to the implementation of their own R&D. However, their cooperation is inadequate. Their development is to a certain extent limited by the need for technological upgrading of supporting and related industries. The problem is also the lack of good human resources in the market associated with brain drain, as well as the lack of cooperation with formal knowledge owners such as public R&D organizations (universities, SAS). Business quality allows for external financing of development projects. Slovakia has an economy with a large industry share, creating good conditions for demand for innovative solutions in line with Industry 4.0. An important determinant of future development is the current subcritical activity of the government in stimulating the development of robotics. Especially cluster collaboration should be the main goal.

Acknowledgement This work was supported by the grant VEGA No. 2/0002/18.

References

1. Fusko, M., Rakyta, M., & Manlig, F. (2017). Reducing of intralogistics costs of spare parts and material of implementation digitization in maintenance. *Procedia Engineering, 192*, 213–218.
2. Arnold, C. K., & Voigt, D. (2016). How the industrial internet of things changes business models in different manufacturing industries. *International Journal of Innovation Management, 20*(8), 1640015.
3. Thoben, K. D. W., & Wuest, S. A. (2017). "Industry 4.0" and smart manufacturing-a review of research issues and application examples. *International Journal of Automation Technology, 11*(1), 4–16.
4. Monostori, L., et al. (2016). Cyber-physical systems in manufacturing. *CIRP Annals—Manufacturing Technology, 65*, 621–641.
5. Gerlitz, L. (2016). Design management as a domain of smart and sustainable enterprise: Business modelling for innovation and smart growth in industry 4.0. *Entrepreneurship and Sustainability Issues, 3*(3), 244–268.

6. Lee, J., Lapira, E., Bagheri, B., & Kao, H. (2013). Recent advances and trends in predictive manufacturing systems in big data environment. *Manufacturing Letters, 1*(1), 38–41.
7. Chiarello, F., Trivelli, L., Bonaccorsi, A., & Fantoni, G. (2018). Extracting and mapping industry 4.0 technologies using Wikipedia. *Computers in Industry, 100*, 244–257.
8. Baur, C., & Wee, D. (2015). *Manufacturing's Next Act*. New York, NY: McKinsey & Company.
9. Reischauer, G. (2018). Industry 4.0 as policy-driven discourse to institutionalize innovation systems in manufacturing. *Technological Forecasting and Social Change, 132*, 26–33.
10. Iivari, M. M., Ahokangas, P., Komi, M., Tihinen, M., & Valtanen, K. (2016). Toward ecosystemic business models in the context of industrial internet. *Journal of Business Models, 4*(2), 42–59.
11. Kamble, S. S., Gunasekaran, A., & Sharma, R. (2018). Analysis of the driving and dependence power of barriers to adopt industry 4.0 in Indian manufacturing industry. *Computers in Industry, 101*, 107–119.
12. Kovacs, O. (2018). The dark corners of industry 4.0—grounding economic governance 2.0. *Technology in Society, 55*, 140–145.
13. Bloss, R. (2016). Collaborative robots are rapidly providing major improvements in productivity, safety, programming ease, portability and cost while addressing many new applications. *Industrial Robot, 43*, 463–468.
14. Fusko, M., Rakyta, M., Krajčovič, M., Dulina, Ľ., Gašo, M., & Grznár, P. (2018). Basics of designing maintenance processes in industry 4.0. *MM Science Journal., 1*, 2252–2259.
15. IFR. (2018). Homepage. Retrieved September 21, 2018, from https://ifr.org
16. Kalaš, V. (2004). Tridsať rokov svetovej robotiky. *AT&P Journal*.
17. Porter, M. E. (1990). *The Competitive Advantage of Nations*. New York, NY: The Free Press.
18. Khamis, A. H., Utembayev, A., Nordin, M.R., Victorero, R., & Abughannam, A. (2012). The Massachusetts Robotics Cluster.
19. Cagáňová, D., Ridzoňová Hlásniková, P., Vraňaková, N., & Chlpeková, A. (2019). Intellectual capital as a key factor in the automotive industry. *ACM Mobile Networks and Applications, 3*, 1–8.
20. Cagáňová, D., Čambál, M., & Weidlichová Luptáková, S. (2010). Intercultural management—trend of contemporary globalized world. *Electronics and Electrical Engineering, 6*(102), 51–54.
21. EC. (2012). Guide to Research and Innovation Strategies for Smart Specialisation (RIS 3), ISBN: 978–92–79-25094-1, https://doi.org/10.2776/65746.
22. Pan, W., Jian, L., Wang, D., &Liu, T. (2017). Patent co-citation Networks for high-tech industrial clusters-a case study of silicon valley high-tech industrial clusters. In *2017 International Conference on Grey Systems and Intelligent Services (GSIS), Stockholm* (pp. 360–366).
23. Burger, P., Baumöhl, E., & Výrostová, E. (2017). Funding structure of the European and North American clusters: Results from an independent questionnaire. *Ekonomický časopis, 65*(6), 485–504.
24. Sung, T. K. (2018). Industry 4.0: A Korea perspective. *Technological Forecasting and Social Change, 132*, 40–45.

Use of Modern Technologies at Baggage Tracking and Its Impact on Airline Revenue

Matúš Bozogáň and Soňa Hurná

1 Introduction

Logistics is likely understood by a couple of approaches whereby its developmental phases progressively change. Logistics constitutes scientific discipline implying processes of effective planning. It includes two basic aspects, namely defining infrastructure and activity supervision. The aspect focused on infrastructure constitutes material-technological basis, technical equipment, and technologies [1]. The importance of logistics is growing along with increasing globalization. All companies are exposed to strong competitive pressure and in this situation play a strategic role. Logistics helps to reduce costs, to achieve higher profits, and to improve customer services [2]. Logistics becomes the most evolving segment and offers an open space for innovation and continuous improvement of its processes [3].

In 2006 approximately 50,000,000 bags around the world had been mishandled during the transportation process within the airline industry. This has caused additional costs and reputation loss for airlines around the world. Together with implementation of baggage fees even in full-fare carriers, customers are demanding seamless service compared to previous situations when checked-in bags were free of charge. Customers usually percept that any paid service should have a higher quality compared to unpaid one. To minimize the impact on passengers as well as on airlines, IATA—International Air Transport Association—introduced Resolution 753 which motivates airlines to implement baggage-tracking system through several checkpoints of baggage transport process. To be compliant with Resolution 753 airlines must have an autonomous system that checks the location of the baggage

M. Bozogáň · S. Hurná (✉)
Department of Business Economics, Mendel University, Brno, Czech Republic
e-mail: xbozogan@mendelu.cz; sona.hurna@mendelu.cz

© Springer Nature Switzerland AG 2020
D. Cagáňová, N. Horňáková (eds.), *Mobility Internet of Things 2018*,
EAI/Springer Innovations in Communication and Computing,
https://doi.org/10.1007/978-3-030-30911-4_6

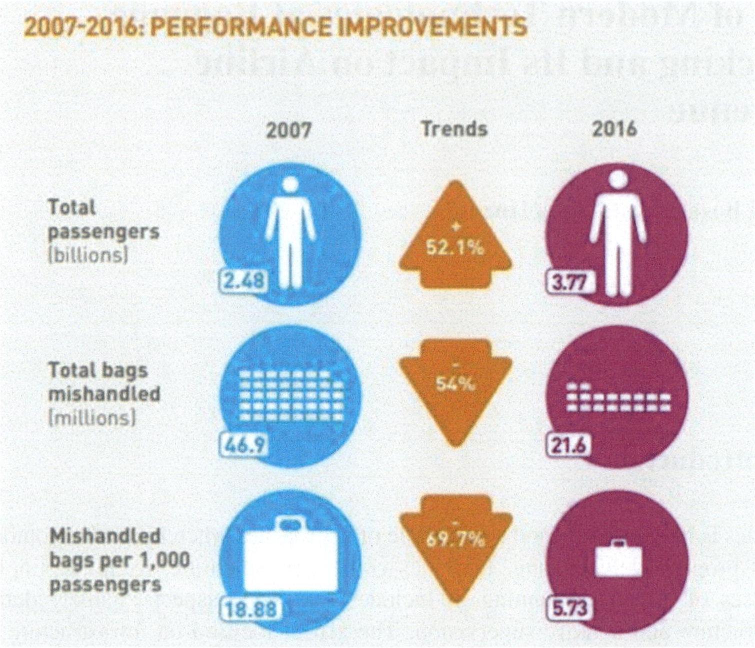

Fig. 1 Performance improvements in baggage handling 2007–2016 [6]

from handing it over at check-in counter or self-baggage-service kiosk all the way until the end of the customer's journey at his/her destination.

The development of large-scale manufacturing system resulted from the development of new technologies which enable the manufacture of products that satisfy unique customer requirements [4]. It raises the need for unification of the basic requirements and regulation for the establishment and operation of an air carrier, air traffic control, passenger check-in construction, and operation of airports and other aviation-related activities [5]. This system should also be able to share information about customers' baggage with interline partners included in the customers' journey (Fig. 1).

1.1 Technology Behind Tracking

To fulfill IATA requirements several levels of tracking can be implemented. Depending on the cost of implementation or available technology airlines choose to use one of the following options for baggage tracking:

- Simple bar code/QR scanning
- Video recognition
- RFID code scanning

Bar Code Scanning

Bar code/QR scanning works on a simple technology which we can find in commercial retail every day. The system scans printed bar code on affixed baggage tag where basic flight data are stored. The system is highly dependent on direct visual contact between scanner and baggage tag. In case the system does not recognize data from bar code, baggage might be loaded on wrong flight or it needs to be manually checked by airline or airport employee for correct assertion. This increases workload as well as possibility of mishandling of baggage.

Video Recognition Software

Compared to bar code scanning video recognition software increases the quality of decoding printed baggage tag data in the form of optical character recognition technology. Thus the system is able to recognize not only bar code-encoded data, but also written basic information on baggage tag. This allows scrapped or partly visible baggage tags to be recognized in the form of combination of visible bar code and written text information. This lowers the quantity of mishandled baggage. However, video recognition is still highly dependable on direct visual contact with baggage tag.

Radio-Frequency Identification

Radio-frequency identification (RFID) is a recent technology used in baggage tracking. RFID works on transmitting of basic information on radio frequency. This reduces the necessity of direct visual contact between scanning technology and baggage tag. Airlines and manufacturers are implementing small RFID transmitter to baggage tag or directly to the baggage to use it. This again improves the quality of recognition and therefore reduces the number of mishandled baggage. RFID was selected over other potential bag-tracking solutions owing to the combination of reliability, maturity, widespread availability, and cost. RFID achieves a read rate of 99–100%, making it the leading technology for ensuring correct bag tracking [7] (Fig. 2).

2 Economic Impact

The price of product consists of cost items that we must reduce. Direct and indirect costs will be reduced by using new methods [9]. Estimated cost for recovery of each mishandled baggage is 100 USD [10]. This amount does not cover goodwill policies and claim handling directly done by customers. With passenger traffic growing annually since 1970 from 310,441,392 customers to 3.97 billion in 2017 demand

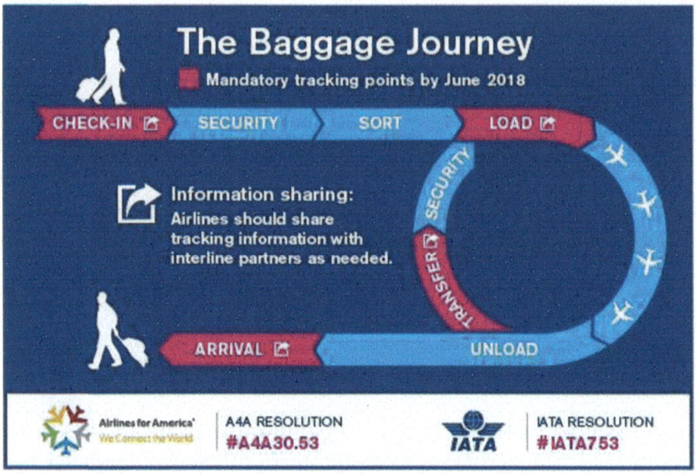

Fig. 2 Mandatory tracking points implemented by IATA Resolution 753 [8]

Table 1 Revenue statistics of selected airlines in the year 2013 [12]

Carrier	Baggage fee policy	12-Month revenue[a]	Per passenger
Spirit	Checked and large carry-on	USD 201 million	USD 19.54
Delta	Checked bags	USD 838 million	USD 8.58
JetBlue	Second checked bag	USD 73 million	USD 2.84
Southwest	No charge for two bags	USD 158 million	USD 1.39

[a] 12-month period ended in September 2013

for baggage transport is growing [11]. According to the SITA baggage report 5.73 bags per 1000 passengers were mishandled in 2016 [6]. This number is expected to decrease with implementation of baggage-tracking system worldwide even though IATA expects continuous growth of passengers' traffic. The cost of individual RFID baggage tag varies from 0.03 to 0.05 USD while the cost of the RFID reader varies from 1500 to 5000 USD depending on manufacturer and implementation. As per SITA this results in approximate average implementation cost of 0.10 USD per customer at the estimated cost saving at 0.20 USD per customer in recovery expenses [6, 7]. At Spirit Airlines almost 18% of airline ancillary revenue came from checked baggage fees. Range of checked in baggage fee varies from airline to airline but average at the amount of 17 USD. This number is acceptable for customers, but at the same time it raises expectation of quality and thus increasing cost of service of the airlines (Table 1).

2.1 Indirect Impact

As this direct cost reduction will help airlines to generate additional profit, the indirect impact can also increase revenue in other areas of airline operations. As the effectivity of RFID tracking ranges from 99% to 100%, additional staff responsible for identification of unreadable baggage tags or incorrect reading will be reduced and thus staff costs will be reduced.

Together with lesser incorrect bag readings, with RFID and autonomous baggage-handling system, loading staff will be able to receive exact information about baggage location during loading process and therefore will be able to make decision to wait or offload baggage based on more actual info. This will reduce possible delays due to safety requirements and baggage of passenger who is not travelling must be offloaded and advanced baggage tracking will show exact loading position to affected loading staff. Delayed flight is increasing the cost of transport for the airline as fees for use of gate or stand are based on actual block time of the aircraft. Together with parking fees delays also generate costs in the form of additional cost for rebooking of customers who missed their connecting flight due to delay.

For commercial airlines transporting combined load (passengers and cargo) advantage of better bag tracking can result in the possibility to transport higher amount of cargo on high-demand routes as in some cases cargo had to be offloaded at the last minute due to transport limits of the aircraft that exceeded due to the fact that in commercial transport passengers' baggage has higher transport priority over cargo shipments. With the availability of exact baggage location for loading and load planning staff at the plane, decision to offload cargo will be based on the most exact data as it will be possible to see if any transfer baggage is missing.

3 Other Ancillary Services

Also other airline auxiliary services are becoming one of the top decisive factors for their customers. As airline market changes, also airlines need and must adapt to such a situation. Additionally, with actual state of the market, with highly competitive products, prices for the air fares are becoming very low and many airlines are struggling to reach positive results at the end of the fiscal year. Therefore, auxiliary services are becoming one of the most important parts in revenue generation within the aviation industry.

3.1 Historical Development

After the Second World War, demand for air travel rapidly increased and together with rapid innovation in technological area airlines could offer very low fares. This resulted in higher potential customer numbers being able to afford an air ticket, which was impossible before. As a result, the necessity to change state of the aviation market became clear.

The first breakthrough came in 1978 when the USA updated 20-year-old law by the new one titled Airline Deregulation Act that deregulated the airline industry. The full title of this act is: "An Act to amend the Federal Aviation Act of 1958, to encourage, develop, and attain an air transportation system which relies on competitive market forces to determine the quality, variety, and price of air services, and for other purposes" [13].

Figure 3 shows causal relationship between liberalization in the market and economic growth. The Act was aiming to encourage the competition in the market. This Act removed the US Government as a controlling body over the aviation market in the USA and created free open market enabling a new competition to enter.

After the successful implementation of US Deregulation Act, other countries followed the example and started with liberalization of their markets step by step. Liberalization in the different markets resulted in rapid development in the aviation industry in terms of passenger's traffic and movements, thus enabling further competition in the market, and therefore again pushed air fares down. In Table 2, the most important events in airline market development and their impacts are noted.

The deregulation of the aviation market enabled new types of competitors to enter the market. After all the adjustments were done, the environment in the aviation was very open and even smaller companies were able to invest in new modern aircraft with guarantee that the legislation will not be in the way.

New low-cost carrier (LCC) model was introduced in the USA by Southwest Airlines, which began the era of LCC around the world thanks to the successful implementation and widely accepted model in the market by the US citizens. This whole model was driven on one thought still advertised by Southwest airlines until today: "If you get your passengers to their destinations when they want to get there, on time, at the lowest possible fares, and make darn sure they have a good time doing it, people will fly your airline" [15]. Their model consisted of several parts, which enabled them to offer very low fares and thus to attract customers who would otherwise not tend to fly:

Fig. 3 The causal relationship between air service liberalization and economic growth [14]

Table 2 Main events of liberalization and their effect [14] (own processing)

Event	Results
US Deregulation, 1978	Emergence of hub and spoke systems Low-cost carriers with nationwide route networks New entrants and integrated cargo carriers
UK Liberalization of Secondary Airports	Growth of international services to Manchester, Birmingham, Glasgow, etc.
Open Skies Agreements for United Arab Emirates	Growth of Dubai as major international hub
Domestic deregulation in India	Development of low-cost carriers and aggressive, expansion-oriented airlines
UK-India Bilateral and Creation of New Frequencies	Growth of capacity New gateways Additional carriers operating UK.-India service
Domestic deregulation in Brazil	Growth of low-cost carrier Gol and others
Single European Market	Growth of low-cost carriers such as Ryanair and EasyJet New services, traffic growth, new gateways throughout European Union

Table 3 Share of capacity of low-cost carriers in European market [14] (own processing)

Year	LCC share of capacity (%)
1996	1.4
1997	2.8
1998	3.7
1999	4.2
2000	6.0
2001	6.4
2002	11.1
2003	20.2

- Fares: Unrestricted and low price
- Network: Point-to-point high-frequency routes
- Distribution: Travel agents and call centers, no tickets
- Fleet: High utilization, same type of aircraft across the fleet
- Airport: Secondary airports with short turnaround times
- Sector length: Short (around 400 nm)
- Staff: High productivity with competitive wages and profit sharing [16]

At later stage, this was widely accepted by other LCC on the market. In Europe after adaptation to Single European Market, this was used by two new airlines which, until today, are still playing the important part in European aviation industry. Ryanair and EasyJet adapted these policies and according to a study of 2006, they were able, together with other smaller LCC, to gain almost 20% of passenger share in just 7 years—from 1.4% in 1996 to 20.2% in 2003 [14]. In Table 3 the development of LCC growth is represented.

Table 4 Pricing comparison between LCC and FSC on similar route as of 19 April 2018 (own processing)

Budapest to Dubai 5 September 2018	Emirates to Dubai DXB	Wizz Air to Dubai DWC
1 × 30 kg baggage	Included	HUF 19,850
Airport check-in fee	Included	HUF 3200
Cabin baggage	Included	HUF 5750
Seat selection	Included	HUF 2250
Refreshment onboard	Included	HUF 2195
Air fare	HUF 102,400	HUF 106,500
Total	HUF 102,400	HUF 139,745
Dubai to Budapest 12 September 2018	Emirates ex Dubai DXB	Wizz Air ex Dubai DWC
1 × 30 kg baggage	Included	HUF 19,850
Airport check-in fee	Included	Included
Cabin baggage	Included	HUF 5750
Seat selection	Included	HUF 2250
Refreshment onboard	Included	HUF 2195
Air fare	HUF 102,200	HUF 30,990
Total	HUF 102,200	HUF 61,035

With the development of LCC, classic network carriers (FSC) gained new challenge in the form of a price competition on similar routes. Where previously they were able to offer premium services (which customers accepted as they had no other choice and they needed to travel), now classic customers were offered much cheaper option to travel. This situation resulted in lowered load factor for classic airlines, which resulted in difficulties to operate profitably on short and medium haul routes. This, together with the Great Recession, led to bankruptcy of several airlines in the USA and Europe. Not only airlines had economic problems, but also customers' ability to buy and travel decreased. To survive such an environment, all airlines needed to focus on profitable operations and therefore they needed to focus on cost saving and generation of additional revenue.

4 Airline Models

To understand the importance of auxiliary services it is also necessary to understand the different approaches by the low-cost and full-service airlines. Even when the core product and main service they provide are the same—transport from point A to point B—they are approaching the pricing differently.

In Table 4, we can see the most common differences between the typical LCC Wizz Air compared to Emirates. Low-cost carriers rely upon a la carte activity by aggressively seeking revenue from checked bags, assigned seats, and extra leg room seating [17].

As with the FSC such as Emirates or Austrian, the customer has all the services included in the fare typically and does not need to worry about any additional payments for the services he/she expects. On the other hand, LCC offers mostly lower fares for customers who would like to travel light—only the basic transport fare is included in the price, and customer needs to pay additionally for services which are typically included in the fare of FSC. Also in the table we can notice that LCC serves only smaller, secondary airports with higher distance to the city center, thus enabling them to offer lower fares as normally landing and handling fees are lower in such airports.

"In the case in which both FSCs and LCCs are operating on the same route, FSCs should prepare their countermeasures carefully against rival LCCs because the cheap airfare of LCCs can be regarded as an attractive advantage for potential passengers who want to use the air transport service with reasonable costs [18]." "The global financial crisis revamped the importance of cost cutting in doing businesses and encouraged the business travelers to choose low cost airlines. In particular, business travelers from small and medium sized enterprises (SMEs) attracted by low-cost airlines compared to the travelers working for large sized enterprises and corporates. As SMEs are price sensitive, the value for money schemes introduced by low cost airlines delivered expected value and benefits [19]."

The big advantage of LCC airlines is the use of harmonized fleet. They normally offer just routes suitable and profitable for the aircraft and do not need to provide any feeder flights; thus, they are able to minimize training, operating, and maintenance cost. Due to the position of network airlines and due to the necessity to optimize capacity according to the need, network carriers use a variety of aircraft that are able to serve different destinations, but for the price of increasing costs of training of the crew and maintenance.

Another difference can be found in customer care approach. Network carriers provide their customers with the ability to contact them directly in several ways, such as social networks, call centers, or email contacts, which are provided nonstop and free of charge. LCC often offer just paid call centers with premium rates or just Internet form with no published direct contact. At the airport FSC does normally have their own ground staff to assist customers in case of any requests or problems, compared to external ground-handling companies contracted by LCC.

If we look at customer comfort, typically FSC offers better onboard service as they use lesser seats on the same aircraft type compared to LCC, thus providing better seating comfort onboard the aircraft.

Table 5 shows different seating capacities of comparable aircraft with different carrier types

The last mentioned difference, however for some customers the most important, is the offered network. FSC offer a possibility to transfer to different flight or partner airline, therefore guaranteeing the customers a higher variety of destinations reachable. In case of any disruption network carrier will get customers to contracted destination by other routing or by use of their partner airlines. Low-cost airlines are normally point-to-point airlines and do not offer any possibility to transfer within one ticket. Even if the customer holds two tickets on the same LCC and because of

Table 5 Seating capacity comparison (own processing)

Seating capacity/airline	EasyJet	Wizz Air	Lufthansa	British Airways	Air France
Airbus A319-100	156	n/a	138	144	142
Airbus A320-200	186	180	168	168	178
Airbus A321-200	n/a	230	205	205	212

the disruption of the first flight he/she misses the connecting flight, LCC will not accept any liability and the customer must find an alternate solution alone and on own costs.

All these differences are enabling LCC to provide lower fares on the similar routes. This approach made it easier for LCC to survive the great recession; however as recession is coming to its end, customers are again willing to pay more for the quality of travel but at the same time are still very price sensitive. Higher transparency and excellent synchronization of systems, big data, and powerful analytics, together with improvements in distribution in line with IATA's New Distribution Capability initiative, create the perfect environment to fine-tune the relationship between airlines and customers and offer them smarter choices and maybe even improve the public perception of revenue and pricing management [20].

4.1 Auxiliary Services

It is necessary to differentiate between three groups of services:

- Core service is a reason why the company is in the market. For an airline, this would be transporting passengers or cargo.
- Enabling service or enabling facilitating service helps the customer to use the core service. For an airline, a check-in service is needed to be able to consume the airline core service.
- Enhancing or supporting services are also auxiliary services, but they fulfill another function than enabling services. Enhancing services are used to increase the value of the service and/or to differentiate the service from that of competitors. Airport lounges and a range of in-flight services are examples of airline-enhancing services [21].

Core Business

"For an airline, the core business is to transport passenger from point of departure to the point of destination" [22].

The core business is an essential part of aviation. Without core business, no auxiliary services would be available as these are connected to core. In general, the typical example of core business-only product is low-cost airlines. They include only transport from point A to point B in the fare and everything else costs extra. If we look at network carriers they prefer so-called all-in-one fares, wherein their main fare and also baggage and seat selection are included. In recent years network carriers realized the change in behavior of potential customers. They are willing to have more flexible options, where not everything is included, so they formed also fares without services included (without checked baggage, without meal onboard, etc.).

Auxiliary Services

The term auxiliary in definition means "secondary or supplementary" [23].

In terms of airline services, it means that service is just an addition to core business service, which is included in the fare. The important part is that auxiliary services are linked to core business services, and cannot be used without the proper main service. Such extras are often considered to be the elementary parts of the service package, which define it and make it competitive. "Service provider develops and manages service offerings, which are based on a thorough analysis and understanding of the customers' needs and expectations from the service" [22].

If we consider the current situation of auxiliary services, it is more than clear that they provide substantial additional revenue to the airlines. According to the research from 2016, which considered the performance of 63 airlines around the world, there is a rise of 7 billion US dollars in average every year in considered airlines together. This shows us the importance of auxiliary services in the current crisis environment in the aviation industry [24].

Types of Auxiliary Services

Before we do in-depth analysis of performance of auxiliary services, it is important that we define them, and explain them how they work.

The list of the auxiliary services consists of:

- Checked baggage and excess baggage
- Pre-order meals
- Buy onboard food and beverage
- Seat assignments
- Sales of commissionable items (hotels, car rentals, insurance)
- Duty-free sales
- Frequent flyer program
- Shuttles
- In-flight Wi-Fi
- Lounge access
- Advertising by the airline (in-flight magazine, ad message at gate or aircraft)
- Call center support

- Priority check-in or screening
- Early boarding benefits
- IFE—in-flight entertainment system
- GDS fees

Checked Baggage and Excess Baggage

In the former days, the checked baggage was almost always included in the free baggage allowance of the fare. With the development of customer categories, the business travelers have shown intention to travel only with cabin bag if it will save costs. This was offered before by low-cost airlines, but business travelers also insisted on better inflight and ground service included in the fare. As a result, network carriers started offering on domestic and continental routes also fares without baggage, which are in general cheaper. Of course, there is an option to additionally buy checked baggage, but the fees would be higher closer to the departure. For example, Austrian Airlines offers three types of so-called first bag offers:

- Paid during sales or reservation
- Paid at the check-in counter
- Paid at the gate

The closer to departure the customer is, the higher the cost for the first checked bag. If we look at excess baggage, every airline has its own rules of transport of baggage. There are two general concepts of measuring free baggage allowance.

Piece concept consists of settled number of pieces allowed to transport by the customer. Zero to three pieces are allowed for transport in general. Every customer has clearly stated on the ticket how many pieces can be transported free of charge. In case this number is exceeded, excess baggage fees apply. It is important to add that every piece of baggage has a limited weight due to economical and safety reasons. In general, it is 23 kg in economy class and 32 kg in higher transport classes. Piece concept is used mostly by EU- and US-based carriers such as Lufthansa, United, Air France, KLM, and British airways.

Weight concept consists of allowed weight for customer. Customers can transport several pieces of baggage free of charge if the total weight does not exceed allowed weight in the ticket. The allowed weight also differentiates according to travel classes and customer status with the airlines. For example, Turkish airlines offer 20 kg in economy class, and 30 kg in business class. This concept is mainly used by Middle East carriers such as Qatar, Emirates, Etihad, or Gulf Air.

"Prior to implementing these new baggage fees, in 2007 the US airlines collected only $467 million in baggage fees, mostly from the excessive weight penalties and the checking of more than the two complementary bags. This revenue stream increased to $1.1 billion in 2008, $2.7 billion in 2009, and has held steady at $3.3–$3.5 billion during 2010–2014 [25]."

In any of these two concepts, if baggage brought by customer exceeds the free baggage allowance, the excess baggage fee applies. These fees differ according to the routing and exceeded quantity. Additionally, airlines at the airport charge also service fee for document issuance at the airport.

It Table 6, we can see the revenue for the bag fees in selected US-based carriers on their domestic routes. We can also see the difference in the amounts charged based on the baggage fee policy. We can clearly see the importance of this service for revenue generation. Spirit Airlines generates the highest revenue per passenger; however airlines had negative perception by customers due to implementation of extra charge also for cabin bags which are taken onboard. This policy was only followed by two major carriers worldwide—Allegiant and Wizz Air. Contritely, Southwest generated the lowest revenue by passenger due to very customer-orientated policy of allowing two checked bags free of charge; however this helped Southwest to transport more passengers in total, as it was in general the first choice by American customer base with the intention to travel with checked baggage.

Pre-order "a la carte" Meals

With the recent downturn in economy and due to very high competition in the airline industry, big airlines decided to follow low-cost carriers with onboard meal offers. Current situation is that even major network airlines offer no or very limited meal service onboard their domestic or continental flights, with the same fare as before the meals were removed from flights. This allows airlines to generate additional revenue as a service of pre-ordered meals. This service is directed to high-value customers, which are defined as passengers motivated to purchase premium services, which generate top margins for an airline. These customers have a high expectation for luxury, comfort, and personal service because they have a prominent level of disposable income [24].

These high-value customers demand premium service; however, they do not want to pay excessive fares for higher classes. Thus, they are ideal candidates for premium economy classes or additional services.

Several airlines cooperate with high-level catering companies or famous restaurants in order to provide competitive a la carte menu with reasonable prices for

Table 6 Baggage fee revenue [24]

Carrier	Baggage fee policy	12 Month revenue[a]	Revenue per passenger
Spirit	Checked and large cabin bag	USD 201,000,000	USD 19.54
Delta	Checked baggage	USD 838,000,000	USD 8.58
JetBlue	Second checked baggage	USD 73,000,000	USD 2.84
Southwest	Two pieces of checked baggage Free	USD 158,000,000	USD 1.39

[a]12-Month period ended in September 2013

Source: Bureau of Transportation Statistics, US DOT, Domestic US Traffic

customer and acceptable margin for the airlines. For example, Air France has a deal with the famous restaurant Maison Lenotre of Paris, or Austrian Airlines has the famous Austrian catering company Do & CO. This service enables customers traveling in economy class to enjoy business class-like meal for the price of economy class plus reasonable charge for the service. Often airlines offer different options to be pre-ordered, with focus on route and with regard to cultural differences.

Buy Onboard Food and Beverage

As stated above, many airlines discontinued food service on most short- and medium-haul flight. As a typical rule alcoholic beverages are not served in economy class. This allows airlines to generate additional revenue in terms of offering snack together with premium drinks or alcoholic beverages onboard during the flight. The prices are higher than during ground sales as also logistic must be taken into consideration. Mostly, for onboard sales, cash is preferred, but in case of payment with credit card, service fee applies, which generates even more revenue for the airlines. This allows customers to buy food only if necessary, but for the airlines, profit per sales is noticeable.

Seat Assignments

Paid seat assignments during reservation are the most recent trend in the airline industry. Previously, seat reservation was free of charge service; however, now it amounts to a significant part of airline revenue.

Typically, the decision factor if seat assignment during reservation will be free of charge is the so-called booking class and passenger status. If the passenger wants to book specific seat in advance in order to have it guaranteed, small fee applies. In most cases, there are several seat categories at the aircraft—window seat, aisle seat, and middle seat—these will cost the least; bulkhead seat and exit row seat with more legroom space will cost more. On some airlines such as United or Delta, also premium economy seats are available. These typically consist of different types of seats with higher seat pitch. These airlines offer, together with premium seat selection, also more auxiliary services, such as free Wi-Fi or better in-flight-entertainment system. As a complementary step, airlines still offer free seat selection during Web check-in, but most of the time, as soon as this is open, the best seats are already sold out to customers willing to pay for guarantee. For the higher value customer, airline waives the payment for the seat assignment. This applies to higher booking classes or customers with high ranking in customer loyalty programs.

Sales of Commissionable Items

Ancillary airline revenue activities also include the commissions earned by airlines on the sale of hotel accommodation, car rentals, and travel insurance. The commission-based category primarily involves sales via the airline's website, but it can also include the sale of duty-free or other consumer products onboard aircraft.

Commission sale is a contract between the airline and other involved party which product airline sells. The revenue mostly consists of percentage from the amount sold. Normally there are several tiers of percentage earned by the airlines.

Table 7 shows the example of travel agency commissions from the sales.

The primary areas of airline sales consist of offering three main types of products: hotel accommodation, travel insurance, and car rental.

Every major network or low-cost airline has a contract with major chains in other forms of businesses. This is a mutually positive situation for both involved parties as these two products' offer comes together. There are travel agencies which focus on all-inclusive experience, in terms of selling the whole package for potential customers, which involves flights, hotel accommodation, and car rental together with travel insurance. The main advantage of this sale is that these products are linked together and it's easier to manage them in case of any disruption to travel plan (for example hotel check-in date change is possibly free of charge due to flight disruption).

Duty-Free Sales

Duty-free sales are part of the commission sales. It is a special category as it is provided onboard the aircraft, in the tax-free or tax-reduced environment. Airlines offer their own magazines onboard with the offers of their contracted partners. Passengers are also offered to prepay selected products via airline website and have them delivered onboard during their flight. This allows customers to have enough time to select the desired product together with the assurance that their product will be available to buy onboard, as some items might be out of stock on the day of flight.

These sales are part of commissionable sale that means from every sales airline do, percentage based on commission level will go to airline as revenue. This style

Table 7 Travel agency commission [26] (own processing)

Sales (USD)	Commission level (%)
<40,000	10
40,000–74,999	11
75,000–124,999	12
125,000–249,999	13
250,000–499,999	14
500,000–999,999	15
1,000,000+	16

of sale is very actively offered onboard low-cost carriers as they are very linked to ancillary revenue, as their fare for the core product is very often low.

To show the importance of duty-free sales we can look at Air Berlin (already ceased operations in 2018) statistics. Onboard revenue from duty-free and inflight sales was €28,431,000 for 2014. This represents 0.7% of the total revenue generated by Air Berlin in 2014, which is a very interesting number compared to possible costs that these sales generate [27].

Frequent Flyer Programs

The system of frequent flyer programs mainly consists of benefits to recurring customers. That means that customers for recognition need to apply for loyalty card, which very often includes free services. Afterwards customers need to travel on this airline (or alliance partner), for which points will be awarded (miles, dollars, points, dots etc.). This allows for the creation of loyalty by customers, which will buy services from the same airline in long term, which means that customers will generate ancillary revenue for the airline.

Credit card rewards with airline miles work differently. Most credit card rewards give you a certain amount of points per dollar spent. Customers can redeem these points for frequent flyer miles, but how much they're worth depends on the credit card. As with any credit card, customer needs to deal with annual fees. Since there is a trend of using credit card more often compared to buying airline tickets, it's usually easier to accrue points through a credit card than it is with just an airline reward program.

The reward of revenue from frequent flier programs is largely tied to co-branded cards. Airlines serving significant populations of credit-worthy customers use cards to tap a stream of retail revenue activity. This allows airlines to create daily transactions with members beyond the two or three trips that consumers normally take each year. Frequent flier programs can also encourage consumers to buy a la carte services.

To show the importance of frequent flyer programs we can consider revenue of Aeroflot for the year 2014, where almost 223 million US dollars in ancillary revenue was generated by their frequent flyer program. This amount stands for 53 US dollars per member of frequent flyer program. Another good example might be Lufthansa Group frequent flyer program called Miles and More in 2013. It generated €31.6 million income (not revenue) for the fiscal year of 2013. In 2014, they more than doubled this amount to €66 million. With around 25 million of members this equals to €2.64 per member [17, 24].

Shuttles and Trains

Ground transport is an essential support for commercial aviation. Buses and trains together with cars and taxi services allow customers to reach their departure at the airport on time. As airlines generate revenue only if seats are occupied on their flights, they start cooperation with ground transport companies.

In some cases, in bigger cities or in very near cities with airports they both operate, airlines offer shuttle services between these airports. The good example of this might be airport pairs in Paris (CDG and Orly—shuttle bus by Air France), Bangkok (Suvarnabhumi and Don Muang—shuttle bus by Thai Airways), or London (Heathrow and Gatwick—shuttle bus operated by British Airways). Besides customers with booked ticket with this airline, who get this service free of charge, there are a substantial number of customers from other airline flights which can use this service against a fee, thus generating ancillary revenue.

Another example of cooperation is cooperation between airlines and train services. An airline can provide a flight number on the train services between selected city pair. Afterwards, sale of this train ticket will generate a provision from sale, as well as provide flexibility for local travelers. This will allow customers from smaller cities, connected to railroad network, to use airline services seamlessly. A good example is cooperation of QBB and Austrian Airlines on route Linz to Vienna, DB Bahn and Lufthansa between several German cities, or the closest one in Slovakia between Bratislava and Vienna, which is provided by Fly Dubai.

The advantage for the customers with booked rail or shuttle transport on the airline ticket is in case of disruption. If in case of train or shuttle delay they miss their connecting flight out of the airport, they will be rebooked free of charge. Normally typical ticket conditions would apply, which might lead to rebooking fees.

In-Flight Wi-Fi

In the current world of duty travels, there is a very high demand of business customers to be able to connect to the Internet during the flight, as it will allow them to work and thus generate revenue from their side. As the technology moved to the current level, airlines are now able to fulfill this request as implementation of in-flight Wi-Fi makes sense.

There are several types of offers through the airlines:

- In-flight free Wi-Fi

This is the most favorite type of service for the customers. Airlines are able to provide this service free of charge as they will use ads which will generate revenue for them. So, the more people connected on network, the more ad click and the more revenue for the airline. Of course, the speed is limited, but for a fee you are able to increase the speed of your connection. This model is used for example by the Norwegian or JetBlue.

- Wi-Fi available for premium classes, paid by the others

This is the most common type of service provided by the airlines. For the higher value customers, the service is provided free of charge; for the rest there is option to purchase in-flight connection onboard with credit card. Thus, airline will generate revenue for the connection and in some cases also for the payment by the card. After payment customers will get Wi-Fi connection through the flight. There is also a possibility to prepay connection in advance for all flights.

- Package model

Package model is the last type of model used. It allows customers to pay for the limited amount of time for the Wi-Fi connection. The airlines' pricing policy mostly offers at least three options including short but expensive 1-h connection, medium-several-hour connection, or full-day model. Pricing is done in the way that if customer gets in touch by the shortest option he/she will get better price to prolong this option. This model is used for example by Lufthansa or United airlines. Due to high demand of this service, in the past year in-flight Wi-Fi became the decision maker deal for customers who needed to work on their trips or for customers who did not want to be limited by the content of in-flight entertainment system.

Lounge Access

Lounge access is the domain of the network carriers; however also low-cost carriers realized that there is a substantial number of customers willing to pay for the comfort before departure at the airport. Typically lounge presents the exclusivity of the airline. Airline offers a variety of high-end products from beverage to catering. But the offer mostly does not stay there. Some of the airlines offer showers, massages, personal bed, or your own personal assistant. In most cases this service is free of charge for higher travel class customers or for the customers with high loyalty status.

For the rest of the customers there is an option to buy access to lounge before entry. Mostly airlines offer different lounges for the paying customers than for the premium customers to avoid premium lounge from being overcrowded and in order not to lose the status of exclusivity. The service in paid lounge is limited compared to premium lounges. However, for the customers with long waiting times between connections, it is more convenient than waiting at the departure gate for several hours. As the tickets with long waiting times are typically much less expensive, the airline is thus able to generate additional revenue from the paying customers.

There are several types of entry options to lounges besides the status entry:

- Day pass
- Lounge annual membership
- Third-party lounge vendor (priority pass, lounge pass)
- Pay-in lounge
- Access with credit card (Diners Club White, American Express Traveler)

Table 8 Advertisement with Spirit Airlines [28] (own processing)

Advertising item	Quantity per quarter	Cost per quarter
Styrofoam cups	250,000	USD 25,000
Aprons	1100	USD 25,000
Air sick bags	150,000	USD 30,000
Boarding passes	1,000,000	USD 30,000
Ticket jackets	400,000	USD 45,000
Clear cups	600,000	USD 55,000
Overhead bins (all bins/all A319s, A321s)	940	USD 225,000
Beverage napkins	3,500,000	USD 60,000
Inflight menu	50,000	USD 60,000
Seatback inserts	70,000	USD 60,000
Tray tables (exclusive, all trays)	5790	USD 160,000
Bulkheads	126	USD 130,000
Window panels	1544	USD 250,000
Plane wrap (6 months)	1	USD 300,000
Plane wrap (1 year)	1	USD 400,000

Advertisements

As airlines transport excessive amount of customers onboard, it is a very interesting part for the third-party companies to affect a wide area of potential customers with advertisements. Airlines thus use several methods for providing space for companies for advertisements. This varies among gates and check-in counters, via the in-flight magazine, and via the in-flight Wi-Fi to physical parts of the aircraft. All types of airlines use advertisements by other companies to generate revenue; it is very common and aggressive for low-cost carriers, as they depend on noncore business income.

Typical example for ancillary revenue aggressive policy is Spirit Airlines, a US-based low-cost carrier, which provides extremely low fares, but you must pay additionally even for the cabin bag. They offer their physical part of the aircraft for advertisement [28].

Table 8 shows quantity and costs of using Spirit distinct parts of their aircraft for advertisements.

However, airline aggressive ad campaign created very intensive discussion in the USA. As Spirit was using personally directed ads to be printed on Web check-in boarding passes, customers felt that it was a breach to their privacy. On the other side, customers decided to travel with airlines known for their aggressive ad culture.

Call Center Support

Typically travel agencies offer their services for a specific service charge for every transaction they are going to perform. Due to economic crisis and very high competition and thus lower income than before, airlines decided to charge for call center support. That means for every minute or service on call with call center you will be charged. Airlines still left other options to contact them via email service center, or via direct visit to their ticketing centers. Charges for service on call vary depending on the carrier. Generally low-cost carriers charge more for calls than network ones.

There are two types of airline call centers in use:

- First one is pay-per-minute when customers call to premium phone number; the part of the money spent will go towards the carrier. Ticketing and service charges still apply. This is a widespread practice in low-cost carrier community.
- Second one is pay-per-service call center, where customer pays only for the service ordered by him/her. Typically, the price for the service via call center is cheaper than with ticketing office or airport office, so customers are motivated to contact call centers.

"Service charges for ticketing with a United representative in the 50 United States and Canada are USD 25 by phone, USD 30 at a City Ticket Office and USD 35 in person at the airport" [29].

Priority Check-In, Boarding, or Screening

In the busy world of commercial aviation, a lot of business travelers prefer to move through typically delaying part of airport process. This also applies with higher value customers who want to enjoy as much airline services as possible. This is the reason why airlines dedicated premium check-in counter and premium security lanes to allow passengers willing to pay for premium service to use them as much as possible, and thus generate revenue.

Typically for low-cost carriers with free seating also priority boarding option applies. It is important to differ priority boarding of network carriers due to customer status from priority boarding of low-cost carriers due to limited overhead bin capacity. Network carrier's priority boarding means higher value customers enjoy more of the service provided by the airline together with security of space in the overhead bins for their personal belongings. Several customers with network carriers also feel privileged to be in front. Almost the same philosophy applies with low-cost carriers. Differently from the network carriers, this service is often paid, but the customer can decide his/her best seat, secure enough space in the overhead bin, and feel privileged to be in the front.

These three services, if paid, often come in package as one product. With the American carriers, the price differs from USD 9 with American airlines to USD 25 with United airlines [30]. Table 9 shows comparable product with several US-based airlines and their prices.

Table 9 Airline priority fees (own processing)

Airline	What It's called	How much it costs
Airtran	Priority boarding	USD 10
Allegiant	Priority boarding	USD 9.99
American	Your choice	USD 9
JetBlue	Even more speed	USD 25 and up
Southwest	Early bird check in	USD 10
United	Premier line	USD 9 and up

Table 10 Tablet rental fees (own processing)

Airline	Offered IFE	Price
Alaska Airlines	Windows 10 tablet	USD 15
Hawaiian	Apple iPad mini	USD 8
United	Samsung Galaxy Tab	USD 10
WestJet	Android tablet	USD 7

In-Flight Entertainment

As the flight times are getting higher due to the possibility to fly to further destinations, it is very important for the airline to provide accurate entertainment during these long flight times. For a very long time on most of the long-haul international flights the so-called IFE—in-flight entertainment—is provided. With the recent necessity to generate as much ancillary revenue as possible airlines decided to offer several levels of IFE.

Typical form of revenue generation via the IFE is the so-called pay-per-view content. In this case there are several options to view free of charge, but if you decide to watch more recent movies or listen to more recent music, you would have to pay for this. As the distribution and cost of this content are very low in digital world, it generates reasonable amount of revenue compared to costs. However, providing screens in the seatbacks is not a very cheap option for the airline, as typical IFE in the backseat may cost up to 10000 US dollars, and airlines must take into account that it will not be the most recent technology in the market.

For this reason, some of the airlines decided to offer possibility to rent a tablet to use for in-flight entertainment. In Table 10 different types of offered tables are mentioned. These tablets are preloaded with the content instead of screens in the seats. This option will allow the airline to save costs on the upgrading of seat IFE, but instead they can just arrange a more recent tablet for only 1/20 of the price of the typical seat IFE. This allows airlines to save costs on upgrading together as a typical seat IFE weights around 6 kilos, but for 200 or more seats it can save a lot of weight and thus a lot of fuel.

As this is the modern technology not very common in EU, we can only see the rent fees in the USA and Canada territory.

GDS Fees

GDS or global distribution system belongs to the travel industry for a very long time. However, in the recent past (2015), one of the major airline groups (Lufthansa) decided to penalize the booking and issuance through GDS with a fee. This fee is not supposed to generate only revenue, but also cover the cost of using GDS as Amadeus, Galileo, Sabre, or Worldspan. Lufthansa realized that it is much cheaper to provide own booking platform and to use it than invest money info GDS systems. Lufthansa also said that the GDSs have limited booking functionality and that "innovative ancillary services and enhanced price options require suitable sales technology" [31].

With the only recent invention of this GDS fee, it is not known the revenue this will generate, but at the end Lufthansa feels that it will lower their costs on using GDS. However, GDS will remain a direct sales channel for the airline, and thus the sole income of primary revenue.

Looking at the analysis of current market of ancillary services of aviation it is necessary to point few important parts of this problematics.

5 Service Analysis

As this new trend is affecting pricing directly or indirectly, customers often act negatively to such a change. Customers were used to have all-inclusive fares, and never thought about additional services. With recent change in pricing and with implementation of a la carte services or branded fares customers have started to feel that airlines are trying to force them to pay additionally for services which were free before.

A recent study shows that customers are accepting various categories of additional payments differently. Extra payment for the excess baggage if often perceived as betrayal by the airline, as reasons behind this are not very understandable by the passengers. Due to common misinterpretation by customers, they tend not to read general conditions and payments at the airport are often followed by discussion and unwillingness to pay for any excess baggage. This is even more often since the wide implementation of a la carte option with FSC as in some fare families there are payments even for first checked bag. This results from improper or inadequate education by the airlines, or more often incorrect information provided by travel agencies in their offices or online [32].

On the other hand, additional services from entertainment group such as in-flight entertainment system or food are accepted more widely. This is mostly because these services are perceived as something offered extra, while excess baggage fees are perceived as additional payment to core service. Reason behind this is that customers are able to bring their own entertainment devices such as tablets, smartphones, or laptops onboard and therefore only customers who prefer this option pay for this [32].

A recent study of customer acceptation of airline fees showed that there is a weak significant correlation between level of acceptance and fee-related anger—the lower the fee-related anger, the bigger the fee acceptance in place. The same study shows that fees for entertainment services and mileage are commonly accepted the most [32].

It is very important to understand that customers expect value for money. Airlines understand that what is for them a way to survive and keep profitability, for customers it might feel as unnecessary additional payments just to get more money. Several surveys were conducted with two different but valid results on how should an airline approach a change in the market. First one says that "it is essential for airlines to emphasize the added value of an ancillary item [33]." Therefore airlines, which choose this way, must work hard and come up with innovative products that customers would accept and would be willing to pay for. So, the approach is to charge only for new and additional services. Another approach according to the study is that "carriers should work on changing, i.e. lowering, customers' expectations [33]."

Second approach is very difficult for FSC as mostly FSC position themselves as premium brands, and this approach is completely opposite to what they have promised and delivered in the past.

Both of the surveys done result into one crucial step for airlines to focus on to raise ancillary revenue—to educate customers as much as possible about the structure of newly introduced fare structures. When customers understand what is behind the new structure and accept the new form of fare structure, there is a preposition that they will be willing to pay for services not previously accepted as additional service.

5.1 Comparison of Models

When we look into the performance in ancillary revenue, we can see that results per passenger are significantly better with LCC due to their aggressive policy in charging for baggage, airport check-in, seat selection, etc. It is not only the way in which LCC are enforcing additional payments, but it is also important to understand that customers of these airlines are normally choosing their services because of availability of low fare, but at the same time they understand that everything above low-cost standard will be charged extra.

Table 11 shows ancillary revenue per passenger as shown and analyzed from annual reports for 2015.

On the reported ancillary revenue per passenger from 2016 we can see that on the top 10 list, there is heavy domination by low-cost carriers, followed by traditional high performers from FSC. Spirit is positioning as an ultralow-cost carrier, where in the basic fare only transportation of customers is guaranteed and transport of any other item is paid. This is enabling Spirit airline to be able to generate the best

Table 11 Ancillary revenue per passenger as reported for 2015 [34] (own processing)

Ancillary revenue per passenger 2015			
USD 51.80	Spirit	USD 38.35	Korean Air
USD 50.84	Jet2.com	USD 36.08	Virgin Atlantic
USD 49.94	Allegiant	USD 35.01	AirAsia X
USD 44.16	United	USD 34.25	Alaska Air Group
USD 42.70	Qantas Airways	USD 32.74	Wizzair

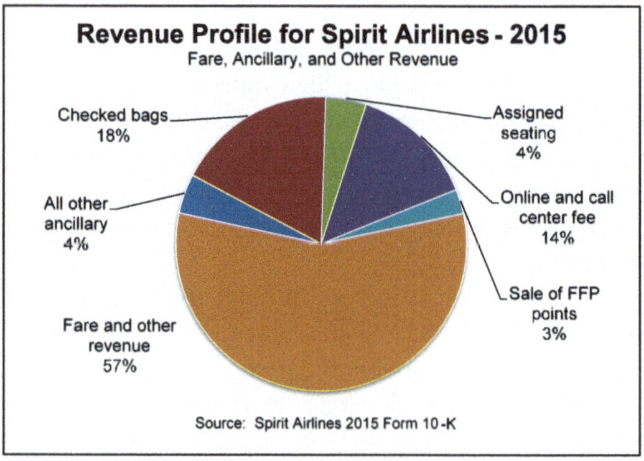

Fig. 4 Revenue profile for Spirit Airlines [34]

ancillary revenue per passenger which is resulting to 43% of total revenue. Their revenue profile is projected in Fig. 4.

The second and third places of the top performers are occupied by LCC Allegiant and Jet2.com. Both airlines are focused on low-cost model in addition to mostly offering leisure destination, where passengers are more likely to be willing to pay for comfort due to the typical nature of their travel and more liberal spending behavior. Both act as holiday package companies in their nature.

Another important note resulting from the table is FSC generation of revenue. United is the only airline whose source of top-performing ancillary service is not solely FFP. They are profiting from long-time-implemented fees for checked baggage on continental routes within the USA, together with offering seat upgrades to their Economy Plus seats and commission earning from sale of miles to Chase bank. The other two FSC in top performer list are Korean and Qantas whose dominating income of ancillary revenue is from FFP. FFP revenue is defined as "sale of miles or points to program partners such as hotel chains and car rental companies, co-branded credit cards, online malls, retailers, and communication services. Direct sale to program members also qualify [34]." These results are showing considerable importance in loyalty programs for premium carriers.

5.2 A la Carte Pricing for FSC

When during the beginning of Ryanair journey Michal O'Leary was thinking about his pricing model, he came up with a solution to sell tickets the same way as restaurants sell their product, the so-called a la carte pricing, which meant that customers will be really paying just for services they prefer, and therefore lower fares would be available. This model was later adopted by Air Canada in 2005 which enabled the company to survive in the most turbulent times more easily than other carriers. With later economic recession customers became more price oriented and premium full-service airlines had a tough time making competition to newly born airlines that preferred low-cost model. The only options were to lower high standard or to adapt prices to the competitive way of LCC. This was only possible by unbundling their all-inclusive product to separate services, where customers have a chance to choose his/her preferred type of fare.

Thus, the era of a la carte pricing begins. According to IdeaWorksCompany a la carte features are defined as "the items on the ancillary revenue menu and consist of the amenities consumers can add to their air travel experience [34]."

Practically the most common features are as follows:

- Onboard sales of food and beverages
- Checking of baggage and excess baggage
- Assigned seats or better seats such as exit row or bulkhead seats
- Call center support for reservations
- Fees charged for payment with credit or debit cards
- Priority check-in and screening
- Early boarding benefits
- Onboard entertainment systems
- Wireless Internet access [34]

These features enabled FSC to come with a new trend of tariffs called "branded fares," where customers have an option to choose his/her preferred level of service from the airline. This way the airline can offer completely basic product for price which is comparable to the price offered by LCC. One advantage is that customers will be able to buy ticket on full-service airline with network model for acceptable price, and a second is that for the airline it enabled the space for additional ancillary revenue generation. Figure 5 shows the preference of travelers from the USA with regard to the type of fare structure they are willing to buy.

If we look at charging fees for checked baggage, with branded fares customers have a chance to choose the option to travel without baggage and therefore they are able to save money on the fare. Therefore, the airline is transporting less weight, which means more fuel-economic flight. Other advantages for customers are that airlines are very flexible and often offer the possibility to pay late for checked baggage even during the check-in. However, this always costs more for the customers than it would during the booking, but the flexibility is important for customers. For airlines that prefer also to offer their capacity for cargo transport,

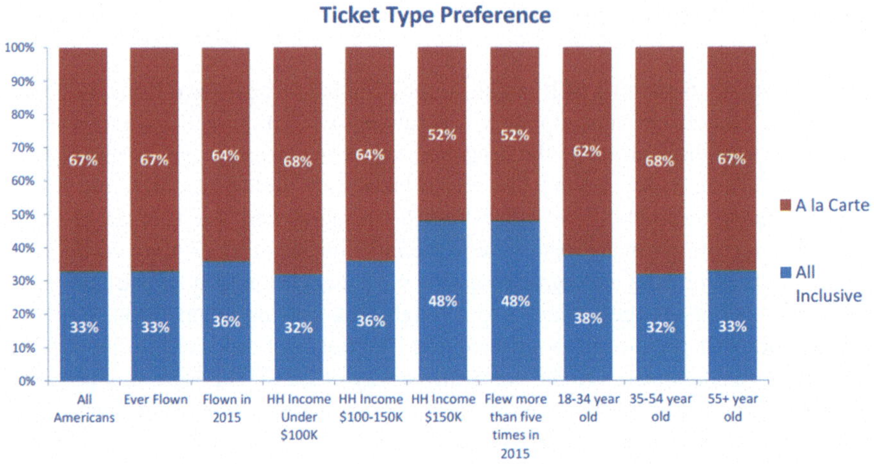

Fig. 5 Preference of US travelers between all-inclusive and a la carte fares in 2015 [35]

this enables more space for cargo transportation which again generates more revenue. The last, but not least, advantage is in staff cost reduction, since with the implementation of checked baggage fees less customers tend to use airport check-in and therefore less staff is necessary and thus cost reduction is implied.

More and more FSC are accepting fare structure of branded fares as it allows their customers to flexibly choose their preference, but at the same time it offers the airline a possibility to generate additional revenue. Figure 6 shows offered fares by Austrian Airlines.

To understand the flexibility in branded fare offers, it is important to describe pricing before and after implementation of these fares. Typical FSC offers two or more travel classes divided into different selling classes indicated by the letter of alphabet. Before implementation of branded fares, as soon as selling class was sold or its time limit for advanced purchase was passed, selling class was sold and customers had an option to buy only higher class with more flexibility and services, but at the same time for higher price. With the implementation of branded fares, selling class remains the same, but always offers different fares according to customer preference. This is enabling customers to choose even with late purchase options how he/she needs to travel and thus enabling him/her to save money, and not to pay for services which are not necessary for his/her journey. On the other hand, customers who prefer early purchase to save money now have an option to choose lower selling class with very flexible conditions in case of unpredicted change of travel plans, which was not possible before as cheap fare always meant very restricted conditions of ticket.

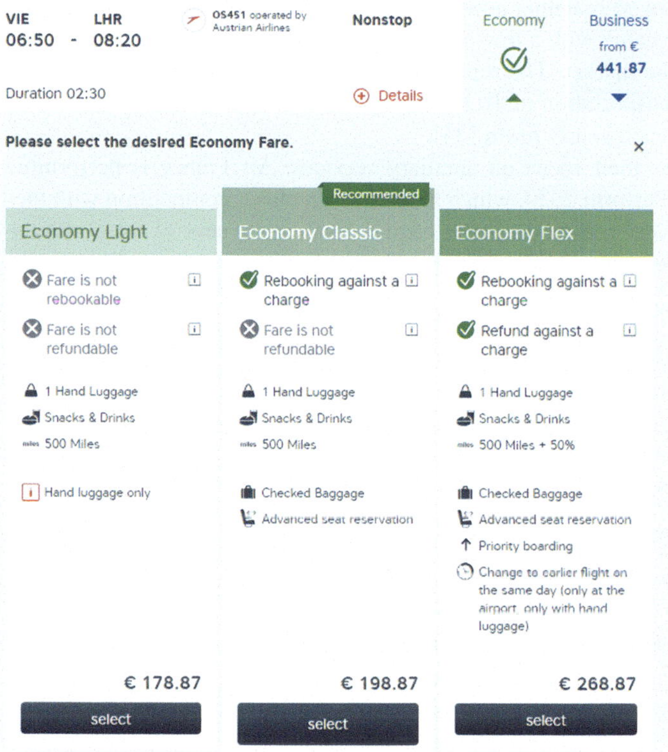

Fig. 6 Branded fares offered by Austrian Airlines [36]

6 Research Cases

To understand the performance of selected ancillary services it is important to analyze current prices and compare it to profitability and usability of service. The next is dedicated to analyzing the current position of the airline in the market, their ancillary revenue, and top-performing services, and possibly prognosing their future development.

6.1 Air France/KLM Group

The integrated group of airlines Air France/KLM is a long-term top performer in the generation of ancillary revenue. Since airlines started reporting ancillary revenue in 2012, steady growth of ancillary revenue of income is in place. Ancillary revenue of Air France/KLM group is sourced to various activities. It presents approximately 7.5% of total revenue reported by the airline. The group expects that ancillary revenue will be reaching a rate of 10% of total revenue until the year of 2020. The group also came with a low-cost initiative to compete with European LCC to reach a part of the market. The reports of airline Transavia, which belongs to the group, showed transportation of 10,800,000 passengers and it is estimated that until 2018 they will reach yearly profit [34].

To show their focus on ancillary services, Air France is performing its long-term plan Perform 2020, which is focused on closer competition with their low-cost concurrence. Their whole experience is divided into several categories that generate ancillary revenue (Table 12):

- Preflight:
  - Time to think
  - Insurance
  - Duty-free online pre-order
  - Surprise gift onboard
  - Flying blue miles for sale
- Baggage:
  - Excess baggage fees
  - Paid baggage with carry-on-only fares
  - Day before drop-off on selected airports
  - Baggage pickup for Paris departure
  - Baggage delivery for Paris arrival
  - Protective case rent for special baggage
- Airport:
  - Lounge access
  - Meet and assist
- Onboard:
  - Economy comfort seat
  - Preferred seat
  - Extra leg room seat
  - A la carte meal
  - Duty free onboard

Table 12 Ancillary revenue reported by Air France/KLM group [34] (own processing)

Ancillary revenue reported by Air France/KLM group				
2012	2013	2014	2015	2016
USD 1,205,727,600	USD 1,714,598,496	USD 2,046,292,309	USD 2,165,996,840	USD 2,586,279,195

Fig. 7 Air France/KLM ancillary revenue growth 2012–2016 [34] (own processing)

- Destination:
  - Hotel
  - Car rental
  - Airport parking
  - Activity booking
  - Airport transfer
  - City passes [34]

The graph in Fig. 7 projects the growth of group ancillary revenue from 2012 to 2016.

6.2 *Wizz Air*

Wizz Air is the fastest growing European low-cost carrier in the market. In 2015, it transported exactly 16,455,669 passengers. From its activities revenue of USD 1,481,381,107 was generated, out of which USD 538,681,419 was reported as ancillary revenue from ancillary services. It represents 36.4% of the total revenue. This resulted in very successful USD 32.74 per passenger [34].

Wizz Air offers a handful of possible add ons to their standard fare:

- Airport check-in
- Airport transfer and parking
- Wizz plus branded fare
- Assigned seating
- Onboard food and beverage
- Call center bookings
- Car rentals, hotels, and trip insurance
- Checked baggage

- SMS service
- Extra leg room seating
- Fast-track security
- Large carry-on baggage
- On-time arrival guarantee
- Paid lounge access
- Priority boarding
- Wizz co-branded credit card
- Wizz discount club
- Wizz flex
- Wizz privilege pass (subscription for flight-related benefits) [34]

Wizz Air is becoming the biggest low-cost player in Europe focused on central and east Europe.

6.3 Spirit Airlines

Regarding ancillary services and ancillary revenue, Spirit is positioning itself as an ultralow-cost carrier, where its core product is completely basic. It is just transport of passengers without anything else onboard. They are also charging fees for even small cabin baggage. This is enabling Spirit to offer extremely low fares, but at the same time they are best performers with regard to ancillary revenue per passenger. In 2016, total revenue reported was USD 2,141,463,000, out of which USD 928,369,000 was from ancillary services. This accounts for 43.4% which is the highest amount in the market. Spirit is also the best performer regarding per passenger revenue, where its last record is estimated as USD 51.80 per passenger [34].

Basically, Spirit is charging extra for every step of their booking process. Breakdown of income from several charges resulted with Spirit in the following numbers:

- Baggage fees—USD 381,386,000
- Passenger usage fees—USD 298,092,000 (PUF is charged for all bookings except those made at the airport)
- Advance seat selection fees—USD 97,786,000
- Cancellation and change fees—USD 43,756,000
- Others—USD 151,105,000
- Considering per passenger statistic reported for 2015 by Spirit:
- Average ticket price: USD 65.25
- Average ancillary price: USD 54.24
- Total average price: USD 119.49

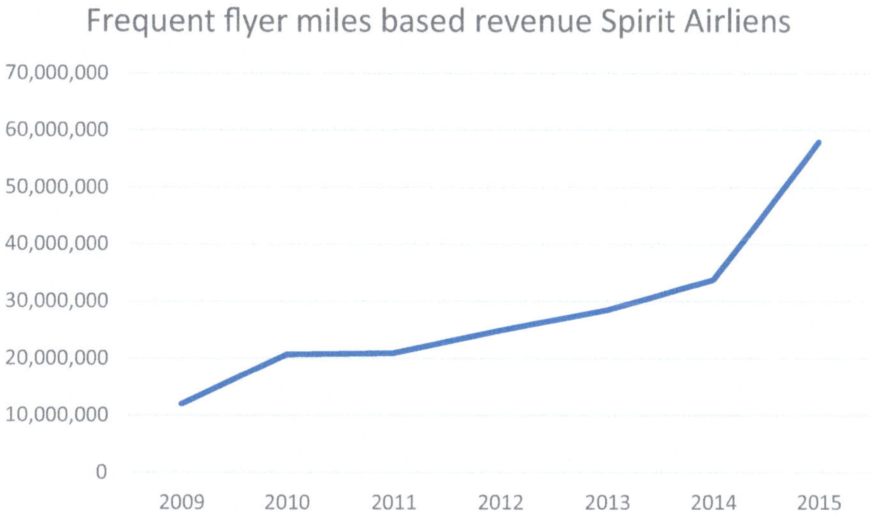

Fig. 8 Growth rate of frequent flyer revenue by Spirit Airlines [34] (own processing)

Besides all information above, Spirit is the airline which does not count sales of their miles and points into ancillary revenue. Cash sales represented for the last few years have a tendency of growth as projected in Fig. 8 [34].

6.4 Cumulative Comparison

To be able to give analysis and estimate for future years it is important to compare data for comparable timeframes. Within aviation the last 2 years, 2015 and 2016, are the most comparable due to nature of development in the market. As more airlines understood the importance of ancillary services, more of them were implemented and later reported by the airlines.

Passenger growth is the principal factor due to a possibility of each customer to buy added ancillary services and therefore generate additional revenue. Comparison of the growth of passengers of 12 selected airlines shows worldwide growth of passenger count due to ending economic recession and stronger position of customers in buying products such as airline transport. Statistics in customer's number also shows the trend of three airlines with statistically advanced numbers. With Air France/KLM this could be the result of implementation of branded fares in their markets, which resulted in more customers being able to afford their services. With Turkish Airlines, significant growth might be the result of their tendency to open new routes in very fast rate. Spirit Airlines also shows a considerable rate due

Table 13 Transported customers and estimated figures for 2017 and 2018 [34] (own processing)

	2015	2016	+/− Growth (%)	2017 (estimate)	2018 (estimate)
EasyJet	64,800,000	68,600,000	**5.86**	72,622,840	76,881,586
Lufthansa	105,988,000	107,679,000	**1.60**	109,396,979	111,142,368
Air France	77,500,000	89,800,000	**15.87**	104,052,129	120,566,209
Virgin America	6,507,000	7,036,000	**8.13**	7,608,006	8,226,515
Hawaiian	10,191,000	10,665,000	**4.65**	11,161,047	11,680,165
Air Canada	38,526,000	41,126,000	**6.75**	43,901,466	46,864,239
Wizz Air	15,200,000	16,455,669	**8.26**	17,815,069	19,286,768
British	42,000,000	43,323,000	**3.15**	44,687,675	46,095,336
Spirit	14,294,000	17,921,000	**25.37**	22,468,325	28,169,502
Turkish	54,674,967	61,248,192	**12.02**	68,611,674	76,860,421
Delta	171,350,030	179,382,874	**4.69**	187,792,296	196,595,949
WestJet	19,652,000	20,281,376	**3.20**	20,930,908	21,601,243
United	138,029,000	140,369,000	**1.70**	142,748,670	145,168,682

to the nature of their product, where they offer very low fares but at the same time possibility to pay for premium services which can attract business travelers.

Figures in Table 13 indicate that the majority of airlines are reporting growth as predicted by IATA (5–8%) for this period. Only airlines mentioned above are overreaching proposed numbers, but it can result in adjusted product.

When comparing revenue reported by the airlines there is a clear indication of turbulent times in the market. Airlines are forced to offer only stable and profiting routes which results in reduction of routes offered, and therefore also reduction of available seat kilometers. It is important to understand that lower revenue does not mean lower profit, as cost reduction might produce more significant saving than lost revenue. An important note from this table is that only US-based carriers are able to report growth in total revenue. This is due to successful implementation of ancillary services to their product. At the same time, Hawaiian services, due to the nature of their location and routes, do not have significant problems with passenger numbers.

Estimates for 2017 and 2018 were calculated with results from previous years adjusted by calculated growth by IATA of 5% (3–8% predicted) annually. Results are processed in Table 14.

Figure 9 shows visual reproduction of total revenue reported by the airlines for 2015 and 2016.

Next measured results were done on the percentage of ancillary revenue compared to total revenue. These results indicate the performance of ancillary services of selected airlines. Result ranges from 0.93% to 78.95% of growth. As all these results are positive, this indicates that airlines are focusing more on ancillary revenue generation, with steady growths. This process is then enabling airlines to have

Table 14 Total revenue for 2015 and 2016, calculated growth and estimated total revenue for 2017 and 2018 with 5% average yearly growth as predicted by IATA [34] (own processing)

	2015	2016	+/− Growth (%)	2017 (estimate)	2018 (estimate)
EasyJet	7,760,957,510	7,318,850,631	**−5.70**	7,267,871,126	7,633,813,658
Lufthansa	33,363,402,072	27,156,989,045	**−18.60**	23,462,970,878	24,820,820,331
Air France	34,104,871,808	28,889,564,819	**−15.29**	25,916,259,285	27,360,737,526
Virgin America	1,489,967,000	1,529,584,000	**2.66**	1,646,733,584	1,723,212,784
Hawaiian	2,314,879,000	2,317,467,000	**0.11**	2,435,931,243	2,551,804,593
Air Canada	12,457,994,160	11,042,328,502	**−11.36**	10,339,648,606	10,891,765,031
Wizz Air	1,573,639,552	1,481,381,107	**−5.86**	1,468,600,593	1,542,669,648
British	20,090,713,731	17,700,498,122	**−11.90**	16,479,674,154	17,364,699,060
Spirit	1,931,580,000	2,141,463,000	**10.87**	2,481,224,767	2,588,297,917
Turkish	11,070,000,000	10,522,000,000	**−4.95**	10,527,227,733	11,053,327,733
Delta	40,362,000,000	40,704,000,000	**0.85**	43,084,097,874	45,119,297,874
WestJet	3,732,659,855	3,208,282,936	**−14.05**	2,917,986,440	3,078,400,587
United	38,901,000,000	37,864,000,000	**−2.67**	38,747,843,737	40,641,043,737

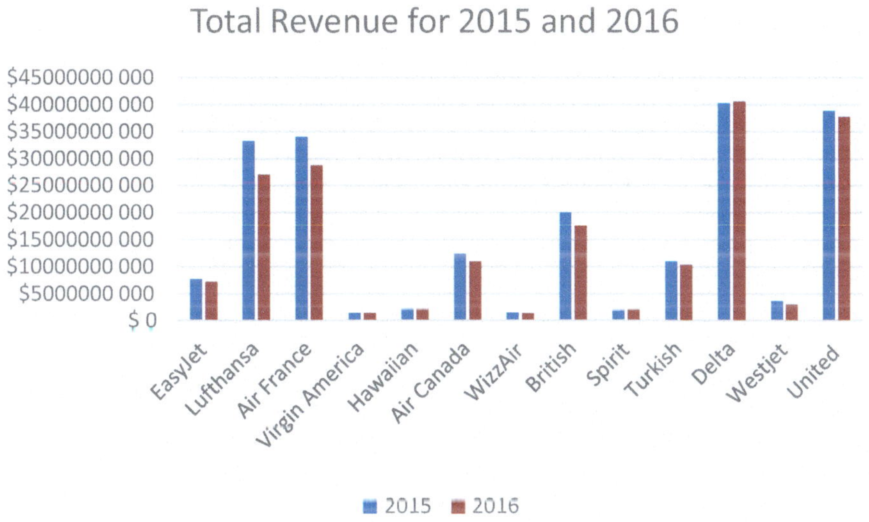

Fig. 9 Total reported revenue of selected airlines and comparison between 2015 and 2016 [34] (own processing)

considerable protection against the fluctuation of fuel prices. The best results were given by the airlines with recent implementation and focus on ancillary revenue generation. Both Hawaiian and British airlines are well proved in their core markets. Another notable comparison is done with results of LCC and FSC. As LCC were using ancillary services as their important part of revenue before, their growth is not

Table 15 Percentage of ancillary revenue from total reported revenue and its prediction for 2017 and 2018 [34] (own processing)

	2015	2016	+/− Growth (%)	2017 (estimate)	2018 (estimate)
EasyJet	18.8	20	**6.38**	21.28	22.63
Lufthansa	4.9	5.5	**12.24**	6.17	6.93
Air France	6	7.5	**25.00**	9.38	11.72
Virgin America	10.8	10.9	**0.93**	11	11.1
Hawaiian	5.5	9.2	**67.27**	15.39	25.74
Air Canada	4.9	6.2	**26.53**	7.84	9.93
Wizz Air	33.7	36.4	**8.01**	39.32	42.47
British	1.9	3.4	**78.95**	6.08	10.89
Spirit	38.7	43.4	**12.14**	48.67	54.58
Turkish	2.1	2.5	**19.05**	2.98	3.54
Delta	8	9.3	**16.25**	10.81	12.57
WestJet	5.4	8.3	**53.70**	12.76	19.61
United	15.1	16.4	**8.61**	17.81	19.35

that significant as with FSC which implemented and focused on ancillary services just recently [34].

Table 15 shows the annual growth of ratio between total revenue and ancillary revenue. Prediction of 2017 and 2018 was calculated with annual growth from 2015 and 2016 with adjustment by the prediction of growth by IATA. IATA predicted annual growth for 2017 and 2018 to be between 3% and 8%, according the core market. This was then averaged to middle value of 5% and calculated accordingly with this number.

Next analysis was done on per-passenger income for ancillary services reported for 2015 and 2016. These results indicate much fluctuated environment of ancillary services, where even the airlines with stable ancillary income are struggling to keep growth on the per-passenger ratio. With negative results, it might be suggested to review their ancillary products that are in place right now and seek opportunities in new and innovative ways of ancillary revenue generation. However, again with British and Hawaiian Airlines, we can see their successful implementation of the new fare structure together with offered ancillary services.

Table 16 analyzes annual per-passenger income from the year before, and estimated results for 2017 and 2018 based on earlier performance. Turkish Airlines are notable due to their stable product and very stable results due to the nature of their product. However, as with the development in the market, it is expected that also premium carriers which do not offer these ancillary services in higher rate will implement such a possibility for their customers as this will enable them to remain profitable and with stable background for future stability and growth.

The last but supposedly most notable analysis was done comparing all changes in different measured sectors and comparing them side by side. This table represents

Table 16 Change in per-passenger ratio for 2015 and 2016, growth calculation and estimate of per-passenger income for 2017 and 2018 [34] (own processing)

	2015	2016	+/− Growth (%)	2017 (estimate)	2018 (estimate)
EasyJet	USD 22.49	USD 21.37	**−4.98**	USD 20.31	USD 19.29
Lufthansa	USD 15.41	USD 13.87	**−9.99**	USD 12.48	USD 11.24
Air France	USD 23.35	USD 24.12	**3.30**	USD 24.92	USD 25.74
Virgin America	USD 24.71	USD 23.70	**−4.09**	USD 22.73	USD 21.80
Hawaiian	USD 12.39	USD 20.20	**63.03**	USD 32.93	USD 53.69
Air Canada	USD 15.99	USD 16.59	**3.75**	USD 17.21	USD 17.86
Wizz Air	USD 34.87	USD 32.74	**−6.11**	USD 30.74	USD 28.86
British	USD 8.91	USD 13.72	**53.98**	USD 21.13	USD 32.53
Spirit	USD 52.35	USD 51.80	**−1.05**	USD 51.26	USD 50.72
Turkish	USD 4.35	USD 4.35	**0.00**	USD 4.35	USD 4.35
Delta	USD 18.75	USD 21.40	**14.13**	USD 24.42	USD 27.88
WestJet	USD 10.34	USD 13.20	**27.66**	USD 16.85	USD 21.51
United	USD 42.26	USD 44.16	**4.00**	USD 45.93	USD 47.77

stability of the airline in the market as were as dependability on their different structures. Notable fact is that even with decrease of total revenue, ancillary revenue is growing steadily on all measured airlines.

Next notable fact is that highest results in growth of ancillary revenue were performed by the airlines with previously negative revenue growth. This indicated their focus on establishment of revenue by means of further focus on ancillary services. In general, the best-performing airlines are the ones with all green numbers. In our case, most stable results are done by Hawaiian Airlines that recently successfully implemented branded fare structure and started focusing on ancillary services.

Our analysis, as projected in Table 17, also clearly resulted in negative average total revenue, which can be indicated in worldwide cost savings done by the airlines, to be able to be sustainable and competitive in the market in the future.

7 Conclusion

As clearly stated above, logistics makes an important factor in baggage transport systems. With increased number of transported passengers and baggage, innovative logistics systems are necessary as well as unification of processes within aviation. With all technologies named and assessed it is safe to state that implementation of baggage-tracking system as per IATA Resolution 753 offers high direct and indirect

Table 17 Analysis of changes in measured criteria [34] (own processing)

	Passengers (%)	Total revenue (%)	Ancillary revenue (%)	An. rev. percentage (%)	USD/pax
EasyJet	5.86	−5.70	6.38	0.60	−4.98
Lufthansa Group	1.60	−18.60	12.24	−8.52	−9.99
Air France KLM	15.87	−15.29	25.00	5.85	3.30
Virgin America	8.13	2.66	0.93	3.71	−4.09
Hawaiian	4.65	0.11	67.27	69.02	63.03
Air Canada	6.75	−11.36	26.53	10.75	3.75
Wizz Air	8.26	−5.86	8.01	1.63	−6.11
British	3.15	−11.90	78.95	58.88	53.98
Spirit	25.37	10.87	12.14	24.08	−1.05
Turkish	12.02	−4.95	19.05	12.02	0.00
Delta	4.69	0.85	16.25	17.50	14.13
WestJet	3.20	−14.05	53.70	31.74	27.66
United	1.70	−2.67	8.61	5.77	4.00
Average	7.79	−5.84	25.77	17.92	11.05

advantages in the form of direct cost savings resulting from baggage recovery and delivery as well as in the form of reduced staff. Implementation has also indirect effect on higher effectivity of loading operations as it offers more exact decision-making of loading supervisor and thus improves usable load ability for other types of load. This approach thus improves customer satisfaction on all levels of transport process.

With regard to abovementioned analysis and prediction, there are several adjustments to be done by the airlines to be in competitive state also in the future. The most crucial step for the full-service carriers is to find a correct and suitable ratio in providing quality but at the same time provide ancillary services for customers. This will enable FSC to generate ancillary revenue which is even more important these days with unstable political situation in more regions, very fluctuating fuel price, and very high demand for premium services for low-cost fares. This is already working with baggage charges where average income per passenger on full-service carrier is resulting in approximately USD 6. On the other hand, it is very crucial for airlines to educate their customers that new fare structures such as a la carte fares are suitable for both frequent travelers and leisure infrequent travelers. Also an important step is to provide branded fares for customers who are willing to pay more for their comfort. This way an airline is able to provide suitable services for larger groups of passengers. Considerable improvement might also be reached with offer of commission-based services. As Frequent Flyer programs are usually working very good on widely established airlines, it is important to offer as much services as possible like car rental, hotels, or insurance on every single touch point with the customer, starting from booking and ending with arrival of the passenger to their destination. This can be reached by higher use of mobile technology, where according to the flight information and location of the passenger they can receive personally targeted offers for specific services.

Low-cost carriers have slightly more difficult stand on the ancillary service market as of today, as they are already trying to pursue steps proposed for full-service carriers. The most crucial step for them is to come with as premium services as possible, but at the same time to offer as low fares as possible. The most notable airlines regarding ancillary revenue generation are the ones serving leisure destinations and at the same time offering holiday packages. This is enabling airlines to act as tour operators who can offer almost the same services as travel agencies who focus mainly on holiday packages.

For both FSC and LCC, there is one step. As current trends show, customers are often choosing more innovative products due to exclusivity even if price and routing of more airlines are the same. Several innovative services were already implemented. For example, Air Baltic implemented a betting system for their departure on time rate. Customers could bet on possible delay time of their flight for a small fee. This is showing the importance of the airline to provide on-time departures, but at the same times it offers their customers a funny way to gamble with their flight. And even when the flight was late, the customer who won the bet with the airline will be less likely to pursue compensation for the delay, as they already received something from the airline. Another notable service is offered by Qatar Airways. They offer a specific fare for transport of the falcons. Falcons are very common animals in the Middle East region, and customers are often willing to pay large amount of money to transport them on the same flight with them, which is normally not possible with other airlines.

Another widely accepted innovative product is auction for upgrade called Smart Upgrade, where customers can propose the amount for upgrade to higher class he/she is willing to pay. Afterwards, according to the rules set by the airline, this offer is accepted or not. This is enabling airlines to generate additional revenue from passengers who normally would not be willing to pay for higher class.

Lufthansa group is also applying their new e-tag for baggage, where customers are able to print his/her baggage tag already at home, and then later to use self-baggage drop-off, which is enabling the airline to reduce staff necessary for baggage acceptance, but at the same time to profit from commission-based sale together with their partner RIMOWA who is crafting baggage supporting these kinds of electronic tags.

As a result, it is very important for all types of airlines to try to offer as much additional services for passengers as suitable. The best proposal for the airline is to try and use new and innovative technologies but at the same time offer services which have well performed until today. For current airlines which are not offering recently implemented a la carte pricing and branded fare structure, a clear recommendation is to deeply consider future implementation due to results.

References

1. Vavrová, A., Špirková, D., & Cagáňová, D. (2018). Impact of legal adjustments of selected countries on logistic of enterprises in the field of transport and forwarding. In *MMS 2018—3rd EAI International Conference on Management of Manufacturing Systems*, Dubrovnik Croatia.
2. Straka, M., Malindžáková, M., Trebuňa, P., Rosová, A., Pekarčíková, M., & Fill, M. (2017). Application of extendsim for improvement of production logistics efficiency. *International Journal of Simulation Modelling, 16*(3), 422–434.
3. Jurik, T. (2016). Concept of service logistics. *Acta logistica, 3*(2), 21–25.
4. Straka, M., Lenort, R., Khouri, S., & Feliks, J. (2018). Design of large-scale logistics systems using computer simulation hierarchic structure. *International Journal of Simulation Modelling, 17*(10), 105–118.
5. Ferencová, J., & Hurná, S. (2017). Logistics technologies in aviation. *Acta logistica, 4*(2), 11–17.
6. SITA. The baggage report 2017. http://comms.sita.aero/rs/089-zse-857/images/baggagereport-2017.pdf. Accessed 17 Sept 2018.
7. IATA. RFID bag tag initiative. https://www.iata.org/pressroom/facts_figures/fact_sheets/Documents/fact-sheet-rfid-bagtag.pdf. Accessed 20 Sept 2018.
8. IATA. (2018). Baggage journey. Infographic. https://www.iata.org/whatwedo/opsinfra/baggage/Pages/baggage-tracking.aspx, last accessed 2018/09/18.
9. Kádárová, J., Teplická, K., Durkáčová, M., & Vida, M. (2015). Target costing calculations and economic gain for companies. *Procedia Economics and Finance, 23*, 1195–1200.
10. Scotti, D., Dresner, M., & Martini, G. (2016). Baggage fees, operational performance and customer satisfaction in the US air transport industry. *Transportation Research part E: Logistics and Transportation Review, 55*, 139–146.
11. World Data Bank homepage. https://data.worldbank.org/indicator/is.air.psgr. Accessed 15 Sept 2018.
12. Bozogan, M. (2016). Analysis and prognosis of development of airline auxiliary services. Diploma Thesis, Faculty of Aeronautics of Technical University of Kosice, Kosice.
13. Public Law. Airline Deregulation Act. U.S Government Publishing Office. https://www.gpo.gov/fdsys/pkg/STATUTE-92/pdf/STATUTE-92-Pg1705.pdf. [Online October 24, 1978]. [Cited: April 19, 2017].
14. InterVISTAS-ga2. The economic impacts of air service liberalization. IATA.org. [Online May 30, 2006]. [Cited: April 19, 2017].
15. http://www.iata.org/whatwedo/Documents/economics/liberalization_air_transport_study_30may06.pdf.
16. King Kelleher, Herb Kelleher. (1967). The mission of Southwest Airlines. Southwest. https://www.southwest.com/html/aboutsouthwest/. [Online]. [Cited: April 19, 2017].
17. Airlines for America. Status of air travel in the USA. airlines.org. http://airlines.org/wpcontent/uploads/2016/04/2016Survey.pdf. [Online April 13, 2016]. [Cited: April 25, 2017].
18. Austrian Airlines. Charming way to fly. austrian.com. https://book.austrian.com/app/fb.fly?pos=AT&l=en. [Online April 17, 2017]. [Cited: April 17, 2017].
19. IdeaWorksCompany. (2018). The CarTrawler yearbook of ancillary revenue. s.l. IdeaWorksCompany.
20. Rajaguru, R. (2016). Role of value for money and service quality on behavioural intention: A study of full service and low cost airlines. *Journal of Air Transport Management, Volume, 53*, 114–122., ISSN 0969-6997. https://doi.org/10.1016/j.jairtraman.2016.02.008. (http://www.sciencedirect.com/science/article/pii/S096969971530096X.
21. Hameed, M. Low cost airlines: A brief history, the current state and the future. AviationKnowledge. [Online September 7, 2011]. [Cited: April 19, 2017].
22. Rautanen, P. (2012). *Airline ancillary services in travel management process*. Helsinki: University of Applied Science.

23. Grönroos, C. (2007). *Service management and marketing: customer management in service competition*. Hoboken: J. Wiley & Sons. ISBN 9780470028629.
24. Marisa, G. U.S. Airlines Lead the World in Ancillary Revenue as Total Tops $38 Billion in 2014. Skift.com. https://skift.com/2015/07/14/u-s-airlines-lead-the-world-in-ancillaryrevenue-as-total-tops-38-billion-in-2014/. [Online Jul 14, 2015]. [Cited: April 27, 2016].
25. Ko, Y. D. (2019). The airfare pricing and seat allocation problem in full-service carriers and subsidiary low-cost carriers, Journal of Air Transport Management, Volume 75, 92-102, ISSN 0969-6997, https://doi.org/10.1016/j.jairtraman.2018.12.003. (http://www.sciencedirect.com/science/article/pii/S0969699717305434).
26. Sorensen, J. Top 10 things you need to know about ancillary revenue and Airlines.IdeaWorksCompany.com LLC. http://info.cartrawler.com/e/32672/airlineseries-en/33mf5/369609110. [Online 2014]. [Cited: April 27, 2016].
27. Lee, S. Explaining (or trying to) travel agent commissions. http://hostagencyreviews.com/. [Online April 18, 2016]. [Cited: April 28, 2016]. http://hostagencyreviews.com/travel-agent-commissions/.
28. Sorensen, J. (2015). The CarTrawler yearbook of ancillary revenue. [http://cartrawler.com/] s.l. IdeaWorksCompany.com LLC.
29. SPIRIT Airlines. Advertise with us. SPIRIT Airlines. http://marketing.spirit.com/mediakit.pdf. [Online April 28, 2016]. [Cited: April 28, 2016].
30. United Airlines. United Customer Contact Center. United.com. https://www.united.com/web/enUS/content/contact/reservations/default.aspx. [Online United, January 01, 2016]. [Cited: April 28, 2016].
31. airfarewatchdog.com. Airlines that sell priority boarding privileges if you're willing to pay. http://www.airfarewatchdog.com/. [Online July 26, 2011]. [Cited: July 28, 2016].
32. http://www.airfarewatchdog.com/blog/8477170/airlines-that-sellpriority-boarding-privileges-if-you-re-willing-to-pay/.
33. DCC—Distribution Cost Charge. Lufthansa Experts. https://www.lufthansaexperts.com/shared/files/lufthansa/public/mcms/folder_3330/file_104217.pdf. [Online July 01, 2015]. [Cited: April 28, 2016].
34. Tuzovic, S., et al. (2014). From 'free' to fee: Acceptability of airline ancillary fees and the effects on customer behavior. *Journal of Retailing and Consumer Services.*, *21*, 98–107.
35. O'Connel, J. F., & Warnock-Smith, D. (2013). An investigation into traveller preferences and acceptance levels of airline ancillary revenues. *Journal of Air Transport Management.*, *I*, 33.
36. IdeaWorksCompany. (2016). The CarTrawler yearbook of ancillary revenue. s.l. IdeaWorksCompany.

Testbed on IT Communication in Smart Grid: A Cybersecurity Demonstration

Yannis Soupionis, Nikoleta Andreadou, and Thierry Benoist

1 Introduction

Connectivity and interoperability are fundamental components in the implementation of the advanced cyberphysical infrastructure due to the large number of technologies, components, and actors involved, from utilities and telecom providers to manufacturers and service providers. This increases the dependence of infrastructures on ICT technologies, which is an emergent area of interest. Events that involve irregular disruptions intentional (e.g., cyberattacks) or accidental (e.g., fiber cable cuts) can cause ominous consequences if operators of critical infrastructure are not able to react in time and efficiently.

Therefore, in order to fulfill preparedness requirements, Joint Research Center (JRC) has advanced the experimentation capabilities in the field of Networked Critical Infrastructures (NIS) by implementing the Experimental Platform for Internet Contingencies (EPIC) [24]. EPIC, as a state-of-the-art cybersecurity testbed, has no difficulties eliciting interest from experts in the field. Moreover, owing to the strategic crosscutting nature of cybersecurity, the EPIC facility finds itself a nexus bridging together a vast array of scientific fields and policy domains such as Energy.

In this paper we have tried to present the complex status of the Smart Grid infrastructure and demonstrate a cyberattack against a specific element of grid, the smart metering.

Y. Soupionis (✉) · N. Andreadou · T. Benoist
European Commission, DG Joint Research Centre (JRC), Directorate E - Space, Security and Migration Unit Technology Innovation in Security, Ispra, Italy
e-mail: yannis.soupionis@ec.europa.eu; nikoleta.andeadou@ec.europa.eu; thierry.benoist@ec.europa.eu

© Springer Nature Switzerland AG 2020
D. Cagáňová, N. Horňáková (eds.), *Mobility Internet of Things 2018*, EAI/Springer Innovations in Communication and Computing, https://doi.org/10.1007/978-3-030-30911-4_7

The complexity is presented briefly by illustrating the Smart Grid Architecture Model (SGAM), which aims to represent an abstract view of Smart Grid architectures and to deliver a structured way for examining and associating different scenario utilizations. The SGAM three-dimensional model offers a clear proof of the major complexity on the communication aspect between entities in the Smart Grid.

The Advanced Meter Infrastructure (AMI), where our use case is focused on, is a main part of the Smart Grid. The future deducts that the customers will not only receive energy from their utilities companies but they will receive information by interconnecting new technologies. These technologies cover areas of electric, communication, and information. They are ready to interact in an automatic way in order to serve demand response notion and customer needs. So AMI is the main element that connects the end user to the main elements of the infrastructure as transmission and distribution network.

The reliability and security of Smart grid [1, 12, 14] and especially the AMI is very important [19]. AMI should offer various aspects of securing Smart Grid and its processes like demand response programs either automated or with human-in-the-loop [15, 18, 26]. The current trend is to offer various and numerous options to users in order to manage their consumption and minimize the cost. All this is succeeded through technology solutions that have to limit electrical outages, have resilient transmission and distribution systems, and to integrate green energy solutions.

In this paper we do not try to answer the complete resilience and security issue for the AMI technology which integrates a diversity of technologies and procedures [22]. The report tries to specify technology independent requirements for implementing secure, resilient AMI solutions, based on a Demand Side Management lab experiment.

Infrastructure interdependencies are more than just a theoretical concern. In order to implement this notion at our lab environment we used specific tools. The two laboratories elements, which consist our cybersecurity testbed [25], are:

- our ICT network emulation laboratory, called EPIC [24], which provides an experimental infrastructure for security and resilience research purposes. The EPIC infrastructure is able to reconstruct network topologies and with special attributes, as latency and loss. This will assist us to develop the communication layer of our test case.
- a rack of Smart Meters, which integrates 2 setups with 1 Data Concentrators and 9 Smart Meters each. We were able to interconnect this rack with the aforementioned other parts of our laboratory.

This paper is organized as follows: Section 2 gives a succinct overview of the EPIC facility. Section 3 briefly describes the current status of Smart Grid reference models and architectures based on the EU directives. Section 4 is dedicated to illustrate the current implementation and demonstration of the proposed testbed into EPIC. Finally, in Sect. 5, we present the conclusions and we suggest our future steps on the specific research area.

2 EPIC Overview and Creation of Emulated Network

In this section we present briefly the EPIC architecture and the steps followed to recreate emulated ICT network topology.

EPIC is a hybrid facility for studying the security and stability of decentralized control systems. It can use simulation for the physical processes and an emulated network system [11, 27] in order to create the cyber/network part of these distributed infrastructures.

EPIC's resources are numerous, as it has 356 experimental nodes, 8 routers only for the experimentation infrastructure, and a few special physical equipment, as Programmable Logical Controllers (PLCs). Based on these computation resources, it is obvious that EPIC could handle simultaneously a lot of experiments by different users. Our EPIC infrastructure contains two control servers, a set of generic servers and other devices for experimental nodes and a pool of switches that connect the nodes (see Fig. 1).

2.1 *EPIC Overview*

The architecture of EPIC suggests the use of an emulation testbed based on the Emulab software in order to recreate the cyber part of NCIs, e.g., servers and corporate network, and the use of software simulation for the physical components,

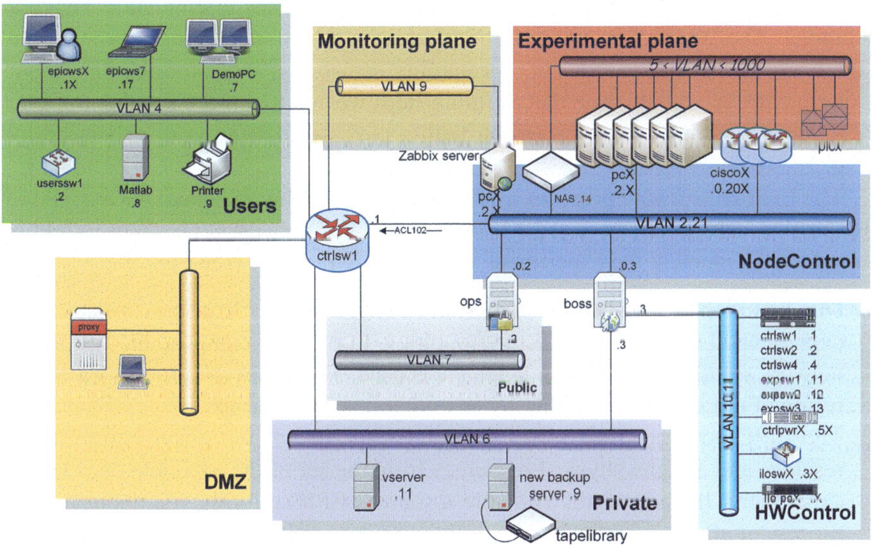

Fig. 1 EPIC's architecture, which includes experimental and control infrastructure

e.g., a chemical process. Figure 1 provides an overview of EPIC together with basic experimentation steps.

By employing an emulation-based testbed we ensure strong fidelity, repeatability, measurement accuracy, and safety of the cyber layer. This approach is well-established in the field of cybersecurity [6] and was chosen in order to overcome the major difficulties that rise while trying to simulate how ICT components behave under attacks or failures.

For the physical layer EPIC uses simulation, since this provides an efficient, safe, and low-cost approach with fast and accurate analysis capabilities.

Although it weakens the fidelity requirement, software simulation enables disruptive experiments on multiple heterogeneous physical processes, satisfying at the same time all other testbed requirements, e.g., repeatability, measurement accuracy, and safety.

2.2 *Recreating Cyber Systems*

Our laboratory testbed, called EPIC is based on a very advanced suite, Emulab [11, 27]. Emulab was created in the University of Utah. Currently, the software is supported and offered to other institutions; therefore, it exists to multiple universities, and private installations throughout the world.

We have formulated in our laboratory an infrastructure for testing by adopting the Emulab software. By implementing EPIC, we can assign physical components and network equipment, e.g., servers, to a logical/virtual topology. The EPIC can set parameters to our physical infrastructures in order to emulate a network topology. The process is made as transparently as possible. This way we gain significant advantages in terms of repeatability, scalability, and controllability of our experiments. We should state that as future steps we aim to replace Emulab to Deter [23], since it is an Emulab extension with a clear specialization in cybersecurity (Fig. 2).

The Emulab software provides a Web interface (see Fig. 3), where we are able to assign physical equipment in order to create emulated networks. The steps to follow in order to implement the experiment are the following ones (see Fig. 2):

1. Our first step is to write our experiment script, which describes in detail our experimental network architecture/topology. In order to compose the script, the network simulator 2 scripting language is used. This eases the process for future re-use of our experiment either by any other researcher or us.
2. The EPIC software then forms/starts the experiment. Based on our script, it reserves and assigns physical resources from the set of the available ones to our experiment. In case of not available resources EPIC informs you about it. The process of allocating resources to someone's experiment is called swap-in.

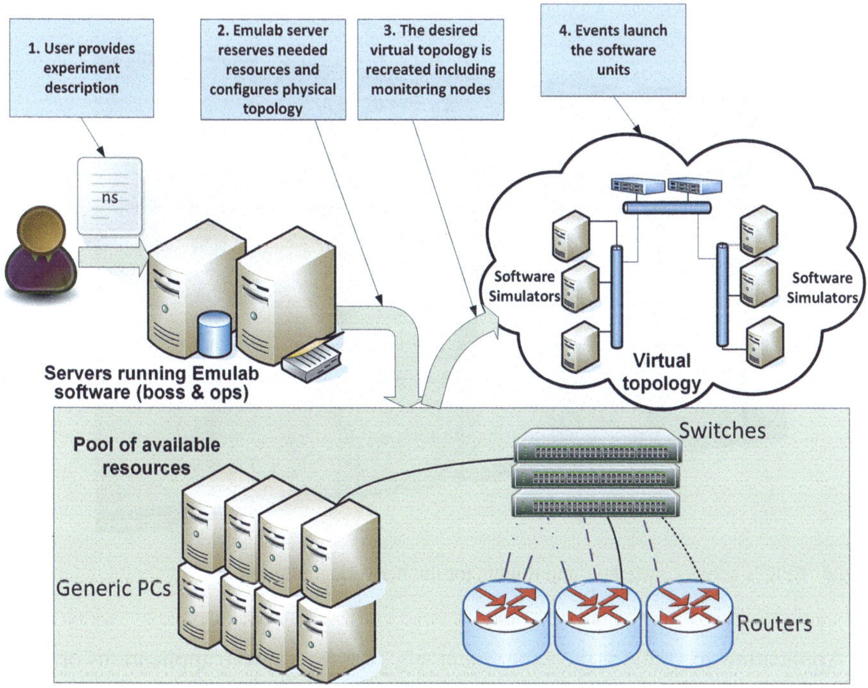

Fig. 2 Architecture of the EPIC testbed: architectural overview and experimentation steps

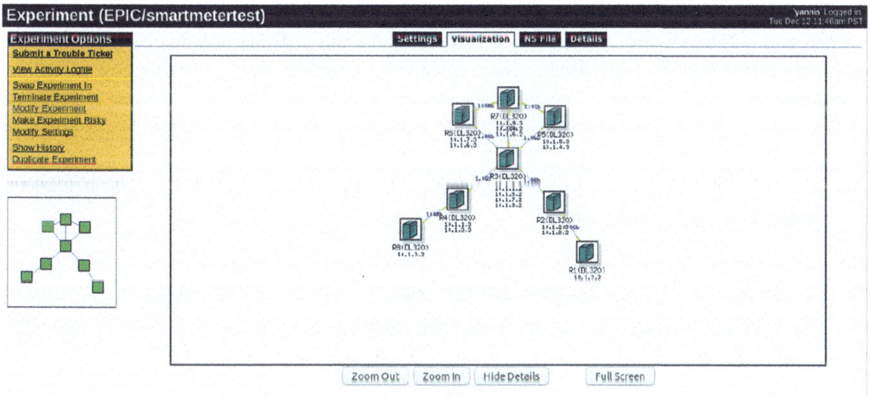

Fig. 3 EPIC's web interface

3. Additionally EPIC creates virtual private networks on the network switches in order to construct the topology by connecting experimental nodes. Finally, the software adjusts other parameters of our network, as latency and packet loss, by assigning additional nodes for those specific purposes.

Fig. 4 EPIC's servers, switches, and routers for the network emulation

4. Application specific for the experiment, e.g., databases, web applications or even dockers, can be initiated either automatically or manually, by logging in to each station.

The EPIC installation at the JRC for the emulation is depicted in Fig. 4, where we can see a part of the 350 nodes. Other part of our infrastructure, as routers and programmable logic controllers, are available as experimental resources in other premises.

3 Smart Grid Overview

In this section we illustrate the current state of Smart Grid reference architectures/models regarding the cyber element and the communication between the entities.

3.1 Smart Grid Reference Architecture

The Smart Grid Reference Architecture was introduced by the European Committee for Standardization (CEN), the European Committee for Electrotechnical Standardization (CENELEC) and the European Telecommunications Standards

Institute (ETSI), based on the EU Smart Grid Mandate M/490, with three main contributions:

- a Dedicated Distributed Energy Domain (DER) domain, which is included in the defined European Conceptual Model,
- the Smart Grid Architecture Model (SGAM), which consists of three dimensions, has been introduced and its main goal is to aid the analysis of Smart Grid use cases, and
- a first fundamental idea of how to comprehend and handle risk of Smart Grid infrastructure, which has been further presented on Smart Grid Information Security (SGIS) reports [7–10, 16].

The EU Conceptual Model has four main domains which include operations, grid users, markets, and energy services, but since we focus on information exchange, we will focus on the Smart Grid Architecture Model (SGAM) [2, 3, 5, 7, 21], which is more close to the information exchange. SGAM consists of (see Fig. 5):

- six zones, which are extracted from the information exchange between main actors and consist of Process, Field, Station, Operation, Enterprise, and Market,

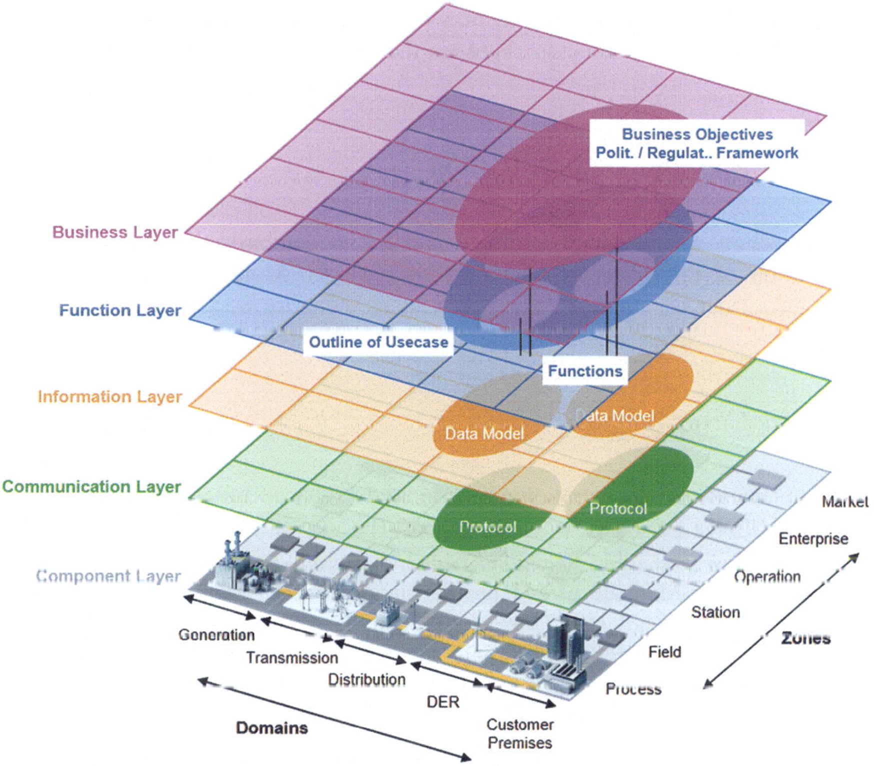

Fig. 5 The proposal of SGAM architecture

- five domains, which are extracted on the energy physical layer and consist of Generation, Transmission, Distribution, Distributed Energy Resources (DER), and Consumer Premises,
- five layers, which are extracted on the view of the stockholders and consist of the following layers: Component, Communication, Information, Function, and Business.

The SGAM assists to crystalize different use cases and identify the involving stakeholders and parties. This will provide not only a common architecture model but also different point of views since more entities will be clearly involved, as information gaps will be revealed. Moreover, this will help processes like standardization into cybersecurity. This issue is a holistic problem, which involves more than one layers. This was clearly presented at Smart Grid Information Security (SGIS) reports by the European Standardization bodies CEN-CENELEC-ETSI [8–10, 16].

The main issue that is more than evident is the amount of different elements and the possible security vulnerabilities for a cyberattack. Therefore, a cybersecurity approach needs more thorough steps and it needs constant optimization.

Based on this architecture we are able to understand the importance of ICT infrastructure and the communication between the various elements [13, 17]. In the demo presented in our next experiment, we are going to infiltrate to a very specific point of the current infrastructure, between the concentrator and the smart meters.

4 Use Cases and Demonstration

In this section we present our testbed and a first demonstration for a possible exploitation of the current communication systems. Moreover we present the requested service capabilities, from which attack scenarios were derived.

4.1 Use Cases

The implemented demonstration/use case in EPIC was based on scenarios of the CEN-CENELEC-ETSI working group. The set of analyzed use cases are presented below (see Table 1) and has been identified to support the functionalities considered under Mandate M/441 by the Smart Grids Coordination Group (SG-CG). As we have already stated the use cases are based on specific wanted/required functionalities for controlling remotely the concentrator and smart meters. The attribute we focused is based on the remote disconnection of the smart grid due to credit exhausted or payment default.

Of course, we should state that the use cases described in the table will vary to the real operational context, where maybe additional security mechanisms are in place.

Table 1 Advanced metering infrastructure use cases

High level functionalities	Low level functionalities	SG-CG use cases
Allow remotely to connect, disconnect, and limit the power supply	Remote connection/disconnection	Connect new customers. Disconnect departing customers
		Disconnect when the payments fails to proceed. Reconnect when a payment agreement is re-established
		Disconnect when load limitation is exceeded or technical accident happens. Reconnect after load limitation is not anymore surpassed or problems repaired
		Allow providers to grant or not remote disconnection (e.g., disable disconnection for protected customers)
		Allow to control power flow status (connected/disconnected)
		Facilitate remote threshold alteration for power limitation per customer or group of customers
		Facilitate prevention and activation of power limitation
	Facilitate power limitation remotely	Facilitate remote threshold alteration for power limitation per customer or group of customers
		Facilitate prevention and activation of power limitation
Provide information about a building's or appliance's status via a specific application interface	Create capabilities for connecting building's communications systems to external information providers	Consumer receives advanced information, such as usage statistics, total consumption, and timely consumption (day, month, etc.)

In Fig. 6 we illustrate the new elements of the demo, apart from EPIC:

- The two concentrators and the smart meters (right part of the figure). Each concentrator has connected on its 9 smart meters.
- A set of 18 light bulbs (left part of the figure). Each lamp is connected to each smart meter. This way we are able to illustrate the disconnection or connection of the smart meter more vividly, by having the lights turned on or off, respectively.

4.2 Demo Description

The demonstrator contains smart meters from different manufactures to show the concept of interoperability. The readings of energy consumption of these smart meters come from actual appliances plugged into this system or might even come from simulated data.

Fig. 6 The smart meter infrastructure setup

On this experiment, we are able to control, monitor, and manage all the meters remotely. Commands can be sent and received to the concentrator which each individual smart meter is connected. We can see the real-time measurement of the energy consumption of each smart meter.

In real life the ICT communication is a circuit that is provided by a local or national carrier that connects on one end the control center and on the other end the different substations/concentrators. There are many technologies applicable to this segment: wireline, wireless, 4G, WiMax, MPLS, point-to-point.

For our demonstration we used the web server installed on both the concentrators. Their use is for remote accessing and customizing the smart meters. We were able to exploit a well-known web-authentication vulnerability and have access as "admin" to the concentrators.

In our experiment in order to implement this communication channels we use EPIC, which emulates different ICT network topologies. In Fig. 7 we illustrate the way we integrated the concentrator and smart meters into the EPIC.

Demo Scenario The proposed scenario for compromising our testbed concentrator-smart meter is as follows:

The control center is able to control the concentrators via a web interface. Both the concentrators have the functionality to disconnect remotely the smart meters and to program them to disconnect on a specific time. The attacker is able to infiltrate to the network and silently listen the communication between the control center and the concentrators. This way is able to obtain the credentials and implement an attack by disconnecting all the smart meters under both the concentrators.

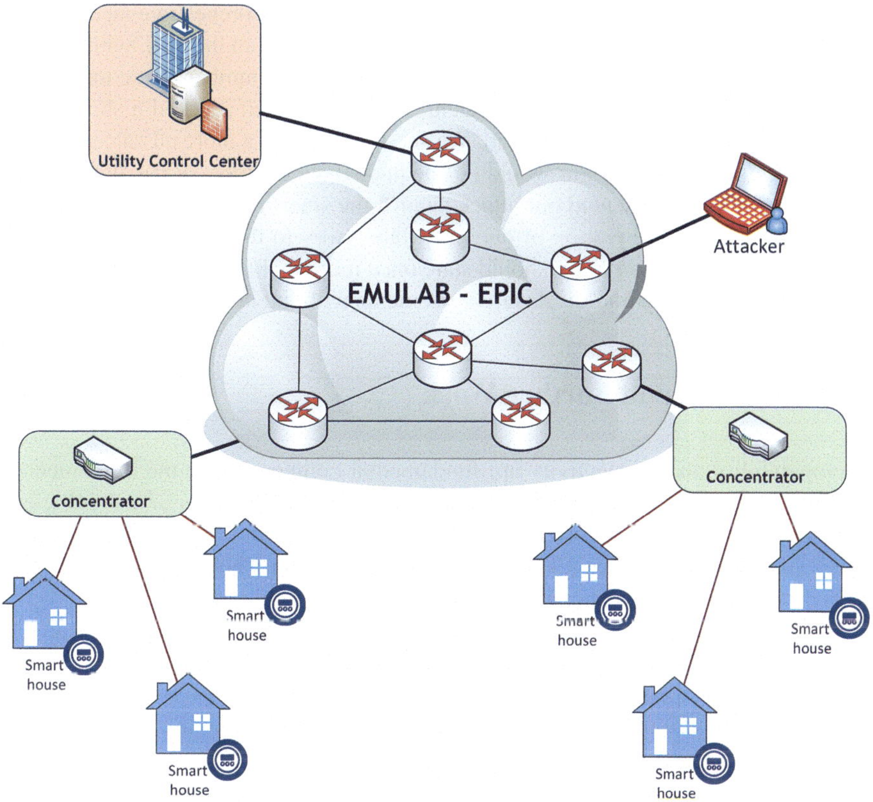

Fig. 7 The logical architecture of our lab test scenario

Demo Implementation In this section we describe briefly our first demonstration of cyberattack against the smart meter testbed. We do not go into details due to confidentiality issues against our industrial partners.

The proposed demo follows the specific steps:

1. The attacker infiltrates the network and silently records and analyzes the data passing through the concentrator and smart meters. This way we were able to:
   - Identify the authentication process and obtain the credentials in order to be able to act on the smart meters. Both concentrators had issue on their web authentication scheme and we were able to introduce ourselves to both system as administrators with the maximum rights.
   - Record all the queries from control center to concentrators and be able to create new commands which were undetectable as compromise or attack.
2. We created fully customizable and nearly fully automated scripts in order to attack and switch off all the smart meters to both the concentrators, by using the credentials we have already extracted from our previous steps.

This way we were able to show that we could be able to orchestrate an extended attack where multiple smart meters will be disconnected from the grid, with serious consciousness to the whole power grid stability. Since the number of the meter was adequate for the extend of our experiment, we were able to succeed a short black out to a part of the created power grid. We should state that a more advanced secure connection to the concentrator would be adequate to tackle most of the attacks.

This section has presented our attack against the smart meter testbed, which is a first step to the full integration of our EPIC laboratory to the extended Smart Grid interoperability laboratory. We should state once more that our presentation is brief in order to respect any non-disclosure agreements with the involving stakeholders.

5 Conclusion and Further Work

Interoperability and connectivity are fundamental components in the implementation of the advanced metering infrastructure due to the large number of technologies, components, and actors involved, from utilities and telecom providers to manufacturers and service providers. They are at the core of the whole product lifecycle, from design to manufacturing and testing.

We illustrated the current state of Smart Grid communication reference architecture and its main elements. One of the important areas is smart meters, where we created a testbed and a specific cybersecurity attack was tested based on the proposed functionalities of the standardization bodies. This infrastructure was integrated into EPIC and our goal was to test the cybersecurity and verify their interoperability with other components [4, 20].

Finally this work will be extended with the collaboration of other partners inside JRC. Our contribution will be to interconnect the Smart Grid laboratory with the EPIC cybersecurity laboratory and emulate the communication channel between the control center and other actors of the Smart Grid.

We need to consider that the Smart Metering demo we presented is only an example of the activities that can be carried out in the laboratory. The same concept we have seen can be extended to other areas relevant to Smart Grids: electromobility, substation automation, grid management, integration of renewables. The aim of this research is to analyze the role of ICT standardization in enabling Smart Grid interoperability, grid management, and distribution automation, where a significant part will be to implement the communication protocols and to study the cybersecurity for the protection of Smart Grid.

References

1. Andreadou, N., Soupionis, Y., Bonavitacola, F., & Prettico, G. (2018). A DSM test case applied on an end-to-end system, from consumer to energy provider. *Sustainability, 10*, 935.
2. CEN/CENELEC/ETSI Joint Working Group Standards for Smart Grids, May 2011. http://www.etsi.org/images/files/Report_CENCLCETSI_Standards_Smart_Grids.pdf

3. CEN/CENELEC/ETSI Smart Grid Coordination Group, Functional reference architecture for communications in smart metering systems, Dec 2011. ftp://ftp.cencenelec.eu/EN/EuropeanStandardization/HotTopics/SmartMeters/CEN-CLC-ETSI-TR50572
4. CEN-CENELEC-ETSI Smart Grid Coordination Group, Smart Grid Information Security (2012)
5. CEN/CENELEC/ETSI Smart Grid Coordination Group, Privacy and Security approach part I, Nov 2013. ftp://ftp.cencenelec.eu/EN/EuropeanStandardization/Fields/EnergySustainabi-lity/Management/SmartMeters/SMCG_Security_and_Privacy_Report_PartI.pdf
6. CEN-CENELEC-ETSI: Smart Grid Coordination Group, Methodologies to facilitate Smart Grid system interoperability through standardization, system design and testing 2014. ftp://ftp.cencenelec.eu/EN/EuropeanStandardization/HotTopics/Smart-Grids/SGCG_Interoperability_Report.pdf
7. CEN-CENELEC-ETSI Smart Grid Coordination Group, Smart Grid Information Security (SGIS). Document for the M/490 Mandate. Dec 2014. ftp://ftp.cencenelec.eu/EN/EuropeanStandardization/HotTopics/Smart-Grids/SGCG_SGIS_Report.pdf
8. CEN/CENELEC/ETSI Smart Grid Coordination Group, Privacy and Security approach part II, June 2014. ftp://ftp.cencenelec.eu/EN/EuropeanStandardization/Fields/EnergySustaina-bility/Management/SmartMeters/SMCG_Security_and_Privacy_Report_PartII.pdf
9. CEN/CENELEC/ETSI Smart Grid Coordination Group, Privacy and Security approach part III, June 2015. ftp://ftp.cencenelec.eu/EN/EuropeanStandardization/Fields/EnergySustaina-bility/Management/SmartMeters/SMCG_Security_and_Privacy_Report_PartIII.pdf
10. CEN/CENELEC/ETSI Smart Grid Coordination Group, Privacy and Security approach part IV, July 2016. ftp://ftp.cencenelec.eu/EN/EuropeanStandardization/Fields/Energy-Sustainability/Management/SmartMeters/SMCG_Sec0108_PartIV.pdf
11. Christos, S., Garcia, A. P., & Genge, B. (2013). On the use of Emulab testbeds for scientifically rigorous experiments. *IEEE Communications Surveys and Tutorials, 15*(2), 929–942.
12. Cintuglu, M. H., Mohammed, O. A., Akkaya, K., & Uluagac, A. S. (2017). A survey on smart grid cyber-physical system testbeds. *IEEE Communications Surveys and Tutorials, 19*(1), 446–464.
13. Commission of the European Communities. (2009). Communication from the commission to the European Parliament, the council, the European economic and social committee and the committee of the regions, Brussels, 30.3.2009 COM 149 final
14. Dong, X., Lin, H., Tan, R., Iyer, R. K., & Kalbarczyk, Z. (2015). Software-defined networking for smart grid resilience: Opportunities and challenges. In *Proceedings of the 1st ACM Workshop on Cyber-Physical System Security (CPSS '15)* (pp. 61–68). New York, NY: ACM
15. Egan, M. (2001, 14 June). *Constructing a European market: standards, regulation, and governance*. Oxford: OUP.
16. ENISA. Appropriate Security Measures for Smart Grids. 06-12-2012. https://www.enisa.europa.eu/activities/Resilience-and-CIIP/critical-infrastructure-and-services/smart-grids-and-smart-metering/appropriate-security-measures-for-smart-grids
17. EU Commission. (2005). Green Paper on a European Programme for Critical Infrastructure Protection COM (2005) 576.
18. Gungor, V. C., Sahin, D., Kocak, T., Ergut, S., Buccella, C., Cecati, C., et al. (2011). Smart grid technologies: Communication technologies and standards. *IEEE Transactions on Industrial Informatics, 7*(4), 529–539.
19. Kammerstetter, M., Langer, L., Skopik, F., Kupzog, F., & Kastner, W. (2014). Practical risk assessment using a cumulative smart grid model. In: SMARTGREENS2014, Barcelona (pp. 31–42).
20. M/491 Smart Grid Mandate Standardization, CEN, CENELEC and ETSI in the field of measuring instruments for the development of an open architecture for utility meters involving communication protocols enabling interoperability. January 2009. http://ec.europa.eu/growth/tools-databases/mandates/index.cfm?fuseaction=search.detail&id=421
21. Robert, W. G., & Langer, L. (2015). *Establishing a smart grid security architecture*. Smart Grid Security (pp. 185–218). Waltham: Syngress, ISBN 9780128021224.

22. Schneider, K., Paul, J.-P., & Liu, C.-C. (2006). Assessment of interactions between power and telecommunications infrastructures. *IEEE Transactions on Power Systems, 21*(3), 1123–1130.
23. Schwab, S. (2010). The DETER project: Advancing the science of cyber security experimentation and test. In *Proceedings of the IEEE International Conference on Technologies for Homeland Security (HST)* (pp. 1–7).
24. Siaterlis, C., Genge, B., & Perez-Garcia, A. (2012). Experimental capabilities of the Network and Information Security laboratory (NISlab), Joint Research Centre, JRC76858.
25. Soupionis, Y., & Benoist, T. (2015). Cyber-physical testbed – The impact of cyber attacks and the human factor. In *2015 IEEE 10th International Conference for Internet Technology and Secured Transactions (ICITST)*, London (pp. 326–331).
26. Uslar, M., Rosinger, C.,& Schlegel, S. (2014, July). Security by design for the smart grid: Combining the SGAM and NISTIR 7628. In *2014 IEEE 38th International Computer Software and Applications Conference Workshops (COMPSACW)* (pp. 110–115). Piscataway: IEEE.
27. White, B., Lepreau, J., Stoller, L., Ricci, R., Guruprasad, S., Newbold, M., et al. (2002). An integrated experimental environment for distributed systems and networks. *ACM SIGOPS Operating Systems Review, 36*(SI), 255–270.

The Platform for Production Capacity Exchange Management as a Support for Implementation of the Industry 4.0 Concept

Sebastian Saniuk

1 Introduction

The marginalization of Europe in world production has prompted German experts, scientists, and industry to develop the concept of Industry 4.0, which will allow them to regain the status of a leader in the industrialization of the countries of Western Europe. The current, close cooperation between Polish enterprises and the German economy raises the need to quickly adapt to the technological requirements of our partners. As part of the proposed concept of Industry 4.0, close connection of physical objects through telecommunications networks is planned. The idea of Industry 4.0 is to create sophisticated business networks, connected by intelligent resources communicating via the Internet, using well-known and already used technologies, including cloud computing, Internet of Things (IoT), big data, etc. This means that today's enterprises, especially small- and medium-sized ones, which want to stay on the market must cooperate within cyber physical systems. Hence there is the need to conduct research related to the assessment of the conditions for the development of Polish enterprises to implement the assumptions of the Industry 4.0 concept and pay attention to areas requiring significant investments [1].

The chapter attempts to identify key areas related to the adaptation process to the functioning of enterprises within modern industrial networks in the face of challenges of Industry 4.0 and the use of the concept of a capacity exchange platform to quickly implement Industry 4.0 has been proposed. The results of the surveys carried out on a selected group of specialist Polish production enterprises in the metal

S. Saniuk (✉)
University of Zielona Gora, Zielona Góra, Poland
e-mail: s.saniuk@wez.uz.zgora.pl

© Springer Nature Switzerland AG 2020

D. Cagáňová, N. Horňáková (eds.), *Mobility Internet of Things 2018*,
EAI/Springer Innovations in Communication and Computing,
https://doi.org/10.1007/978-3-030-30911-4_8

sector and the results of research conducted within the "Modeling virtual production networks of SMEs" grant are shown in this chapter. The project was financed by the Polish National Science Center based on the agreement no. 4039/B/H03/2011/40.

2 The Concept of Industry 4.0 and the Cyber Industry Network (CIN)

Strong competition and growing customer expectations in modern markets means that along with an increase in production efficiency, product customization is also taking place. There is now a product manufacturing model in which the customer decides about the product configuration. At the same time, the price of the product should be close to the price of products produced in mass production. This is exactly what the Industry 4.0 concept provides, which assumes the creation of a fully integrated system of suppliers, producers, and clients creating the so-called cyber physical systems (CPS), which are open sociotechnical systems, capable of implementing a number of functions and actions imposed by production, logistics, or management [2].

Industry 4.0 means the use of intelligent mechatronic CPS products (machines, devices, robots, means of transport, etc.) throughout the product development chain from appearance of product concept, creation of virtual documentation, product creation in a virtual production environment, printing of 3D models, laboratory and industrial tests, simulation, and transition from virtual production environment to the real environment. Then, there is the implementation of computer-aided and computer-controlled production, logistics, sales, service, and recycling of products. It means the ability to meet customer expectations while maintaining high profitability of the production process thanks to the dynamic adaptation of autonomous modules of the entire process of preparation, production, and delivery of the product to the customer using IoT and information recorded in big data and cloud computing [2–4]. The Industry 4.0 concept is oriented to a significant improvement in production efficiency due to the use of production and human resources of cooperating network partners who have unused production capacities and expect the benefits of such cooperation. This means that there is a need to conduct research on the development of cooperation network forms of enterprises operating as a smart factory 4.0 [1].

The modern, dynamic business world is characterized by rapid change, intense growth of competition, and a rise in globalization, meaning that the development of new technologies is essential for enterprises to remain competitive in the face of these increasingly challenging requirements [5]. Noteworthy is the other perception of the company in the Industry 4.0 concept. Each company is perceived as an intelligent module for use in the production and delivery of the product to the customer. The size of the enterprise, which has been significant so far, is no longer relevant. This means that in future enterprises must focus on the growth of the used technology, the level of highly qualified staff, and openness to unlimited communication, e.g., communication networks. Cooperating partners in the network will be able to offer more complex, innovative products and services tailored to the clients' needs [6, 7]. Such actions result in shorter product life cycle and an increased range of products, as well as a change of processes to processes of high efficiency along with the change of devices and machines to more flexible devices and machines, taking into consideration the necessity to constantly assure the assumed quality of products [8].

Moreover, functioning in an enterprise network is significantly influenced by gaining experience, know-how, and new knowledge based on mutual relations between cooperating enterprises [9]. The development of network forms of cooperation and solving many problems related to the formation and management of collaboration is an excellent basis for the rapid implementation of the Industry 4.0 concept, especially in the SME sector [6].

The idea of a production network called the cyber industry network (CIN) means the manufacturing of joint production orders using fully automated processes of individual network partners, in which communication takes place via the Internet, and the necessary data is stored in the cloud (cloud technology) [6]. Such a network organization using modern communication technologies ensures that all partners have access to the necessary information from anywhere in the world. This allows for the development of a partnership consisting of the combination of core competencies and a better orientation to the growing expectations of the client and thus effective gaining of a competitive advantage in the market. Unfortunately, despite the obvious benefits of enterprise participation in the network, the process of establishing a network is not easy and raises many fears especially in SMEs. The process of partner selection is very difficult. A number of features required by the network organization, such as production capacity, possessed technology, level of quality of offered services, financial condition, experience, and communication skills, should be taken into account when selecting a network partner [10]. Each area has a fundamental impact on the results of joint implementation of tasks. An equally important problem is trust between partners, need to invest in information technology, and problems with its use. This particularly affects small- and medium-sized enterprises, which due to the high costs of implementation and servers are only able to use advanced information technologies to a limited extent.

The level of application of the newest technologies in manufacturing enterprises is very low. This is particularly evident in the latest report of the International Federation of Robotics—IFR, which presents the differences between implementations in the field of automation and robotics in Poland and other countries. In 2016, there were on average 74 industrial robots per 10,000 employees employed in the global production sector, while in Poland the average density of robots was 32 robots. In comparison to other countries in the region, Polish enterprises have the weakest positions, e.g., in the Czech Republic 101 robots, in Slovakia 135, and in Germany the robotization density in 2016 amounting to 309 robots. For comparison, in the most automated countries in the world, such as South Korea, the robotization density index in 2016 amounted to as much as 631 robots [11].

Polish industrial enterprises are still characterized by a low level of automation and robotization of production. In 2015, only about 15% of enterprises in Poland were fully automated, which corresponds to the stage of development of industrial production in technology 2.0. About 76% of enterprises were characterized by partial automation (research carried out by Astor in 2015). The level of robotization and digitization of production 3.0 is characterized by constantly low level of advancement. Polish industrial enterprises are still at the beginning of the third industrial revolution [12], which does not give good prospects for the rapid implementation of Industry 4.0. The financial barrier is one of the most frequently mentioned, especially in the case of SMEs [6].

An important barrier limiting the rapid application of the Industry 4.0 assumptions is the constant lack of climate for the cooperation of Polish enterprises, especially in the group of industrial SMEs. Preliminary research carried out on 100 enterprises operating in the industrial sector of the Lubuskie Voivodeship shows a number of barriers related both to the experiences of previous cooperation and to difficulties in creating various forms of network cooperation.

The level of delays and the scale of non-adjustment to the fourth industrial revolution can also be demonstrated by the declared problems associated with the sale of unused capacity in the network, which often entails the need for its accurate estimation and valuation. Most respondents indicate a relatively short time to determine the availability of the machine/device. This means a short time to respond to an inquiry, in most cases, no more than a day (78% of respondents). However, determining the costs of the used resource is a big problem. Only 61% of respondents declared that they can accurately determine the cost of using the resource offered on the Web. The surveyed enterprises also identified a number of other problems arising from networking. Finding the right network partner is often associated with the need to solve a number of problems related to both the risk of cooperation with an unknown enterprise and the method of searching for a cooperator. Based on respondent answers, today's key problems of enterprises which form networks are as follows:

- Lack of IT platforms for the exchange of production capacity (90% of respondents)
- Untimely completion of projects (73% of respondents)
- Disloyalty of partners involved in the network (45% of respondents)
- Problems with finding a network partner (41% of respondents)
- Low quality of services (products) by potential partners (36% of respondents)

Every enterprise is seen as a specific intelligent module which offers different possibilities which can be used in the entire logistics chain in the Industry 4.0 concept. The size of the enterprise does not matter. Meanwhile, the role of technology used, the level of highly qualified staff, and openness to unlimited communication using increasingly common technologies, such as cloud computing, big data, and Internet of Things, are constantly growing dynamically (Fig. 1).

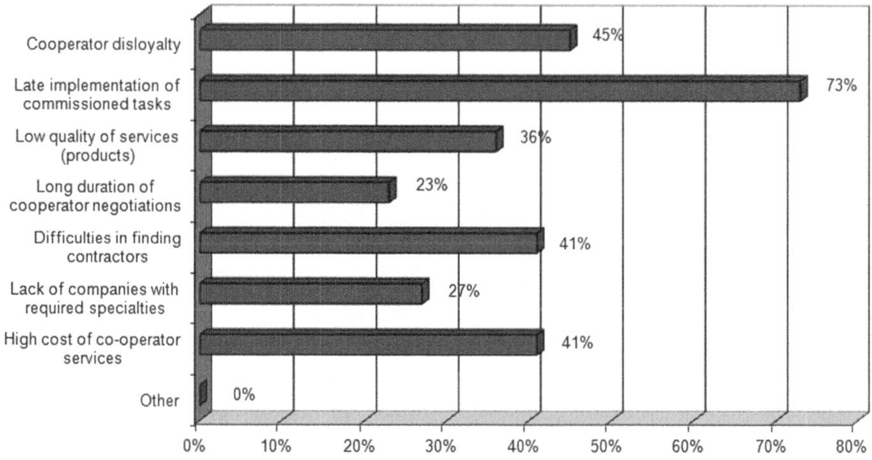

Fig. 1 Key problems associated with cooperating

A condition for the development of network forms of cooperation is the development of a model of company cooperation by creating future cyber physical systems (CPS). CPS should ensure data collection, processing, and impact on physical processes within the entire production network due to unlimited network connections of intelligent, mechatronic resources that communicate with each other (machines, devices, robots, means of transport, etc.), with a negligible human participation performing only the functions of a supervisor and/or a coordinator. Therefore, building a model requires developing a set of conditions related to the way the company operates in the production network, establishing temporary networks oriented towards a joint production undertaking, planning the load of geographically dispersed resources, and controlling the production or financial settlement of partners providing resources for production. Accession of the enterprise to such a network operating within the framework of the Industry 4.0 concept requires first of all an initial assessment of the technological potential of the company, the know-how possessed, the employees' competencies, and the ability to cooperate by providing resources. This process is also aimed at manufacturing in the enterprise [13]:

- The ability to reduce the technological gap, including ensuring the so-called technological readiness
- Adequate socio-technological potential
- Internal capacity to absorb and adapt innovations

3 The Platform of Production Capacity Exchange

The authors propose the use of a computer system for production network planning that can help in the creation and management of the cyber industry network for small and medium enterprises. Because at the moment there is no possibility of full automation of data flow between partners in the proposed approach, the key person is a broker who is responsible for supervising data collection on new production projects and enterprise opportunities (offers from enterprises).

The platform is an excellent brokering tool that on the one hand allows tasks planned for the new project to be presented and, on the other hand, allows the collection of offers from companies wishing to participate in the project. On the basis of the collected offers, the algorithm responsible for generating a set of production network variants capable of timely project execution starts. At the same time, a detailed resource utilization schedule is generated for each of the designated admissible variants of CIN. Each variant of the network is a set of selected enterprises for the execution of individual tasks that can guarantee the timely execution of the project while meeting the condition of the project's budget. The broker can choose the best variant and designate the network manager who will be responsible for the project execution. Because data on enterprise resources and their offers for tasks (cost of resource use, availability) are made available only to a trusted broker, there is no general access to confidential data, which increases the system's security against unfair competition. Each of the companies registered on the platform has access only to information about orders and planned tasks [14].

The broker is responsible for entering information about the project which is divided into tasks, as presented in Fig. 2. The following data for the new production project is introduced [14]:

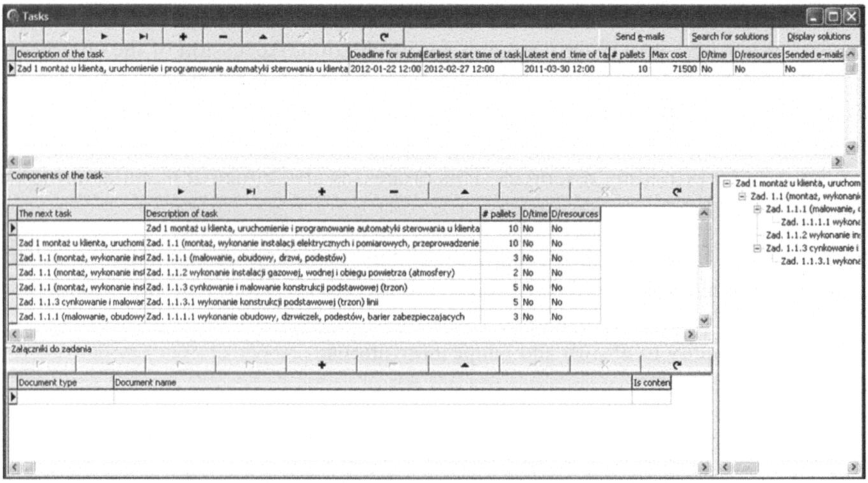

Fig. 2 Information on the data of the planned (new) project (own work)

- Deadline for submitting bids for the new production project
- Client's required project completion date
- Number of pallets required for delivery
- Total project budget (total cost)
- Technological documentation, drawings, etc.

Once the above has been completed, the enterprises can enter information about the cost of utilization and availability of resources to the platform. Similarly, they can enter information about the availability of transport means (Fig. 3). The data from enterprises will be transferred to the platform automatically from machines and devices using big data and the Internet-of-Things technology.

In addition, all information necessary for planning transport between partners, e.g., the cost of using the vehicle, permissible load capacity, and availability, is entered into the platform.

After taking into account all the information about the resources of enterprises available in real time during the planned production project, the platform can start the procedure of acceptable variant planning of the cyber industrial networks to ensure the execution of the production project. The proposed procedure is based on checking of sufficient conditions that can guarantee the production project execution on time with assumed cost (budget) [6].

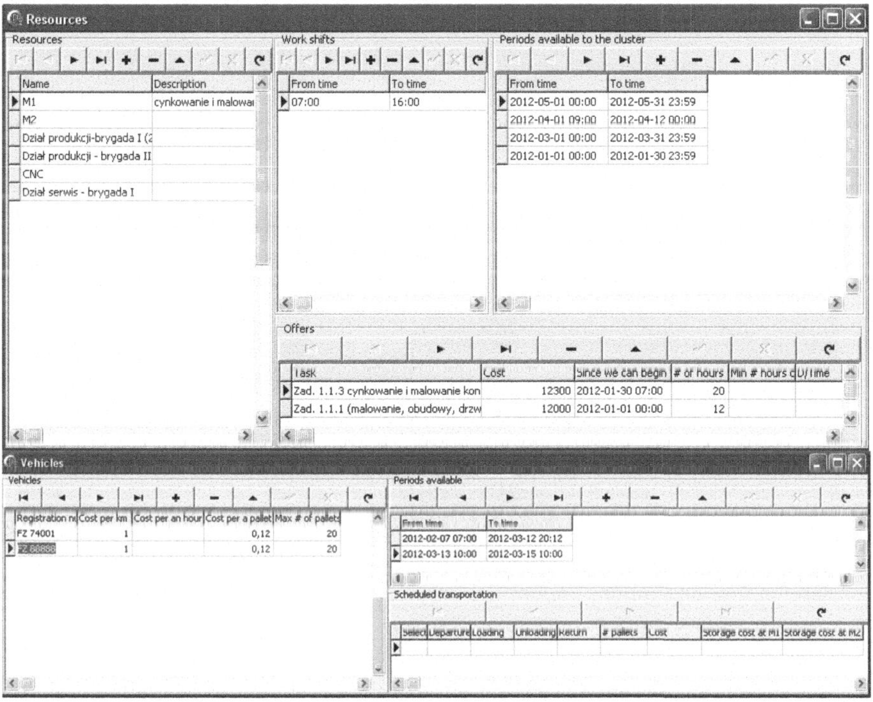

Fig. 3 Information on the availability of production resources, means of transport of the potential partners (own work)

Each variant is characterized by selected companies and information about used resources, transport between partners, and total cost of project execution. A set of acceptable solutions of CIN for the planned new project is presented in Fig. 4.

In addition, for each acceptable variant of CIN, a schedule of the production flow is proposed (Fig. 5). The detailed schedule can support the efficient completion of the project and help coordination of transport and manufacturing operations between partners. It can also support monitoring and control of the progress of project execution in the network of enterprises. In the future, all information from the platform and between each enterprise will exchange automatically using Internet-of-Things technology. Currently, a serious limitation of the proposed solution based on new, automated communication technologies is the lack of adequate equipment in enterprises that are able to create cyber industrial networks.

The capacity exchange platform allows resources of various enterprises that have available production capacity and wish to participate in the joint execution of new projects to be connected. The cyber industrial networks (CIN) are formed for those enterprises whose available resources guarantee timely execution of the entire

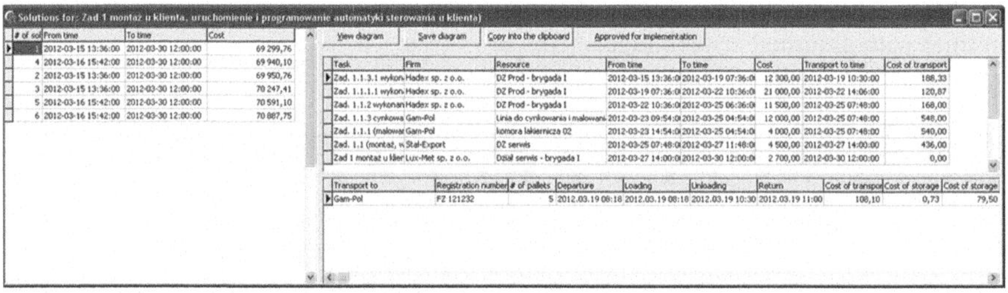

Fig. 4 Results screen—a set of acceptable solutions (own work)

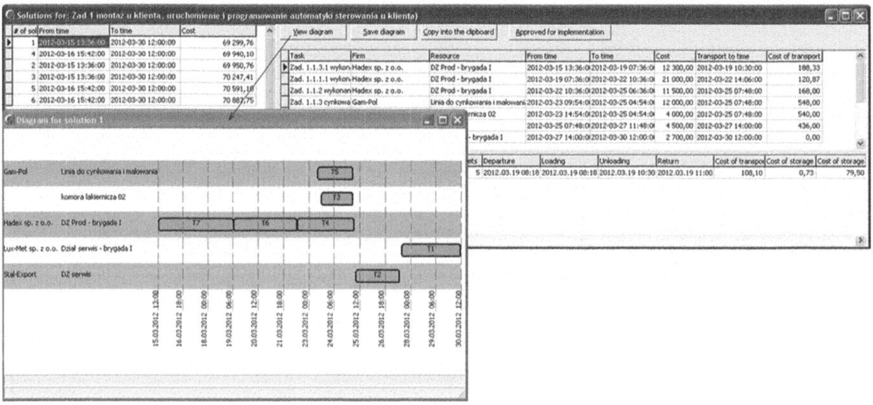

Fig. 5 A schedule of production flow for the specified variant of CIN (own work)

project while not exceeding the total planned budget. The proposed tool makes it possible to establish cooperation in the form of a production network at the level of coordination of information flows and physical flows between network partners, especially through the use of telecommunications networks and new technologies related to the Industry 4.0 concept. Such platforms, consisting of enterprises from the SME sector, provide excellent support for the implementation and development of the Industry 4.0 concept and building together smart factories 4.0 [14].

4 Conclusion

The presented approach to building cyber industry networks offers a solution to the current problem that small and medium enterprises have extremely limited potential in forming smart 4.0 factories independently. Such a network supports better use of enterprise production capacity, e.g., machinery, employees, equipment, and tools, and connects different companies with narrow specialization. An important advantage is also increasing resource utilization and productivity of partners' systems. Practical functioning of cyber industry networks allows a computer system supporting the prototyping of such a network, which is illustrated in this chapter. It should be emphasized that the creation of such smart factories from many small and medium enterprises specializing in a given activity or process enables the creation of CITs without large financial outlays. It requires only a simple interface and the ability to share through a Web service which means low implementation and maintenance costs.

In this chapter, key problems of Polish enterprises from the metal sector related to the process of adaptation to the challenges of Industry 4.0 and the new idea of creating production networks are presented. The idea of creating production networks provides for the use of a platform operated by a broker. This solution is dedicated especially to the group of small- and medium-sized production enterprises. The platform will provide a facility for intelligent resource communication, where enterprises within a network can automatically exchange real-time information on available resources for the purpose of a given project.

One of the significant advantages of the network planning system proposed in the chapter is the ability to create dynamic temporary networks of enterprises that can be simultaneously involved in many different projects (networks). Thanks to this, enterprises can easily establish cooperation ties and obtain a high level of utilization of available resources. During the development of the Industry 4.0 concept, the proposed solution will integrate resources of various enterprises through the use of information from intelligent mechatronic products such as machines, transport vehicles, and devices.

References

1. Saniuk, S., & Saniuk, A. (2018). Industry 4.0—technical-economic development perspective for the metallurgical production. In *26th International Conference on Metallurgy and Materials—METAL 2017. Brno, Czechy, 2017* (pp. 2288–2292). Ostrava: Tanger Ltd..
2. Olszewki, M. (2016). Mechatronizacja produktu I produkcji—przemysł 4.0. *Pomiary Automatyka Robotyka, 3*, 13–28.
3. Chui, M., Löffler, M., & Roberts, R. (2010). The internet of things. *The McKinsey Quarterly, 2*(47), 1–9.
4. Mrugalska, B., & Wyrwicka, M. K. (2017). Towards lean production in industry 4.0. *Procedia Engineering, 182*, 466–473.
5. Grabowska, S. (2016). Business model metallurgical company built on the competitive advantage. In *25th Anniversary International Conference on Metallurgy and Materials. Brno, Czech Republic, May 25th–27th, 2016. Conference proceedings* (pp. 1800–1807). Ostrava: Tanger.
6. Saniuk, S., & Saniuk, A. (2018). Challenges of industry 4.0 for production enterprises functioning within Cyber Industry Networks. *Management Systems in Production Engineering, 26*(4), 212–216.
7. Walters, D., & Buchanan, J. (2001). The new economy, new opportunities and new structures. *Management Decision, 39*(10), 822–823.
8. Gajdzik, B. (2013). World class manufacturing in metallurgical enterprise. *Meta, 52*(1), 131–134.
9. Corvello, V., & Migliarese, P. (2007). Virtual forms for the organization of production: A comparative analysis. *International Journal of Production Economics, 110*, 5–15.
10. Franke, U. (2001). The concept of virtual web organizations and its implications on changing market conditions. *Electronic Journal of Organizational Virtualness, 3*(4), 43–64. www.virtual-organization.net. Accessed 21 May 2018.
11. International Federation of Robotics. (2018). Robot density rises globally, Frankfurt. Retrieved February 07, 2018, from https://ifr.org/ifr-press-releases/news/robot-density-rises-globally. Accessed 1 Jun 2018.
12. Gracel, J. (2016). Industry 4.0—kluczowe pytania i odpowiedzi. *Automatyka, Podzespoły, Aplikacje, 6*(10), s36–s39.
13. Soszyńska, E. (2012). Konwergencja technologiczna a potencjał społeczno-technologiczny krajów Unii Europejskiej. *Metody Ilościowe w Badaniach Ekonomicznych, 13*(3), 191–201.
14. Saniuk, S., & Saniuk, A. (2017). Decision support system for rapid production order planning in production network. In *Advances in Intelligent Systems and Computing* (Vol. 637, pp. 217–226). Berlin: Springer.

Bike Sharing: As Urban Transport Solution in a Small City

Mária Holienčinová, Zdenka Kádeková, Tomáš Holota, and Ľudmila Nagyová

1 Introduction

At the beginning of our study, we gave attention to the factors which define a smart city. We know that an integrated transport and communication network saves time, energy, and money. Digitization doesn't just upgrade cities: it improves quality of life. The definition of a smart city is continuously evolving as discoveries emerge that improve our standard of living.

We discovered that such a city should be digitalized first and foremost—with 4G, plentiful Wi-Fi hot spots, and high smartphone usage. Transport and mobility should be knowledge based, with smart parking, traffic sensors, and car sharing apps. A smart city is sustainable, with a focus on clean energy and environmental projection. In addition, there is excellent online access to governmental services and a high level of citizen participation [1]. To make a city "smart," we need to build upon the strategies which could mitigate a city's current issues and avert future problems which are driven by the urban society and its rapid expansion [2].

The factors that define intelligent cities and form the basis for analysis and Smart Cities Index creation are as follows (Table 1).

EasyPark team has created the Smart Cities Index 2018. It represents more sophisticated analysis with new criteria and updated findings; the index provides granular level data on behavioral habits and government investments from cities around the world. This index is made up of 24 factors on the basis of which 500 cities worldwide were analyzed and then ranked the top 100 to determine the cities that manage their assets and resources most efficiently. Each factor is scored from 1

M. Holienčinová (✉) · Z. Kádeková · T. Holota · Ľ. Nagyová
Slovak University of Agriculture in Nitra, Nitra, Slovak Republic

© Springer Nature Switzerland AG 2020

D. Cagáňová, N. Horňáková (eds.), *Mobility Internet of Things 2018*,
EAI/Springer Innovations in Communication and Computing,
https://doi.org/10.1007/978-3-030-30911-4_9

Table 1 Some of the factors creating Smart City Index [1]

Smart parking	Smart building	Urban planning	Wi-Fi hot spots
Car-sharing services	Waste disposal	Education	Smartphone penetration
Traffic	Environment protection	Business ecosystem	Living standard
Public transport	Citizen participation	4G LTE	How the city is becoming smarter
Clean energy	Digitalization of government	Internet speed	Cyber security

to 10; the higher the score, the better. For our study, we chose the factor transport as one of the key points of the smart city platform in the condition of the city of Nitra.

Figure 1 shows the situation where individual cities are ranked according to the "traffic" factor from the transport and mobility area by the best score. The best position was reached by the cities of Nordic countries that show very good results not only within this category but also in other smart solutions of the city. The first place belongs to the Danish town of Koge which reached the highest possible score of 10 as well as the Finnish town of Hämeenlinna. In the list of top 10 cities according to the "traffic" factor, as mentioned above, there were a total of eight cities from the Nordic countries (4× Finland, 2× Denmark, 2× Norway). Slovenia took the third place with its capital city of Ljubljana and Manama in Bahrain was placed as ninth. Bratislava, our capital city of Slovakia, was ranked 36th according to the "traffic" rating, but Bratislava was ranked 96th in the Smart Cities Index 2018 in the overall rank/score ranking.

According to Mauritz Börjeson, CBDO of EasyPark Group, "Big Data has changed the face of the world as we know it, because it allows us to create better solutions to real world problems. Without better solutions, global urbanization would lead to problems such as traffic congestion, housing shortages and pollution—by using Big Data, we can help tackle these important global issues. In our case, we target mobility and help reduce the footprint of driving traffic while creating a much more welcoming experience for drivers." Every city in the final index should be praised for their efforts as it showcases not only those cities which are at the forefront of smart urban growth but also those which are making impressive strides towards a digitalized infrastructure. Therefore, cities at the bottom of the ranking should not be interpreted as "not smart," but rather that they are newly emerging smart cities with room to grow and improve [3].

			TRANSPORT AND MOBILITY				SUSTAINABILITY					
#	CITY	COUNTRY										
1	Koge	Denmark	3.10	2.95	10.00	2.88	8.47	6.78	8.18	6.85	8.58	9.
2	Hämeenlinna	Finland	2.13	2.88	10.00	6.70	7.86	3.63	5.24	4.08	9.33	4.
3	Ljubljana	Slovenia	3.40	4.83	9.85	2.35	4.97	1.83	4.25	6.70	7.83	3.
4	Tampere	Finland	4.83	1.98	9.85	6.70	7.86	3.63	5.24	4.68	9.33	4.
5	Vantaa	Finland	4.00	5.88	9.85	6.70	7.86	3.63	5.24	3.25	9.33	3.
6	Bergen	Norway	8.13	6.18	9.63	7.90	9.92	1.75	8.18	6.10	9.70	7.
7	Espoo	Finland	2.88	5.80	9.63	6.70	7.86	3.63	5.24	4.53	9.33	5.
8	Trondheim	Norway	6.18	4.30	9.48	7.90	9.92	1.75	8.18	6.25	9.70	7.
9	Manama	Bahrain	1.00	2.35	9.48	2.28	1.00	1.00	2.13	2.05	1.08	6.
10	Odense	Denmark	4.75	2.80	9.48	2.88	8.47	6.78	8.18	6.55	8.58	9.

Fig. 1 Top 10 cities according to the "traffic" factor [1]

1.1 Traffic Congestion and Cycling Commuting/Bike Sharing

Traffic congestion is a vexing problem felt by residents of most urban areas. Despite centuries of effort and billions of euros worth of public spending to alleviate congestion, the problem appears to be getting worse. Despite the exasperation that traffic congestion causes, most people know surprisingly little about it or what can be done about it, and much of what is stated in the media is oversimplification [4].

Since 2009, more people are living in cities than in rural areas, and 7 out of 10 people are expected to live in cities by 2050. Rapid urbanization is associated with issues such as pressure on energy, waste management, and transport networks as well. This calls for cities to actively anticipate, act, and mitigate these (future) risks, which expand to having attention for economic development and improving quality of life of residents [5]. With bike-sharing systems popping up all over the world, it's about time we look critically at the role these systems can play in a city's urban fabric and transportation system. While bikes have been an integral part of the modal mix in many cities for years, they served a similar purpose to automobiles: exclusively personal mobility. Bike sharing has altered that paradigm, essentially creating a new mode of public transit [6].

Using a bicycle for commuting to school, work, or errands (even if just short distances) can improve physical and mental fitness, reduce problems with parking/traffic congestion, and lower carbon emissions [7].

Commuting, the journey to and from work, makes a key contribution to economic prosperity, and is an important aspect of travel behavior in modern societies. The fact that commuting is mandatory for most employees, and occurs between fixed points at fixed times, however, means that commuter trips are vulnerable to traffic congestion [8]. Over short distances, the bicycle presents a good alternative to motorized transport, can reduce these negative effects, and can even help to improve public health levels and combat obesity. The city of Nitra has got the ideal conditions for the development of bicycle transport on a daily basis as it is not a very large city by square kilometers and the marginal residential areas of the city are not several kilometers far from the center. Public bicycle is a solution that improves not only the city logistics in the field of traffic management, but in itself it is also a logistically challenging endeavor. Planning the network, managing large number of bikes and stations in the city system including the transport of bikes between stations, as well as providing logistics solutions for equipment servicing are a few of many challenges [9].

Whether cycling facilities are available in a workplace reveals an employer's attitude towards modes of commuting. The presence of cycling facilities, such as secure storage, showers, and changing rooms, makes cycling more attractive [10].

In the Netherlands a tradition and positive attitude towards cycling, as well as good bicycle facilities, have led to the highest bicycle rate in the world. In addition, the national government encourages further bicycle commuting by providing tax benefits and enhanced facilities such as "bicycle highways" [8].

Many studies have been carried out aimed at commuting, in particular the role of the car [11, 12]. However, bicycle commuting has so far got only limited attention. Available research found that the weather conditions and climate, socioeconomic aspects, travel distance, and attitudes towards cycling explain individual's bicycle mode choice [13]. There was conducted a longitudinal study into the effect of the introduction of a free public transport card on attitudes and mode choice. They found that the introduction affected the attitude and social norm towards bus use and increased the use of public transport. Although the introduction did not affect the attitude towards bicycling significantly, the intention to use the bicycle decreased. A similar increase occurred in the Netherlands, where after the introduction of a free public transport card for students, cycling rates dropped [14].

Rates of cycling to work vary significantly from one urban area to another. Many factors influence the choice to cycle to work. Some are part of geography, such as terrain and weather. In one study, research executives considered the cycling choices for each home-to-workplace pair at the level of neighborhoods. The model results highlight the importance in joining up network-level interventions, for instance to reduce both route circuitry and on-road stress. The results also highlight the importance of landscape amenities along commuter routes, and the role of city-specific cycling culture after controlling for network design and geography. The insights indicate further need for closer collaboration between promoters of commuter cycling and wider urban disciplines to shape effective, low-stress routes in the heart of cities [15].

As more and more cities add programs, it is important to more concretely define bike sharing as a mode of public transportation. Acceptance of bike sharing as a public transit mode will also help it to be considered more seriously by cities looking to expand their transit services. Cities around the world are cash-strapped, leading many to a lose-lose dilemma of higher quality service covering a smaller portion of the population versus lower quality service covering a larger portion. One of the beauties of bike sharing as public transit is that it is indescribably cheap [6].

2 Material and Methodology

This chapter is structured as follows. In the first section, introductory part, a short literature review is provided, followed by some examples from practice. Then the data collection and research design are described. The last two sections present the research results and conclusions.

The aim of the chapter is to point out the opportunities and limitations that result from the concept of shared bicycles in the conditions of city of Nitra. The Arriva bike stations are located mainly in the city center and thus their location has the greatest use especially for tourists. Through marketing research, we want to find out what people think of the use of the Arriva bike system and gain important information for the full use of bicycles by the residents themselves.

In order to achieve the stated aim of the chapter primary and secondary sources of information had been collected and used. Underlying secondary data were obtained from available literature sources, i.e., from professional publications from domestic and foreign authors and organizations. When data processing and formulating conclusions of the chapter were used methods of analysis, synthesis, induction, deduction and the comparative method.

In order to meet the objectives of the chapter, marketing research was conducted. Marketing research was conducting in the period from May 2018 to September 2018.

The questionnaire was evaluated by contingency tables, which were prepared in MS Excel; graphic representations of the obtained data were created subsequently. As we were interested in a deeper analysis of the results, the following assumptions were tested:

- *Assumption 1*: There exist the preferences for most frequently used kind of transportation in the city, and these preferences vary by registration status.

Table 2 Characteristics of respondents

	Category	Absolute frequency	Relative frequency
Gender	Female	237	55%
	Male	194	45%
Age group	15–19	68	16%
	20–25	106	25%
	26–35	119	28%
	36–45	69	16%
	46–55	48	11%
	56 and more	21	5%
Economic activity of respondents	Employed	271	63%
	Unemployed	3	1%
	Student	143	33%
	Maternity leave	6	1%
	Retiree	8	2%
Education	Primary education	61	14%
	Secondary education	177	41%
	University education	193	45%

- *Assumption 2*: There do not exist the differences in the reasons for absence of using Arriva bikes between students and employees.

The questionnaire was processed in Google Forms and people were asked to fill in on social networks and via emails. Some questionnaires were filled in printed form by personally meeting. Finally, in the research outcomes were involved 431 citizens of Nitra of which 57% was non-registered citizens and 43% registered citizens (Table 2). Some of the processed questions are presented in the chapter.

3 Results and Discussion

3.1 City of Nitra, History, Geography, and Demography

Nitra is situated on seven hills on both sides of the river Nitra in the south-west of Slovakia. Because of its strategic geographic location, it became one of the most important centers for business, culture, and education in Slovakia in the past times as well as today [16].

Nitra was first time mentioned in 871–873 AD and is known as the place where the first Christian church was built in the year 828 on the present-day Slovak territory. Cyril and Methodius, the patron saints of Europe, preached here and in 880 the first Slavic diocese was established here. During the Great Moravian Empire, one of the largest settlements in Central Europe, it remained an important center of commerce until the late Habsburg Empire. The old town and the castle which was the largest one in Slovakia introduce us to the local history; scores of sacred monuments reflect the Christian history. Nitra is the seat of various institutions, universities, theatres, and museums and a well-known location of fairs as well as a modern center of industries. Every year a wide range of social, cultural, and sports events attract a multitude of visitors [17]. Nitra lies at an altitude of 190 m (623 ft) above sea level and covers an area of 100.45 km^2 (38.8 sq. mi). It is located in the Nitra river valley in the Danubian Lowland, where the bigger part of the city is located. Nitra has a population of 78,559 (as of January 2, 2019). Nitra is currently the fifth largest city in Slovakia.

3.2 Travelling in Nitra

City of Nitra was the first city in Slovakia that offers to citizens and visitors discovering the new adventure in the city—the green Arriva bike. It is a new, modern, and environmentally friendly way of transport around the city thanks to the first bike-sharing service in Slovakia [18]. City of Nitra has launched bike sharing as a supplement of public transport. The bike sharing project is a joint project of City of Nitra and company Arriva which operates public transport in Nitra. Arriva put €200,000 into this project. In this amount is included the cost of buying not only bicycles, but also licenses, concessions, stands, GPS, and applications. The city of Nitra, which built bicycle locations, supported the project with a sum of €27,000. City of Nitra signed with Arriva a cooperation agreement in which both parties undertook to promote city mobility with an emphasis on cyclotourism and bicycle sharing. Both expect shared bikes to make cities more accessible, improve the traffic situation, and increase passenger interest in urban public transport [19].

Arrive bike appeared in the city below the Zobor mountain on July 28th 2017. From that time residents of Nitra and visitors of the city could use 70 shared bikes at seven stations [20]. Figure 2 shows individual station locations (green circles) that are located in the very center of the city. We marked the center area with a red circle. In the marginal residential areas of the city there are no stations which could be fully utilized by residents on the way to the city center. The blue circle captures the marginal parts of the city. This picture displays a lot of reserves in the coverage of Arriva bike stations.

All bicycles are thoroughly serviced during their winter sleep from December till February, so they can continue to serve faithfully during the season from March to November. At any time during this period, day and night, users can enjoy the cycling around the smart city.

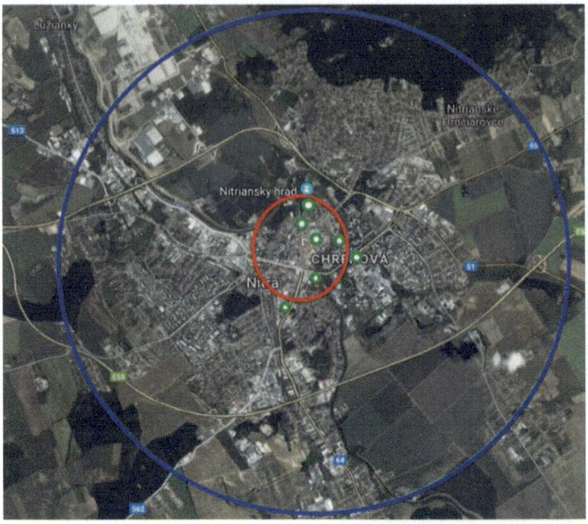

Fig. 2 Stations of shared bicycles in Nitra. Source: Own processing

Fig. 3 Renting a shared bicycle [21]

3.3 Registration and Renting a Shared Bicycle

Registration for using shared bike is free of charge. Users can register via the web page www.arrivabike.sk or via mobile app Arriva bike. After the registration it is possible to use all shared bicycle schemes in the world operated by nextbike. Bicycle rental is available via mobile app or directly via bike computer. Rental is possible through the mobile app or smart card (issued by Arriva Nitra, nextbike) when up to four bikes can be rented at a same time. Citizens and visitors have been so far able to rent bicycles through the bicycle computer or mobile app available on Google Play and the App Store. A new feature of the year 2018 is the opportunity to use Arriva Nitra public transport cards to rent shared bicycles (Fig. 3). To rent bicycle via the public transport card, registration is required at Arriva bike web page or through the app Arriva bike. The payment is executed automatically via credit card, after finishing the rental [21].

Thus, to rent a shared bike is easy. It is needed to arrive at one of the seven stations, where the Arriva bike is located, and choose one of them. The information about free bikes at the selected location is available in the Arriva bike mobile app. Each registered user can use not only bike-sharing system of Arriva bike in Slovakia, but also all the schemes of bike-sharing system operated by nextbike all around the world [21].

There is no registration fee for registering via smartphone or online via the operator's website. If the rental is subject to paying a fee and is not free of charge (first 60 min of system use), it is necessary to input payment card details before the rental begins. The operator charges a fee of €1 for performing a credit check on the customer. The fee of €1 shall be credited to the account of Arriva bike. Subsequently, it shall be cleared during the following payment for bike rental. Based on the service used, the operator is eligible to charge regular rental fees according to the current price list, which can be found online at www.arrivabike.sk. Only persons older than 16 years of age may become registered customers. Regular fare of Arriva bike in Nitra is €0.50 for 60 min and €3 for 24 h. Annual subscription is €25. This subscription fee includes the first 60 min free of charge; the following 60 min is charged with a valid tariff. Tariff after surpassing 24 h is €6 for 60 min. The payment is executed automatically via credit card, after finishing the rental. Bicycles can only be returned at official bike stations in Nitra. In order to return the bike, the user has to lock the bike at one of the seven stations and confirm return via bike computer [22].

Bike sharing in Nitra is interesting and popular. One year since its launch, it has got almost 8700 users and 20,000 bicycle rentals. Nevertheless "Boom" registration of new users still continues. Before this year's biking season, 1800 new users were registered in April 2018, and the bikes were rented approximately 140 times a day, with the most popular stations being the City Park (Fig. 4) and Trieda Andreja Hlinku. The first anniversary of the project, which was on July 28th 2018, allowed cyclists to actively recall the project's success by renting a bicycle free of charge at any station for 2 h [17].

Also thanks to bike-sharing system, the interest in cycling in Nitra is unprecedented, and Nitra has also been involved in the cycling competition. In 2018, 48

Fig. 4 Shared bicycles at City Park station [22]

registered teams and 161 contestants competed for the city of Nitra. There were 2482 journeys made, with a total of 19,266 km and thanks to all involved for up to 5257 kg less carbon dioxide emissions in the air [23].

3.4 Questionnaire Research Outcomes

In the introductory part of the questionnaire, we asked the respondents what kind of transportation they use most often in the city. We noticed certain differences in responses between registered and non-registered citizens (Fig. 5). The highest percentage of registered citizens use a bus to move around the city, while the bus is used by only 17% of unregistered citizens. Registered citizens as the second most frequent kind of transportation use the bicycle (21%), followed by car (19%). The car is the most popular kind of transportation in the city of Nitra, used mainly by unregistered inhabitants, as 62% of respondents marked this option. Only 6% of unregistered users use the bicycle. In the case of taxis (10%, 9%) and other kinds of transport (5%, 4%) there are no significant differences between registered and unregistered citizens. Respondents were given the opportunity to mark and give their own response to the most commonly used kind of transportation, where the majority marked the motor scooters and electric scooters.

In relation to the assessment of this question, we have formulated *Assumption 1*, where we assume that there exist the preferences for the most frequently used kind of transportation in the city, and these preferences vary by registration status.

H_0: There do not exist the preferences for the most frequently used kind of transportation by registration status.
H_A: There exist the preferences for the most frequently used kind of transportation by registration status.

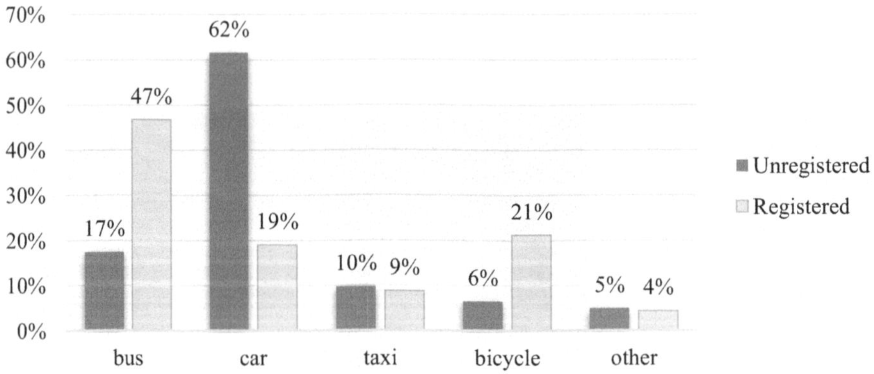

Fig. 5 What kind of transportation do you use most often in the city of Nitra? Source: Own research and processing

Table 3 Results of Kolmogorov–Smirnov test

Kolmogorov–Smirnov test	
D-stat	0.351
D-crit	0.169

D-stat > D-crit
The result is significant at $p < 0.05$
Source: Own processing, XLStat

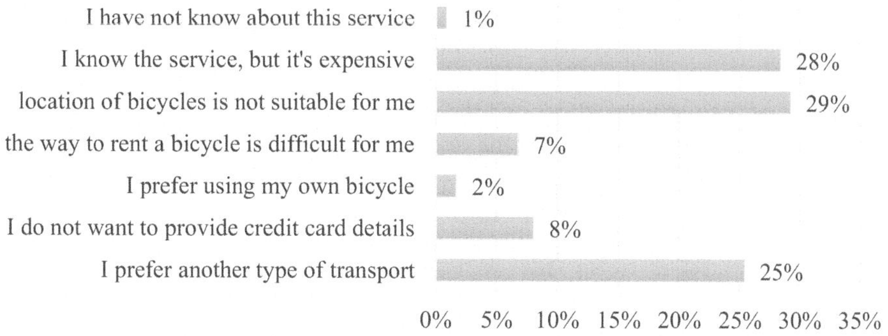

Fig. 6 Why haven't you used the system of shared bicycles—Arriva bikes—yet? Source: Own research and processing

For evaluation the nonparametric Kolmogorov–Smirnov test was used. This test is used in situations where a comparison has to be made between an observed sample distribution and theoretical distribution. We carried out the analysis, and the results are shown in Table 3.

Since the calculated value is greater than the critical value, we reject the null hypothesis and conclude that there exist the preferences for some kind of transportation.

Following the common questions for both registered and unregistered people, we have asked the unregistered citizen the question: Why haven't you used the system of shared bicycles—Arriva bikes—yet? The asked question has given us the answers on which areas we need to focus on in order to achieve greater use of bike sharing (Fig. 6). Regarding this question, respondents had the option to mark 1–3 different responses. The most important reason why the service is not used by respondents is the high price, which accounts for up to 28% of the respondents' answers. Inappropriate location of bicycle stations was marked by 29% of respondents and 25% of marked responses represent the use of another kind of transport. The least marked answer, only 1%, represents the lack of information about the abovementioned service, which tells us about the well-spread information about bike-sharing system among the citizens of Nitra. The survey also showed us that 2% of responses identified the use of their own bicycle as a reason. A complicated way of using bike sharing was marked by 7% of respondents and only 8% of respondents see the problem in providing the credit card information for payment.

Table 4 Results of Kruskal-Wallis test

Kruskal-Wallis test	
K (observed value)	*2.853*
K (critical value)	*5.865*
DF	*1*
p-Value (two-tailed)	*<0.063*
Alpha	*0.05*

Source: Own research and processing

By setting a more favorable price for bike-sharing service as well as more logical placement of bike stations all around the city and enhancing the number of these stations, we will contribute to reducing or eliminating two of the three reasons why respondents do not use the service of bike sharing in the city of Nitra. The most difficult reason to influence is the priority of citizens to use the other kind of traffic. Here it is suitable to focus on greater awareness among the citizens, mainly about the ecological aspects, speed, and efficiency of such transport as well as its side effect in the form of reduced traffic density in the city.

For a deeper analysis of the obtained results, an *Assumption 2* was set out: "There do not exist the differences in the reasons for absence of using Arriva bikes between students and employees." We wanted to prove that there are common reasons for absence of using regardless of the economic category of respondents. If our assumption is confirmed, we can conclude that the most often marked reasons are crucial.

The analysis of most common reasons for not using Arriva bikes was carried out by the Kruskal-Wallis test, the results of which are given in Table 4. In connection with the evaluation of this question, we wanted to find out whether the choice of the particular option (the main reasons) is different in the economic category of respondents, exactly between students and employees (*Assumption 2*). For this reason, the following hypotheses were formulated:

H_0: *There do not exist the differences in the reasons for absence of using Arriva bikes between students and employees.*

H_A: *There exist the differences in the reasons for absence of using Arriva bikes between students and employees.*

Based on the theoretical level of significance, the zero hypothesis of the same choice of answer by economic category is accepted. In other words, we can assert that there are no statistically significant differences in the choice of student- and employee-based reasons at the level of alpha significance.

When questioning the use of Arriva bikes, respondents have the option to mark up to three different answers. The most frequent answer was that up to 50% of all responses use the shared bicycles in the city of Nitra during their leisure time (Fig. 7). Regarding the age of the respondents, up to 25% of them marked the option of using the shared bicycle as a way of transportation to school. When using the Arriva bike for active sport we received only 12% of all possible answers. Only 7% of responses consisted of using bike sharing on the way to work and the least

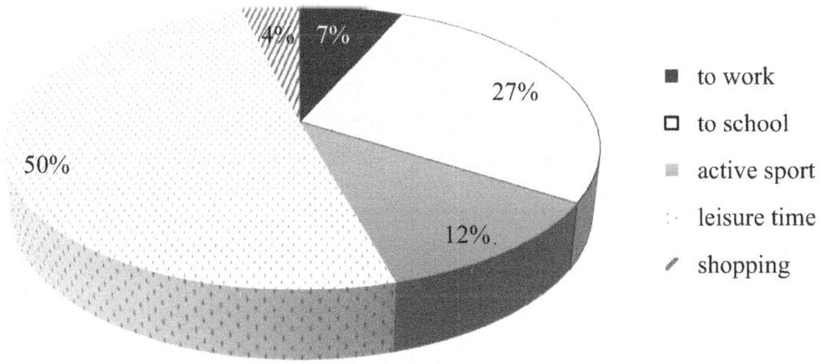

Fig. 7 For what purpose do you use Arriva bikes? Source: Own research and processing

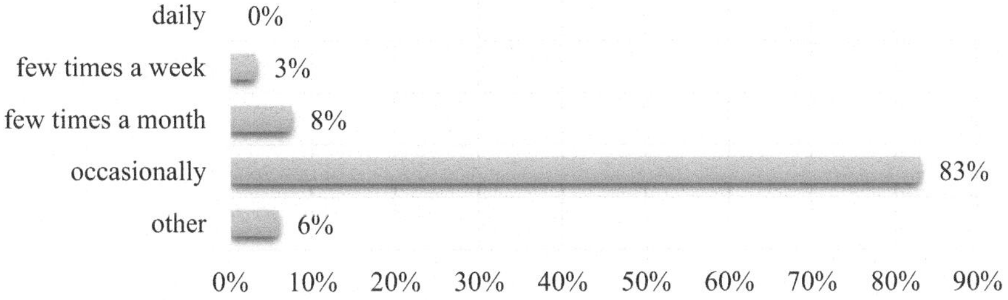

Fig. 8 How often do you use the shared bicycles—Arriva bikes? Source: Own research and processing

marked answer, i.e., 4%, was that the respondents use the Arriva bikes on the way to the store. From all answers regarding the question of purpose to use the Arriva bikes, a tremendous potential for using the shared bicycles for riding to work and to school could be seen. The abovementioned lower number of responses is derived from the previous question, where 66% of respondents said that they were using a bus and car as the most preferred kind of transportation.

The following question provides really important information for our research, as we want to find out the frequency of using Arriva bikes by the citizens of Nitra (Fig. 8). The results are not optimistic as the use of bicycles is occasional for the majority of registered respondents (83%). Only 8% of respondents use bicycles a few times a month and 3% respondents a few times a week. As a daily kind of transportation in the city, using Arriva bikes was marked by 0% of respondents. The "other" option was marked by 6% of respondents, most responding that they have tried this service just once. This question is related to the answers of unregistered users and reasons why they do not use the service of shared bikes (Fig. 6), which outlined three most important reasons: expensive service, no suitable locations, and preference of another type of transport that formed up to 82% of all the abovementioned reasons.

Respondents could, once again, mark 1–3 options when answering what are their reasons for using Arriva bike system (Fig. 9). This question offered us an

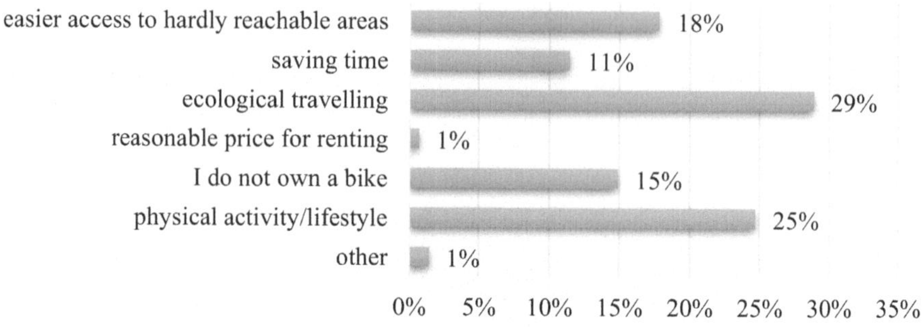

Fig. 9 Why do you use Arriva bike system? Source: Own research and processing

Fig. 10 Would you like to enhance the number of shared bicycle stations? Source: Own research and processing

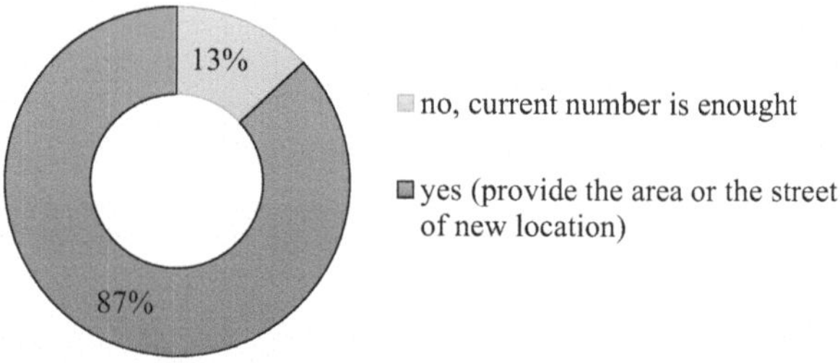

important answer about motivation of people to use bike-sharing system. The collected findings will help us to state a further strategy of disseminating and promoting this kind of transportation. The most common answer, up to 29% of all responses, was the ecological side of shared bikes. Physical activity and healthy lifestyle were accounted for 25% of all possible answers as the most important reason for using the shared bikes. The favorable price and other unspecified reason accounted for only 1% of all possible answers.

The final question was related to enhancing of the number of shared bicycle stations, providing us the specific suggestions to improve the current location of bicycles and thus their actual use. More than 87% of all respondents would welcome to enhance the current number of bicycle stations (Fig. 10). This result can be considered as the most important priority in redefining the current strategy in connection with addressing the urban transport issue. By summarizing the free answers and following analyses on the map of the city of Nitra, we can say that the majority of respondents have largely marked the residential areas in the suburbs of the city of Nitra as the most strategic places for bike-sharing stations.

4 Conclusion

The number of urban smart bike sharing programs is growing in cities in Slovakia. This concept is still being improved and enriched by further technical innovations that offer new functionalities for both customers and operators. This chapter focuses on the city of Nitra, which is currently struggling with the issue of critical transport—the situation is more than serious and is getting even worse.

Based on a marketing survey which included the opinions of Nitra citizens we wanted to point out the opportunities and constraints arising from the concept of shared bicycles in the conditions of the city of Nitra. The results showed that among the most important reasons why the service is not used by respondents are the high price, which accounts for up to 28% of the respondents' answers, and inappropriate location of bicycle stations (29%).

If we want this system to contribute to improving the traffic situation in the city, it is necessary to reassess and extend the location of bike stations to the marginal residential areas of the city where most of the population is inhabited. In our survey more than 87% of all respondents would welcome to enhance the current number of bicycle stations that can be marked as the priority in redefining the current strategy in connection with addressing the urban transport issue. By strategic placement into residential areas of the city, bicycles would have much better use (to work, to school, for shopping). Nitra's advantage is that the peripheral parts of the city are not very distant from the center, which means that the journey by bicycle from marginal part to the center cannot last very long. We suggest for the beginning stage that by splitting existing stations into smaller stations, which decreases availability but increases the accessibility, ridership can be increased by high percentage—with the same amount of bicycles.

The results connected with the frequency of using the Arriva bikes by the citizens are not optimistic as the use of bicycles is occasional for the majority of registered respondents (83%).

For a deeper analysis of the results of marketing research, the following assumptions were tested:

Assumption 1: There exist the preferences for most frequently used kind of transportation in the city, and these preferences vary by registration status—based on the confirmed Kolmogorov-Smirnov test.

Assumption 2: There do not exist the differences in the reasons for absence of using Arriva bikes between students and employees—based on the confirmed Kruskal-Wallis test.

We live in a society where congestion for political and social reasons is labeled as "a major problem to be solved" but adopting the most effective solutions is still missing. We actually prefer the problem to the solutions. Of course, replacing every car with a (shared) bicycle overnight is not a realistic scenario. Barriers for bicycle adoption include a lack of safe parking spaces, vandalism and theft, and inconvenience and cost of owning and maintaining a bike. These issues are addressed by bike-sharing systems. Nonetheless, the success of bike-sharing systems depends on other aspects, such as station accessibility (how far a commuter has to walk to reach a station) and bike availability (the likelihood of finding a bike at the station). These findings again underline the call for optimizing bike-sharing system in the condition of the city of Nitra in order to develop a reliable urban transportation infrastructure with minimization of traffic congestion. The development towards smart cities promises interesting opportunities for sustainable innovation and collaboration between public and private parties.

References

1. EasyPark Group. Retrieved January 3, 2019, from https://easyparkgroup.com/smart-cities-index/
2. Bawa, M., Cagáňová, D., Szilva, I., & Špirková, D. (2016). Importance of internet of things and big data in building smart city and what would be its challenges. In A. Leon-Garcia et al. (Eds.), *Smart City 360°. SmartCity 360 2016, SmartCity 360 2015. LNICST* (Vol. 166, pp. 605–616). Cham: Springer.
3. Venture Smarter. Retrieved January 3, 2019, from https://www.venturesmarter.com/resources/easypark-smart-city-index
4. Wachs, M. (2002). Fighting traffic congestion with information technology. *Issues in Science and Technology, 19*(1), 5–15.
5. Finch and Break. https://www.finchandbeak.com/1108/smart-cities-smart-transit-bike-shares.htm
6. The City Fix. Retrieved September 13, 2018, from http://thecityfix.com/blog/bike-sharing-the-newest-mode-of-public-transport/
7. Keene, D., McDonald, P. (2018). Bicycle commuting: Improving personal and global well-being–one revolution at a time. Retrieved January 5, 2019, from https://digitalcommons.spu.edu/common_learning/dcl_2018/dcl_2018_events/1/
8. Heinen, E., Maat, K., & Van Wee, B. (2013). The effect of work-related factors on the bicycle commute mode choice in the Netherlands. *Transportation, 40*(1), 23–43.
9. Cichosz, M. (2013). IT solutions in logistics of smart bike-sharing systems in urban transport. *Management, 17*(2), 272–283.
10. Abraham, J. E., et al. (2002). Investigation of cycling sensitivities. In *Transportation research board annual conference* (pp. 1–12). Washington, DC.
11. Dargay, J. M., & Hanly, M. (2007). Volatility of car ownership, commuting mode and time in the UK. *Transportation Research. Part A: Policy & Practice, 41*(10), 934–948.
12. Susilo, Y. O., & Maat, K. (2007). The influence of built environment to the trends in commuting journeys in the Netherlands. *Transportation, 34*(5), 589–609.
13. Parkin, J., Wardman, M., & Page, M. (2008). Estimation of the determinants of bicycle mode share for the journey to work using census data. *Transportation, 35*(1), 93–109.
14. Rietveld, P. (2000). Non-motorised modes in transport systems: a multimodal chain perspective for The Netherlands. *Transportation Research. Part D: Transport and Environment, 5*(1), 31–36.
15. Cervero, R., Denman, S., & Jin, Y. (2019). Network design, built and natural environments, and bicycle commuting: Evidence from British cities and towns. *Transport Policy, 74*(1), 153–164.
16. Bárta, V. (2016). *Nitra: Ako si Ťa pamätáme* (1st ed.). Slovenská Ľupča: AB Art Press.
17. City of Nitra Homepage. Retrieved September 27, 2018, from https://www.nitra.sk/
18. Arriva Homepage. Retrieved September 28, 2018, from https://arriva.bike.sk
19. Nitra 24 Homepage. Retrieved September 28, 2018, from https://nitra.dnes24.sk/v-nitre-uz-funguje-zdielanie-bicyklov-stala-sa-tak-prvym-mestom-na-slovensku-275985
20. Záborský, J. (2018). Prvé dojmy z bikesharingu. *Trend, 39*(27), 52–53.
21. Arriva Homepage. Retrieved September 28, 2018, from https://arriva.bike/en/this-season-brings-a-new-feature/
22. Arriva Homepage. Retrieved September 28, 2018, from https://arriva.bike/portfolio-item/arriva-bike-stanoviste-c-5305-mestsky-park/
23. MyNitra Homepage. Retrieved September 28, 2018, from https://mynitra.sme.sk/r/143/spravodajstvo.html

Aspects Affecting Emission Production from Small Heat Sources

Jozef Jandačka and Michal Holubčík

1 Introduction

The world's population is rising, and this growth affects an expanding consumption of energy in all its forms [1]. To enhance energy self-sufficiency, a great attention is paid to optimisation, and improvement efficiency of heat sources, and use of renewable energy sources as well [2, 3]. The biomass is one of the renewable energy sources derived from organic matter such as wood, crop waste or garbage.

The energy stored in the biomass is mainly utilised by combustion process. The wood can be combusted in various forms, such as logs, wood chips, briquettes and pellets. The straw of cereals and oilseeds can also be modified in the mentioned forms. Good fuels for combustion are materials rich in hydrogen and carbon, called hydrocarbons. Such fuels include natural gas, coal, diesel, gasoline, propane, wood, agricultural residues and municipal solid waste. Ideally, all hydrogen and carbon would split off and combine with the oxygen in the air to create water vapour, carbon dioxide and heat [4]. The combustion of biomass does not guarantee its best use at low emission concentration [5, 6]. Since biomass fuels are primarily composed of carbon, hydrogen and oxygen, the main products from burning biomass are carbon dioxide and water. Flame temperatures can exceed 2000 °C, depending on the heating value and moisture content of the fuel, the amount of air used to burn the fuel and the construction of the furnace. Bad settings of combustion conditions during biomass combustion can have negative influence on the efficiency and high concentration of emissions [7].

J. Jandačka · M. Holubčík (✉)
Faculty of Mechanical Engineering, Department of Power Engineering, University of Zilina, Zilina, Slovakia
e-mail: michal.holubcik@fstroj.uniza.sk

© Springer Nature Switzerland AG 2020
D. Cagáňová, N. Horňáková (eds.), *Mobility Internet of Things 2018*,
EAI/Springer Innovations in Communication and Computing,
https://doi.org/10.1007/978-3-030-30911-4_10

According to the Slovak Environmental Agency (SAZP), energy, especially small heat sources, has a significant share in the production of emissions, especially solid pollutants (particulate matter, PM). Households, in particular by using small heat sources for solid fuels, produce 77.2% of PM up to 10 μm (PM10), and 87% PM up to 2.5 μm (PM2.5). Households also account for a 25% share of carbon monoxide (CO) production, a 17.7% share of NO_X, a 3.3% share of SO_2 and a 23.3% share of non-methane volatile organic compounds (NMVOCs). According to these data, concentrations of pollutants in the air are very often exceeded, especially in the regions of northern and central Slovakia, especially in the Žilina Region. This is directly related to the high representation of the use of small sources of heat for solid fuels in this area, namely according to a survey by the Statistical Office of the Slovak Republic. In these areas the share of domestic solid fuel heating is around 60–80% while the average is around 21.43%.

Energy production based on combustion of solid fuels always produces greenhouse gas emissions and pollutes the environment, which weakens air quality and can harm the environment and human health. Large and medium heat sources are currently relatively strictly regulated and use effective air pollution control to reduce emissions [3]. However, small heat sources are currently not controlled and theoretically they can produce an unlimited amount of pollutants.

Given the increasing share of small heat sources for solid fuels and worsening emissions solutions need to be sought to make the use of solid fuels more efficient in small heat sources while minimising environmental impact. The greatest potential for reducing emissions from small combustion plants is heat-treated heat sources such as fireplaces, stoves and boilers with manual operation. Automatic sources of heat for combustion of solid fuels offer lower emission abatement options according to statistics and previous research. However, it is well known that the use of old technologies, older boiler designs, choice of poor quality or incorrect fuel type, improper handling or maintenance negligence brings diversity, which leads to an increase in the negative impact on the environment [4]. The use of modern technologies should help to make more efficient use of chemically bound energy in solid fuels and reduce emissions. However, the period of modernisation of heat sources is relatively long and longer in the case of central boilers than in the case of fireplaces or stoves. New sources of heat for solid fuels should be tailored to new eco-design regulations, emission limits for new plants for the future and requirements for installing a stand-alone emission control facility, and sales should be supported by a ban on the sale of old and inefficient technologies and information campaigns for the public.

The design of heat sources with manual stoking has a simple construction due to its low cost. The most common fuel burning in the boilers is lumber and coal. The classical and most widespread design of these heat sources is a large-volume outbreak (with over-fire or with downdraft combustion) in which large amounts of fuel can be loaded for as long as possible. Freshly added fuel is gradually heated and dried and then volatile flammable fuel is released, the ignition of which initiates burning. The process of combustion proceeds at different speeds. Fuel will be degassed with variable control, and released fume will be utilised, but a large grate area is required to burn the flammable fuel [7].

Manually operated heat sources are mostly used in heating facilities with a heat output range of up to 50 kW. With these heat sources it is very hard to regulate the combustion process. It is possible only by changing the fuel stoking in combination with the regulation of the amount of combustion air, its redistribution. The primary air supply influences the release of the gaseous fuel component and the secondary air combustion of released flammable gases. Combustible wood combs usually have a simple manual control of the combustion air supply, so the air supply is usually more or less constant (the flow can vary depending on the change in the draft in the chimney). Since it is not possible to provide the currently required amount of combustion air with simple, reliable and inexpensive regulation, this type of heat source does not allow the emission of the environment to be reduced.

The most common technology used in boilers with a manual household appliance is the over-fire boiler, boiler with downdraft combustion or gasification boiler.

The application of high-quality technologies can lead to effective utilisation of biomass as a fuel with low-emission production [6]. The quality of biomass combustion in a heat source can be affected by the following aspects: type of fuel, amount and distribution of combustion air, method of combustion, design and regulation of combustion appliance, etc. [7]. These factors affect the performance and emission parameters of a boiler or fireplace. Emission is the term used to describe the gases and particles which are put into the air from the biomass combustion or emitted by various sources. In Europe, the most problematic pollutants in terms of harm to human health are particulate matter (PM), ground-level ozone (O_3) and nitrogen dioxide (NO_2) [8, 9]. To maintain the environmental friendliness of biomass, states issue measures in the form of emission limits, which have to be respected [10].

This work describes various factors affecting the performance and emission limits of the boiler for combustion of wood pellets. The experimental measurements were focused on the verification of the following influences: the amount and temperature of combustion air, relative humidity of combustion air, fuel feeding and fuel moisture on thermal performance and emission concentrations in automatic boiler.

2 Methodology of Boiler Testing

2.1 The Experimental Boiler

Currently, more and more automatic boilers are used for solid fuel combustion that use a self-supplying fuel system to the furnace. In the case of biomass or coal of a suitable fraction it is almost an exclusively different combination of the screw conveyor. Automatic fuel transport to the boiler ensures the stability of the combustion process, stable performance, high combustion efficiency, low pollutant production and, above all, user comfort. With theoretically continuous fuel supply there is a problem regarding when it is necessary to transport a very small amount of fuel continuously to the focus. Therefore the fuel is virtually always delivered cyclically and the whole process is automated with the fuel supply being the output temperature of the flue gas, the heat transfer medium, or the conveyor is run periodically with the specified dosing and standing time. Automatic heat sources for combustion of solid fuels can use grate combustion or special burners in the combustion chamber. Always combust the minimum amount of fuel that is necessary to achieve the desired heat output. A relatively small area of the furnace is needed which leads to rapid thermal decomposition of fuel and burning of solid residues. The hopper space must be large enough due to the length of the flame usually equipped with a refractory lining which allows the perfect combustion of the flammable fuel. The automatic heat sources have a control unit that controls all systems and communicates with the operator. The degree and precision of automation are the main reasons for the differences in boiler prices with the same power category.

A hot water boiler was applied for testing of abovementioned factors. The fully automatic wood pellet boiler can reach nominal performance of 18 kW. The combustion is carried out in cast iron, high-quality, retort burner and fuel is transported from the fuel tank into the combustion chamber using a screw feeder. The combustion air is supplied into the chamber by a fan. The combustion of waste products takes place near the cast iron rectifier of flue gas. The boiler control system offers unprecedented management of combustion air, operation time and standing time of the screw feeder for the fuel supply.

2.2 *Used Experimental Fuel*

In order to minimise measurement error, the fuel with the same properties was applied in all tests. Technical parameters of wood pellets made of spruce sawdust are presented in Table 1.

Table 1 Technical parameters of wood pellets

Technical parameters		
Diameter	6 mm	STN EN 14961-2 [11]
Length	40 mm	STN EN 14961-2 [11]
Moisture	9.98%	STN EN 14774 [12]
Ash content	0.452%	STN EN 14775 [13]
Mechanical resistance	98.67%	STN EN 15210 [14]
Content of fine particles	0.239%	STN EN 15210 [14]
Calorific value	17.58 MJ/kg	STN EN 14918 [15]
Bulk density	650 kg.m^{-3}	STN EN 15103 [16]

The moisture was measured on fuel sample submerged to a fraction size below 1 mm. It was dried at 105 °C ± 2 °C until constant weight (weight change did not exceed 0.2% of total weight loss over the next 60 min). The thermogravimetric analyser was used for determination of fuel moisture.

The ash content of fuel sample was measured by heating at a temperature of 550 °C until constant weight (weight change did not exceed 0.2% of total weight loss over the next 60 min). The thermogravimetric analyser was used for determination of fuel ash content.

Mechanical resistance and content of fine particles were determined by the device Lignotester which blows on a sample of pellets in the airstream with prescribed pressure for prescribed time. The weight loss of the pellets is determined.

The gross calorific value was measured by the calorimeter LECO AC 500. The measurement of gross calorific value is based on the law of conservation of energy, which states that the heat released by the investigated substance shall be equal to the heat that is absorbed by water and the calorimeter.

The calorific value of wood pellets was calculated based on the gross calorific value obtained from the calorimeter, H_2 content from elemental analysis and measured moisture content by the equation:

$$Q_{\mathrm{i}} = Q_{\mathrm{S}} - 2453.(M_{\mathrm{ar}} + 9\mathrm{H})\left(\frac{\mathrm{MJ}}{\mathrm{kg}}\right) \tag{1}$$

The density determination procedure was as follows: 10 samples of each test fuel were randomly selected. Each sample was weighed on an analytical scale to the nearest 0.1 mg. Further the sample was immersed in a suitable fluid and the volume of the sample was determined from the increase in fluid volume. Subsequently the sample density was determined on the basis of the equation:

$$\rho = \frac{m}{V}\left(\frac{\mathrm{kg}}{\mathrm{m}^3}\right) \tag{2}$$

2.3 *Experimental Measurement of Heat Power of Boiler*

Thermal characteristic and measurement of the boiler heat power were carried out on an experimental setup (see Fig. 1). This scheme consists of the automatic boiler, device for regulation of heat produced by the boiler, flue gas analyser, particulate matter analyser, data logger for recording the measured data every 20 s and a computer for evaluation of the results.

The heat power of the boiler was measured and calculated in accordance with STN EN 303-5 [17], which is based on the calorimetric method. Equation (1) for calculation of the performance includes mass flow measured by a magnetic

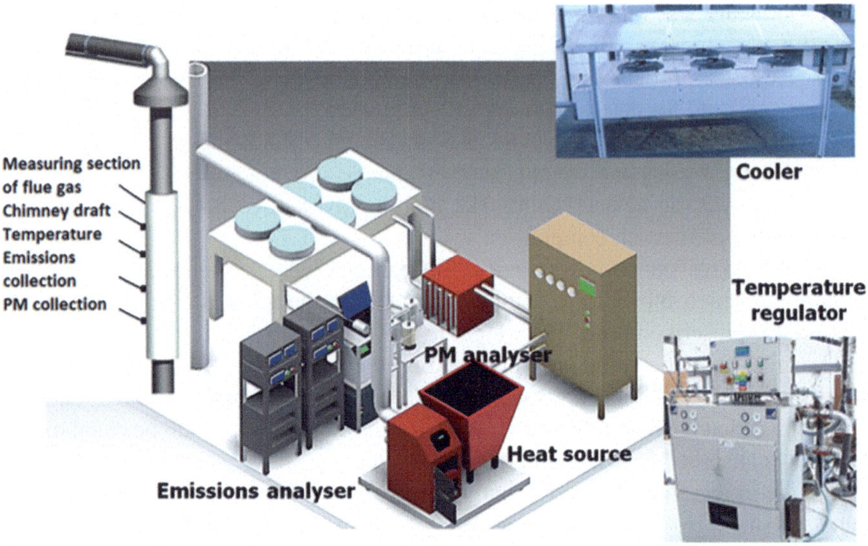

Fig. 1 Experimental device for testing heat sources

induction flow meter multiplied by the specific heat capacity, and temperature difference between outlet and inlet of heat transfer medium in the boiler [18]:

$$Q = \dot{m}.c_{\text{p}}.(T_2 - T_1) \quad (3)$$

where Q is the thermal power of heat source [kW], m is the mass flow of heat transfer medium at the inlet and outlet [kg/s], c_{p} is the specific heat capacity of heat transfer medium [J/kg.K], T_1 is the temperature of heat transfer medium at the outlet of heat source [K] and T_2 is the temperature of heat transfer medium at the inlet of heat source [K]. In order to determine the performance, the temperature gradient was maintained by the temperature regulator. This device consists of a heat exchanger, two pumps, two valves, and control, regulatory and safety devices. The constant chimney draft of 12 ± 2 Pa was ensured with a fan, which was regulated by the frequency converter.

The temperature of the heat transfer medium at the inlet and at the outlet of the heat source was measured with the correlated PT100 metal resistance resistors with an accuracy of ±0.4% and a range of 0–150 °C. The principle of resistive temperature sensors is based on a change in the electrical resistance of metals with a temperature. The flow rate of the heat transfer medium was determined by the YOKOGAWA ADMAG AXF magnetic induction flowmeter. The flowmeter works on the principle of Faraday's law, does not produce any pressure loss and achieves an accuracy of ±0.35%.

During the determination of the heat output it was necessary to maintain the set temperature gradient by means of the temperature controller. It is a device that detects and regulates the boiler water temperature. The device consists of a heat exchanger, two circulation pumps controlled by frequency converters, two control valves, and control and safety device.

The heat produced by the heat source was passed through the heat exchanger temperature controller to the secondary circuit of the heat transfer medium. The heat from the secondary circuit was transferred to a dry automatic cooler located outside the heat exchanger via a heat exchanger. The cooler was equipped with six fans, the speeds of which were regulated according to the temperature of the heat transfer medium.

2.4 *Measurement of Emission Concentration*

The collection of the flue gas samples was done in measuring section (see Fig. 1) which is located in the chimney. The lower edge of the flue gas measuring section was 1 m from the outlet axis of the flue gas duct of the heat source into the flue. The measurements of flue gas temperature-chimney temperature, chimney vacuum-chimney draft and sampling for gaseous emissions in flue gas were done in the flue gas measuring range.

The concentration of CO and NO_X was recorded by the flue gas analyser ABB AO 2020 with infrared spectroscopy. The device records the emission per unit of parts per million (ppm). The units were to be converted to mg/m^3. The conversion process is as follows:

$$Y = X.\frac{M}{22,41}.\frac{T}{273,15}.\frac{p}{101325} \tag{4}$$

where Y is the calculated emission output (mg/m^3), X is the measured emission (ppm), M is the molar mass (g/mol), 22.41 is the normal molar volume (dm^3/mol), T is the flue gas temperature and p is the absolute flue gas pressure (Pa).

The normalised oxygen concentration ($O_{2\ \text{standardised}}$) in the flue gases from the central heat source is in accordance with STN EN 303-5 [17, 19] value of 10%. The values of emission production were calculated based on the oxygen content in the flue gas according to the following equation:

$$Y_{(10\%O_2)} = Y.\frac{21 - O_{2\text{standardised}}}{21 - O_{2\text{avg}}} \tag{5}$$

where Y is the concentration of the emission [mg.m^{-3}], $O_{2\text{standardised}}$ is the normalised oxygen concentration [%] and $O_{2\text{avg}}$ is the average oxygen concentration in the flue gases [%].

3 Methodology for Various Aspects

The following aspects were experimentally tested:

- Combustion air amount
- Combustion air temperature
- Combustion air relative humidity
- Fuel moisture
- Fuel feeding

3.1 *Combustion Air Amount*

Used experimental boiler disposes the regulation of combustion air amount by the fan speed controller in the range of 10–100%. Combustion air mass flow rate reaches values from 45 kg/h to 142 kg/h. The speed of fan is controlled by the regulation system of boiler. All other parameters during determination of the impact of combustion air amount were constant, mainly fuel feeding, chimney draft, temperature and relative humidity of combustion air.

The supplied combustion air was divided into primary and secondary combustion air. The primary combustion air was supplied under the fuel layer and the secondary combustion air was supplied over the burning wood pellet layer. The combustion air mass flow was not measured directly. The anemometer was used for measuring the airflow velocity at combustion air inlet. The mass flow was calculated through air density.

Combustion has three requirements—fuel, air and heat. If any of these three are removed, burning stops. When all three are available in the correct proportion, combustion is self-sustaining, because the fuel releases excess heat to initiate further burning. Complete combustion of biomass requires a certain amount of air. Air consists of 21% oxygen and about 79% nitrogen. Therefore, the product of a stoichiometric combustion of biomass in air will include carbon dioxide, water vapour and nitrogen. This reaction will generate heat. When oxygen (O_2) is found in the flue exhaust, it usually means that more air was supplied than needed for complete combustion to occur. When too little oxygen is supplied to the burner, carbon monoxide (CO), a highly toxic gas, forms in the flue gas. Not only improved fuel efficiency but also decreased soot generation will occur when the correct amount of air is supplied to the combustion process [7].

3.2 *Combustion Air Temperature*

For fuel-fired industrial heating processes, one of the most potent ways to improve efficiency and productivity is to preheat the combustion air going to the burners. The source of this heat energy is the exhaust gas stream, which leaves the process at elevated temperatures. A heat exchanger, placed in the exhaust stack or ductwork, can extract a large portion of the thermal energy in the flue gases and transfer it to the incoming combustion air. Recycling heat this way will reduce the amount of the purchased fuel needed by the furnace. There are two types of air preheaters: recuperators and regenerators. Recuperators are gas-to-gas heat exchangers placed on the furnace stack. Internal tubes or plates transfer heat from the outgoing exhaust gas to the incoming combustion air while keeping the two streams from mixing. Recuperators are available in a wide variety of styles, flow capacities, and temperature ranges. Regenerators include two or more separate heat storage sections. Flue gases and combustion air take turns flowing through each regenerator, alternately heating the storage medium and then withdrawing heat from it. For uninterrupted operation, at least two regenerators and their associated burners are required: one regenerator is needed to fire the furnace while the other is recharging [20].

The temperature of combustion air was set to the desired level before entering the experimental boiler. This setting was realised by using of heat exchanger (see Fig. 2) that was connected by the pipe. The heat exchanger was cooling or heating the combustion air. During the experiments temperatures between −5 °C and 40 °C gradually increased in steps of 5 °C. For all measurements the same conditions were used, mainly fuel (18-s feeding and 25-s pause for the screw feeder), the same combustion air amount and the same chimney draft.

The experimental heat exchanger for changing the combustion air temperature was located in front of experimental boiler and it was in square cabinet made from steel sheets. The main part of the heat exchanger (see Fig. 2) was made from copper tubing. For intensification of heat transport corrugated aluminium ribs with regular spacing of 5 mm were used. The heat transfer medium-antifreeze mixture with water was circulated through copper tube. This liquid flowed through circulating thermostat Julabo FP40 that also regulated the temperature of the liquid.

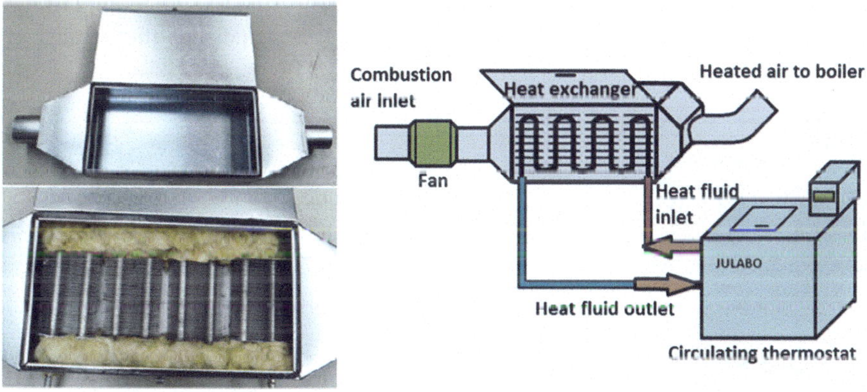

Fig. 2 The scheme of the device for combustion air temperature changing

3.3 Combustion Air Relative Humidity

From previous work [21] results it is seen that changing ambient temperature and pressure changes the usable heat slightly, while air humidity does not change it significantly. Furnaces are usually operated with an excess air ratio of $\lambda \geq 1.5$ to allow a complete combustion. Due to the difference between flue gas and ambient temperature and due to the air humidity which is usually below saturation, flue gas losses increase with increasing excess air ratio [21].

Regulation of relative humidity of combustion air was realised before the inlet to experimental boiler by using polystyrene chamber (Fig. 3). This chamber with the size of 100 cm × 100 cm × 150 cm with extra walls is used for better mixing of combustion air with water vapour. The vapour was generated by steam humidifier. It was regulated by controller of relative humidity in the inlet of combustion air to experimental boiler. The ambient air had 23 °C and 49% relative humidity. During the experiments relative humidifies of combustion air in the range of 50–100% gradually increased in steps of 10%. Mollier h-x diagram was used for controlling the amount of added water vapour. All other settings of boiler or ambient conditions were constant during all measurements.

3.4 Fuel Moisture

The efficiency of heat production from the wood biomass depends on both the design of the heat generator and the energy properties of wood biomass. Energy properties of the wood biomass are significantly dependent on its humidity content, which adversely affects not only the basic energy properties of biomass, such as combustion heat and heating value, but also the combustion process in the furnace, namely flame temperature, volume of the produced flue gas and temperature of the dew point of flue gas. The use of flue gas heat content—the rate of cooling of the flue gases prior to their diversion into the atmosphere—and thus the volume of

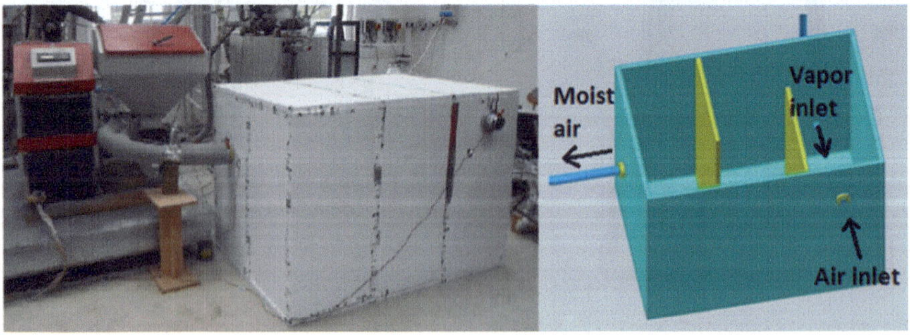

Fig. 3 The principle of combustion air humidification

the flue gas loss are dependent on the design solutions of the exchanger of heat generator. Nowadays, the boilers for combustion of wood biomass with medium output reach the thermal efficiency of 75–85% and modern wood-fired boiler units with guaranteed energy properties reach the efficiency of 92%. The standard boiler heat losses are flue gas loss, loss in the form of a volatile matter and combustible in solid residues (which are also called the chemical and mechanical unburned products), loss by thermal radiation and convection of heat from the surface of the boiler and finally operating losses. The flue gas loss in boilers utilising combustion of fossil fuels and biofuels is the biggest loss [22].

Fuel moisture change was realised of standard wood pellets. In standard conditions they have 9.98% moisture. To achieve lower moisture (5% and 0%) the wood pellets were dried in drying oven at a temperature of 105 ± 2 °C. For increasing the fuel moisture (10–30% gradually increased in steps of 5%) water was added as needed. These wood pellet samples were burned in experimental boiler during standard conditions with nominal settings of regulation.

3.5 *Fuel Feeding*

Fuel feeding influences the quality of combustion because in combination with air supply the production of emissions, mainly CO, is influenced. Air supply and oxygen concentration changes in flue gas have an important role in the combustion process. The minimum amount of CO and the highest efficiency of the boiler could be achieved at the concentration of oxygen in the flue gas of 6–7%. Increasing or decreasing the concentration of oxygen in the flue gases promotes growth of the CO concentration and a decrease in the efficiency of the pellet boiler [23].

The experimental boiler contains the regulation of fuel supply by screw feeder. The regulation controlled the electric motor with the gearbox. The regulation is of two positions—start or stop—screw feeder supplies wood pellets or not. In the control algorithm it is possible to set the time of feeding and time of pause of feeder. The standard setting based on previous measurements [24] was 18 s for feeding and 25 s for pause.

During the experiments three series of testing were realised. In the first the pause time of 25 s was constant and the feeding time was changing (18 s, 15 s, 12 s and 9 s). During the second series the feeding time was 18 s constant and the pause time was changing (25 s, 28 s, 31 s and 34 s). The last series of measurements was realised with constant time difference of 7 s between the feeding time and pause time (15 s/22 s, 18 s/25 s, 20 s/27 s and 30 s/37 s). All other conditions, mainly the amount and temperature of combustion air, were constant.

4 Results

4.1 The Results of the Impact of Amount of Combustion Air

Figure 4 shows the average values of heat power of heat source, CO and NO_X concentration depending on the amount of combustion air which was set by fan speed. The combustion air mass flow rate increased proportionally to the increasing fan speed up to 60% of rated speed. A smaller increase in fan mass flow rate was recorded between 60% and 100% of the nominal speed. The results show that the heat power of heat source increased proportionally with the increasing amount of combustion air or fan speed up to 60% of rated speed—mass flow rate was approximately 108 $kg.h^{-1}$. Further increase of fan speed caused decreasing of heat power of heat source. The production of CO was the lowest during burning of wood pellets with 30% of rated speed of fan setting—mass flow rate was approximately 90 $kg.h^{-1}$. Minimal fan speed setting caused several times higher CO production. It was caused due to incomplete combustion process with lack of oxygen in some areas of combustion chamber. NO_X concentration was not too affected by the amount of combustion air except 10% of rated speed of fan setting and reached values around 170 $mg.m^{-3}$.

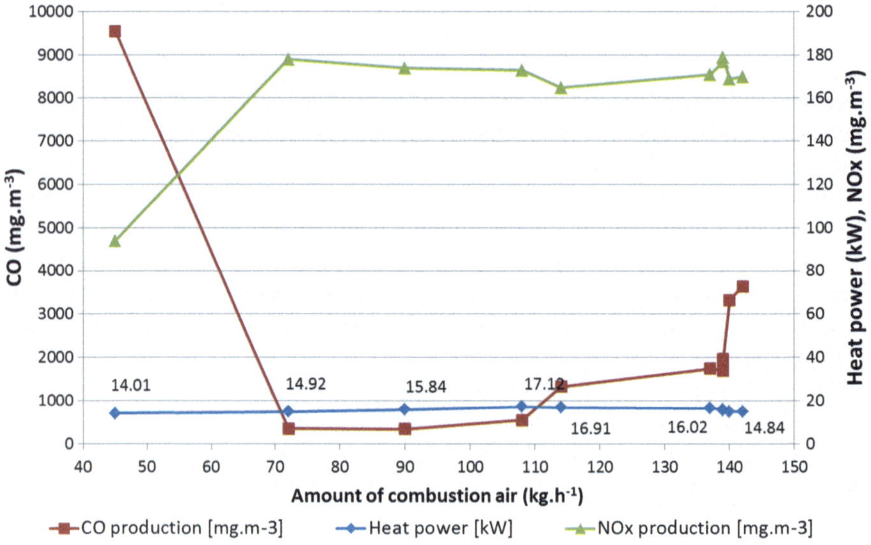

Fig. 4 The impact of combustion air amount

4.2 *The Results of the Impact of Combustion Air Temperature*

The results of experiments with average values of heat power, CO and NO_X concentration with various combustion air temperatures are shown in Fig. 5. The heat power of tested heat source increased, CO concentration decreased and NO_X concentration increased with the increasing temperature of combustion air. The higher heat power was due to energy content in air—warmer temperature has higher energy content. This can also help CO production that was the lowest during the highest temperature of combustion air when the highest NO_X concentration was also observed. This NO_X concentration increase was caused by the flame temperature increase when NO_X is formed in higher extent [25, 26].

4.3 *The Results of the Impact of Combustion Air Relative Humidity*

Experimental results of heat power of tested boiler, CO and NO_X production during wood pellet burning with various relative humidity of combustion air are presented in Fig. 6. Higher water content in combustion air causes lower flame temperature and lower temperature in combustion chamber of boiler. This was reflected by decreasing of heat power of boiler, increasing of CO production and decreasing of NO_X concentration in flue gases. This is due to the reduction of the flammability amount to the detriment of water content in the combustion process. The result is

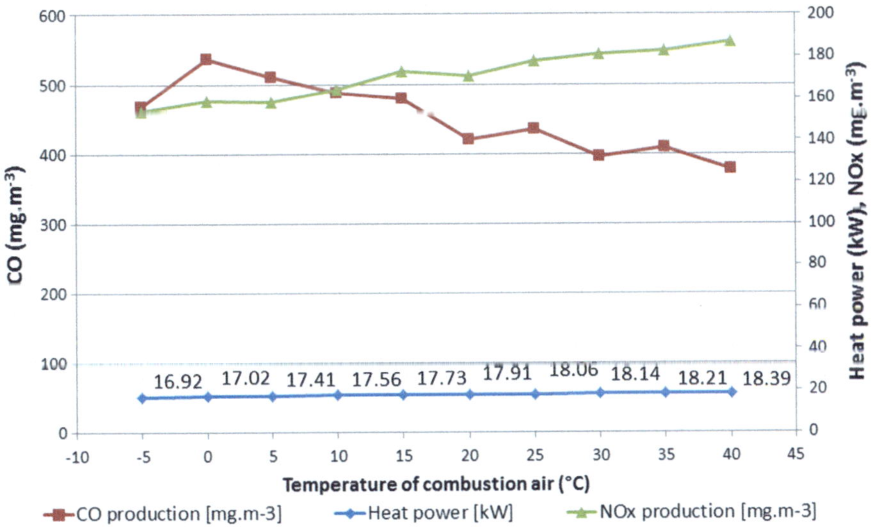

Fig. 5 The impact of combustion air temperature

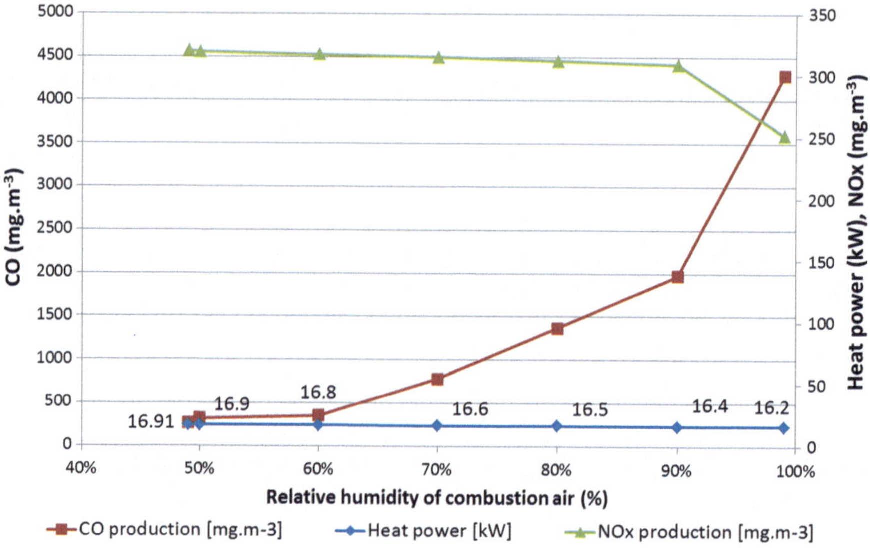

Fig. 6 The impact of combustion air relative humidity

therefore lower energy released during wood pellet burning. The increasing relative humidity of combustion air from 49% to 99% caused 4.2% decrease of heat power and seven times higher CO production.

4.4 The Results of the Impact of Fuel Moisture

In Fig. 7 are shown the results of the impact of fuel moisture in thermal and emission parameters of wood pellet boiler. The fuel moisture had the highest impact on combustion process in comparison with other tested aspects. The highest heat power was observed during combustion of wood pellets without water content. In this case the lowest average CO concentration and the highest average NO_X concentration in flue gases were also recorded. With increasing fuel moisture the heat power decreased, CO production increased and NO_X production decreased. The heat power has decreased by more than 50% and CO production has increased more than 12 times in comparison with the results during burning of standard wood pellets with 10% moisture. These negative consequences arise for lower flammability amount with incomplete combustion process with lack of oxygen in some areas of combustion chamber.

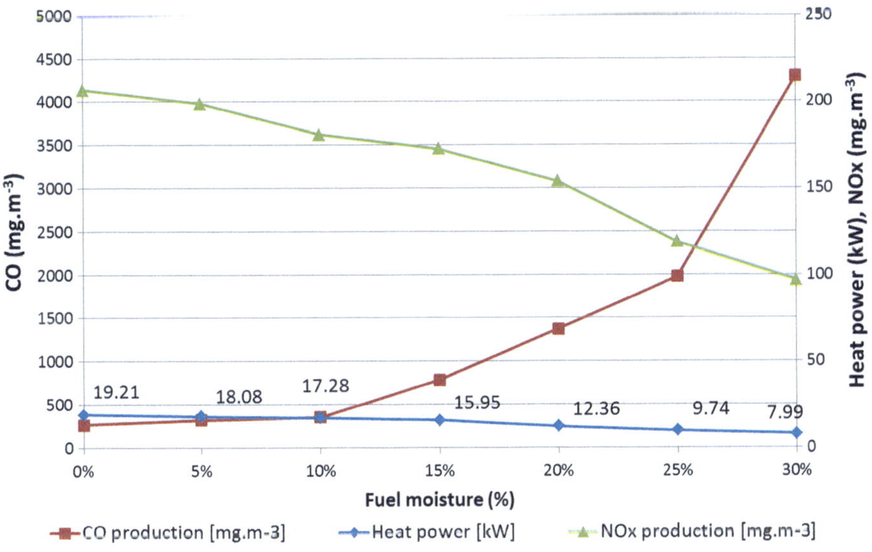

Fig. 7 The impact of fuel moisture

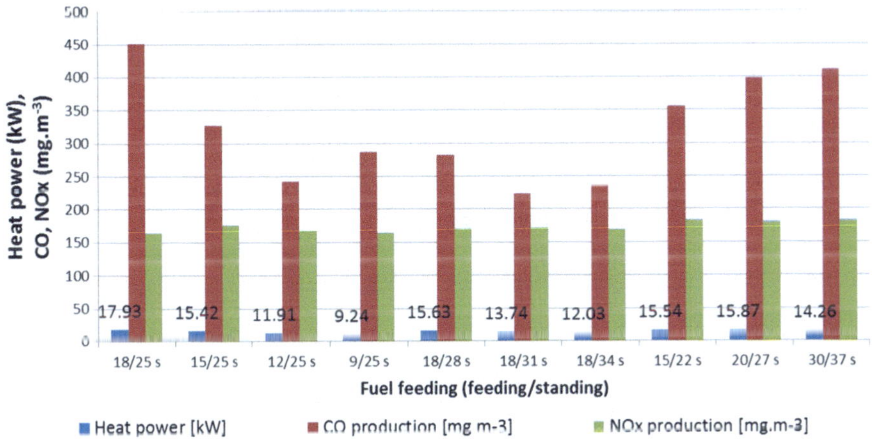

Fig. 8 The impact of fuel feeding

4.5 *The Results of the Impact of Fuel Feeding*

Average values of heat power of boiler, CO and NO_X production depending on fuel feeding are shown in Fig. 8. The results of the measurements show that the combination of feeding time and pause time has a major impact on the heat power of boiler and production of emissions. Decreasing the time of wood pellet feeding in combination with increasing the pause time has a linear impact on the heat power decrease. The highest observed heat power was measured during the setting of 18 s/15 s. During this setting the highest CO concentration in flue gases was also measured. The CO production was dependent on the combination of fuel and combustion air fuel. If there was a higher fuel amount (higher wood pellet feeding) the highest was the production of CO.

5 Conclusions

The measurement results showed that it is possible to influence the process of wood pellet combustion by many factors. These aspects have a major effect on the heat power of the heat source and also on the production of emissions, mainly formation of CO and NO_X.

The obtained results showed that fuel moisture has the highest effect on the combustion process. This is the reason for the need to use the driest fuel in real conditions. Most wood pellets meet this requirement but other solid fuels are mostly wet and need to be naturally dried before burning. This improvement of fuel can also decrease the CO production.

The combustion air amount and its properties has the most significant impact on heat power and production of emissions. For optimal combustion of biofuels like wood pellets it is necessary to ensure appropriate excess air given by the combustion stoichiometry and also the low relative humidity and high temperature of combustion air. This state is shown in Fig. 9 where A area indicates the optimal combustion conditions. These conditions allow effective burning of standard solid

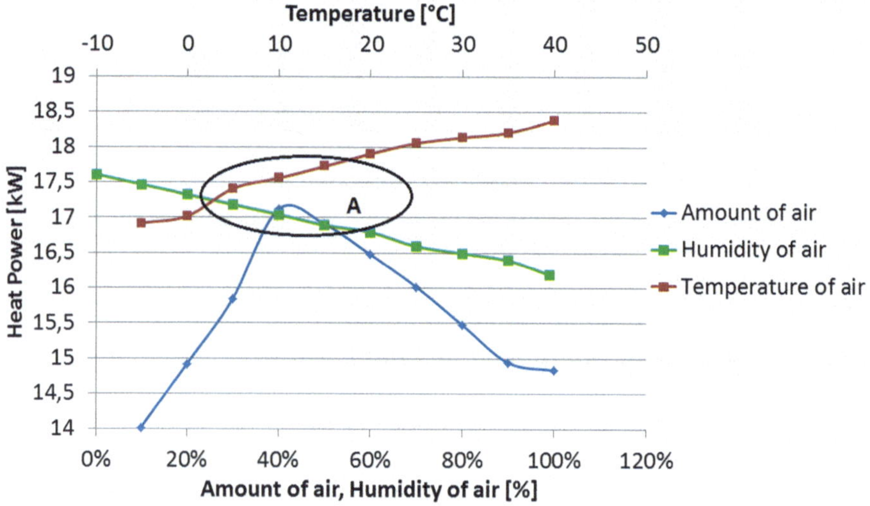

Fig. 9 The dependence of heat power on various aspects

fuels with low emission production. For wood pellets the temperature should be approximately 20 °C and relative humidity about 40–60% with optimal excess of air.

Based on the observed results it could be recommended to operate the heat sources in accordance with the manufacturer's instructions at optimal settings with quality fuel and standard combustion air.

Acknowledgement This work was created under the projects KEGA 033ŽU-4/2018 "The heat sources and pollution" and APVV 15-790 "Optimization of combustion of biomass with low temperature melting behavior of ash".

References

1. Lacinák, M., & Ristvej, J. (2017). Smart city, safety and security. *Procedia Engineering, 192*, 522–527.
2. Ristvej, J., Zagorecki, A., Holla, K., Simak, L., & Titko, M. (2013). Modelling, simulation and information systems as a tool to support decision-making process in crisis management. In Modelling and Simulation 2013—European Simulation and Modelling Conference, ESM 2013 (pp. 71–76).
3. Pilat, P., Patsch, M., & Malcho, M. (2012). Solar heat utilization for adsorption cooling device. *EPJ Web of Conferences, 25*, 01074.
4. Nosek, R., Jandacka, J., & Szlek, A. (2011). Simulation of coal combustion process in small boiler. *AIP Conference Proceedings, 1337*, 43.
5. Deliiski, N., Dzurenda, L., & Brezin, V. (2013). Calculation of the heat energy needed for melting of the ice in wood materials for veneer production. *Acta Facultatis Xylologiae Zvolen, 55*(2), 21–32.
6. Zagorecki, K., J. Ristvej, & K. Klupa. (2015). Analytics for protecting critical infrastructure, 2015, In: *Communications—Scientific Letters of the University of Zilina*, 17(1), (pp. 111–115).
7. Sadaka, S., Johnson, D. M. (2014). Biomass combustion, agriculture and natural resources, 1156.
8. Vitazek, I., Vitazkova, B., & Ploth, J. (2013). Production of gas emissions from biomass heat source. *Engineering Mechanics, 20*(3/4), 289–298. ISSN 1805-4633.
9. Taus, P., & Tausova, M. (2009). Economic analysis of FV power plants according to installed performance. *Acta Montanistica Slovaca, 14*(1).
10. Dzurenda, L. (2004). Emission of NO_2 from the combustion process of wet wood and bark. *Drvna Industrija, 55*(1), 19–24.
11. STN EN 14961-2. (2011). Solid biofuels. Fuel specifications and classes. Part 2: wood pellets for non-industrial use.
12. STN EN 14774. (2010). Solid biofuels. Determination of moisture content. Oven dry method.
13. STN EN 14775. (2010). Solid biofuels. Determination of ash content.
14. STN EN 15210. (2010). Solid biofuels. Determination of mechanical durability of pellets and briquettes.
15. STN EN 14918. (2010). Solid biofuels. Determination of calorific value.
16. STN EN 15103. (2010). Solid biofuels. Determination of bulk density.
17. STN EN 303—5. (2012). Heating boilers. Part 5: heating boilers for solid fuels, manually and automatically stoked nominal heat output of up to 500 kW. Terminology, requirements, testing and marking.
18. Nemec, P., Caja, A., & Malcho, M. (2013). Mathematical model for heat transfer limitations of heat pipe. *Mathematical and Computer Modelling, 57*(1-2), 126–136.

19. Horak, J., Jankovska, Z., Straka, F., Buryan, P., Kubesa, P., Hopan, F., & Krpec, K. (2013). Problems of determination of characteristic temperatures of biomass ash fusibility (in Czech). *Chem, 107*(6), 502–509.
20. Industrial Technologies Program. (2004). Improving process heating system performance—sourcebook for industry.
21. Neuenschwander, P., Good, J., Nussbaumer, T. H. Combustion efficiency in biomass furnaces with flue gas condensation, biomass for energy and industry, 10th European Conference and Technology Exhibition June 8–11 1998, Würzburg.
22. Dzurenda, L., & Banski, A. (2015). Dependence of the boiler flue gas losses on humidity of wood biomass. *Archives of Thermodynamics, 36*(4), 77–86.
23. Kirsanovs, V., Žandeckis, A., Veidenbergs, I., Blumbergs, I., Gedrovičs, M., & Blumberga, D. (2014). Experimental study on optimisation of the burning process in a small scale pellet boiler due to air supply improvement. *Agronomy Research, 12*(2), 499–510.
24. Werle, S. (2014). Impact of feedstock properties and operating conditions on sewage sludge gasification in a fixed bed gasifier. *Waste Management & Research, 32*(10), 954–960.
25. Rimar, M., Fedak, M., Dudinak, P. (2007). Process of nitrogen oxides generation in pressure burners, manufacturing and industrial engineering. issue 3, pp. 81–84.
26. Adamczyk, W., Werle, S., & Ryfa, A. (2014). Application of the computational method for predicting NOx reduction within large scale coal-fired boiler. *Applied Thermal Engineering, 73*, 341–348.

The Importance of Housing Companies in the Development of Sustainable Housing and Increase in Competitiveness

Janka Babelová and Jarmila Machajdikova

1 Introduction

An essential part of market economy is a functional real estate market, the mechanism of which is based on market principles. It should create conditions to utilise area potential as well as to attract qualified workforce into urban area. Housing market is part of real estate market. The development of flat market in Slovakia and an enormous interest of foreign developers to enter this market after the year 2000 influenced mainly the influx of direct foreign investments [1], growth of gross domestic product (GDP) and offers for investment financing from commercial banks. Those were mainly available investment sources for developers for building of flats determined for ownership as well as loans and possibility to gain subsidies from the state for citizens. The building of rental housing fund was forgotten as well as the creation of suitable conditions for origination of housing companies, which play an important role in the development of rental housing [2].

We consider housing companies as an important institutional condition for the development and functioning of housing market—prevailingly in rental flats. They are part of institutional environment while a vital role is played by partnership of the state with villages, towns and organisations providing rental housing.

Foreign experience shows that sustainable development of housing including rental housing, which is considered a remarkable phenomenon, might be provided by housing companies on non-profitable principle as a new type of a developer.

Housing organisations create a particular category of non-profitable organisations and at the same time are important institutions that function mainly in the field of rental housing. Their activity is the most important in the European countries as

J. Babelová (✉) · J. Machajdikova
Institute of Management, Slovak University of Technology in Bratislava, Bratislava, Slovakia
e-mail: jarmila.machajdikova@stuba.sk

© Springer Nature Switzerland AG 2020

D. Cagáňová, N. Horňáková (eds.), *Mobility Internet of Things 2018*,
EAI/Springer Innovations in Communication and Computing,
https://doi.org/10.1007/978-3-030-30911-4_11

Austria, France, Great Britain, the Netherlands, Denmark and Sweden. For example, public housing companies in Sweden have no specific social housing sector. The problems solved by social housing in other countries are in Sweden addressed by the municipal housing sector (in this text referred to as public housing). Swedish municipalities are responsible for providing housing to their residents. Municipality-owned dwellings emanate from a long tradition in the welfare state [3].

Several common signs are typical for non-profitable organisations acting in these countries. Basic principles of their structure and functioning are similar—their legal status, membership in them, their main activity, financing, renters and rent. From the legal point of view, non-profitable housing organisations have legal status, as well as independent legal entities. Their form is either a cooperative or a particular type of trading company. Mainly they pursue public interest, so they belong to the public sector. Besides legal entities (e.g. charity organisations) also citizens who are touched by housing or want to join welfare activities might be members of non-profitable organisations. The main activity and aim of these organisations are to provide affordable rental housing. To achieve affordability in standard level flats, a non-profitable principle is needed, on which these organisations function—they do not try to have profit, but essential is economic (cost) principle. Besides this main activity, they might also have side activity, which is housing building for sale. Their main activity is financed from income received from the sale of these flats [4].

From historical point of view, the first housing organisations in Europe that have a character of non-profitable organisations originated at the turn of the nineteenth and twentieth centuries. In this period the need for housing was initiated by the growth of population in towns caused by industrialisation of the society, foundation of many factories and interest of their owners to create suitable housing conditions for their employees.

2 Housing Companies and Sustainable Housing

Housing institutions do not function in vacuum, but for their effective functioning there are institutional conditions in developed economies. Housing societies are in the majority of countries financially supported by citizens and the central government. The most frequent forms of support are accessibility of public building sites for the building, investment subsidies and tax relief.

As an example, we might present Austria, which is the first from the view of realisation of environmentally and energetically saving housing. It is thanks to well-created and effectively functioning system of housing policy and well-controlled housing societies. In housing building, they use quality building materials, in no small extent renewable sources of energy—solar, wind, geothermal, water energy and energy from biomass.

The reason why we present Austria as an example of sustainability of housing is the amount of state subsidies resulting from the orientation of the country as a whole in ecology. Regarding the fact that heat production carries with itself increase in air emissions, building of energetically non-demanding houses/flats with the aim to use renewable sources of energy has a strong support in Austrian legislative.

It needs to be emphasised that Austria has a leading position in the international role in realisation of environmental and energy-saving housing. Low energy standard of houses (including rental) is the minimum standard for newly built buildings in all federal countries of Austria. Some federal countries implemented even stronger rules for building and reconstruction of buildings, the so-called passive standard. It is a building with the need of heat for heating lower than 15 kWh/m^2 of useful area, and at the same time need of primary energy is lower than 120 kWh/m^2 [5].

The fact that housing companies coordinate its activities with the towns might be visible also in smaller assumption of socially excluded locality.

The sites determined for residential building are localised mainly in stabilised areas of towns and villages. Agricultural soil is not taken for housing building, and it also prevents the origination of hot islands [6, 7].

In exchange for the help the non-profitable societies receive from the state, they are obliged to keep the requirements of public administration in the standard of built flats and invested costs. It usually lowers the price of rental houses compared to the price of other flats offered in the market.

3 Materials and Methods

This chapter aims the analysis and synthesis of knowledge available from professional publications of authors dealing with non-profitable housing sector with their determination up to contribution of housing companies for the development of rental housing sector in the frame of chosen European states. The statistical data processed in this chapter was obtained from the EUROSTAT database, the Statistical Office of the Slovak Republic, the Ministry of Transport and Construction of the Slovak Republic.

4 The Importance of the Development of Housing Companies in the Slovak Republic

Foundation of non profitable organisations, including housing ones, in Slovakia, was allowed by legislative since 1997 based on the Act No. 213/1997 Coll. on non-profitable organisations providing generally beneficial services [8]. A non-profitable organisation is a legal entity based on this act that provides generally beneficial services determined in advance and for all users with same conditions, and its profit must not be used in favour of founders or members of organisation and nor its employees, but it must be used entirely for the provision of generally beneficial services. It means that created profit must be reinvested back to the subject of company activity. The mentioned act in nine points generally defines beneficial services, while the last is 'the provision of housing, administration, maintenance and renewal of housing stock'.

In 2004 it also allowed for the origination of the first two non-profitable housing organisations in Slovakia. In 2004 in September the society for the development of housing in Bratislava, n.o., originated. The founders were the capital city Bratislava with 90% share in the registered capital and Istrobanka, a.s., with 10% share. Subsequently, in January 2005, another one originated in Martin under the name of Matra, n.o., as 100% daughter company of the town Martin. The stated non-profitable housing companies presented a new approach in the provision of rental housing with focus on complex provision of housing—building of rental houses, flats into the ownership as well as their subsequent administration and maintenance. Foundation of the non-profitable housing company in Bratislava was due to significant financial possibilities as experience of workers of founding bank as well as positive approach and willingness of the management of the capital city Bratislava [9]. In Martin, an important contribution to the foundation of the company was the experience of Dutch professionals as well as the grant project of the Dutch kingdom 'Possibilities of development of non-profitable sector in Slovakia'. Primary intention of the given project was the support of the development of rental non-profitable sector in Slovakia with the usage of know-how in the administration of municipal flats in the Netherlands.

4.1 Relationship Between the Housing and Labour Mobility

Mobility in the labour market reflects the ability to adjust the labour market, which has an impact on the flexibility of the economy [10]. We distinguish the spatial and structural mobility. The spatial mobility depends on commuting of employees to work; the structural mobility reflects the ability of the labour force to exchange jobs between the sectors or to adjust to a new work and qualification.

Generally accepted thesis of a low labour motility in the Slovak labour market is probably caused by low price of labour and high direct costs for the travel to work. Further, it might be assumed that historical orientation of Slovakia to agriculture conditioned the unity of people with soil; then the consequence is the low volume of sales of properties and overall low level of moving.

Non-existence or just minimal possibilities of rental housing, which at the same time induces its high price in parity to the height of salaries, is one of the obstacles for higher labour mobility. The price of rent also correlates with economic power and attractiveness of the region linked with the height of average salaries. The negatives influencing labour mobility in comparison with foreign countries (especially countries of the old EU) include the ratio of costs of travelling/salaries, worse services in the field of transport, and more and bigger rural area.

The underlying problem of competitiveness of the present Slovak economy is lack of qualified labour force. On the other hand, sufficient potential of qualified labour force is a condition for sustainable growth of economy as well as growth of the share of industrial production on GDP. However, it needs creation of conditions for stabilisation of workplaces in the industry, via which it is possible to contribute significantly also to balancing of considerable regional disparities in Slovakia. The Slovak labour market has one of the highest fluctuations of labour work in the EU, and low internal geographic mobility of workforce contributes to it [11]. The factors which hinder mobility are, besides the others, high fare and accommodation cost and mainly insufficiently developed market with rental housing (see Fig. 1). According to the data from Eurostat database, the share of people living in rented dwellings in 2016 was more as 30.0% in EU member states. By contrast, in Slovakia only 10.7% lived in rented dwellings (with reduced price 1.7% and market price 9.2%).

We consider adequate rental housing as a significant source of employment, as it has close links to national economy. Housing availability influences the labour market as it mitigates the pressure on salary increase. In this sense, it might influence export ability of branches in which less qualified workers work. In case of job loss, it is important that a citizen does not have to move to another flat which complicates the availability of several potential possibilities for a job. This fact is especially important under the conditions of globalisation and low employment security.

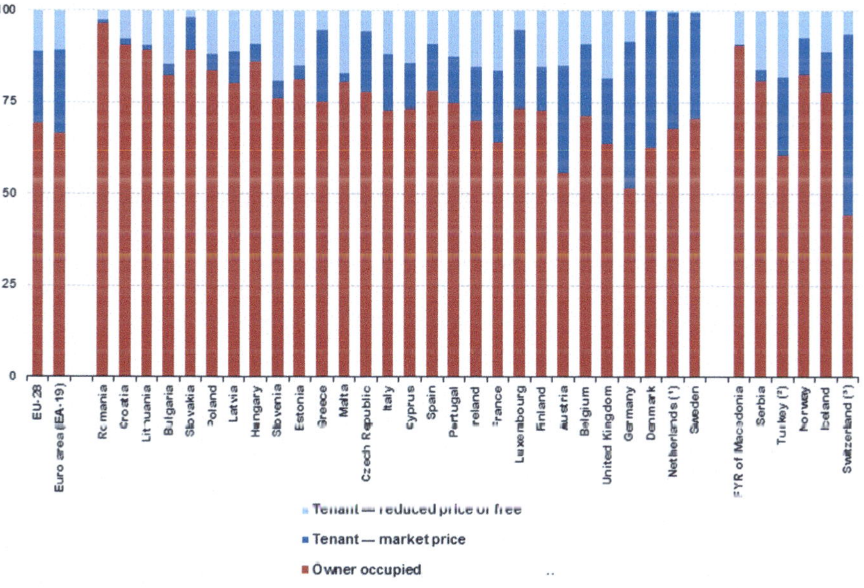

Fig. 1 Distribution of population by tenure status, 2016 (% of population). Source: [12]

4.2 Housing Mobility

Functioning of the market with flats is closely related with the housing mobility, which is closely linked with desirable and undesirable work mobility. Residential and geographical mobility contribute to the efficient matching of jobs and the allocation of human resources within the labour market [13]. We remark that in the Slovak Republic many people have become 'prisoners' of their flats, due to lack of offer of rental flats but also due to economic reasons. The requirement of increased mobility of housing (in the form of rental housing) is a condition for higher mobility of workforce and so also for better options for finding employment in the labour market.

The internal migration rate for Slovakia, measured as gross regional outflow as a percentage of working-age population, is significantly lower than in most EU countries and other advanced economies as well as neighbouring countries [14].

With the aim to provide competitiveness of all regions and support labour mobility, it is needed to diversify the offer of housing, not only with various types of the ownership and utilisation, but also with variability of the offer from the view of costs for its provision in such a way that quality housing will be available in proportion to financial sources of households.

Immediate availability of housing for employees is a necessary condition for the development of dynamic labour market; on the other hand the lack of rental houses blocks the employment development.

5 Conclusion

In the developed economies there is the effort to support spatial accordance between housing and the labour market. Regarding foreign experience, especially from Anglo-Saxon countries, it might be said that there is a positive impact of housing policy on the employment and mobility. The function of housing societies (or investment companies on the non-profitable principle) is the provision and administration of rental (as well as social) housing stock. The term social housing is generally used in the European countries; even this term is not always understood as the same. In the USA, for example, they prefer the term low- to moderate-income housing. In all countries, however, it is the facilitation of the access to housing for families with lower up to moderate income and at the same time the increase of mobility of inhabitants from the view of employers' needs.

In conclusion, it might be assumed that the creation of investment societies on the non-profitable principle 'housing societies' with the majority share of the town and setting of suitable conditions would provide flats for companies for gaining and stabilisation of their employees as well as for employees working in the state administration (besides others also for graduates from universities, who continue in their PhD studies).

We also assume that their creation in the Slovak Republic would also generate other effects—for example state subsidies would return via the following:

- Contributions of employees (who the companies would employ).
- Investment housing society (as an owner of rental house) would pay to the given town/village taxes from properties, which creates a substantial income item in the accountancy statements of the towns and villages.
- It is assumed that the company would also gain employees from other countries, who would come with their families and pay contributions and taxes (they would not work 'illegally').

Acknowledgments This chapter was supported by Slovak Research and Development Agency under the grants VEGA No. 1/0604/18 Economic Aspects of Sustainable Construction within Intelligent Urban Structures.

References

1. Sevela, M. (2016). Role of foreign direct investment stock in business cycle synchronization. In *19th international conference enterprise and competitive environment (ECE). Procedia social and behavioural sciences* vol. 220 (pp. 506–512). Brno, Czech Republic.
2. Ivanicka, K. (2006). Social housing sector and the role non-profit housing organizations. Nehnuteľnosti a Bývanie (property and housing).
3. Magnusson, L., & Turner, B. (2008). Municipal housing companies in Sweden—social by default. *Housing, Theory and Society, 25*(4), 275–296.
4. Spirkova, D., & Fulopova, C. (2016). The causal relationship between the housing and labor mobility. In *27th international-business-information-management-association conference* vol. SI–VI (pp. 2826–2832). Milan, Italy.
5. Janiss, R. (2014). Housing policy tools to support the sustainable development of housing stock. Dissertation work. Slovak University of Technology in Bratislava.
6. Spirkova, D. (2015). The causal link between land grabbing and its implications. In *26th international-business-information-management-association conference* vol. SI–VI (pp. 2723–2729). Madrid, Spain.
7. Golej, J. (2012). *Land and development process*. Bratislava: Slovak University of Technology.
8. Zákon č. 213/1997 Z. z. v platnom znení o neziskových organizáciách poskytujúcich všeobecne prospešné služby (Act No. 213/1997 Coll. on non-profit organisations providing services of public benefit. Retrieved from http://www.epi.sk/zz/2012-56
9. Spirkova, D., Zubkova, K., & Machajdikova, J. (2017). *Housing policy. Monography*. Slovakia: Slovak University of Technology in Bratislava.
10. Vagac, L. (2013). Internal labour mobility in Slovakia. European Employment Observatory.
11. Jurcova, D. (2010). *Labour migration in the Slovak Republic. Demographic research*. Bratislava, Slovakia: INFOSTAT.
12. Eurostat. Housing statistics. Retrieved from https://ec.europa.eu/eurostat/statistics-explained/index.php?title=Housing_statistics#Tenure_status
13. Henley, A., Disney, R., & Carruth, A. (1994). Job tenure and asset holdings. *Economic Journal, 104*(423), 338–349.
14. Paci, P., Tiongson, E. R., Walewski, M., Liwinski, J., & Stoilkova, M. M. (2007). *Internal labor mobility in Central Europe and the Baltic Region. World Bank Working Paper No. 105*. Washington, D.C.: The World Bank.

Smart Surveillance Systems and Their Applications

Andrei Braicov, Ivan Budanaev, Marco Cosentino, Walter Matta, Alessandro Mattiacci, Carlo Maria Medaglia, and Mircea Petic

1 Introduction

Information technology develops at a fast pace, and its progress influences the evolution of many other fields of science. It is difficult to say nowadays if there are any aspects of our modern world which have not yet been affected in some way by the IT revolution. One such area of today's society is the urban surveillance and security. Although security is paramount in its core concept, the importance of this area has substantially increased since the launch of the Global War on Terrorism military campaign in 2001. This can easily be observed with the shift of military technologies into domestic applications and the confluence of internal and external security [1].

Modern surveillance systems consist of many individual components, and represent a very complex architecture. Most common problems that arise when building such products are related with data transfer speed and reliability, data analysis, and automation of processes. The widely spread high bandwidth mobile networks and attack resistant cryptosystems provide solutions to the first two of the above requirements. Computer vision algorithms permit to analyze streams of data and detect predefined patterns to spot hostile or abnormal behavior and automatically send signals to supervisors at the monitoring stations.

This paper addresses this specific area of modern security—surveillance systems. This article is written in the framework of the Wide InTegration of sensor Networks to Enable Smart Surveillance (WITNESS) project which has the aim to design

A. Braicov · I. Budanaev · M. Petic (✉)
Tiraspol State University, Chisinau, Republic of Moldova

M. Cosentino · W. Matta · A. Mattiacci · C. M. Medaglia
Link Campus University, Rome, Italy

© Springer Nature Switzerland AG 2020
D. Cagáňová, N. Horňáková (eds.), *Mobility Internet of Things 2018*,
EAI/Springer Innovations in Communication and Computing,
https://doi.org/10.1007/978-3-030-30911-4_12

a system for urban surveillance and security to help detect, prevent, and give an efficient response to terrorist threats and attacks.

The aim of this paper is to describe the methodology of the research in the project, starting with the architecture of the surveillance system.

This article is structured as follows: First, we start with the state of the art of similar research topics. Then we will describe our main aspects of the system architecture. More details are given in the following sections, concerning sensor models.

2 Related Works

The topic of smart surveillance is very popular nowadays. Its idea came from the practical issues where video surveillance was used to monitor municipal buildings, banks, train stations, etc. Surveillance becomes more and more important because of increasing number of exceptional situations that require a high attention to the people and society [2]. Smart surveillance is the use of automatic video analysis technologies in video surveillance applications [3]. Automatic video analysis technologies usually take into account both GPS and telemetry data. As video data also contains noise, video sequences are usually divided to be processed into smaller pieces. A separate analysis of smaller video sequences allows one to create an overview of the whole video.

The current available smart surveillance infrastructure allows to configure and implement the algorithms for recording and filtering video data streams from interconnected objects on the Internet. One of the examples of state-of-the-art infrastructure are RFID frameworks. They are elaborated in such a way that they provide the functionality to monitor space (2D or 3D), identify suspicious events, and react by generating appropriate responses to the situation. RFID frameworks have been developed as a result of researchers' efforts [4] and commercial demands. Examples of open-source RFID frameworks include Mobitec [5], AspireRFID [6, 7], and the Fosstrak project [8, 9] that provide free infrastructure deployments.

Another approach would be to use Wireless Sensor Networks (WSNs). These were initially used as surveillance in military conflict zones. The first implementations of WSNs used distributed sensor networks (DSNs) technology. Just as the first sensors were quite large, their applicability was reduced as well as due to their limited wireless connectivity. Currently the sensors are significantly smaller and cheaper. This led to the implementation of sensor networks for monitoring apartments, the environment, and the use of body sensors. WSN is considered one of the most prospect technologies of the present century [6, 10].

The variety of WSNs platforms is great. There is a platform that only addresses the system as a network of sensors. Other platforms work with devices and other sensor networks connected to the WSNs. There are WSN development and monitoring systems that have limited extensibility, for example, Moteview [11] and [12]. The following tools provide development and/or programming environments for WSNs systems: Hourglass, SenseWeb, jWebDust [13], and GSN [14]. A more detailed description of the architectural particularities of the WSNs systems can be found in [13].

The effectiveness of WSNs in the surveillance process is acknowledged. However, there are approaches that seek to improve the use of WSNs by combining them with unmanned aerial vehicles (UAVs) in surveillance. UAV is a solution in situations where it is necessary to fly over dangerous areas without endangering people's lives.

Paper [15] presents a project example describing the interaction between WSNs technology and UAV tools used for border surveillance. The UAV in the given case is considered a quadcopter.

Research focused on the use of quadcopter in terrestrial surveillance is focused on identifying cost-effective solutions and preserving the same functionality. Usually a quadcopter is driven by the means of proprietary framework APIs from a laptop or a PC. A quadcopter is useful in reading the altitude to the ground, as well as in measuring air temperature, humidity, and gas composition [16].

3 Use Case Scenarios

WITNESS proposes an innovative technological solution to incidents and accidents that may occur in an unpredictable urban scenario characterized by crowded scenes with potentially complex structured man made surroundings. The typical scenarios WITNESS will cater for are those where an incident or accident has caused disruption in the normal 24/7 operation of a public space (for instance, a metro, a railway, or a bus station). In such environments a nominal flow of people can be expected and, therefore, normal behavior can be predicted. We also envisage that the public spaces of interest will be monitored by fixed cameras and by police forces.

Some examples of possible case studies are as follows:

- Natural disasters: A disaster is a serious disruption of the functioning of a community or a society involving widespread human, material, economic or environmental loss and impacts. Disasters caused by natural or technological reasons are identified as NaTech.
- General disruption of individuals: Public areas witness the presence of intoxicated individuals, usually disruptive in small groups. Such events may have an effect on the normal flow of people, for instance, in a metro or railway. Such individuals may start pushing one another and other people in the surroundings.
- Public events related to holidays, manifestations, or protests where there is an abnormal accumulation of people with high density. As noticed from latest terrorist attacks, these types of scenarios are a very attractive target for terrorists [17, 18].
- Accident: This describes a category of events that may be caused by the failure of electrical power, for instance, delaying metro or trains, or by physical accidents

happened to individuals, including suicide attempts. In such case an entire station may be closed and events may more or less slowly affect the entire area.
- Incident: This could be caused by a terrorist threat, even including hostage situations.

Specific use case scenarios of WITNESS system applications would include monitoring of the sport events, large-scale peaceful demonstrations, and violent protests. During these events, the abnormal behavior under study is sought to be well-organized groups of people who aim to destabilize public order by taking hostage among civilians, using guns, fumigants, or explosives. Depending on the event type, the size of monitored area can span between 10,000 and 300,000 m^2. The average occupancy of the monitored area is projected to be up to 20,000 people.

4 WITNESS Architecture

WITNESS implements a distributed multi-layered architecture to satisfy the operational requirements of a situational awareness and decision-making system. This approach facilitates building a flexible and pervasive enough product, ready to be automatically reconfigured and quickly redeployed when needed. The cornerstone of the system is the data, which is collected by a predefined set of sensors—wearable by police forces deployed on the grounds or sensors installed on UAVs and police vehicles. In this way, the layer responsible for collecting data in WITNESS system represents a sensor fusion grid that leverages data from multiple heterogeneous sensor nodes, including cameras, microphones, and drones. A schematic diagram of the system's architecture is depicted in Fig. 1.

Fig. 1 WITNESS architecture

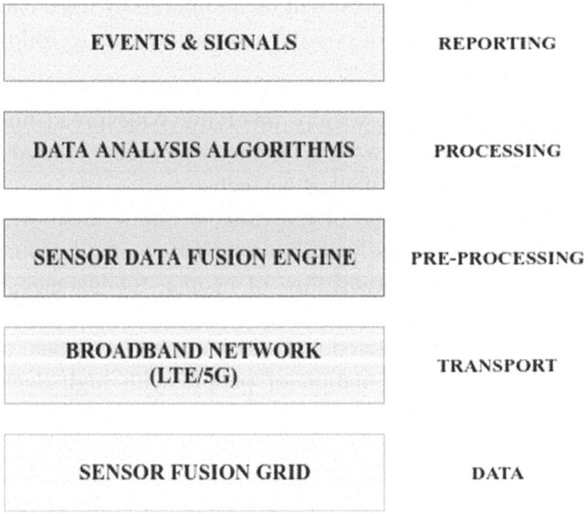

Data originating from different sources dispersed in multi-dimensional space (police forces on the field, quadcopters in the air, lags and losses during data transfer synchronization) needs to be gathered and pre-processed before it is submitted for analysis. This is a difficult task to address, since information received scales exponentially in terms of location, space, time, and means (multi-source information). Recent advances in ICT technologies can boost the efficient acquisition, fusion, and integration of information from the above sources. Sensor data fusion component will be implemented to address the above problem. Data fusion technologies involve the fusion of multi-sensory data to estimate the position, speed, attributes, and identity of the detected and flagged targets, e.g., a person, a vehicle, or an object in the operation area.

It is widely accepted that the data transfer between system components of the above architecture requires a dedicated broadband communication infrastructure (e.g., LTE, 5G technology) to be deployed in a very short time, so as to be promptly used to support the communications among security forces deployed on the field and used to gather and to process awareness data coming from monitoring devices (e.g., wearable sensors or mobile nodes) and to perform the command and coordination of the forces. In this regard, the underlying infrastructure plays a critical role in the security, processing, flow, supporting information requirements throughout the operational forces.

In particular, the proliferation of multi-purpose sensors provides ample room for sensing the physical world. For example, a host of visual processing algorithms can provide credible information about the context of a given actor (e.g., location, behavior). At the same time Wireless Sensor Networks (WSNs) provide the means for autonomic continuous information collection, in the scope of large-scale heterogeneous environments. However, even with these technologies at hand, there is need for the fusion of information captured by multiple sensors and modalities to the end of identifying situations, especially in the scope of highly distributed heterogeneous and volatile environments where people and entities (i.e., people, sensors, vehicle, etc.) may dynamically join and leave.

Within the proposed methodology, WITNESS ensures that data and information are delivered to the right place on time and optimally encoded for use by their intended recipients to take the appropriate actions at the right time. The security and safety of this information will be granted by the dedicated LTE cell. This architecture is a key enabler of Net-centric Enable Capability (NcEC) and is essential for "information superiority" and "decision superiority": it will automatically enable the security units deployed in the area of interest to communicate with whom of interest and get an immediate perception of who and where suspects and security forces are.

WITNESS will adopt a breadboard architecture enabling plug and play integration of the various components. The architecture also enables the integration of third-party components. Furthermore, this architecture will give the possibility to assign a task to a UAV and to reallocate it to another one when the first drone will have low battery or any other malfunction scenario. This need arises for the known critical issue of the battery duration of the UAVs.

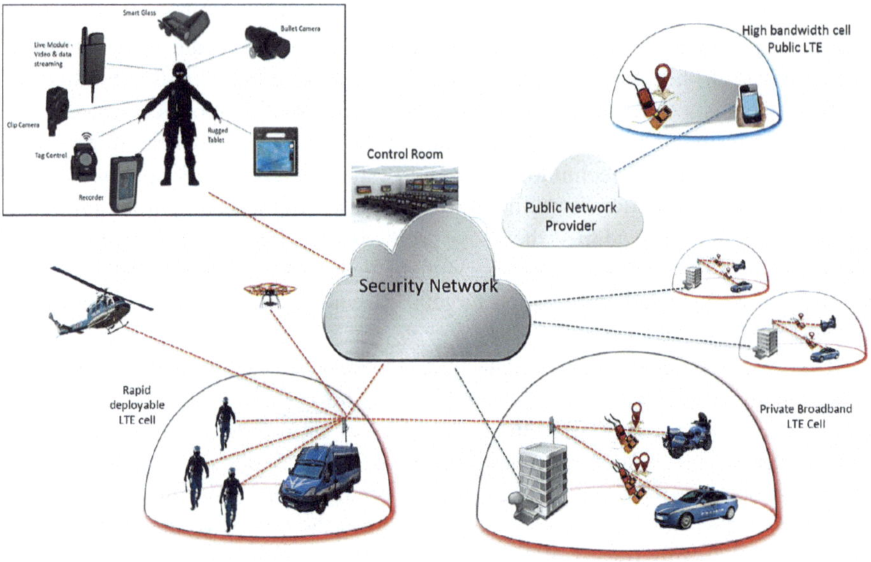

Fig. 2 WITNESS concept

This project aims to explore technological concepts and approaches already proposed for military operational fields for use in civilian security. Those concepts, and in particular the Network Enabled Capabilities (NEC), have practically never been applied in civilian applications, and are new even to the military sectors. At a global level, however, there are already focus groups questioning on how to apply the NEC philosophy to the security applications (especially in the USA), thus is imperative to start the development of a European blueprint on this topic. One of the technological impacts of WITNESS will be the generic Internet-of-Things approach towards creating a civil security C2 situation assessment and decision aiding framework (Fig. 2).

5 Methodology

Taking into account the current approaches in this field and the use cases for the system to be built the project WITNESS research will follow the scheme presented in Fig. 3.

The stages included in Fig. 3 are referred to the whole project. In this paper we refer to the following stages:

- Reference Scenarios and Operational Requirements,
- Architectural Design and Functional Description.

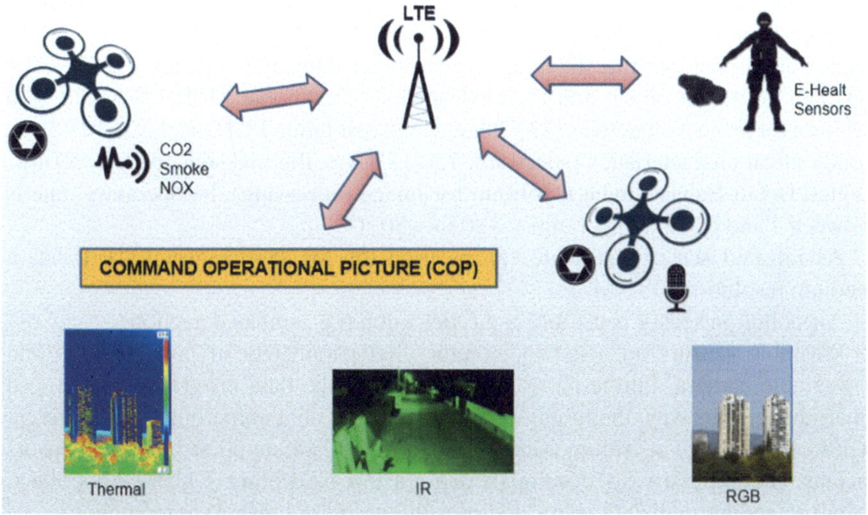

Fig. 3 WITNESS architecture details

Table 1 Parameters and devices

Parameters (sensors)	Devices
Optical sensor (video analysis)	RGB camera/thermal camera/infrared camera
Sound sensor	A microphone
Air quality sensor	SNS-MQ135

Each of the above scenarios was discussed in the corresponding sections of this paper. Below we will describe the solutions for architectural design and functional descriptions. The key elements of the system are quadcopters and different types of sensors.

In order to obtain a more accurate result we need a complete supply of important parameters for the proper determination of the instant situation. Therefore, the parameters to monitor are presented in Table 1.

The UAV solution for the architecture described can be a quadcopter, i.e., DYS D800 X4 Professional Multi-Rotor [19]. It has a maximum payload of up to 6.5 kg, which allows porting of several types of sensors. Flight time—up to 15 min. A platform NVIDIA Jetson [20] can be used to build the Artificial Intelligence application that will process the data provided by the sensors.

Another UAV solution candidate is Phantom2 [21]. Having a lower price, it provides higher flight time—up to 25 min. However, its maximum payload is only 1 kg.

Third candidate is a piloted quadcopter, with the A3-PRO flight controller [22]. Its flight time is 22 min, maximum payload is 1 kg, and its max speed is 22 m/s. Its advantages are the different available development platforms (Linux, ROS, QT) and the different programmable functionalities available.

The thermal camera is useful for detecting people in low visibility conditions (total darkness, fog, or smoke). It can see through smoke or light mist, it does not require additional lighting, the image being obtained due to temperature differences between the target object and the environment. The model FLIR LS-X can be a solution for project objectives [23]. It has a high resolution LCD display (640×480), good optical characteristics (spectrum: 7.5–13.5 μm, thermal sensitivity: <50 mK, Digital Detail Enhancement algorithm for image processing). Its operating time is between 4 and 6 h, at temperatures −20 to +50 °C.

As infrared sensor candidate can be used the ML8540 sensor [24]. It has a medium resolution (48×47).

Air pollution sensor is useful for gas detection (i.e., smoke detection).

Wearable sensors are used to monitor vital parameters of police agents and a wearable camera. Furthermore, system can use the data provided by a rugged smartphone. This way, the monitoring system relies on a numerous set of different types of sensors. The volume and complexity of data supplied by these sensors dictates the necessity to develop a data fusion algorithm, which will collect, serialize, and normalize the data, and pass it forward into the system pipeline.

6 Conclusions

WITNESS provides an opportunity for companies to increase the competitiveness of the WSN (wireless sensor network) industry by developing novel sensors so far never exploited in this field.

One of the project objectives is the definition of tools, technologies, and methods that will facilitate the countering to attacks and critical situations. This study will lead to some conclusions about the state-of-the-art resources and, hopefully, to the definition of new methods that will improve the ability to develop a monitoring solution in short time. The results obtained will be used in further stages of the WITNESS project.

Acknowledgements This research is sponsored by the NATO SPS Programme under the funding NATO SPS G5437—WITNESS—Wide InTegration of sensor Networks to Enable Smart Surveillance.

References

1. Wilson, D. (2012). Military surveillance. In *Routledge handbook of surveillance studies*. London: Routledge Taylor & Francis Group.
2. Sayantani, S., & Sarmistha, N. (2014). A case study on smart surveillance application system using WSN and IP webcam. In: *Proceedings of Applications and Innovations in Mobile Computing (AIMoC)*, Kolkata. ISBN: 978-1-4799-3880-3.

3. Hampapur, A., Brown, L., Connell, J., Pankanti, S., Senior, A., & Tian, Y. (2003). Smart surveillance: Applications, technologies and implications. In: *Fourth International Conference on Information, Communications and Signal Processing, 2003 and the Fourth Pacific Rim Conference on Multimedia, Proceedings of the 2003 Joint*, Singapore. ISBN: 0-7803-8185-8.
4. Prabhu, S., Xiaoyong, S., Harish, R., Chi-Cheng, C., & Rajit, G. (2006). WinRFID a middleware for the enablement of radio frequency Identification (RFID) based applications. In: R. Shorey, M. C. Chan, W. T. Ooi, & A. Ananda (Eds.), *Invited chapter in mobile, wireless and sensor networks: Technology, applications and future directions*. New York: Wiley.
5. *MobiTeC-open source RFID middleware 1.0. Mobile technology center*. http://mobitec.ie.cuhk.edu.hk/rfid/middleware/. Accessed 24 Sept 2018.
6. *AspireRFID OW2 project*. https://projects.ow2.org/view/aspire-rfid/. Accessed 24 Sept 2018.
7. Dimitropoulos, P., & Soldatos J. (2010). RFID-enabled fully automated warehouse management: Adding the Business Context. *International Journal of Manufacturing Technology and Management, 21*(3/4), 269–288.
8. *Fosstrak: Open Source RFID software platform*. http://www.fosstrak.org. Accessed 24 Sept 2018.
9. Floerkemeier, C., Roduner, C., & Lampe, M. (2007, December). RFID application development with the Accada middleware platform. *IEEE Systems Journal, 1*(2), 82–94.
10. Shu, Y., et al. (2014). *International Electrotechnical Commission: White Paper: Internet of Things: Wireless Sensor Networks: Wireless Sensor Networks Project Team*, Geneva (p. 78). http://www.iec.ch/whitepaper/pdf/iecWP-internetofthings-LR-en.pdf. Accessed 24 Sept 2018.
11. *MoteWorks getting started guide. Revision G* (2012, March). *Massachusetts: MEMSIC*. http://www.memsic.com/userfiles/files/User-Manuals/moteworks-getting-started-guide.pdf. Accessed 24 Sept 2018.
12. Ritter, H. (2018). *Short presentation of ScatterWeb software and hardware for next generation wireless networks*. http://files.messe.de/cmsdb/001/14916.pdf. Accessed 24 Sept 2018.
13. Chatzigiannakis, I., Mylonas, G., & Nikoletseas, S. (2005). jWebDust: A java-based generic application environment for wireless sensor networks. In: *Proceedings of the First International Conference on Distributed Computing in Sensor Systems (DCOSS 05)* (pp. 376–386).
14. Aberer, K., Hauswirth, M., & Salehi, A. (2006) *The global sensor networks middleware for efficient and flexible deployment and interconnection of sensor networks*. LSIR Report 2006-006.
15. Berrahal, S., Kim, J., Rekhis, S., Boudriga N., Wilkins, D., & Acevedo J. (2016, March). Border surveillance monitoring using Quadcopter UAV-Aided wireless sensor networks. *Journal of Communications Software and Systems, 12*(1), 67–82.
16. Borah, D. R., Debnath, L., & Gogoi, M. (2016). A review on Quadcopter Surveillance and Control. *Journal of Engineering Technology, 4*(1), 116–119. ISSN: 2348–7305.
17. *Country reports on terrorism 2016*. https://www.state.gov/j/ct/rls/crt/2016/. Accessed 24 Sept 2018.
18. *Global terrorism index 2016*. http://economicsandpeace.org/wp-content/uploads/2016/11/Global-Terrorism-Index-2016.2.pdf. Accessed 24 Sept 2018.
19. *DYS D800 X4 professional multi-rotor package for aerial photography and heavy lift*. https://www.unmannedtechshop.co.uk/dys-d800-x4-professional-multi-rotor-package-for-aerial-photography-and-heavy-lift-pnf/. Accessed 24 Sept 2018.
20. *NVIDIA Jetson TX2 module*. https://developer.nvidia.com/embedded/buy/jetson-tx2. Accessed 24 Sept 2018.
21. *Phantom 2 specifications*. https://www.dji.com/phantom-2. Accessed 24 Sept 2018.
22. *A3-PRO flight controller*. https://www.dji.com/a3. Accessed 24 Sept 2018.
23. *Tactical Handheld thermal monocular FLIR LSX model LS-X*. https://www.flir.com/products/lsx-r/. Accessed 24 Sept 2018.
24. *Lapis semiconductor infrared image sensor (IR sensor) ML8540*. http://www.lapis-semi.com/en/semicon/sensor/ir-sensor.html. Accessed 24 Sept 2018.

Enhancing Resilience by Understanding Smart City and Safe City Concepts and Their Common Systems

Jozef Ristvej and Maroš Lacinák

1 Introduction

To become a smart city is a vital goal of all modern cities. Today, in the world there are already many cities, successfully reaching smartness in some fields. In the Slovak Republic, some cities, such as the capital city of Bratislava, Košice, Poprad and later also Žilina and others, are advancing in reaching this goal as well, but a common understanding of what does it mean to be a smart city and what are its systems and features is still not achieved. Also, there is often lack of importance given to the concept of safe city, even though safety is with no doubt an important quality of a city.

After this introduction, the second section of this chapter describes the structure and relationships of smart city and safe city concepts as we understand them with brief explanation of every structural level and of dividing the systems into layers. Those two concepts are interconnected and systems, presented below, belong to both systems.

The third section then continues with defining concepts of smart city and also safe city. The fourth section summarises all the systems of the smart city concept, describes their smart and safe layers and further summarises features, needed in every system.

J. Ristvej (✉) · M. Lacinák
Faculty of Security Engineering, University of Žilina, Žilina, Slovakia
e-mail: Jozef.Ristvej@uniza.sk; Maros.Lacinak@fbi.uniza.sk

© Springer Nature Switzerland AG 2020

D. Cagáňová, N. Horňáková (eds.), *Mobility Internet of Things 2018*,
EAI/Springer Innovations in Communication and Computing,
https://doi.org/10.1007/978-3-030-30911-4_13

2 Structure of Concepts and Systems

For better understanding of the concepts, a structure of relations within the concepts was set up, breaking them into smaller groups. Those groups are systems, layers and features. Components are the individual solutions, the smallest parts of the hierarchy. The relationships between components, systems and layers are illustrated in Fig. 1. In this chapter, the structure and relations are explained. The actual definitions will be presented in the following chapters.

2.1 Concepts

Concept, as this term is used in our work, is a complex vision about the future condition from the point of view of specific philosophy—in our cases philosophy of smartness of the city, or its safety. Development and implementation of the concept are enacted in individual systems of the concept. Concepts of smart city and safe city are equal; they partially blend together and consist of the same systems. While identifying the systems, we were influenced by the strategies of the Finnish cities of Tampere and Helsinki [1–3] and personal communication with Mr. Haapakangas [4].

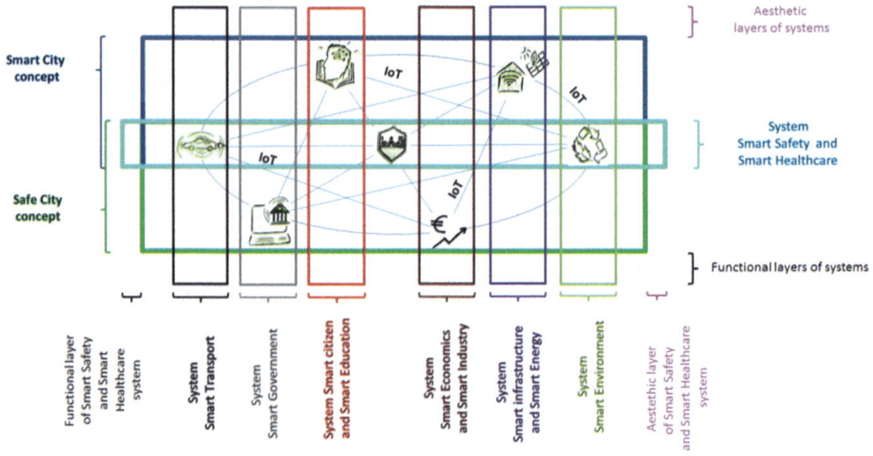

Fig. 1 Structure and relations between concepts of smart city and safe city, their common systems and separate layers (own creation)

2.2 Systems and Layers

First, there is a group of systems. Every system belongs to both concepts, is divided into four layers and consists of many features that belong to one or sometimes more layers. Layers are illustrated in Fig. 2, and smart and safe layers of every system are defined in Chap. 3.

The first layer is a functional one. It consists of components that serve for the functionality of a given system, without further addressing the safety or smartness of solutions.

The second layer is a safety layer. It includes components and parts of components that primarily serve for complex protection of users, components of the systems and their surroundings.

The third layer includes intelligent components. The research of the smart cities deals mainly with these layers of systems. As was mentioned before, concepts of smart city and safe city partially blend together, and the same can be said about some components that serve for the safety and smartness of the system at the same time.

The fourth layer covers the aesthetics of the system components. Design sometimes also belongs to a safe or smart layer, but often pursues only the aesthetic function without another meaning.

In this layout it can be seen that systems in certain ways exceed borders of concepts. The intersection of a given system with the smart city concept creates the smart layer, and the intersection of the system with the safe city concept creates the safe layer. Specific example is the system of smart safety and smart healthcare. This system itself is an intersection between smart and safe city concepts and it is also a cross-sectional system that covers safe layers of every other system.

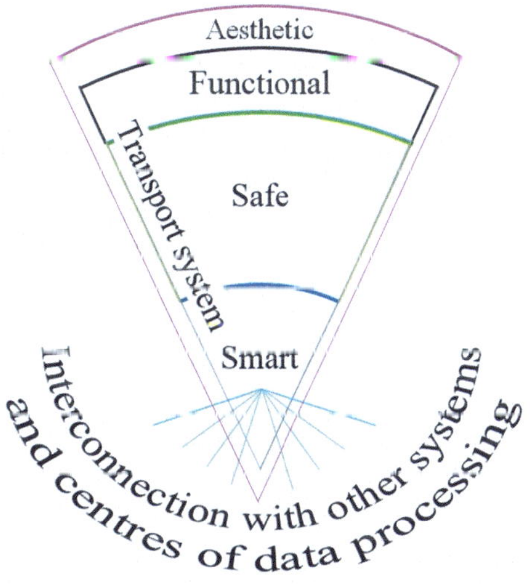

Fig. 2 System of smart transport and its layers (own creation)

2.3 Features and Components

The features are subgroup of the systems. While the systems represent different fields of research within the systems, features have direct relation with the focus of its system.

Components are individual, particular solutions for a given concept, system and feature. For example, for the concept of safe city, system of smart transport and feature "tools for prediction of impacts of various crisis phenomena on traffic and planning of evacuation and backup routes", the simulation program VR Forces® can be presented. It is used in the Laboratory of Modelling and Simulation of Crisis Phenomena in Transport in the University Science Park of the University of Žilina. This component belongs to the safe layer of the system.

Features and components are not finally an enclosed issue with fast development of technologies. In every layer, new features, that we did not yet made provision for, can emerge. Of course, features can be further divided into smaller groups, but those are not the essential subject of the research and so the chapter does not deal with deeper dividing.

3 Defining of Concepts

Our research works with two concepts: smart city and safe city.

The smart city itself was defined in our work as the city that, by the interconnection with the safe city concept and integration of technology and natural environment, enhances the effectivity of processes in every field of the city functioning to achieve sustainable development, safety and health of citizens in order to increase the life standard of citizens of the city and its region.

In some of our previous articles [5] the term safe city was used as one of the systems within smart city concept. Later during our work, we came to an opinion that this perception does not cover the whole scope of the urban safety. That is why we think that an extension to the given system is needed.

As such an extension, this chapter presents the safe city as a concept, equal to the concept of the smart city. The two concepts are partially blended together. Their intersection is a system, originally introduced as the safe city system. Later on, we changed its name into the smart safety and smart healthcare system, to make it more coherent with other systems. Its cross-sectional character illustrates that all the systems of the smart city and safe city concepts have to be safe. Systems of smart city therefore have their roots in the concept of safe city; they belong to both concepts. While the smart city deals with possibilities of component usage within systems, safe city ensures protection of society, property, environment and these components, and their informational, cybernetic and physical safety. In a way, safe city is a term that not only deals with the intelligent innovations, but also uses older useful solutions, not yet depending on the intelligent technologies.

The term safe city itself is used in the professional literature, even though it has not been so frequently and deeply addressed as smart cities. While most resources focus on reducing crime by using security cameras [6, 7] Fedorov et al. (2012) have a broader insight. In their work, creation of a unified plan of response to major crises and improvement of safety and quality of life by reduction of crime, urban violence and terrorist threats and by acceleration of responsiveness to crisis phenomena are considered as the goals of a safe city [8].

According to Finka et al. (2016), smart concept of the urban safety includes a wide range of aspects and activities linked to public spaces, from crime prevention, physical protection of the environment and accessibility to institutional and organisational aspects [9].

We consider the safe city concept as one of the basic, necessary conditions for the creation of a smart city, and in this chapter the two are defined in a similar way.

The concept of safe city by the interconnection with the safe city concept and by the use of physical, organisational, hardware and software means of protection increases the effectiveness of processes in the field of urban safety and health by minimisation of crime and terrorist threats, in order to allow citizens life in a healthy and sustainable environment and a simple access to quality healthcare and in order to reach high levels of preparedness, resilience and fast reaction ability to threatening or arising crisis phenomena.

4 Defining of System Layers

When it comes to defining the individual systems of these concepts, this chapter rather takes a closer look on respective layers. Smart layers and safe layers of every system were defined. In general, it can be said that if we want to define the system itself, the definition would always go down to satisfying all the definitions of each respective layer.

Defining the system layers was approached with three basic aspects in mind:

- Tools (What does the system use within this layer?)
- Outputs (What does the system create or what it does by the use of previously identified tools?)
- Aims (How do outputs change the city and the life of citizens? Why should we want it?)

The final definitions try to cover all of those aspects and questions. To maintain compact form of definitions, many details had to be contained in a generalised description. Except for the system of intelligent safety and intelligent healthcare, that is a cross-sectional system, every other system has defined its smart layer and safe layer separately.

Prior to every definition, many scientific and practical sources were studied to cover a wide range of previously defined approaches. Those are usually mentioned before the definition itself in a separate clause. Afterwards, the results were consulted with the members of Smart Tampere Team, as Tampere is considered as one of the smartest medium-sized cities in Europe [10].

4.1 Smart Safety and Smart Healthcare System

This system embraces all the other systems of smart city and safe city concepts, because every system has to be safe. It is also the conjunction between smart city and safe city concepts. It means that both respective definitions are partially applicable for this system.

The system of smart safety and smart healthcare by the use of smart technology, natural environment, physical, organisational, hardware and software means of protection increases the effectiveness of processes in the field of urban safety and health by minimisation of crime and terrorist threats, in order to allow citizens life in a healthy and sustainable environment and a simple access to quality healthcare and in order to reach high levels of preparedness, resilience and fast reaction ability to threatening or arising crisis phenomena and to increase the safety of usage and consequences of every system of smart and safe city concepts.

Features of the system are not divided into smart and safe layers, because of the specific, cross-sectional character, that means that the whole system belongs to the smart and also safe layer at the same time. Therefore, there are only three layers in this system: the functional layer, the core layer (smart and safe character) and the aesthetic layer. Identified features are (Fig. 3) as follows:

Because of already wide and still rapidly growing pool of possibilities, identified features are not definite, and they can be further divided into more specific fields, but it is not the concern of this chapter to go too deeply in every single system.

Fig. 3 Features within safe city concept (own creation)

4.2 Smart Transport System

Smart transport system is one of the most discussed systems. It affects all citizens and administrative and business bodies within the city as well as has many impacts, such as pollution, street congestion and travel times across the city. Its importance can also be seen from the fact that inadequate transport systems constrain a city's economy and vitality and can cause a loss of valuable space. Therefore dealing with transport needs to be tied with the housing and land-use policy. It also needs to take into account the density of the city. The most dense points are becoming the centres, where it is most convenient to gather shops and services, so that no one has to walk or drive great distances to get what is needed. From this point of view, one may say that a smart and sustainable city should be dense [11].

The most important sources that helped to define its smart and safe layer were references [11–16].

Smart transport layer is part of the smart transport system that, by the integration of smart technology, planning and construction, increases the effectiveness of transport and movement of citizens and goods in the city and surrounding region, in order to reduce transport times, noise, costs, unnecessary economic charges and traffic congestion to minimise air pollution and to allow simple access to any public area through the city, even to the people with special needs.

As well as the other systems of safe city, smart transport system too partially belongs to the concept of safe city with its safe layer.

Safe layer of the smart transport system by the use of technology, planning, construction, traffic rules and supervision minimises the amount of various traffic accidents in order to increase the safety of all traffic users and means of transportation and safety and integrity of traffic infrastructure.

The layers of the smart transport system should include the following features:

- **Functional layer:** quality roads and other transport lines, bridges and tunnels, means of transportation, road signs, relevant data
- **Safe layer:** safe parking systems, data monitoring sources from critical parts of infrastructure (supervision management), tools for prediction of impacts of various crisis phenomena on traffic and planning of evacuation and backup routes, traffic rules
- **Smart layer:** urban mass transportation (UMT) systems: UMT preference traffic, with effective back-to-back connections and possibility to adjust services in order to meet changing commuter needs, preference and support of emission-free vehicles, bicycles and pedestrians, smart parking systems, effective road signs and traffic light system with real-time adjustments, real-time informational tools for traffic consumers, intelligent tools for collection and work with data, planning and prediction of results of changes in the traffic, drone management, solutions for shortening everyday travel needs
- **Design and other features**

4.3 Smart Government System

Studied articles that helped to define layers of the smart government system were references [17–22].

Smart layer of the smart government system by the integration of smart information systems and communication networks creates transparent governance and effective digital government and with innovative policies fights financial, environmental and service challenges of the city in order to achieve long-term benefits and smart development of the city.

Safe government layer by the use of software, construction, organisational solutions and tools increases the cybernetic safety of used systems and connections; the safety of government executives, objects and facilities; as well as the safety of people and their possessions in an administered region in order to ensure the resilience of government functioning and to raise the law abidance and the trust of citizens towards the government.

Digital government should digitalise city services and help with communication within government and its departments and also between government and citizens, helping both sides to better understand the reasoning and requirements of the other and enabling participation of the citizens in the city planning. Transparent governance and quality and accessibility of services and information are important factors of smart government.

Services should be improved by mineralisation of bureaucratic delays and increasing the efficiency of bidirectional interactions while taking into account also the needs and experiences of tourists.

Features of smart government system include the following:

- **Functional layer:** the law, hierarchical and functional system of executives, government objects, facilities and offices, relevant data
- **Safe layer:** secure Internet connection and communication systems (including intranet), law enforcement, justice system
- **Smart layer:** smart development policies, user-friendly and effective digital government, effective communication systems, open data for the public
- **Design and other features**

4.4 Smart Citizen and Smart Education System

In order to create a smart city, we must realise that even the most advanced technology will not answer its purpose, if the citizens are unable to understand the vision of a smart city and to embrace its possibilities. Current education needs to be adjusted, so that it will let students grow into smart citizens, aware of their possibilities, rights and duties. In defining this wide system, the most important sources were references [23–28]. The definition of this system is a bit different from the others, because rather than using technologies and tools to create something new, it focuses more on the attitude, behaviour, knowledge, information sharing and ability of citizens to use available information.

Smart layer of the smart citizen and smart education system presents a cooperation and relationship between government, smart city stakeholders, teachers and citizens and students, where in the welcoming atmosphere teachers are to guide the students through all-round theoretical and practical knowledge: knowledge about all the smart systems and deeper about students' individual strengths and interests for a future job and daily decent, cultural and healthy life, raising a smart citizen, who is knowledgeable about the reason and the use of smart technologies to their full potential, interested in participation in the city planning, takes smarter decisions based on real-time data and by the use of smart devices participates in sustainability of development and gives feedback for further improvement.

For the most part, the safety of smart citizen and smart education system is handled by the safe city concept. Safe layer of the smart citizen and smart education system contributes to overall safety by raising citizens with a strong moral base, educated in first-aid actions, in behaviour during crisis situations and in survival basics, and in some cases also by raising experts with higher education in safety-oriented fields of study.

Features of smart citizen and smart education system are:

- **Functional layer:** students, teachers, citizen, schools, knowledge and relevant data, culture
- **Smart layer:** teaching and learning methodologies, contents of education, smart teaching and learning applications and serious games, citizens' smart habits, smart applications, technologies and devices for everyday citizen use, government, e-Government, smart city stakeholders, companies cooperating with students
- **Safe layer:** education topics (morality, first aid, crisis behaviour, survival basics, physical education), safety-oriented higher education, possibility to learn self-defence, helpline and psychologist services
- **Design and other features**

4.5 Smart Economy and Smart Industry System

Industrial companies are an important part of the development and they are needed to participate in technological progress. Most important sources for defining the smart economy and smart industry system were references [29–31].

Smart layer of smart economy and smart industry system by the use of circular, sharing and low-carbon economies, smart infrastructure and energy, smart production technologies, industrial digitalisation, artificial intelligence, virtual and augmented reality and collaboration between the academic world and companies creates an economy with effective, sustainable ways of production and flexible labour market that offers diverse economic opportunities to create innovative products for handling maintenance and operational issues, in order to save costs and resources, supporting tourism and striving for sustainable economic growth of the city, its citizens and region.

Safe layer of smart economy and smart industry system by the use of the circular economy, laws, terms of treaties, smart production technologies, safety tools like switches, aids and devices, safety protocols, automatisation, safe Internet connection and audits creates an environment, where people can develop their business, invest, buy and sell goods without the fear of being scammed or unjustly deprived of their property, and environment, where the risks of health issues for workers are minimised in order to support and stimulate the economic growth of the city, its citizens and region in the secure way.

Features of the smart economy and smart industry system are:

- **Functional layer:** companies, shops, banks, start-ups, law and commercial code, employers, employees, customers, goods, relevant data
- **Smart layer:** trade policies, sustainable production policies, smart systems for communication, accounting, energy, business opportunities, networking opportunities, tourist opportunities, requalification opportunities
- **Safe layer:** secure Internet and intranet connections, secure communicational systems, safety protocols, safety devices, switches, tools and aids
- **Design and others**

4.6 Smart Infrastructure and Smart Energy System

The infrastructure and the power to operate it are the subjects of this system. It has specific place among the systems of smart and safe city concepts, because technologies and tools, previously often working on a mechanic principle, are turning online and into electricity-powered ones. The importance of energy resilience and the infrastructure to deliver it is rising. The most important sources for defining layers of this system were references [32–36].

Smart layer of the smart infrastructure and energy system cooperates with smart environment system and by the use of renewable energy and smart energy systems enhances grey infrastructure of the city; turns buildings into responsive, energy, digital, sustainable and comfortable; provides reliable telecommunication infrastructure, smart grid and effective and sustainable energetic services; and enables centralised control of energy and heating systems, in order to achieve energetically neutral or positive infrastructure, capable of restoring waste energy, and carbon neutrality, to make the infrastructure collaborative as a whole and to conform to the needs of users.

For the most part, the safety of infrastructure of the city is again handled by the safe city concept. Safe layer of smart infrastructure and smart energy system contributes to infrastructure safety by complying to building codes, equipping the infrastructure with sensors, being able to alert responsible personnel about threats and creating backup energy systems in order to improve the resilience of infrastructure and energy systems and strengthening their ability to work continuously in harsh conditions.

Features of the smart infrastructure and smart energy system:

- **Functional layer:** grey infrastructure (buildings, energetics, heating and cooling systems, telecommunication infrastructure), relevant data
- **Smart layer:** smart energy systems, smart grid, power plants of renewable energy, energy-saving and -controlling applications, smart housing systems, waste energy-restoring systems
- **Safe layer:** building codes, sensors, monitoring the state of the infrastructure component, protective measures for critical parts of infrastructure and energy systems, backup energy systems
- **Design and others**

4.7 Smart Environment System

When it comes to an environment in the smart city concept, people mostly tend to see it as green fields and parks within the cities. Those are surely important, but not the only part of the system. The future on smart technologies is about to fill the environments, green and grey alike, with technologies that will help us to effectively interact with the environment. On defining the system and its layers, we worked mainly with references [37–41].

Smart layer of smart environment system by the use of smart devices such as sensors and communicational technologies and green infrastructure creates an enjoyable, interactive, automatic, data-generating and -evaluating environment with enough leisure public spaces in order to make the environment itself support decision-making, to allow effective control and use of the environment, suitable landscaping and development in line with the philosophy of a sustainable development, where people have enough space for their leisure activities, relaxation, sports and hobbies.

Safe layer of smart environment system by the use of smart devices such as sensors and communicational technologies creates automatic data-generating environment in order to allow the support of decision-making in response to a crisis phenomenon and by the use of natural features of green infrastructure supports nature sanitation, in order to protect ecological system (nature, biodiversity and resources) and thus creating sustainable, clean, healthy and enjoyable ecosystems that enhance physical and psychical health of citizens.

Possible sources of further information about safety within our environment are works of Nosek and Holubčík et al. [42–44].

In a wider, regional point of view, safe layer cares also about the well-being of animals and about the use of land, preserving the natural diversity and ensuring food safety of rural agricultural output.

Features of the smart environment system:

- **Functional layer:** green infrastructure, blue infrastructure (water and wastewater management), grey infrastructure (shared with smart infrastructure system), community (shared with smart citizen system), animals, relevant data, land and agricultural output management
- **Smart layer:** smart devices, smart connections, centres for controlling the parts of environment (flats, buildings, city regions ...), leisure playfields, landscaping
- **Safe layer:** nature sanitation features, rainwater management, heat and emission regulation, decision-making support systems, health-monitoring, parks, food safety management
- **Design and others**

5 Research in Progress: Methodological Procedure of Implementation of Safe City Solutions

The ways of improving safety and transforming the city into smart and safe city vary.

Some cities, like Finnish city of Tampere, are looking for new ideas by the cooperation with university students and local companies, to whom the city provides live testing areas for new ideas to be implemented. Safety is not viewed as a separate field of development, but a value that belongs to everything else and needs to be consulted with every step taken in other fields [45].

In Slovak city of Žilina, the solutions come from city officials or directly from citizens. Citizens can submit their proposals on the web page of the city, via the mobile application or directly by email. City officials are regularly attending conferences, competitions, seminars, etc. to keep in touch with the newest possibilities, trends and practices in the world and to be able to identify possibly needed and achievable solutions for the city [46]. When it comes to choosing the most suitable solution, it all boils down to the knowledge and involvement of citizens and city officials. The pool of considered solutions is limited by the awareness and willingness to step up to raise a proposal. That is why there is strong need to support continual learning of the executives and collaboration with other Slovakian but also foreign cities.

In attempts to build up concepts of smart city and safe city in their whole complexity, a methodological approach is needed. Our current research aims to establish the methodological procedure of safe city concept implementation. It consists of more phases that are summed up in Fig. 4, along with the visualisation of their connections.

For region analysis and assessment of risks, elaborated methodological procedures for the Slovak Republic are used by administrative bodies. The in-depth descriptions of those procedures, that influenced my own understanding of the processes, can be found in publications of Zánická Hollá et al. [47] and methodology of assessment of selected risks on a national level was described by Jánošíková and

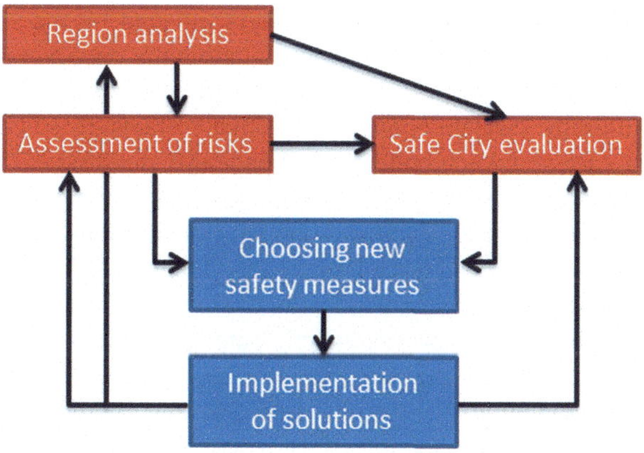

Fig. 4 Scheme of methodological procedure of safe city concept

Hudecová [48]. Those are the most influencing sources for the phases of region analysis and assessment of risk in methodological procedure in progress.

Safe city evaluation phase refers to various indexes, used to create safe city rankings around the world. Currently, there is no such complex index in use within the Slovak Republic. To create one in the third phase, I collected practices from some abroad-used indexes: Safe City index of the Economist Intelligence Unit [49], Personal Security Index of Canada by Canadian Centre for International Statistics [50] and online interactive Better Life Index by OECD [51]. Methodology for Assessment of Safety of Citizens in Regions of the Slovak Republic is also taken into account. This methodology was created by Kováč [52] and despite not becoming widely used to evaluate Slovak cities yet, it does provide valuable insight into Slovak circumstances.

Our research indicates that the least amount of studies so far was dedicated to the phase of choosing new safety measures. We see possibilities of addressing this phase with today's technologies in order to deliver effective support for decision-making, allowing the executives to choose the most suitable solutions for increasing the safety of the city.

6 Conclusion

For further development of smart cities in the Slovak Republic, definitions for the smart and safe layers of every system of the concept were presented. By this summarisation, we want to set up a base ground for understanding tools, actions and aims of systems that belong to both smart and safe concepts. For the development of crisis management and resilience in Slovak cities, solutions that will merge the philosophies of smart and safe city concepts need to be implemented and for successful implementation the relationships within a clear structure of concepts, systems, their layers and components need to be understood.

Resilience of the city is strongly affected by the smartness and safety of city systems. To create high resilience within the city, every system needs to work with high efficiency and in connection with other systems. As well as the relationships within the city are connected through systems, these connections need to be taken into account when building resilience in a smart way.

In addition, we need to highlight that in all the fields, the development of technology must be followed by the education of citizens in the use of technology. Even the most advanced smart city failed to fulfil its purpose, if the feature of smart citizen and smart education was missing. And that might become a task more challenging than one might anticipate.

The task for the next research is to complete methodological procedure of implementation of safe city solutions that should be based on above-mentioned sources with application of knowledge about smart city and safe city concepts, summarised in this chapter. The phases of the procedure should be elaborated as one, continuous process. To help to achieve this, the whole methodology will be supported by web page guide, providing aid in the process, sheet to store information and support for decision-making of the executives. In-depth elaboration on understanding smart city and safe city concepts with connection with the upcoming methodological procedure should allow effective transformation towards smart and safe cities by the implementation of the most suitable components.

Acknowledgments "With the support of the project VEGA 1/0628/18 Minimizing the level of experts' estimations subjectivity in safety practice using quantitative and qualitative methods".

"This work was supported by the research project VI20152019049 "RESILIENCE 2015: Dynamic Resilience Evaluation of Interrelated Critical Infrastructure Subsystems", supported by the Ministry of the Interior of the Czech Republic in the years 2015-2019".

"This work was jointly supported by the Slovak Research and Development Agency under the contract No. SK-CN-2017-0023".

"This work was supported by the Internal Grant Scheme of Faculty of Security Engineering, University of Žilina from the grant No. IGP201805".

The views expressed, however, are solely those of the authors and not necessarily those of the institutions with which they are affiliated or of their funding sources. The authors are solely responsible for any errors or omissions.

References

1. Blomqvist, T. (2018). *Smart Tampere* (p. 15). Tampere: Studio Libeskind.
2. Smart Tampere Ecosystem. Retrieved May 7, 2018, from http://smarttampere.fi/en
3. Helsinki Smart Region. Retrieved May 15, 2018, from https://www.helsinkismart.fi/
4. Haapakangas, V. (2018). *Concepts Smart City, Safe City and cybernetic security*. Tampere: Personal communication.
5. Lacinák, M., & Ristvej, J. (2017). Smart City, Safety and Security, In: 12th International Scientific Conference of Young Scientists on Sustainable, Modern and Safe Transport, TRANSCOM 2017. *Procedia Engineering, 192*, 522–527., ISSN: 1877-7058. https://doi.org/10.1016/j.proeng.2017.06.090.

6. Axis Communications. (2015). *A smart city is a city, where people feel safe*. Brožúra: Axis Communications.
7. Huawei: Huawei Technologies Co., Ltd. (2013). *Huawei smart city solution*. Brožúra: Huawei Technologies.
8. Fedorov, V., Ana, R., & Terekhov, A. (2012). "Safe City"– an Open and Reliable Solution for a Safe and Smart City. In: ZAO IskraUralTEL. Ekaterinburg. 6s., zo dňa 15.3.2018, dostupné na internete. Retrieved from https://pdfs.semanticscholar.org/fff6/d861abe7dc3d022c5e7b1d4f25d87d2850e8.pdf
9. Finka, M., Ondrejička, V., & Jamečný, Ľ. (2016). *Urban safety as spatial quality in smart cities*. Bratislava: Slovenská technická univerzita.
10. Giffinger, R., Fertner, Ch., Kramar, H., Pichler-Milanovic, N. & Meijers, E. Europeansmartcities 3.0 (2014) - Ranking. Retrieved May 12, 2018, from http://www.smart-cities.eu/?cid=3&ver=3
11. The Worldwatch Institute. (2016). *Can a city be sustainable? (state of the world)* (p. 448). Washington, DC: Island press,. ISBN 13: 978-1-61091-755-1, ISBN 10: 1-61091-755-3.
12. Benevolo, C., Dameri, R. P., & D'Auria, B. (2015). Smart mobility in smart city. In T. Torre et al. (Eds.), *Empowering organizations, lecture notes in information systems and organisation* (Vol. 11, pp. 13–28). https://doi.org/10.1007/978-3-319-23784-8_2.
13. Çiğdem, Ç. (2016). The key role of the smart traffic applications in smart city systems, International Conference on Advances in Science ICAS, Istanbul, https://doi.org/10.13140/RG.2.2.24617.39527.
14. Docherty, I., Marsden, G., & Anable, J. (2018). The governance of smart mobility. *Transportation Research Part A*. in press.
15. Mrityunjaya, D.H., Naveen, K., Laxmikant, S.A. & Kelagadi, H.M. (2017). Smart Transportation, International Conference ON I-SMAC 2017 (IOT in Social, Mobile, Analytics and Cloud) (I-SMAC), IEEE, ISBN 978-1-5090-3243-3.
16. Zhili, L., Xudong, J., & Wen, C. (2012). Solving the last mile problem: Ensure the success of public bicycle system in Beijing. *Procedia - Social and Behavioral Sciences, 43*, 73–78.
17. Hadden, D. What is the Smart in Smart Government. Retrieved October 25, 2018, from http://freebalance.com/open-government/what-is-the-smart-in-smart-government/
18. Lunde, K. Nordic Smart Government. Retrieved October 25, 2018, from https://nordicsmartgovernment.org/
19. Mutiara, D., Yuniarti, S., & Pratama, B. (2018). Smart governance for smart city, IOP Conference Series: Earth and Environmental. *Science, 126*, 1–10. https://doi.org/10.1088/1755-1315/126/1/012073.
20. Smart Tampere; Smart Government & Citizen. Retrieved October 25, 2018, from http://smarttampere.fi/en/smart-ecosystem-themes/smart-government-and-citizen
21. NEC: Smart Government. Retrieved October 25, 2018, from https://www.nec-enterprise.com/solutions/Smart-Government-229
22. Rochet, C. (2015). *Urban lifecycle management: A research program for smart government of smart cities* (p. 33). Marseille: Institut de management Public et Gouvernance Territoriale.
23. ByeongGuk, K. Daily edventures. Retrieved April 23, 2018, from http://dailyedventures.com/index.php/2013/03/11/korea/
24. Capdevila, I., & Zarlenga, M. I. (2015). Smart City or smart citizens? The Barcelona case. *Journal of Strategy and Management, 8*(3), 266–282.
25. Cardullo, P., & Kitchin, R. (2017). *Being a 'citizen' in the smart city: Up and down the scaffold of smart citizen participation* (p. 24). Maynooth: The Programmable City Working Paper 30.
26. Bull, R. (2016). Smart citizens for smart cities: Participating in the future. *Proceedings of The Institution of Civil Engineers – Energy, 169*(3), 93–101.
27. Zhi-Ting, Z., Ming-Hua, Y. & Riezebos, P. *A research framework of smart education,* Smart Learning Environments. Retrieved October 26, 2018, from https://slejournal.springeropen.com/articles/10.1186/s40561-016-0026-2

28. Kresin, F. (2016). Smart cities value their smart citizens. In V. Mamadouh & A. Wageningen (Eds.), *Urban Europe. Fifty Tales of the City, chapter: 22* (p. 6). Amsterdam: Amsterdam University Press. https://doi.org/10.26530/OAPEN_623610.
29. Smart Tampere: Smart Industry. Retrieved October 26, 2018, from http://smarttampere.fi/en/smart-ecosystem-themes/smart-industry
30. Investopedia, Sharing Economy. Retrieved October 26, 2018, from https://www.investopedia.com/terms/s/sharing-economy.asp#ixzz5IJNu1iSI
31. Kumar, T.M.V. & Dahiya, B. (2017). Smart economy in smart cities, In: *Smart Cities, Local Community and Socio-economic Development: The Case of Bologna*, 1st ed, Chapter 1, (pp. 3–76).
32. Christiansen, H. Ch. & Kongsmark, E. *Defining a Smart Energy City*. Retrieved October 26, 2018, from http://urbantransform.eu/about/smart-energy-city/
33. Tokody, D. Schuster, G. and Papp, J. (2015). Smart City, Smart Infrastructure, Smart Railway, *International Conference on Applied Internet and Information Technologies: ICAIIT 2015, At Zrenjanin, Serbia*, (pp. 231–234).
34. Meier, M. *Smart Energy: What is the Risk?* Retrieved October 26, 2018, from http://www.remotemagazine.com/main/articles/smart-energy-what-is-the-risk/
35. Lund, H., Østergaard, P. A., Connolly, D. & Mathiesen, B. V. *Smart energy and smart energy systems*. Retrieved October 26, 2018, from https://doi.org/10.1016/j.energy.2017.05.123
36. Cormia, R.D. *Getting Smart About Smart Energy*. Retrieved October 26, 2018, from https://www.slideshare.net/rdcormia/getting-smart-about-smart-energy
37. IGI Global: What is Smart Environment. Retrieved October 26, 2018, from https://www.igi-global.com/dictionary/smart-environment/27178
38. SlideShare: Smart environment. Retrieved October 26, 2018, from https://www.slideshare.net/franson0605/smart-environment
39. Tóth, A. (2017). Green infrastructure - its elements, areas and systems. *Enviro Magazine, 22*(3), 8–9.
40. Mikulecký, P. (2012). Smart Environments for Smart Learning, DIVAI 2012 - 9th International Scientific Conference on Distance Learning in Applied Informatics, Štúrovo, ISBN 978–80–558-0092-9.
41. Lovecek, T., et al. (2016). Level of protection of critical infrastructure in the Slovak Republic, In: *Conference: International Conference on Engineering Science and Production Management (ESPM)*, (pp. 163–168).
42. Nosek, R., et al. (2017). Analysis of paper sludge pellets for energy utilization. *BioResources, 12*(4), 7032–7040.
43. Holubčík, M., et al. (2012). Optimization of the production process of wood pellets by adding additives. *International Journal of Energy Optimization and Engineering, 1*(2), 20–40.
44. Holubčík, M., et al. (2015). Factors affecting emission concentrations in small heat sources. *Communications - Scientific Letters of the University of Zilina, 17*(3), 18–24.
45. Blomqvist, T. (2018). *Process of finding and implementation of a Smart City solutions in Tampere: personal communication.*
46. Slebodník, L'. (2019). *Process of finding and implementation of a Smart City solutions in Žilina: personal communication.*
47. Zánická Hollá, K., Šimák, L., & Ristvej, J. (2010). Systematic method of risk assessment in industrial processes. In *Risk analysis VII: Simulation and hazard mitigation* (p. 43). Southampton: WIT Press. https://doi.org/10.2495/RISK100111. ISBN 9781845644727. ISSN 1743-3517.
48. Jánošíková, G., Hudecová D. (2013). *Methodology of Assessment of Selected Risks on a National Level*, Ministry of Interior of Slovak Republic, (p. 30).
49. The Economist - Intelligence Unit: Safe Cities Index 2017. Retrieved January 8, 2019, from https://dkf1ato8y5dsg.cloudfront.net/uploads/5/82/safe-cities-index-eng-web.pdf
50. Centre for International Statistics at the Canadian Council on Social Development: Personal Security Index. (1998). CSLS Conference on the State of Living Standards and the Quality of Life in Canada. 19 p. In: M. Kováč (Ed.), *Possibilities of Measuring of Safety of the Citizens*

in the Conditions of Regions of Slovak Republic, Dissertation thesis, Faculty of Security engineering, University of Žilina, pp. 126. Retrieved January 8, 2019, from http://opac.crzp.sk/?fn=docviewChild000D4083

51. The Organisation for Economic Co-operation and Development (OECD). (2017). *Better life index*. Better life index. Accessed 8 Jan 2019.
52. Kováč, M. *Possibilities of measuring of safety of the citizens in the conditions of regions of Slovak Republic, dissertation thesis* (p. 126). Žilina: Faculty of Security engineering, University of Žilina. Retrieved January 8, 2019, from http://opac.crzp.sk/?fn=docviewChild000D4083.

Industry 4.0: The Need for Rethink Strategic Planning

Anna Saniuk

1 Introduction

As a result of a strong and very dynamically developing competition of Asia, European countries are looking for new opportunities to gain a competitive advantage in the market. Industrial enterprises need new concepts, innovative methods, and solutions that revolutionize production and increase the flexibility and efficiency of industrial enterprises.

The Industry 4.0 is one of such concepts, which is gaining increasing popularity. This concept was created in 2011 in Germany and based on high-tech strategies. It gives the opportunity to significantly increase productivity and production efficiency. The result of its implementation is very modern, fully automated factories of the future, in which machines and devices communicate with each other and exchange data using the Internet-of-Things (IoT) technology.

The new concept of Industry 4.0 offers great opportunities to develop industrial enterprises, but it is also a huge challenge and involves overcoming many fears and solving many emerging problems. New technical and technological solutions require the selection and purchase of modern machines, devices, robots, and software, which are associated with high costs. At the same time, their economic use is fast with such dynamic technical and technological development. Thus, there is a fundamental question about the profitability of production based on the Industry 4.0 concept and the need for a completely new approach to costing, planning, budgeting, and overall business management.

A. Saniuk (✉)
University of Zielona Gora, Zielona Gora, Poland
e-mail: a.saniuk@iizp.uz.zgora.pl

© Springer Nature Switzerland AG 2020

D. Cagáňová, N. Horňáková (eds.), *Mobility Internet of Things 2018*,
EAI/Springer Innovations in Communication and Computing,
https://doi.org/10.1007/978-3-030-30911-4_14

The large costs of automation and digitization cause the need for their constant monitoring and control. The indirect costs of the enterprise increase significantly and the role of depreciation costs in the total costs of the enterprise increases. This results in the need for very accurate settlement of indirect costs according to various settlement keys on many different products, tailored to individual customer needs (personalization of production).

The chapter presents the results of research on the aspects of business management related to the implementation of the Industry 4.0 concept in industrial enterprises. The aim of the research was to build a procedure of the implementation of the concept Industry 4.0 based on the most important problems of an implementation process and main areas of needed, identified changes. The novelty of the research described in the manuscript is a new procedure of implementation of the concept Industry 4.0 which helps to reduce the effects of the most important threats identified in the chapter and also connect the company's strategy with the introduction of the concept Industry 4.0.

2 Theoretical Background: Benefits of an Implementation of the Industry 4.0

The Industry 4.0 is a kind of a new trend which leads mainly to automation and data exchange in manufacturing technology. This concept includes especially cyber physical systems, Internet of Things (IoT), cloud computing, cognitive computing, etc. All these physical objects have their own unique Internet protocol (IP) addresses which allow them to communicate and exchange information among themselves and to any other external devices over the same network [4, 7].

The Industry 4.0 concept gives enterprises completely new opportunities to gain a competitive advantage in the market. The factories 4.0 must be incomparably more flexible than traditional industrial companies [1]. They have the ability of mass customization and mass personalization, which requires modern production lines. Mass customization of products to customer needs is called the long tail strategy and consists of offering and delivering nonstandard products customized to specific customer needs and requirements, sometimes to a large extent modified. They usually have a higher price resulting from higher production costs, but also many supporters for whom a product adapted to their needs is important [6].

The main aim of the Industry 4.0 solution implementation is to significantly increase efficiency and reduce the costs of the entire supply chain. Machines and devices communicate with each other online, thus enabling a quick analysis of many data received in real time [2, 3]. A detailed analysis of the benefits of the methods and technological solutions used in Industry 4.0 is presented in Table 1. The analysis presented in this table was made on the example of the metallurgical industry.

Table 1 The most important technological solutions and the resulting benefits of the introduction of the Industry 4.0 concept [8]

Implemented solution	Main benefits
Autonomous robots	– Quick (few seconds) customization of the production line to produce the next item of the product – Reduce costs associated with operating the production lines
Simulations	– Dramatic reduction time of device configuration – Improve product quality
Vertical/horizontal software integration	– Full integration of all IT systems – Fully automated value chain
Internet of Things	– Improve communication and interaction between devices and elements in progress – More centralized control – Real-time decision-making
Additive manufacturing	– Reducing the cost of materials used in production – Reduction of machine downtime – Weight reduction of components
Cyber security	– Increased communication reliability – Advanced user identification systems that provide access to devices
Cloud	– Extensive data exchange – Increased cloud efficiency – Drastic decrease in response time (up to several milliseconds)
Big data and analytics	– Real-time decision-making – Increased energy efficiency – Optimizing product quality – Improved maintenance and service
Augmented reality	– Supporting different activities – Accelerating decision-making processes

3 Research Results: Needs of Changes in Enterprise According to the Industry 4.0 Implementation

Implementation of the Industry 4.0 concept requires many different changes and also a well-prepared plan which minimizes the risk of many significant and dangerous events for the company. On the basis of the observations carried out in years 2016–2017 in the manufacturing enterprises of western Poland and the interview technique, four areas of changes were distinguished, as shown in Fig. 1: technical infrastructure (demand for machinery and equipment); automation of device operation and communication between them; employee competences (the skills of employees needed to control an automated manufacturing system); and collaboration with other companies which means building relationships and competencies together. Especially small and medium enterprises cannot build the factory 4.0 independently, so they have to organize it with other companies what allows joint manufacturing of products as a network of companies [9].

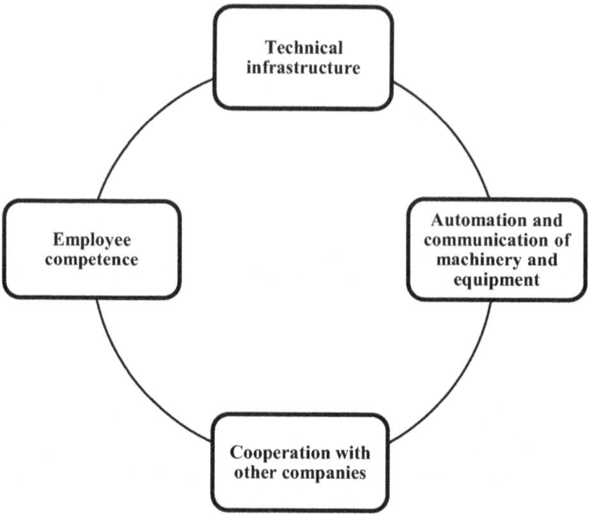

Fig. 1 Areas of company changes in the Industry 4.0 [9]

First and foremost, the companies must expand the technical infrastructure. Machines and devices are needed that can communicate in real time via the Internet. Besides, employees' competences are necessary. They have to control an automated manufacturing system. The research shows that collaboration with other companies is also important. Especially a sector of small and medium enterprises should build relationships and competencies with other companies which help to manufacture products together in the network.

A well-prepared implementation process of Industry 4.0 can reduce or sometimes completely eliminate threats and increase the chances and benefits of using this concept. The chapter presents the results of research on the aspects of business management related to the implementation of Industry 4.0 in industrial enterprises.

Based on the analysis of possibilities to implement the Industry 4.0 concept into manufacturing enterprises of western Poland (research conducted in years 2016–2017 [9]), the most important, selected threats resulting from the implementation of the Industry 4.0 concept have been identified, which can have a key impact on company management. The results are shown in Table 2.

The implementation of the Industry 4.0 concept primarily requires large expenditures on technical infrastructure, such as modern machines, devices communicating with each other via the Internet, new technologies, modern software, databases enabling the collection and processing of a huge amount of data in real time, and cloud computing technologies. Thus, indirect costs increase in the structure of the company's total costs and generate the need to accurately account for many different products. Traditional costing methods used so far are insufficient because they use only few keys to account indirect costs for a product. They do not show the actual, objective cost of a specific product variant, but the cost is generally averaged.

Table 2 The key selected threats resulting from the implementation of the Industry 4.0 concept which have an impact on company management (own study)

Cause	Effect	Needs
High expenditures for machines, devices, software, technologies, ...	Significant increase in indirect costs	Change of costing method
High expenditures for machines, devices, software, technologies, ...	The risk of losing liquidity	Monitoring and control of liquidity ratios
A large number of differing products	Various conversion keys for indirect costs	Change of costing method
Many variants of one product	Different costs	Change of costing method
Fast technological progress	Fast economic consumption of fixed assets	High costs of depreciation of fixed assets
Frequent changes of strategy and strategic goals	Budgeting too static, too time consuming to make changes	A method that allows frequent, less time-consuming changes

Large expenditures on infrastructure cause the risk of losing a financial liquidity, which is one of the key threats when implementing the Industry 4.0 concept. There is a strong need for continuous monitoring and control of financial liquidity. In addition, traditional budgeting has to change due to its overly static nature. The collection and analysis of data from machines and devices in real time generate the need for rapid changes of strategy, strategic goals, and their quick implementation.

Modern technologies, machines, and devices needed in Industry 4.0 are subject to rapid economic consumption resulting from fast technical and technological progress. The infrastructure must demonstrate very high efficiency of functioning. In accordance with this concept, downtime and failures should be solved before the actual occurrence, based on data sent by machines and devices, hence the need to accumulate in the form of amortization write-off funds for the replacement of used machines, devices, software, etc.

4 Research Results: The Proposed New Procedure of the Implementation of the Concept Industry 4.0

As a result of the conducted research, a procedure was designed which the main aim is to implement the Industry 4.0 concept dedicated for industrial enterprises. This procedure consists of five stages shown in Fig. 2 and is called the Procedure of the Implementation of the Concept Industry 4.0 (PICI).

In the first stage, the company takes the most important decisions regarding the implementation of Industry 4.0:

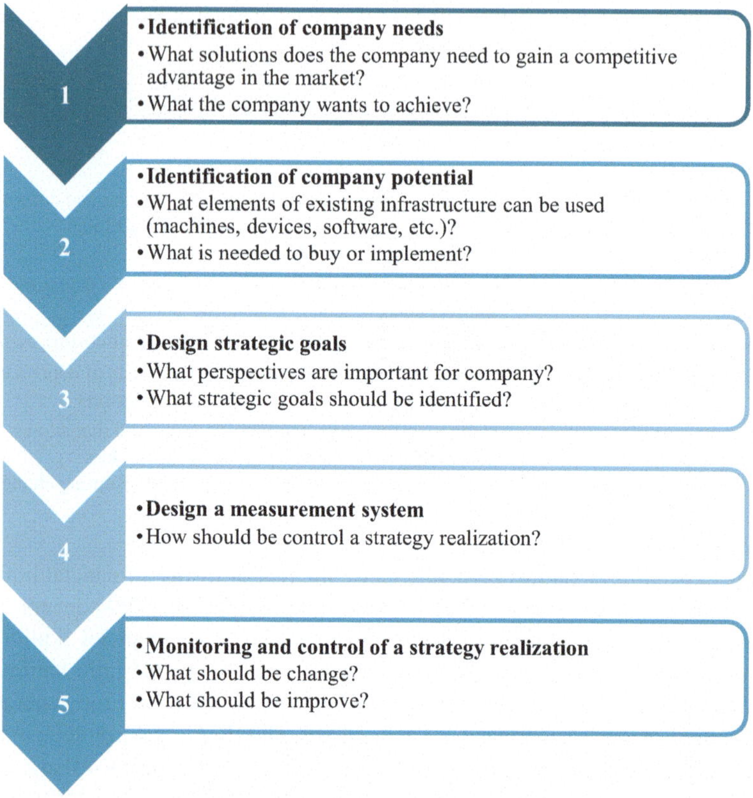

Fig. 2 Procedure of Implementation of the Concept Industry 4.0 (PICI)

- What solutions does competition have?
- In which direction does the competition develop?
- What is able to achieve competition?
- What solutions does the company need to gain a competitive advantage in the market?

The second stage is an identification of a company potential. The infrastructure should be detailed analyzed for adaptation to the solution used within Industry 4.0 (shown in Table 1). Based on this analysis a list of useful machines, devices, software, etc. should be made. The company decides what is needed to buy, improve, and implement. The cost of that is calculated. As part of this stage, decisions are also made regarding where will the financial resources come from to finance their purchase and sources of financing.

Information taken in previous stages is the basis for the development of a new enterprise strategy (third stage). The author of the chapter suggests to use the balanced scorecard method (BSC). Thus, the company decides what perspectives are important for its functioning. According to the BSC approach, four perspectives are fundamental and necessary: financial, customer, internal processes, and learning and knowledge [5]. It is needed to establish what areas are also significant. Next strategic goals for each perspective are designed and the strategy map is made.

In the fourth stage a measurement system is designed. The strategy must be implemented quickly. For this reason, continuous monitoring and control of the implemented strategy are needed. For each strategic goal defined in the strategy map, at least one indicator is designed which is called key performance indicator (KPI). KPIs help to measure and assess a level of strategy realization. In other words, they allow a progress toward organizational goals to be defined and measured and also management of enterprise is easier. The essence of KPIs is to select the most important indicators to rationalize and simplify decision-making, paying attention to the strategically most important areas of the organization. In this stage the target value of each KPI is established and planned time of realization, too.

Finally, the fifth stage is the analysis of a level of KPI which shows if the strategy can be implemented on time. The actual values of KPIs are compared to those planned. If not, corrective actions are planned, which accelerate the implementation of a specific strategic goal.

The application of the procedure proposed by the author enables to implement the Industry 4.0 at the same time taking into account the most important threats identified in Table 2 and enabling their effects to be reduced. In addition, it ensures simultaneous control of the most important areas of enterprise and helps to connect financial aspects with customer needs, requirements for improving the processes and applying new technologies, as well as need for continuous employee development and gaining competitive advantage and also general company development.

The results of conducted research presented in Chap. 3 showed among others that enterprises must control financial liquidity. It is connected with high costs associated with the purchase of modern machines, devices, robots, software, etc. The new proposed PICI also has the advantage that allows continuous monitoring and control of the company's financial liquidity. A financial liquidity control should be included as one of the strategic goals in the developed strategy map in the financial perspective stage 3. In the fourth stage a liquidity radio should be used to control.

Research results also indicate the need to change the approach to traditional financial planning in the form of a budget. Traditional budgeting is highly time consuming and cost consuming within conditions of such rapid technical and technological progress and so often strategy changes. Nowadays, there are an increasing number of critical opinions about the use of traditional budgets. Detailed financial planning is outdated. It is necessary to respond to market changes quickly. Traditional budgets are not generally matched to the continuous changes in the structure and processes in the organization. They do not minimize costs because they do not consider creativity. The balanced scorecard can be used instead of traditional budgeting as the basis of planning. It allows strategies in the enterprise to be implemented effectively and quickly.

5 The Proposed Costing Method

Manufacturing of many variants of different products according to customers' requests (mass customization and mass personalization) generates a large problem of cost calculation. More complex technology requires a greater number of ancillary activities related with the special conditions of production, requirements of quality, supervision, technical requirements, or storage. Products have different lead times, quantity, and kinds of operations involved in manufacturing, manufacturing times, setup times of operations, costs of works, transport, etc.

The results of the conducted analysis clearly show an increase in indirect costs in the implementation of Industry 4.0 solutions. The costs of depreciation of fixed assets and intangible assets (e.g., depreciation of the software) are mainly growing. This means the need for a cost calculation (costing method) which allows indirect costs very precisely to many different products to be allocated. In traditional cost calculations average manufacturing costs are calculated. It is completely unsuitable for settling high indirect costs on many different products.

In this situation the author of the chapter proposes to use the activity-based costing method (ABC). This is applied in several very different bases of accounting indirect costs called cost allocation key costs. The ABC method uses not only financial data on the consumption of economic resources (cost data), but also information on consumption expressed in physical terms, occupied area, time, resources, etc. The use of many different cost allocation keys of indirect costs significantly improves the accuracy of cost estimation. The activity-based costing consists of five stages [10]:

1. Identification of activities existing in the organization
2. Measurement of the costs of each activity
3. Indication of the activity drivers and activity output measures
4. Creation of cost objects responsible for the costs from each place of the generation costs, in which the activity is
5. Activities of cost assignment between the products

In the first stage the business process analysis (BPA) is conducted. Primary and secondary activities are identified in a company. The primary activities are directly involved in the manufacturing process. The secondary activities are associated with the maintenance of the company, the customer, the product, or the order. In the second stage the activity costs are calculated depending on the extent to which the product is participated in the activity. Then in the third stage activity drivers (cost drivers) and activity output measures are set to each activity. The cost driver shows what activity cost depends on. An activity cost is divided by activity output measures which are defined for a given period (e.g., month, year) and it is counted activity unit cost action (stage four). Finally, in the last stage activities of cost are assigned between the products.

6 Conclusions

Currently, enterprises must introduce new, innovative solutions and follow or even stay ahead of the competition to gain a competitive advantage on the market. The Industry 4.0 concept includes many solutions that will revolutionize the industry and can significantly increase production efficiency. However, this is a big challenge for companies that have to overcome many fears and constraints and solve many emerging problems.

The process of implementation the Industry 4.0 concept must be carefully planned and efficiently carried out. It requires the use of specially developed implementation procedures, which will reduce the risk of failure and reduce the greatest threats. As shown by research carried out in industrial enterprises in western Poland, the greatest threat is loss of liquidity. The costs of new, modern machines, devices, robots, etc. communicating via the Internet in real time are high and cause a significant increase in indirect costs in the structure of total company costs, which requires a different, more accurate settlement of indirect costs on many different products.

The proposed procedure for implementing the Industry 4.0 concept PICI is dedicated to industrial enterprises and based on the balanced scorecard approach. It enables to connect financial goals and customer requirements with requirements for process improvement and modernization as well as employees and whole enterprise development. The system of measurement of the implementation of the strategy follows continuous monitoring and control of the most important aspects of business management. The financial liquidity of the company is also controlled. The purpose of further research is to build a strategy map model dedicated to companies implementing the Industry 4.0 concept.

References

1. Bawa, M., Caganova, D., Szilva, I., & Spirkova, D. (2015). Importance of internet of things and big data in building smart city and what would be its challenges. In *First EAI international summit, Smart City 360°* (pp. 605–616). Bratislava, Slovakia and Toronto: Revised Selected Papers.
2. Brodny, J., Alszer, S., Krystek, J., & Tutak, M. (2017). Availability analysis of selected mining machinery. *Archives of Control Sciences, 27*.(LXIII(2), 197–209.
3. Fortuna, C., & Mohorcic, M. (2009). Trends in the development of communication networks: Cognitive networks. *Computer Networks, 53*(9), 1354–1376.
4. Jasiulewicz-Kaczmarek, M. (2013). Role and contribution of maintenance in sustainable manufacturing. In: 7th IFAC Conference on Manufacturing Modelling, Management, and Control. 2013 in N. Bakhtadze, K. Chernyshov, A. Dolgui, V. Lototsky, (eds.), Series Title: Manufacturing Modelling, Management, and Control, Vol. 7, Part 1, (pp. 1146–1151). International Federation of Automatic Control.
5. Kaplan, R. S., & Norton, D. P. (2007). *Using the balanced scorecard as a strategic management system*. Boston: Harvard Business Review.

6. Kramarz, M., Kramarz, W. (2014). The flexibility and specialization of resources—competitive strategies of materials decoupling points of a network supply chain of metallurgic products. In: 16th EWGT Meeting (Euro Working Group on Transportation), Portugalia, Porto, In: Procedia—Social and Behavioral Sciences, 111/2014, (pp. 741–750).
7. Pawar, S. P. (2015). *Smart citied with internet of things (sensor networks) and big data*. Chinchwad, Pune: ASM'S IBMR.
8. Saniuk, S., & Saniuk, A. (2017). Industry 4.0—Technical-economical development perspective for the metallurgical production. In *26th international conference on metallurgy and materials—METAL 2017* (pp. 2288–2292). TANGER: Brno.
9. Saniuk, A., & Saniuk, S. (2017). Management of the metallurgical enterprise according to the industry 4.0 concept. In *26th international conference on metallurgy and materials—METAL 2017* (pp. 2034–2039). TANGER: Brno.
10. Więcek, D., & Więcek, D. The influence of the methods of determining cost drivers values on the accuracy of costs estimation of the designed machine elements. In Z. Wilimowska, L. Borzemski, & J. Świątek (Eds.), *Information systems architecture and technology: Proceedings of 38th international conference on information systems architecture and technology—ISAT 2017. ISAT 2017* (Advances in intelligent systems and computing) (Vol. 657, pp. 78–88). Cham: Springer.

Intelligent Hardware-Software Platform for Efficient Coupling of Water-Energy Nexus in Smart Cities: A Conceptual Framework

Naran Pindoriya, Markus Duchon, Pragya Kirti Gupta, Venkatesh Pampana, S. N. Singh, Jakob Giza, Bastian Hackenberg, Arvind Kumar Rajput, and Janki Jethi

1 Introduction

Energy efficiency, sustainable energy, energy storage devices, and real-time information and control are of prime importance to manage the smart city network efficiently while reducing its carbon footprint. International Telecommunication Union (ITU) [1] highlights the role of information and communication technology (ICT) integration in building smart sustainable city infrastructure. Mohanty et al. [2] emphasize that the two closely related emerging technology frameworks are the IoT and big data analytics, making smart city infrastructure efficient and responsive. A smart city relies on widely distributed smart devices (collectively forming a massive IoT network) to monitor the urban environment in real time, collects increasing volumes of new data for intelligent decision-making, and facilitates various services

N. Pindoriya (✉)
Indian Institute of Technology Gandhinagar, Gandhinagar, Gujarat, India
e-mail: naran@iitgn.ac.in

M. Duchon · P. K. Gupta · V. Pampana
fortiss GmbH, Munich, Germany
e-mail: duchon@fortiss.org; gupta@fortiss.org; pampana@fortiss.org

S. N. Singh
Madan Mohan Malaviya University of Technology, Gorakhpur, UP, India
e-mail: snsingh@iitk.ac.in

J. Giza · B. Hackenberg
sonnen GmbH, Wildpoldsried, Germany
e-mail: j.giza@sonnen.de; b.hackenberg@sonnen.de

A. K. Rajput · J. Jethi
Gujarat International Finance Tec-City, Ahmedabad, Gujarat, India
e-mail: arvindkumar.rajput@giftgujarat.in; janki.jethi@giftgujarat.in

© Springer Nature Switzerland AG 2020
D. Cagáňová, N. Horňáková (eds.), *Mobility Internet of Things 2018*,
EAI/Springer Innovations in Communication and Computing,
https://doi.org/10.1007/978-3-030-30911-4_15

to improve the quality of urban living. The distributed network (mostly known as wireless sensor network) of intelligent sensor nodes, as well as data centers/cloud where sensor data are stored and shared, constitutes a smart city infrastructure. Participatory sensing plays an indispensable role in emerging initiatives of a smart city, which retrieves sensor data from groups of people or communities and other critical infrastructure of water and electricity within the city. Real-time information management from subsystems of water-energy network, knowledge-based systems, and intelligent decision support is an integral part for effective energy and water infrastructure management in smart sustainable city.

Water-energy nexus is most intrinsically linked to the core of environmentally sustainable smart cities as shown in Fig. 1. Clean and sustainable water supplies and low carbon energy access are the essential building blocks for economies, health, and quality of life. It is possible to address the water-energy nexus by utilizing the advanced information and communication technologies (ICT) and the distributed monitoring and optimal utilization and control of distributed energy resources such as solar PV and energy storage along with technological advancements in wastewater treatment, solid waste management, district cooling system, etc. Developing distributed renewable energy generation and battery energy storage are the major interventions besides the energy efficiency and demand electrification. The pervasive deployment of the IoT and advanced ICT, especially the smart metering, will generate big energy data in terms of volume, velocity, and variety.

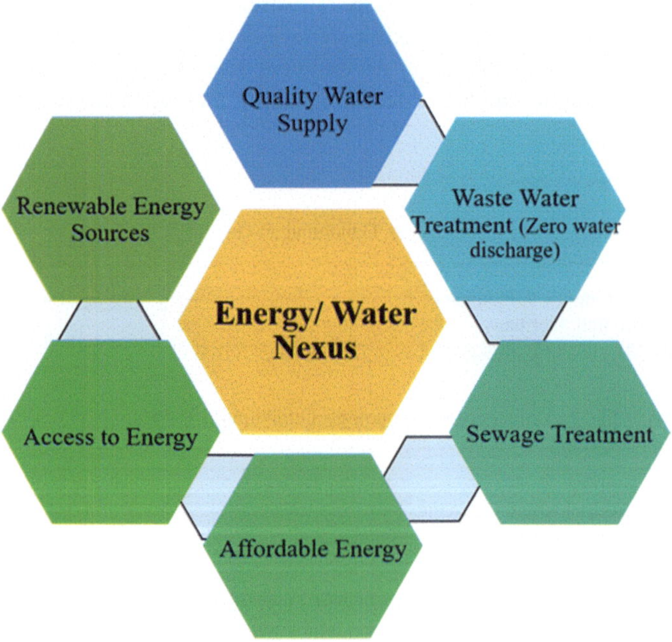

Fig. 1 Water-energy nexus: Building blocks for smart cities

This can bring huge benefits to the energy efficiency and management; however, it must be processed and communicated in an energy-efficient manner.

Water-energy nexus topic is a widely researched topic and the sustainability issues surrounding the availability of clean water and energy have drawn the attention to the problem from all around the world. Specifically, in several countries like China, Middle East and North Africa (MENA), Spain, and Brazil, the water-energy nexus is already considered crucial for economic growth. A study by Kahrl et al. [3] conducted in 2007 on water-energy nexus in China found that there is a small fraction of total energy consumption used in providing nonagricultural water supply. It further found that the water-energy nexus was of little relevance to policymakers. Finally, while designing the water pricing and conservation, energy investment plays a big role. Siddiqi et al. [4] conducted a study in the Middle East and North Africa (MENA) region. They observed that there is a weak dependency of energy systems on freshwater, and rather a strong one on the water abstraction, purification, and treatment. They also concluded that although the energy savings is inevitable with water efficiency, the energy implications must be considered while controlling or stabilizing water demand in the region. Hardy et al. [5] evaluated the water-energy nexus in Spain. They focused on two aspects of the nexus, namely, energy used in the water sector and the water required in running the energy sector. They approached the problem considering Spain's vision on biofuel production by 2020. They identified the water-use cycle and recommend that energy audits must be carried out at each stage of the water-use cycle. They also highlighted that the maintenance phases are important and must be considered while designing optimization techniques. Evaluating the electricity use for water production and supply in Brazil, Vilanova [6] identified the potential regions in Brazil suitable for hydraulic and energy efficiency measure in water supply systems. They identified potential regions in Brazil where energy generation, energy wastage, and water loss indicators were similar even though the demographics, economic, and biodiversity of these regions were different. Smart water grid system [7] implemented in Singapore manages the water supply network efficiently and ensures 24/7 quality water supply to communities.

The present-day water management and energy management are mainly centrally controlled with limited information access. As the distribution network complexity increases, the centralized control is often prone to downtime due to network latency, connection losses, damages, etc. A refinement or reconfiguration mostly requires shutdown of the whole system and leads to overall degradation of performance. Betterment in both water and energy management requires distributed monitoring and controlling of the loosely coupled distributed infrastructure. A distributed control with a centralized coordination can yield a localized and robust control among loosely coupled infrastructure with a platform for more information gathering and analysis.

The effort has been put into research and technical development for pathway towards energy-efficient sustainable water and clean energy access to the communities. On the one hand, water management mainly focuses on the following aspects: (1) improving the distribution by enacting faster decisions and control actions and

(2) conserving the available resources by increasing the efficiency and recycling the wastewater. On the other hand, clean energy access focuses on optimal utilization and advancement in distributed energy resources.

The energy demand should be covered mainly by renewable energy sources among others for water treatment or street lighting. Here the use of storage technologies is essential since such energy is not available at all times. By energy storage technologies, surplus energy can be stored and made available again, when demand exceeds supply. Towards distributed controlling, software applications [8–11] are designed in such a way that the physical infrastructure is considered a loosely coupled software entities. Such applications enable the integration of energy-efficient techniques. Literatures [7, 8, 12, 13] discuss the strategies to improve the energy efficiency of the water supply systems. Analysis of energy efficiency in supply systems based on modeling and optimization has been presented in [14]. From the magnitude and complexity of the problems around the water-energy nexus arises the need for optimization and analytical tools for better schedule and control algorithms using computational intelligence and machine learning techniques that would help in improving both energy efficiency and water management. The work on the energy efficiency [8–10] will be taken into consideration for the efficient operation of the water infrastructure.

Pilot-scale project on wastewater treatment (WTP), smart energy, and solid waste management has been initiated at Gujarat International Finance Tec-City (GIFT). This chapter presents the systematic framework for demonstrating the intelligent hardware-software platform for the identified test bed. The test bed is equipped with WTP, STP, and street lighting system at the smart city site of GIFT, Gujarat. The combination of water and smart energy infrastructure highlights the underlying commonality of various technologies. Not only does the supply of cheap and green energy reduce costs, but there is also scope for better planning for energy generation and sharing of various technologies. Better planning approaches can be designed using common knowledge, mathematical models, optimization, and analytical tools. The intellectual merit of this chapter is to develop a hardware-software framework for deployment in smart sustainable cities, which enhances energy efficiency in water (WTP, STP) and energy (streetlight) infrastructure.

This chapter is organized as follows—Sect. 2 describes the design of a conceptual framework for intelligent energy management of water and energy infrastructure in smart city. GIFT city along with technical details of use cases (WTP, STP, and street lighting system) at test bed, the proposed framework, representative scenarios, and corresponding simulation results are presented in Sect. 3. Finally the concluding remark is presented in "Acknowledgements."

2 Conceptual Description

A multidisciplinary research is required to develop a technological unified framework for hardware-software architecture to integrate advanced technologies such as IoT and ICT for addressing the challenges of efficient management of water and energy in smart cities. Online monitoring and resource management system is being developed for improving reliability and quality of a water-energy network in smart sustainable city.

The following aspects are looked into, under the unified framework:

- *ICT interface and IoT devices* for online monitoring and management system for improving reliability and quality of a water-energy network in smart city
- *Integration of on-site distributed clean energy resources and energy storage devices* for better network efficiency and reduced carbon footprint
- *Optimization and decision-making algorithms* provide load and energy storage scheduling based on the energy price, demand, load profiles, and energy storage status for water-energy network efficiency
- *Hardware and software platform* to improve energy efficiency and water management—optimize water-energy usage by real-time monitoring and demand-side resource management

Figure 2 shows the interaction of physical entities with the software controller through a robust and reliable communication network. It shows a generic view of the technologies and physical entities that are part of the smart city landscape.

(a) *Physical layer:* It consists of advanced metering and IoT sensors connected at the major process equipment in the critical infrastructure like WTP, STP, and street lighting system. Electrical distribution network with RE-based

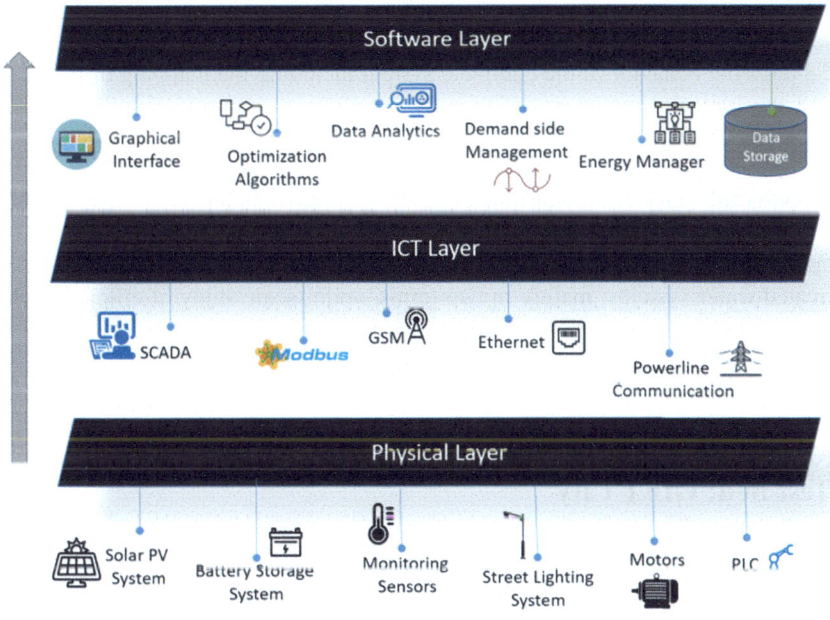

Fig. 2 Physical entity interaction with the software controller through communication network

generation—solar PV, etc.—and battery energy storage (BES) are covered in this layer. These physical assets are to be operated in the most efficient and reliable manner in real time. This can be achieved by real-time monitoring of critical decision variables at the plants and thereafter communicate them to software layer (application layer) via appropriate communication protocol in communication layer for the optimization, real-time decision making and control purpose.

(b) *ICT layer:* ICT includes various aspects of information handling such as the communication channels, ability to manipulate data, and information transmission. Various modes of transmission (wired Ethernet, power-line communication, or wireless networks like GSM, Wi-Fi, Zigbee) would be used depending on the location of nodes and their accessibility. The physical infrastructure for the communication channel must be robust and reliable. It must be capable of manipulating data in order to transfer the complete information from one to another. In the context of smart grids, power-line communication (PLC) and supervisory control and data acquisition (SCADA) could also be considered as alternatives to the conventional communication channel.

(c) *Software layer:* From software engineering perspective, defining the data models and modeling the real entities into the software are crucial for controlling. A scientific work should be carried out both at the physical and software layers, while communication layer will support in data provisioning. To combine these objectives, a component-based architecture, where functionalities can be added as new loosely coupled software components, is proposed in this section. The interoperable functionalities are organized in the form of interoperable software services that belong to either water or energy infrastructure. The architecture captures the behavior of the complete system including the mapping of process and the logical components. The software platform comprises optimization, decision-making, and multilevel control algorithms for improving water and energy efficiency for smart city as a whole. The proposed software platform supports full stack from real-time controlling to the data analytics capabilities. For instance, street-lighting system is considered for the optimal energy management. On the other hand, the focus is on improvement of distribution of treated water systems, mainly in two terms—pilot-scale study of efficient leak detection system to check for unaccounted for water (UFW)/nonrevenue water (NRW) losses, and residual chlorine sensor/oxidation reduction.

3 Test Bed: GIFT City

The framework of concept is proposed at water and electricity infrastructure in GIFT city—Gujarat International Finance Tec-City of Gujarat, India. The GIFT city envisions a smart city infrastructure with efficient water and energy distribution networks in a distributed manner. Water infrastructure in GIFT is designed to provide "potable water in all taps" with total water requirement of 60 million liters per day (MLD). The source for water is Narmada Canal, from where water is pumped to WTP for treatment and then distributed across the city. Water distribution in GIFT is via pipelines in utility tunnel where the twin-distribution pipeline is interconnected with each other to maintain 24 × 7 water supply, which drastically reduces the UFW loss. The present process of WTP consists mainly of filtration process—dual-media filters (made up of sand and gravel) and microcartridge filters. Moreover, wastewater [5] is emerging as an affordable and reliable alternative water supply. In line with sustainable infrastructure plan, GIFT city aims for "zero liquid discharge." Hundred percent of wastewater generated will have various uses (HVAC blowdown, horticulture, landscaping, car washing, road cleaning, flushing of sewers, flushing at building level, etc.), which will minimize freshwater demand.

The test bed comprises STP, WTP, and streetlight system as shown in Fig. 3. The use cases for identified locations are presented in Tables 1 and 2. The total sewage generation in GIFT city is approx. 50 MLD. The sewage generated from each building will be collected through gravity in sewage collection pipeline, which will be treated in STP. The present STP is based on a biological process in which the main energy-requiring component is aeration (air aeration blowers) to have sufficient oxygen available for microorganisms and to allow proper mixing in the tank. GIFT city has developed solid waste management system consisting of automated waste collection system (AWCS) through pipelines, segregation, and

Fig. 3 Test bed and use cases at GIFT city

Table 1 Use case of aeration blower at STP

Title	Dissolved oxygen (DO) level-based optimal operation and PV + battery integration on aeration blower at STP
Description	Development of an algorithm which controls the variable frequency drive (VFD) of air blower used for aeration process in MBBR tank according to the DO level in sewage and schedules the battery charging and discharging time to minimize the cost of electricity purchased from gird
Basic flow of events	• Collect DO sensor data, load trend of air blower, real-time battery status, PV forecasts, weather forecasts, energy costs, status of grid supply • Send all data to iEMS • iEMS generates control signal for VFD and battery using the optimization algorithm • Send control signals to 4diac • Control of VFD and battery switches through 4diac • Controlled air injection in MBBR tank and optimized battery switching

Table 2 Use case of hypo-dosing pump and air compressor at WTP

Title	PV and battery integration on hypo-dosing pump and air compressor at WTP
Description	To maximize the use of solar PV system and battery to support the load of hypo-dosing pump and air compressor in WTP and minimize the overall cost of energy purchased from grid
Basic flow of events	• Input data: Development of load trends using historical data for hypo-dosing pimp, real-time battery status, PV forecasts, weather forecasts, energy costs, status of grid supply • Send the input data to optimizer • Get the output from optimizer (duration and quantum of energy from PV and battery) • Send control signals to switches

treatment of waste to avoid human intervention as a best possible way and to reduce health and environmental issues. For treatment and disposal of solid waste, a pilot-scale plant of plasma pyrolysis of capacity is installed and working at GIFT city. The efficiency of system has been increased by using the gases generated in primary chamber, to burn (decompose) the waste in secondary chamber. To online monitor the data of gas emission from the chimney of the plant, a gas (CO_2 and CO) sensor can be installed in chimney. The inlet feed of plasma pyrolysis consists of two fractions—sewage sludge and solid waste. The sewage sludge is added in the system to increase the calorific value of the waste.

Development of environmentally sustainable electricity supply network would integrate a range of technologies like renewable generation, battery energy storage system, gas-insulated switchgears, substation automation, and thermal storage system at a single location which enhances the quality of power, reduces power loss, saves money, and creates business for people in community. Such a network will not only maximize the benefits to the consumer by providing low carbon generation but also facilitate better scheduling of energy storage systems. Further, intelligent street lighting also has significant potential in energy saving for sustainable smart city. Looking into the aspects of high reliability, high performance, fast maintenance, accurate energy monitoring, and less intervention of human scheduling, a centralized monitoring and control system with wireless-based communication can be developed [7, 8] to take appropriate energy consumption reduction measures through power conditioning and control.

To implement an optimal scheduled water-energy coupled network, an advanced energy storage system in a distributed manner understanding the cross-commodity interactions is also necessary. The sonnenBatterie, manufactured by Sonnen GmbH, is comprised of state-of-the-art lithium iron phosphate (LiFePO4), which makes it unique from other lithium-ion batteries as it is widely considered to be a more stable and safer compound. In addition to the production of intelligent energy storage systems, Sonnen GmbH builds the sonnenCommunity. The sonnenCommunity connects people who produce their own electricity into a worldwide energy-sharing network consisting of more than 30,000 sonnenBatteries. The exchange of energy exactly as required enables the members of the sonnenCommunity to supply 100% of their own energy. Sonnen GmbH also manufactures the sonnenCharger for intelligent charging of electric vehicles. This enlarges the pool of advanced storage systems that can be used as a virtual power plant.

The water-energy nexus with advanced energy storage and use of renewable energy resources has not yet been available. This chapter presents the framework to implement a software platform with progressive optimization algorithms to interconnect different infrastructure and enable their real-time monitoring and control. Fostering the utilization of solar PV generation advanced storage technologies will be integrated which allows for a modular and distributed operation of infrastructure. The system proposed envisages an intelligent distributed architecture with a smart centralized coordinator. The test bed in GIFT city is distributed into grids, where each grid has a water and/or energy management system. In this section, the proposed framework of hardware-software platform has been presented.

As shown in Fig. 4, the pumps and streetlight are already connected with information network and sensors. So the information about the amount of energy required and the daily operational duration of pumps and streetlights is known. In the proposed framework, an additional software layer is added, which is capable of (1) collecting and analyzing data from various sensors, (2) making smart decisions based on the advanced optimization techniques and data analysis, and (3) controlling pumps and streetlight based on the smart decisions. Towards distributed monitoring and real-time controlling, the 4diac framework is used. It implements the IEC 61499 standard for industrial control applications. It enacts the distributed control and monitoring of various energy resources. Using 4diac, heterogeneous control devices can be reconfigured and adapted with nearly no downtime. 4diac RTE will be used in conjunction with the intelligent energy management system (iEMS) framework. This framework provides high-level capabilities such as data analysis, scheduling, and monitoring interface with bidirectional data flow to 4diac application. Multiple instances of iEMS will be used to receive data from the distributed infrastructure in the test bed. The use of multiple instances of iEMS enables the possibility of both centralized and localized controlling. The distributed grids in the test bed are controlled and monitored with Eclipse 4diac with iEMS, where 4diac provides real-

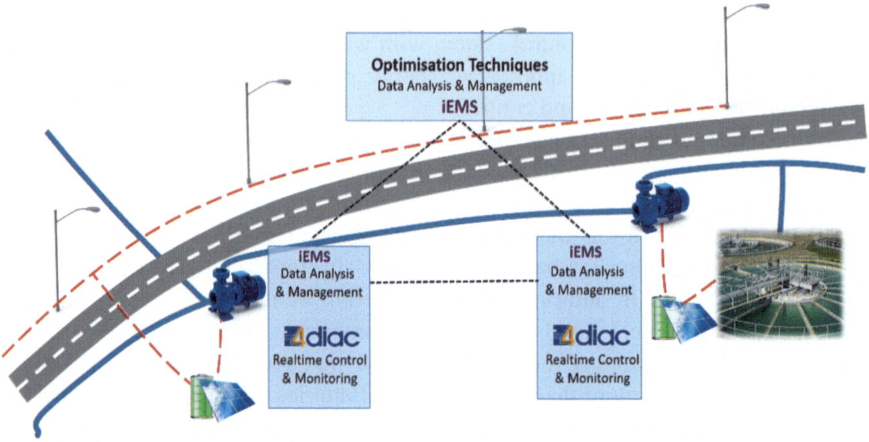

Fig. 4 The hardware-software platform for real-time monitoring and controlling of WTP, STP, and streetlights

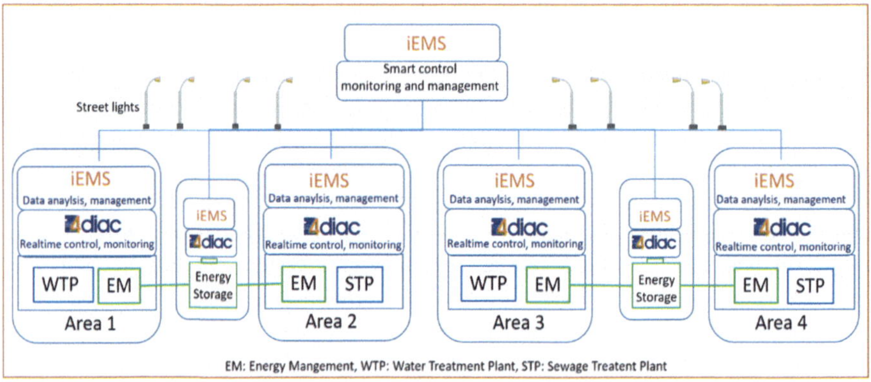

Fig. 5 Block diagram of energy management and real-time control

time control and field data for analysis in iEMS. At the edge, iEMS uses the data and analyzes it to make distributed systems intelligent by efficiently distributing the available energy and water resources and utilizing the water-energy nexus.

Towards making the smart decisions, iEMS executes optimization algorithms to provide intelligent decision commands to 4diac. Figure 5 shows the distributed systems connected with energy storage systems, which provide a modular energy storage hardware, which uses 4diac for real-time plugging and unplugging of new storage modules. On the top, a centralized coordinating controller is used for higher level monitoring and decision-making with iEMS. Since the proposed system uses a distributed architecture providing intelligence at each grid, the latency, interruption, and damage in communication infrastructure are eliminated and an efficient and uninterrupted water and energy distribution network is foreseen.

Towards water-energy nexus, the methodology needs to be developed to maximize the use of renewable energy to power the entire physical infrastructure. The power consumption appliances that meet the water and energy requirements comprise both water pumps and lighting systems. To address the water-energy nexus, the optimization of PV module and battery storage integrated with STP, WTP, and streetlight system is done. The single-line diagram for the STP powered by both the grid and photovoltaic system (PV) is provided in Fig. 6. Arrows depict the power supply to the aeration blower, which can be powered from the grid (bus_grid), PV, and batteries. All the three power sources are connected with the feeder bus (bus_feeder) to ensure the required supply to the aeration blower at all times. In Fig. 7, the single-line diagrams for WTP and streetlights are presented. Similar to the STP connections, the air compressor and hypo-dosing pump for WTP and streetlights are powered by three sources (grid, battery, and PV). The water-energy nexus is handled as sharing the same PV source for meeting the energy demands of WTP and streetlights. The battery backup system is separate for WTP and streetlight.

The initial simulation results using the open energy modelling framework (OEMOF) [15] are presented in Figs. 8, 9, and 10, which incorporate the PV module and battery energy storage integration to STP, streetlight cluster, and WTP. With OEMOF complex cross-sectoral energy systems can be modelled and optimized based on a generic graph-based description of the system [16]. The optimization objective of the simulations is to minimize the electricity purchase from the grid using solar PV generation by controlling the battery energy dispatch. Table 3 shows the logic to regulate the blower speed in order to maintain microorganism in the sewage tank. Based on the power requirements for the various blower speeds, the battery schedule is calculated.

In Fig. 8 results of integration of 15 kWp solar PV and 30 kWh battery energy storage at STP are shown. It is observed that with coordinated and optimized

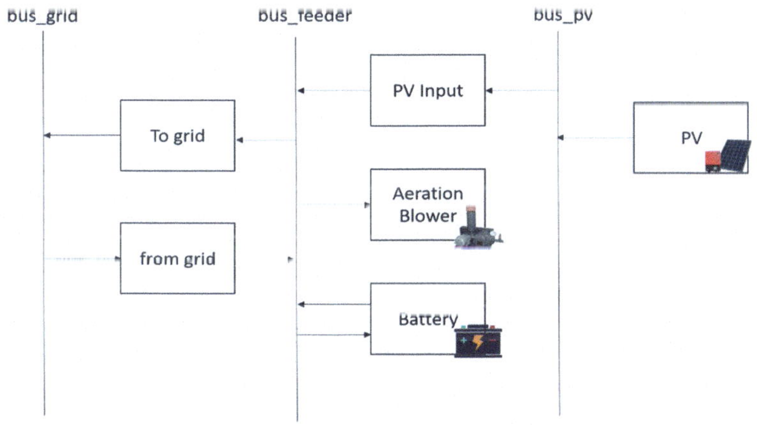

Fig. 6 Single-line diagram for STP use case

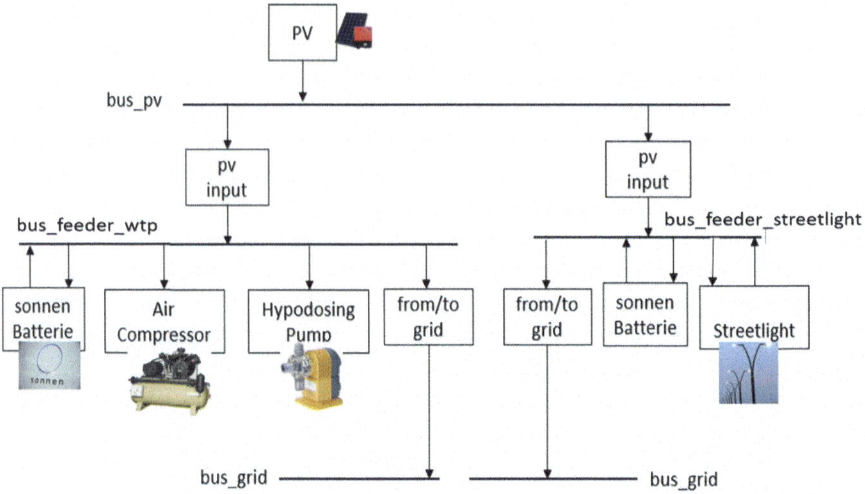

Fig. 7 Single-line diagram for WTP and street lighting use cases

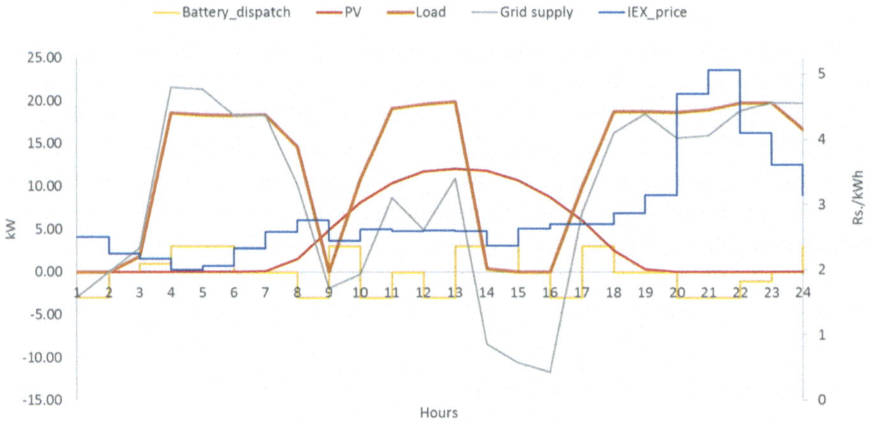

Fig. 8 Simulation result of PV and battery energy storage integration at STP

charging and discharging of battery storage, grid dependency for plant operation can be reduced by 16%. Simulation results of solar PV and storage integration for streetlight operation are shown in Fig. 9. Results show that with integration of three Sonnen batteries (each of 10 kWh rating) and dimming control of smart streetlight system, 25 streetlights of 4 pillars can be operated directly without having dependency on grid supply. Figure 10 shows the simulation results for WTP use case with PV installation of 10 kWp and one Sonnen battery (10 kW) and connected load of one hypo-dosing pump and seven air compressors. The results show that, by prioritizing the use of PV and utilizing as much as energy available in the battery, the grid dependency can be reduced by 50%. These merits would

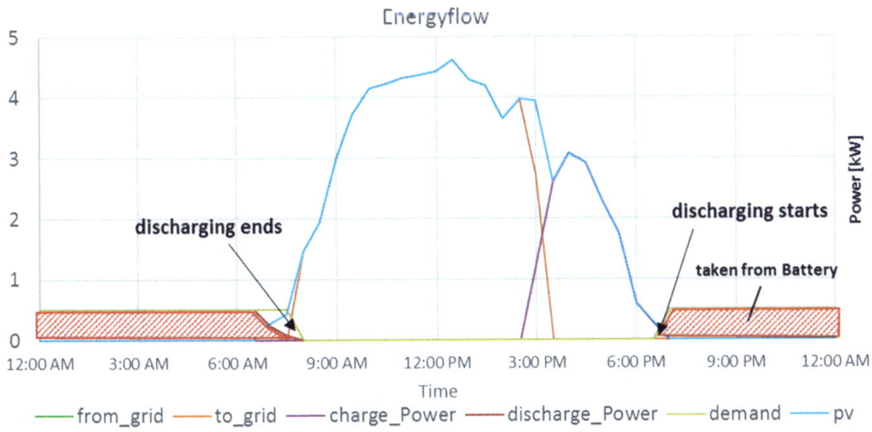

Fig. 9 Simulation result of PV and battery integration at Gyan Marg street lighting system

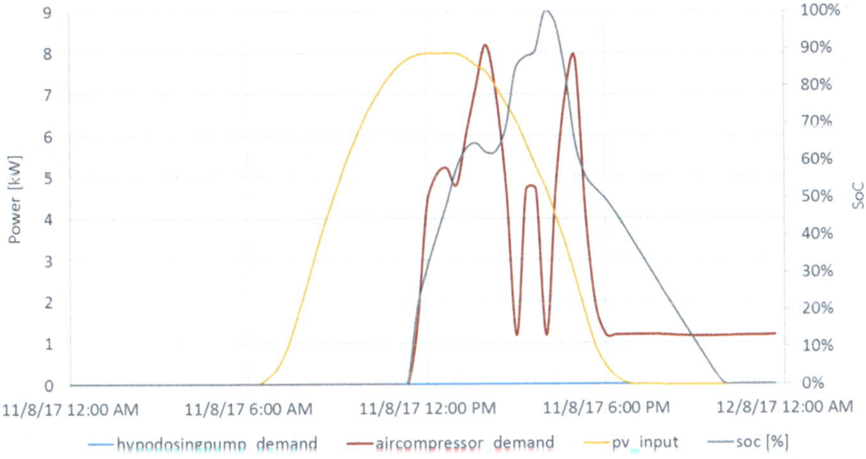

Fig. 10 Simulation result of PV and battery energy storage integration at WTP

Table 3 Pseudo code for blower speed and battery energy dispatch

Pseudo code for aeration blower at STP	Pseudo code for battery energy dispatch
Function: Dissolved Oxigen_level() start procedure input("Dissolved Oxigen level") calculate(optimum blower speed) output("blower speed") end procedure	Function: battery_dispatch() start procedure input("SoC, load, PV_{gen}, grid price ") calculate(optimum battery scheduling) output("battery dispatch") end procedure

increase when dynamic electricity prices are introduced in the network. It is worth to emphasize that integration of PV and storage system has decreased the dependency of fundamental processes like STP and streetlight on grid supply and enhanced the reliability of smooth and secure operation of city.

The deployment of various smart and innovative technologies in GIFT will enable effective use of resources, i.e., water and energy by reducing wastage. The development of water infrastructure and microgrid with smart street lighting at GIFT is unique of its kind, not involving any known IP issues. Moreover, the pattern of development of utilities is acting as a pioneer/model for upcoming smart cities in India. All the components of water infrastructure, ranging from treatment and storage to supply of potable water 24 x 7, is fully automated with philosophy of water neutral and any tap drinking water. The innovation is especially generated in the field of real-time measurement and data fusion in the different domains simultaneously. This data fusion will allow advanced forecasted algorithm and techniques to improve the overall energy efficiency significantly.

4 Conclusion

Coupling of cross-commodity infrastructure and integration of energy resources is a challenge for smart cities. In this chapter, we presented a conceptual hardware-software platform to manage the water and energy infrastructure within GIFT smart cities using ICT. As a result, the test bed identified for this study is STP, WTP, and streetlight clusters attached to WTP which are energized by solar PV, BES, and utility grid. A detailed description of the test bed is also presented, which serves as the evaluation platform for the proposed methodology. Then, the use cases from the test bed have been identified which best explains their functional requirements. Two-layer architecture has been proposed to address the energy management and real-time control of the use cases where a distributed controller and monitoring system is proposed for each use case level. In addition to it, a centralized coordinating controller is used for higher level monitoring and decision-making. The proposed framework has the advantage of eliminating the latency and any interruption or damage in communication infrastructure which will result in efficient and uninterrupted water and energy distribution within GIFT city.

Preliminary results on energy management of these use cases with fixed and dynamic energy pricing models were presented and it can be confirmed that a significant reduction in the power drawn from the utility grid can be found. As a result, the considerable reduction in operating cost can be achieved which is due to efficient management of solar PV and BES systems. It was also noticed that software controller, communication infrastructure, and optimization techniques serve as the basis for solving the challenges in water-energy nexus.

Acknowledgements This work is being carried out for ongoing ECO-WET research project, under the flagship of IGSTC (Indo-German Science and Technology Centre). Authors would like to thank Federal Ministry of Education and Research (BMBF, Germany) and Department of Science and Technology (DST, India) for funding the research and development activities of the project.

References

1. *The Promise of Smarter Cities—Pathway for Smart Sustainable Cities: A guide for city leaders.* http://www.itu.int/en/itunews/Documents/2016-02/2016_ITUNews02-en.pdf.
2. Mohanty, S. P. (2016). Everything you wanted to know about smart cities: The internet of things is the backbone. *IEEE Consumer Electronics Magazine, 5*(3), 60–70.
3. Kahrl, F. (2008). China's water-energy nexus. *Water Policy, 10*(1), 51–65.
4. Siddiqi, A., & Anadon, L. D. (2011). The water-energy nexus in Middle East and North Africa. *Energy Policy, 39*(8), 4529–4540.
5. Hardy, L. (2012). Evaluation of Spain's water-energy nexus. *International Journal of Water Resources Development, 28*(1), 151–170.
6. Vilanova, M. (2015). Exploring the water-energy nexus in Brazil: The electricity use for water supply. *Energy, 85*, 415–432.
7. Public Utilities Board Singapore. (2016). Managing the water distribution network with a smart water grid. *Smart Water, 1*, 1–13.
8. Duchon, M, Gupta, P.K., et al. (2014). Advancement of a sensor aided smart grid node architecture. In Cyber-Enabled Distributed Computing and Knowledge Discovery (CyberC), International Conference on (pp. 349–356), IEEE.
9. Bytschkow, D., Zellner, M., et al. (2015). Combining SCADA, CIM, GridLab-D and AKKA for smart grid co-simulation. In *Innovative Smart Grid Technologies Conference (ISGT), 2015 IEEE Power & Energy Society* (pp. 1–5).
10. Gupta, P.K., & Schaetz, B.. (2016). Constraint-based graceful degradation in smart grids. *2nd International Workshop on Software Engineering for Smart Cyber-Physical Systems* (pp. 8–14). ACM.
11. Gupta, P. K., Gibtner, A. K., Duchon, et al. (2015). Using knowledge discovery for autonomous decision making in smart grid nodes. In *IEEE International Conference on Industrial Technology (ICIT)* (pp. 3134–3139).
12. Amare, E., Kebede, F., et al. (2017). Wastewater confronting realities for sustainable livelihood in developing countries: Case Study Mekelle University, Ethiopia. *Water Conservation Science Engineering, 2*(1), 21–30.
13. Bhansali, G., Singh, B. P., et al. (2016). Strategic optimisation of microgrid by evolving a unitized regenerative fuel cell system operational criterion. *International Journal of Sustainable Energy, 35*(8), 722–733.
14. Diniz, A. M. F., Fontes, C. H., et al. (2014). Dynamic modeling and simulation of water supply system with applications for improving energy efficiency. *Energy Efficiency, 8*, 417–432.
15. oemof developer group. (2018). *oemof—open energy modelling framework (v0.2.2) (Version v0.2.2)*. Zenodo. https://doi.org/10.5281/zenodo.1302372.
16. Hilpert, S., et al. (2018). The open energy modelling framework (oemof)-a new approach to facilitate open science in energy system modelling. *Energy Strategy Reviews, 22*, 16–25.

Scheduling Heuristic to Satisfy Due Dates of the Customer Orders in Mass Customized Service Industry

R. Sudhakara Pandian, V. Modrak, Z. Soltysova, and P. Semanco

1 Introduction

In the mass customization, the major concern is to satisfy all customers with varying outputs and with different due dates. The flow shop is quite traditional in meeting customers' demand and supply from the manufacturer in right time. The flow shop is restricted to make a variety of products whereas mass customization makes one step further in manufacturing different models with more variety. The objective of mass customization is to provide customized services or produce customized products at low cost and acceptable delivery time [1]. Mass customization approaches were not viable with conventional manufacturing practice, since traditional manufacturing strategies can efficiently produce especially standard products [2]. Requirements of clients of municipal services usually differ from one person to another person and they can require different delivery times of the services. For that reason, new approaches are demanding from service providers in mass customized environment based on smartphone apps and tools of Internet of Things. Naturally, effectiveness of potential smartphone apps depends also on supporting software tools. In this chapter, a novel method using simple heuristics for mass customization to select appropriate batch sizes that meet the due dates prescribed for various customers is proposed. The suggested heuristics algorithm is performing to be better. The Gantt chart is used to clearly explain the scenario of proposed method.

R. Sudhakara Pandian (✉)
Vellore Institute of Technology [VIT], Vellore, India

V. Modrak · Z. Soltysova
Technical University of Kosice, Kosice, Slovakia

P. Semanco
Lear Corporation, Presov, Slovakia

© Springer Nature Switzerland AG 2020

D. Cagáňová, N. Horňáková (eds.), *Mobility Internet of Things 2018*,
EAI/Springer Innovations in Communication and Computing,
https://doi.org/10.1007/978-3-030-30911-4_16

2 Literature Survey

In this chapter an exhaustive review of literature is presented for scheduling of jobs with appropriate batch sizes in mass customization. Zhang et al. [3] proposed an ant-based algorithm suitable for assembly line balancing used in mass customization and authors present that the line balancing problem in mass customization could be optimized using metaheuristics algorithm, the so-called ant colony algorithm for the type I method. Noorul Haq et al. (2006) [4] presented an approach based on a hybrid genetic algorithm for balancing of assembly lines as novel approach in a mass production. Boysen et al. [5] presented an idea of mixed model assembly lines in the classification of physical resequencing that shifts workpieces to a new sequence during assembly. Multiobjective mixed model created by Rahimi-Vahed et al. [6] for sequencing problems of assembly lines was solved by authors by using scatter search. Related complexity problems of assembly processes were studied by Modrak and Marton [7, 8]. Franz et al. [9] proposed to eliminate the amount of the overload situations by heuristic approach. Their problem is to insert unblocked orders into a given area of an existing sequence; therefore the resequencing is not considered. A combined optimization for rescheduling workpieces is given by Gujjula and Günther [10]. Valero-Herrero et al. [11] focused their research on resequencing problem by using several rules for placing and releasing workpieces. Gaia Nicosia et al. [12] described the problem of assigning operations to an orderly sequence of nonidentical workstations, which also took into account relationships of precedence and cycle time restrictions. The main purpose of this work was to minimize workstation costs by using a dynamic program-based algorithm to reduce the number of states and present several fathoming rules. Two-index vehicle flow formulation was suggested by authors Tabue and Minner [13] for the vehicle routing problem and they also presented non-explicitly numbered mixing banks with symmetry reduction. There are also existing models that show how several downstream sequences can be optimized in one upstream sequence. About standardized components with batch production were not included in their paper. They consider JIS's supply of finished workpieces after assembly rather than JIS's supply of necessary assembly materials. Even though above- and below-mentioned works come from the manufacturing industry, those scheduling and capacity balancing problems and solutions are also applicable in the service industry.

For similar workpieces of a single model and for the combination of models of different products produced on the same line, the problem of balancing the assembly line is linked to a problem of sequencing, which must decide on the sequence of assembly of the model units (Merengo et al., [14]). The sequence is important for the efficiency of a line, as the task times can vary significantly between the products. Depending on the type of intermixing of the units, two variants arise: A mixed-model line produces units of different models in an arbitrarily intermixed sequence (Bukchin et al., [15]), while a multi-model line produces a sequence of batches each containing units of only one model or a group of similar models with intermediate setup operations. Balancing and sequencing are linked to a lot sizing problem (Burns and Daganzo, [16]).

In batch scheduling problems solved by heuristics, usually dispatch rules that can build a batch sequence quickly. Due to the simple implementation and short calculation time, dispatching rules have been used for decades (Blackstone et al. [17]) and are still being applied today. Heuristics hardly find an optimal result in complex problems; they only try very quickly to find a reasonably good solution. Metaheuristics means generic algorithms that can be applied to different types of problems because their search logic is independent of the problem solved. They are usually iterative procedures that seek better solutions by changing those that have already been found (such as GA and SA). An optimal result cannot be guaranteed by metaheuristics, but using them it is possible to find good solutions in short computational times. They are usually much faster than precise methods and achieve better results than simple heuristics. Artificial intelligence was also used to solve the batch processes' short-term scheduling. These methods impersonate human reasoning and learning and apply the knowledge acquired to solve new problems.

Mane and Ghadle [18] consider a total penalty for n works, a one machine scheduling problem in which all works have a common due date. The main aim is to determine the optimal value of this due date and an optimal sequence in order to minimize the total penalty. By considering a static one machine problem, Baker and Bertrand [19] analyzed the procedures for selecting the due date. The common due date was one of the procedures for assigning the due date involved in their study.

Kim et al. [20] presented a genetic algorithm that minimizes the number of stations in an ALBP on two sides. Castillo and Gazmuri [21] considered dispatch rules for the heuristics used in batch scheduling problems and drew up a batch sequence schedule. Other researchers (see [22, 23]) explained the flow shop scheduling algorithm to minimize the completion of n-jobs of m-machines and suggested an alternative heuristic algorithm for the problem of flow shop scheduling. Dispatching rules have been used for a long time since the simple implementation and reduced calculation time are available. In complex problems, heuristics will hardly find the best result; they will find a better solution.

From the above literature review it is inferred that there is a scope for mass customized problems. In this chapter it is well addressed in the coming sections that the heuristics algorithm is useful in the categorization and optimization of different types of patients' throughput time in a mass customized environment.

3 Problem Description

In a mass customization the customers are selective in ordering of their required design rather than getting the mere supply. There are many similar models with varying simple features required to manufacture with no major deviation in manufacturing setup. So the customers bother about time for their cost to be spent on delivery of the orders placed after accepting due dates.

The sequencing of jobs and the selection of batch size in a mixed-model flow shop problems play a vital role in order to meet due dates. When selecting the batch size, the makespan should not be increased for the particular sequence. Taking all this into consideration a heuristics model is developed in this chapter. The model description is as follows.

3.1 Assumptions Made

There are ***n*** number of clients visiting the service providers and ***m*** number of operations to be performed. All the operations of different clients are processed in various workplaces (WP). The total number of workplaces is ***s***. In the example illustration the total number of clients considered is three and the total number of workplaces is eight (see Fig. 1).

The setup time and movement time between workplaces are not considered.

The batch size is variable for all the clients, the advantage of mass customization. Batch size varies from 1 to Q.

The due date for each batch of clients plays a vital role to fix the batch size.

Heuristic Steps:

Step 1: As discussed earlier, 8 workplaces and 3 clients with different requirements will be considered as shown in Fig. 1.
Then, operation times in given WPs are summarized and denoted as processing times.

Step 2: The processing times (in seconds) for all the three clients are transformed into mxn matrix (see Table 1).

Step 3: The sequence in which the clients are to be sent for various operations is calculated using benchmark heuristics like FIFO, SPT, MAKESPAN, and LPT.

Step 4: After finding out the best sequence for the clients to be processed in various workplaces, the makespan is calculated using the algorithm given by Modrak and Pandian [22].

Step 5: Subsequently, Gantt chart is drawn as shown in Fig. 2.

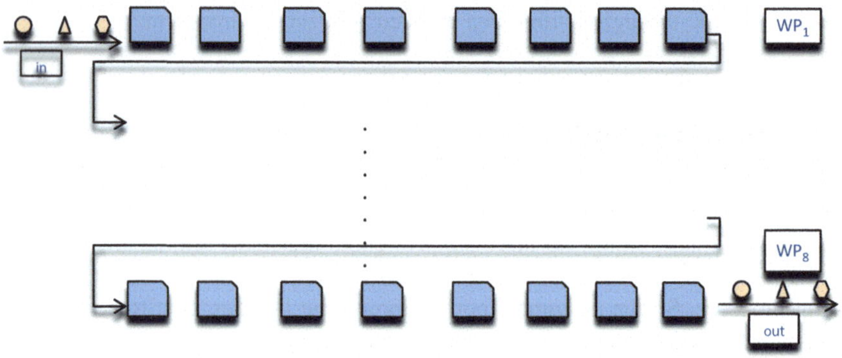

Fig. 1 Flow shop for mass customization environment

Table 1 Matrix *mxn* with processing times (seconds)

Client operation	J1	J2	J3
O1	30	20	30
O2	20	30	20
O3	20	30	20
O4	20	30	20
O5	30	20	30
O6	20	30	20
O7	20	30	20
O8	20	20	30

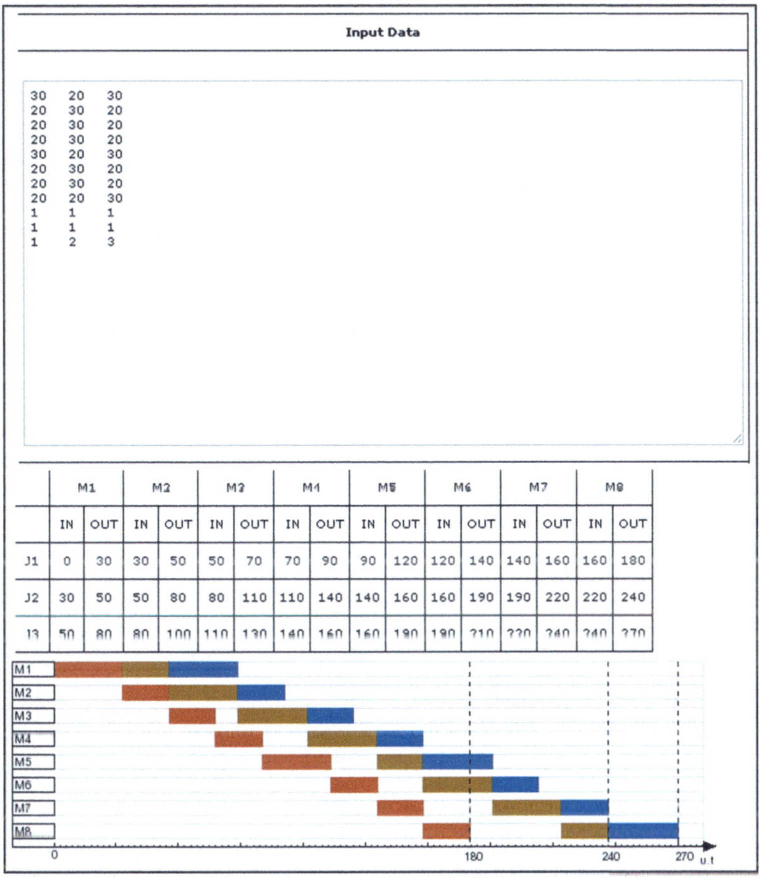

Fig. 2 Gantt chart with makespan determination

Step 6: In this step the final calculation for satisfying the due dates is done with respect to the requirements of all different clients. The proprietary software is used to find out the exact due date and corresponding batch size of the clients for further processing.

4 Illustration with Example

In our example we consider that there are three different clients to be processed on eight workplaces. Each workplace consists of a number of operations. All three clients will visit all eight workplaces for their various operations. The processing time for all clients is given in Table 1. Firstly the sequence of three clients is calculated for the given processing time. In this case, it is obtained as 1 → 2 → 3. Then the makespan is calculated as mentioned in step 4. The makespan for different batch sizes is calculated as given in Table 2.

In this table, we found that if the due date is less than 3.5 days the batch size of 350 for each client is selected which is the maximum size possible in column 2 of the table, S. No. 7, showing "select." So the appropriate batch size is selected for recommendation to managerial decision-making.

Initially the orders are collected from clients that are given in Table 3.

From the orders, there are two different cases considered before calculation of makespan and selection of due date.

Table 2 Selection of batch size corresponding to makespan

S. No.	Selection of batch size	Batch for client J1	Batch for client J2	Batch for client J3	Makespan (in days)
1	Select	50	50	50	0.47
2	Select	100	100	100	0.94
3	Select	150	150	150	1.41
4	Select	200	200	200	1.88
5	Select	250	250	250	2.34
6	Select	300	300	300	2.81
7	Select	350	350	350	3.28
8	Not select	400	400	400	3.75
9	Not select	450	450	450	4.22
10	Not select	500	500	500	4.69
11	Not select	550	550	550	5.16
12	Not select	600	600	600	5.63
13	Not select	650	650	650	6.09
14	Not select	700	700	700	6.56
15	Not select	750	750	750	7.03
16	Not select	800	800	800	7.50
17	Not select	850	850	850	7.97
18	Not select	900	900	900	8.44
19	Not select	950	950	950	8.91
20	Not select	1000	1000	1000	9.38

Table 3 Orders from clients

	Monday	Tuesday	Wednesday	Thursday	Friday	Total
J1	100	150	100	200	150	700
J2	150	100	100	150	200	700
J3	150	100	100	150	200	700

Case 1:

The orders collected on the first week from Monday to Friday will be started for processing on the second week from Monday. Here the following options are considered for calculation of makespan:

(a) Considering full amount of total orders in one flow (batch size 700)
(b) Considering half amount of total orders in first batch (batch size 350)
(c) Considering 50 numbers of orders in first batch
(d) Considering 10 numbers of orders in first batch

Case 2:

The orders collected on the first week from Monday to Friday will be divided into two. In our problem the orders collected from Monday to Wednesday will be summed up and grouped into first category. Then the orders from Thursday and Friday will be grouped into second category. Then the first category will be started for processing on the first week from Thursday and the second category will be started for processing from the second week from Monday.

Here the following options are considered for calculation of makespan:

(a) Considering half amount of total orders in first batch (batch size 350)
(b) Considering 50 number of orders in first batch
(c) Considering 10 numbers of orders in first batch

The Gantt charts showing (in Fig. 3) results of two cases are given below for interference. It is understood that case 1 (a) consumes more time, 6.5 days, and case 1 (d) takes lesser time. It is understood that if we reduce the batch size further in case 1 (d), it is possible to complete all the client orders within 1 day.

5 Results and Discussions

In order to analyze achieved results by using different cases and different batch sizes, calculated makespans are tabulated in Tables 4a and 4b.

As it is clear from Tables 4a and 4b, the best makespan has been achieved in case 1, S. No. (d). However, from customer point of view total waiting time is important that has to be calculated by including time from the time of order registration. Then, the best solution is achieved in case 2, S. No. 3. Finally, it

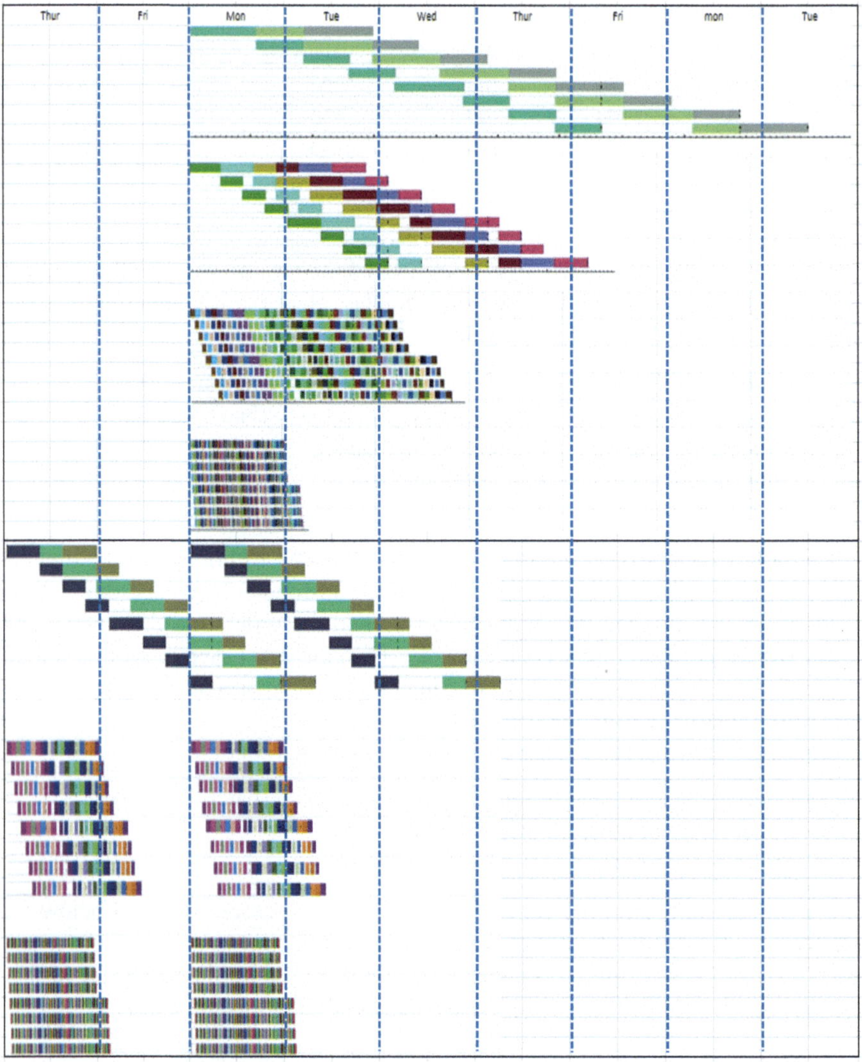

Fig. 3 Gantt chart (in days)

Table 4a Case 1

S. No.	Batch size	Makespan (days)
a	700	6.5
b	350	4.5
c	50	2.5
d	10	1.3

Table 4b Case 2

S. No.	Batch size	Makespan (days)
1	350	5.3
2	50	3.5
3	10	3.2

can be stated that there are two effective approaches for optimization of due date satisfaction. The first is via reducing batch size, and the second by determining the time intervals (in days) for completion of orders.

6 Conclusion

In this study a novel approach was discussed and analyzed to satisfy and to meet various clients' requirements in order to compete with the market demands and mass customization environment with day-to-day changing needs. In particular the service industries are facing challenges to take customized orders from various customers and thereafter to complete the services in acceptable time with required quality. For such situations, our approach offers a flexible solution to make the customers to feel satisfaction. Moreover, by this approach higher service productivity can be achieved. This method is technically still being developed for more factors to be included and we will get the feedback from various sectors about the approach. In future, the model can be extended with setup time, transport time, bottleneck operations, and labor assignment to make it more suitable for real-world problems.

Acknowledgement The authors gratefully acknowledge the contribution of the Scientific Grant Agency of the Ministry of Education, Science, Research and Sport of the Slovak Republic-VEGA under the grant No. 1/0419/16.

This publication is the partial result of the project that received funding from the European Union's Horizon 2020 research and innovation program under the Marie Skłodowska-Curie grant agreement number 734713.

References

1. Davis, S. M. (1987). *Future perfect*. Boston: Addison-Wesley.
2. Phanindra Kumar Chadalavada University of Kentucky Master's Theses Graduate School. (2006). Minimum flow time schedule genetic algorithm for mass customization manufacturing using minicells.
3. Zhang, Z., Wenming, C., Lin, S. & Qingliang, Y. (2009). An ant based algorithm for balancing assembly lines in a mass customization environment IEEE 1–4.
4. Noorul, H. A., Jayaprakash, J., & Rengarajan, K. (2006). A hybrid genetic algorithm approach to mixed-model assembly line balancing. *International Journal of Advanced Manufacturing Technology, 28*, 337–341.

5. Boysen, N., Scholl, A., & Wopperer, N. (2012). Resequencing of mixed-model assembly lines: Survey and research agenda. *European Journal of Operational Research, 216*, 594–604.
6. Rahimi-Vahed, A. R., Rabbani, M., Tavakkoli-Moghaddam, R., Torabi, S. A., & Jolai, F. (2007). A multi-objective scatter search for a mixed-model assembly line sequencing problem. *Advanced Engineering Informatics, 21*, 85–99.
7. Modrak, V., & Marton, D. (2012). Modelling and complexity assessment of assembly supply chain systems. *Procedia Engineering, 48*, 428–435.
8. Modrak, V., & Marton, D. (2014). Configuration complexity assessment of convergent supply chain systems. *International Journal of General Systems, 43*(5), 508–520.
9. Franz, C., Hällgren, E. C., & Koberstein, A. (2014). Resequencing orders on mixed-model assembly lines: Heuristic approaches to minimise the number of overload situations. *International Journal of Production Research, 52*(19), 5823–5840.
10. Gujjula, R. & Günther, H.O. (2009). Resequencing mixed-model assembly lines under just in-sequence constraints. In: International conference on computers and industrial engineering (CIE 2009). IEEE; (pp. 668–73).
11. Valero-Herrero, M., Garcia-Sabater, J. P., Vidal-Carreras, P. I., & Canos-Daros, L. (2014). Solving the car resequencing problem with mix banks. *Dirección y Organización, 54*, 36–44.
12. Gaia, N., Pacciarelli, D., & Pacifici, A. (2002). Optimally balancing assembly lines with different workstations. *Discrete Applied Mathematics, 118*, 99–113.
13. Taube, F., & Minner, S. (2018). Resequencing mixed-model assembly lines with restoration to customer orders. *Omega, 78*, 99–111.
14. Merengo, C., Nava, F., & Pozetti, A. (1999). Balancing and sequencing manual mixed-model assembly lines. *International Journal of Production Research, 37*, 2835–2860.
15. Bukchin, J., & Rubinovitz, J. (2002). A weighted approach for assembly line design with station paralleling and equipment selection. *IIE Transactions, 35*, 573–585.
16. Burns, L. D., & Daganzo, C. F. (1987). Assembly line job sequencing principles. *International Journal of Production Research, 25*, 71–99.
17. Blackstone, J. H., Phillips, D. T., & Hogg, G. L. (1982). A state-of-the-art survey of dispatching rules for manufacturing job shop operations. *International Journal of Production Research, 20*, 27–45.
18. Mane, J. K., & Ghadle, K. P. (2012). Common due-date assignment and job scheduling on single machine and parallel machines. In *The bulletin of Society for Mathematical Services and Standards*. Online: 2012-03-01 ISSN: 2277-8020 (Vol. 1, pp. 6–13).
19. Baker, K. R., & Bertrand, J. W. M. (1980). *A comparison of due-date selection rules, industrial engr. Report 1980/02*. Eindhoven: University of Technology.
20. Kim, Y. K., Kim, Y., & Kim, Y. J. (2000a). Two-sided assembly line balancing: A genetic algorithm approach. *Production Planning & Control, 11*, 44–53.
21. Castillo, F., & Gazmuri, P. (2015). Genetic algorithms for batch sizing and production scheduling. *International Journal of Advanced Manufacturing Technology, 77*, 261–280.
22. Modrak, V., & Pandian, R. S. (2010). Flow shop scheduling algorithm to minimize completion time for n-jobs m-machines problem. *Tehnicki Vjesnik-Technical Gazette, 17*(3), 273–278.
23. Semanco, P., & Modrak, V. (2012). A comparison of constructive heuristics with the objective of minimizing makespan in the flow-shop scheduling problem. *Acta Polytechnica Hungarica, 9.5*, 177–190.

Possibility of Accelerating Power Signal Calculation in a System of Internet of Things

O. N. Choporov, A. P. Preobrazhenskiy, and I. Ya. Lvovich

1 Introduction

Nowadays we can observe rapid development of the Internet of Things in wireless communication systems in urban and suburban areas.

Among network technologies, the most rapidly growing area, you can select the wireless communication system. Despite the relatively short period of operation wireless data networks of this kind are actively developing on the basis of accepted standards governing. In addition to cost and ease of installation, wireless communication systems provide essentially inaccessible cable network services in the form of mobile access.

Building a wireless access system includes a series of steps, starting from writing the terms of reference indicating the area where the system will be used, and ending with the start-up.

Because this step is very important the simulation is used which includes the main parameters of the wireless access systems, such as the service area and the signal quality. This is due to the fact that the design process is time consuming and expensive, so the use of different kinds of modeling tools eliminates errors that may occur during the operation of wireless access systems, as well as the optimal location of access points.

Special need for design automation wireless communication systems occurs in urban areas, which creates a specific radio propagation environment. The most significant contribution to the propagation of electromagnetic waves is made by

O. N. Choporov
Voronezh State Technical University, Voronezh, Russia

A. P. Preobrazhenskiy · I. Y. Lvovich (✉)
Voronezh Institute of High Technologies, Voronezh, Russia

© Springer Nature Switzerland AG 2020

D. Cagáňová, N. Horňáková (eds.), *Mobility Internet of Things 2018*,
EAI/Springer Innovations in Communication and Computing,
https://doi.org/10.1007/978-3-030-30911-4_17

factors such as multiple reflections from the walls of buildings and diffraction at the edges and corners of buildings.

To assess signal propagation a variety of methods based on statistical and deterministic analysis of the nature of signal propagation are currently used [1–3]. In this chapter we consider an approach for accelerating calculations in the computation of the characteristics for signal propagation in wireless communication. To describe the signal of electromagnetic waves in the course of their propagation, we use the ray-path model [4–6].

2 The Use of Ray Method

Deterministic methods provide higher accuracy of calculation of electromagnetic wave scattering characteristics in comparison with statistical ones, and modern communication systems operating with small spatial scales are the most effective, but they have high requirements for the accuracy of the given model of the medium.

Among these methods we can mention the method of geometrical optics method, geometrical theory of diffraction, methods of solutions of the parabolic equation, and corresponding numerical methods of solution of problems of electrodynamics.

If we use a quasi-deterministic three-dimensional model of wave propagation in urban areas and the corresponding calculation algorithm, then the main drawback will be a large amount of calculations, as it is necessary to carry out calculations at each point of space where the electromagnetic wave will propagate.

The waveguide model is based on the representation of the urban environment of the signal propagation in the form of a waveguide; it does not take into account such factors as the height of the building (in this case, one level is taken) and forest park zones.

Among the deterministic methods can be distinguished ray-tracing method, which most accurately describes the phenomenon of propagation of electromagnetic waves in space.

This method is universal and accurate, but the disadvantage of its use is a large amount of calculations, because when calculating by this method calculations are made at each point of space.

This method is based on the decomposition of the electromagnetic field into rays and the search for rays that connect the source of electromagnetic waves and the receiver, taking into account the heterogeneity of urban development.

Depending on the algorithm used, the rays are taken into consideration, which when reaching the receiver are reflected in the mirror or diffused from the walls of buildings and/or diffracted at the corners of buildings.

The model of scattering is expressed as a function of the two-beam light distribution (bidirectional reflectance distribution function, BRDF) that is the most common way of representing the surface reflection properties of the material (without taking into account the internal dispersion).

BRDF for each point on the surface of an object determines the energy transfer coefficient between any pair of directions (direction of incidence and direction of reflection) at that point.

In general, it depends on the properties of the material, the wavelength (i.e., color) of the incident electromagnetic wave, its polarization, etc.

In general, BRDF allows to model materials with complex optical properties quite accurately.

In those, for example, in which fabric, paint, etc. are applied when compared with the more familiar objects, it is possible to compare the BRDF texture.

Only the texture determines the change in color with the change in position on the surface of the material, and BRDF—the change in color with the change in the direction of illumination or observation.

Understanding BRDF in the sense of the function from two directions in one point, you can set it in different ways.

The simplest and most well-known method is to set BRDF as a simple function from a mathematical point of view with some parameters that determine the properties of the material.

With simple parameterized models, however, it is not possible to specify complex optical properties of surfaces. Sometimes, when realism is needed, more complex functions are used.

They can model more complex properties, but are not suitable for cases where you want to use real measured BRDF.

In some cases, the calculation method is applicable to a quasi-three-dimensional environment; that is, it is assumed that the height of the walls is much higher than the height of the transmitter, so that the diffraction from the roofs can be neglected.

All possible ways of distribution due to multiple reflection from the walls and diffraction at the corners of buildings can be found, systematized, and combined.

The algorithm is based on the assumption that in the propagation of radio waves in urban environments, there are only three types of rays: and emanating from the transmitter, reflected, and received as a result of diffraction.

At the first stage, the beam propagation tree is built, with the beam coming from the transmitter in the root and each point of reflection or diffraction generates additional rays; as a result, the process becomes recursive. The recursion depth is set as the initial parameter.

Then, based on the beam propagation tree, all propagation paths for a given receiver location are found. In this case, the tree does not need to be calculated again, since it was built regardless of the location of the receiver.

Further, there is a transformation of two-dimensional beam propagation into three-dimensional, and using the formulas of the geometric theory of diffraction beam is calculated taking into account the reflections and diffraction. A large number of models are designed to calculate the field distribution, designed for parallel-perpendicular type of building.

We shall research the proposed approach for different positions of the mobile station (MS_1, MS_2, MS_3, MS_4, Fig. 1) and measure the total power consisting of

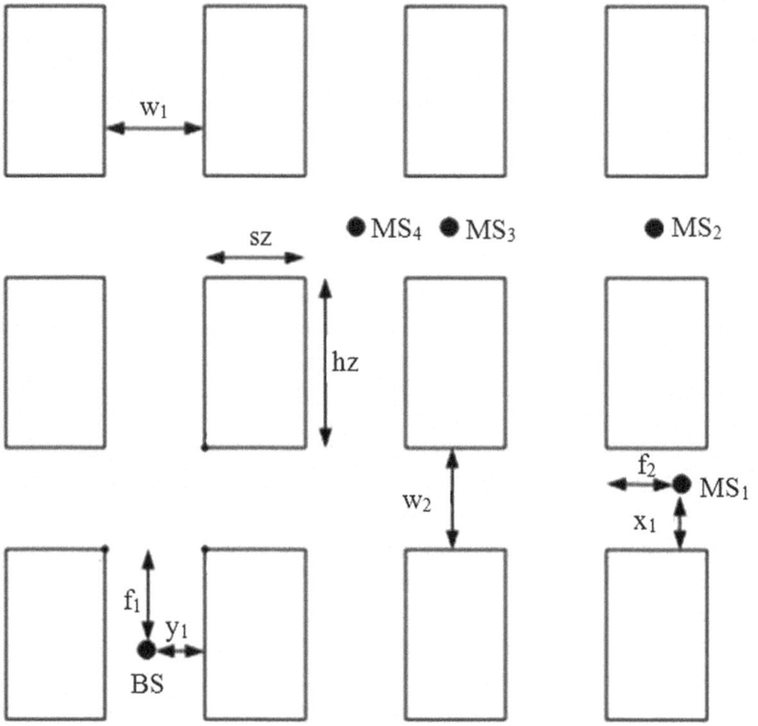

Fig. 1 Example of urban street construction

the sum capacities contributed by each beam depending of the base station (BS) removal from the first crossing (the parameter f_1, Fig. 1).

First, let us consider:

- The impact of the width of the main and secondary streets (w_1 and w_2, Fig. 1) on the signal level at the receiving point, when it is removed from the BS.
- The impact of the change in the MS position on a secondary street (change of k_2) on the signal level.
- The impact of the signal level change in the MS position on a secondary street (change in x_1).

After having made calculations, we can conclude that by changing the location of BS along the main street (that is, in case of the BS approaching to the intersection with a secondary street where the MS is located, Fig. 1) the signal level at the receiving point increases (and the attenuation decreases).

We may also notice that the number of rays that determine the signal level decreases.

The greatest change in the parameters at the admission point is observed when the MS is placed at the first secondary street.

According to obtained data, it could be concluded that by changing the width of the main street (w_1 reduction, given a fixed value of the width of the secondary street) the signal level and the number of rays at the receiving end decrease.

If we change the width of the secondary street (w_2 reduction, given a fixed value of the width of the main street) the signal level and the number of rays in the receiving point increase.

By changing the MS position along the secondary street (distance from BS), signal strength and number of rays at the receiving point decrease.

The largest contribution to the total power is made by the rays, which have the smallest path (i.e., the number of reflections from the building walls is low).

In this regard, there is a possibility to take into account an error in the determination of the signal level Δ (about 1–2 dB), which will reduce the number of considered rays that contribute to a minimum signal level, and accordingly reduce the time for calculating the signal level for the current position of the MS.

To perform these operations, an algorithm of calculation of the ray number is used.

In so doing, the principle of the maximum correlation coefficient [7, 8] in the parameters of the polynomial is used, describing how the signal level is distributed for the rays depending on how the BS and MS are located.

According to the fact that at the reception point a large number of rays can propagate, in order to reduce the amount of stored data in the database (DB), they should be stored as a polynomial of N-th degree. It provides opportunities for considerably reducing the volume of stored data.

As input data in calculation of the number of ray algorithm the values of the BS and the MS coordinates are used.

Based on the BS and the MS coordinates, we can calculate the distances in main and secondary streets to the reception point.

In the calculation of the correlation coefficient the following symbols are presented:

α_1 is a vector where each element is the power of the transmitted signal, which is a function of the distance among the BS and MS, that, in turn, corresponds to the input parameters.

α_2 is the vector containing the appropriate correspondences in the database.

k is a vector, containing the correlation coefficient.

As the output result of the calculation algorithm the ray number connecting the BS and MS will be used along with other parameters in the calculation algorithm of the signal level.

This algorithm works within the model based on the method of ray tracing [9]. As additional input parameters to the ray number calculation algorithm the error Δ is accepted by the signal power relative to the maximum level. MS_2, MS_3, and MS_4 are in the centers of their respective rectangles. Parameter values were as follows: $w_1 = 30$ m and $w_2 = 20$ m, sz = 20 m, hz = 40 m, $(hz/2) \le f_1 \le hz$, $(w_1/2) \le y_1 \le w_1$, $(sz/2) \le f_2 \le sz$, and $(w_2/2) \le x_1 \le w_2$.

The functioning of the algorithm calculating the number of rays is carried out in two stages:

- Using correlation coefficients, where the polynomial is calculated according to the total power of the power contributed by each ray
- On the basis of the found polynomial where the number of rays is calculated by taking into account the specified margin of the error

3 Use of Correlation Method for Accelerating Calculation

Taking into account that the case of sufficiently dense urban development is practically the most interesting, it is reasonable to carry out an assessment for the worst conditions of distribution, i.e., for reception on cross streets (perpendicular to the line connecting the points of reception and transmission).

In such a model, we assume that for the walls of buildings we have absolute absorption and there is still no influence (as well as the surface of the earth) on the field strength at the receiving point.

The largest contribution to the total power of creating the rays, which are the smallest path (i.e., the number of reflections from the walls of the buildings is low), in this regard, perhaps the addition of error in the determination of the signal level (of the order of 1–2 dB), which will reduce the number of considered rays that contribute a minimum signal level, and accordingly reduce the time spent calculating the signal level for the current position in MS.

To perform these operations, the module for calculating the number of rays is part of the module models and calculates the signal propagation.

The principle of maximum correlation coefficient in the parameters of the polynomial describing how the signal level for the rays is distributed depending on how the BS and MS are located is applied for the operation of the ray number calculation unit.

Let us present the sequence of the proposed algorithm implementation:

1. Input of initial parameters
2. Calculation of the number of polynomial in the database and the correspondences of the ray number by the distance from the BS to MS by means of the correlation coefficient
3. Calculation of the total number of rays for selected polynomial
4. Calculation of the number of rays for the selected polynomial taking into account the error Δ
5. The output of ray number

The process of searching the number of rays N is reduced to performing the following procedure for finding the maximum of the correlation coefficient [2]:

$$k_{\max} = \frac{\int_{S_1,\ldots S_L} \alpha_{2\max}(\gamma_1,\ldots,\gamma_L)\,\alpha_1(\gamma_1,\ldots,\gamma_L)\,d\gamma_1\ldots d\gamma_L}{\int_{S_1,\ldots S_L} \alpha_{2\max}(\gamma_1,\ldots,\gamma_L)\cdot\int_{S_1,\ldots S_L} \alpha_1(\gamma_1,\ldots,\gamma_L)\,d\gamma_1\ldots d\gamma_L} \tag{1}$$

where $S_1, \ldots, S_L$—the limits of parameters $\gamma_1, \ldots, \gamma_L$ variation.

For the vector of correlation coefficients by means of standard procedures we determine maximal elements for which the numbers are corresponded to the optimal numbers of polynomials in the database for the given input parameters.

For approximating the given curve we will use the method of least squares (MLS) [3].

This problem is linked with the construction of weighted approximation by MLS. Basic functions can be placed in function in a nonlinear manner, and evaluation function can be replaced by the maximum deviation, or we can use any other evaluation function.

On the basis of this method of approximation under this evaluation function it is impossible for us to determine the best coefficients of the approximation with a finite number of operations.

The problem is reduced to solving a system of linear equations. The main peculiarity of the model described above is that when the distance of the MS from the BS is increased, the number of rays at reception point increases, and, therefore, the order of the polynomial increases.

4 Results

According to the chosen polynomial degree, the error between the experimental and theoretical values is equal to 5%.

Figure 2 presents two curves: experimental (based on the values obtained in the course of calculation) and theoretical (based on approximation using MLS).

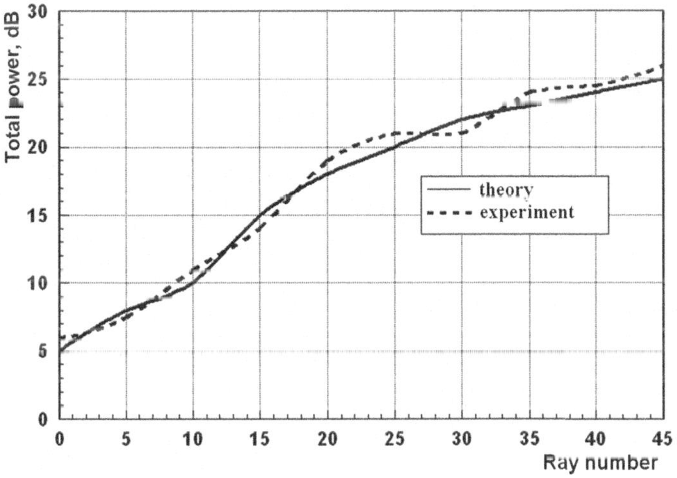

Fig. 2 Approximation of the multimode power correspondence on the ray number

For a given error (6%) the polynomial degree is equal to 5; therefore, when storing in the DB the coefficients of the polynomial instead of the experimental values, it allows to reduce the amount of information.

According to the algorithm, on the base of the selected polynomial the number of rays is calculated with the introduction of error Δ, relative to maximum level. For example, for the graph depicted in Fig. 2, with an error $\Delta = 1.5$ dB, the number of rays taken into account at the reception point is reduced from 45 to 14.

5 Conclusion

The proposed approach which is based on correlation coefficients and MLS approximation usage allows to reduce the number of iterations in the calculations of the signal level by means of the method of ray tracing. For $\Delta = 1.5$ dB, the number of iterations is reduced by 2–3 times (depending on the remoteness of MS).

References

1. Sarkar, T. K., Ji, Z., Kim, K., Medour, A., & Salazar-Palma, M. (2003). A survey of various propagation models for mobile communication. *IEEE Antennas and Propagation Magazine, 45*(3), 51–82.
2. Ichitsubo, S., Furuno, T., Taga, T., & Kawasaki, R. (2000). Multipath propagation model for line-of-sight street microcells in urban area in. *IEEE Transactions on Vehicular Technology, VT-49, 2*, 422–427.
3. El-Sallabi, H. M., & Vainikainen, P. (2003). Radio wave propagation in perpendicular streets of urban street grid for microcellular communications. *Progress In Electromagnetics Research, PIER, 40*, 229–254.
4. Hunt, W. (2015). Virtual Reality: The Next Great Graphics Revolution. Keynote Talk HPG 2015, Oculus Research, (pp. 1–2).
5. Schettino, D. N., Moreira, F. J. S., & Rego, C. G. (2007). Efficient ray tracing for radio channel characterization of urban scenarios. *IEEE Transactions on Magnetics, 43*(4), 1305–1308.
6. Wen, J. A correlation coefficient of belief functions 2017. https://arxiv.org/abs/1612.05497v2.
7. Zhang, W. (2000). Fast Two-Dimensional Diffraction Modeling for Site-Specific Propagation Prediction in Urban Microcellular Environments. In: IEEE Transactions on Vehicular Technology, VT-49, 2, March (pp. 428–436).
8. Bure, V. M., Parilina, E. M. (2013). Probability theory and mathematical statistics (p. 416), St. Petersburg, Russia.
9. Liu, Z.-Y., & Guo, L.-X. (2011). A quasi three-dimensional ray tracing method based on the virtual source tree in urban microcellular environments. *Progress In Electromagnetics Research, 118*, 397–414.

Concept of the Factories of the Future in Slovak Industrial Companies

Jozef Hercko, Miroslav Fusko, and Lubica Kotorova Slusna

1 Introduction

Current industrial factories are facing new challenges related to emergence of fourth industrial revolution, also referred to as Industry 4.0. Industry 4.0 represents new dimension of implementation of technological solutions in industrial factories' environment. Factories are thus exposed to the requirements of their own transformation into modern Factories of the Future. Industry 4.0 represents machines connected into collaborative community [1]. The concept Factories of the Future is based on modularity, collaboration, IoT, mass customisation, modelling and prediction, and distributed control [1]. Industry 4.0 is becoming one of the megatrends, which is being gradually transformed into formulation of strategic innovative directions at both corporate and sector levels [2].

The concept Factories of the Future, which is part of the broader concept Industry 4.0, is not fully defined, while it is subject to constant changes and replenishment depending on the development of key technologies. Nonetheless, EFFRA defined in its strategic documents [3] six strategic technologies and activators for the Factories of the Future, namely advanced manufacturing processes, mechatronics for advanced manufacturing systems, information and communication technologies, manufacturing strategies, knowledge worker modelling, simulation and forecasting. For comparison, the Italian cluster of intelligent factories defined in its publication Fabbrica Intelligence [4] eight supportive technologies for the Factories of the Future, namely advanced production processes, mechatronics for advanced man-

J. Hercko (✉) · M. Fusko
University of Žilina, Univerzitna 1, Žilina, Slovakia
e-mail: jozef.hercko@fstroj.uniza.sk; miroslav.fusko@fstroj.uniza.sk

L. K. Slusna
Strategos, s.r.o., Klincova 37, Bratislava, Slovakia

© Springer Nature Switzerland AG 2020

D. Cagáňová, N. Horňáková (eds.), *Mobility Internet of Things 2018*,
EAI/Springer Innovations in Communication and Computing,
https://doi.org/10.1007/978-3-030-30911-4_18

ufacturing, methods and tools for simulation, planning and forecasting, ICT for manufacturing, strategies for manufacturing management, production and deployment of innovative materials, technologies for sustainable manufacturing, and technologies and methods for human-centred manufacturing. Main innovative element of the Factories of the Future is enhancement of information and communication technologies, enabling full application of the concept of company management, such as computer-integrated manufacturing (CIM), computer-integrated construction (CAD) and computer-integrated technical preparation of production (CAPP), which have grown into the concept of digital factories. As part of the Factories of the Future concept, there are many other concepts such as dynamic manufacturing network providing agile infrastructure for networking development of products [5], or end-to-end manufacturing between FoFs [6].

It is important that modelling of development of objects in factories (products, processes and production sources) has mastered a wide range of data. These data come from a variety of sources, such as system tracking, production planning and performance evaluation tools. It is important that data having physical characteristics (such as location of objects) and abstract properties (for example, state transitions) are consistent with their static representation [7]. Modelling requires effective software tools which must be optimised and customised to be able to provide all necessary features, which will help to integrate already existing tools and processes and bring them into line with strategy [8]. In this context, intercultural management is crucially [9] connected with competencies of employees [10].

Intelligent manufacturing is characterised by vertical and horizontal integration as well as end-to-end integration for end-to-end connections, i.e. integration of IT system in different horizontal levels of CPS networks [11]. Industry with vertically integrated intelligent manufacturing is called an intelligent factory. Such factory meets the following characteristics: there is a connection of data and network resources, high level of interconnection, IoT and edge computing [12, 13].

Modelling and simulations significantly contribute to streamlining the decision-making process. Through them, virtual manufacturing and virtual factories can be realised [7].

At present, it is necessary to address the problems and narrow places in the field of informatisation and manufacturing. One of the possible solutions is usage of sets of technologies, and therefore the connection through cloud manufacturing is being used. It includes many technologies such as cloud, IoT, service-oriented technologies and other powerful technologies [14].

Due to the complexity of the topic of the Factories of the Future the factories are facing considerable uncertainty that affects their willingness to implement new comprehensive solutions. These uncertainties are reflected in different sectors of industry and countries, i.e. textile industry in Germany [15]. Developing to this model is one of the challenges for effective managing of transformation of the economy and broad implementation of the Industry 4.0 principles. An effective model should ideally be applicable also to higher level of the sectors, eventually clusters, which became part of the industrial development of many countries [16]. Clusters will play a significant role in pushing of the Industry 4.0 concepts and the

Factories of the Future into real life, and thereby influencing their competitiveness. Recently, there have been several approaches to determine the readiness of the factories to transform into Factories of the Future.

The aim of this chapter is to verify the methodology by which factories will be able to identify the current state of implementation of each key component of the concept of the Factories of the Future and consequently define areas with the highest potential to make progress in managing the company. It will test the capabilities of two industrial factories operating in Slovakia where industry produces up to 25% of GDP.

2 Testing of the Readiness of Factories

Proposed model integrates core manufacturing processes and key pillars of company's growth which in proposed concept are people, technologies and standards related to safety. Basic processes within this model are pre-manufacturing processes, main manufacturing processes, handling and supporting manufacturing processes and finally sales processes. Model is also complemented with strategy and corporate culture, which co-create the overall environment and impact the competitiveness of factories. Within the model a set of 130 questions of model were prepared. Model is based on all key trends in modern factories as pull systems, advanced industrial engineering, digital and smart factories, Factories of the Future, Industrie 4.0, digitalisation, etc. All parts of this concepts are implemented into the model presented (see Fig. 1).

One of the primary goals of each company is to increase the effectivity of the particular processes by using simulation. Simulation enables to imitate process in production area, logistics, assembly, etc. [17]. Simulation forms the basis for digital factory. Digital factory represents the most progressive approach to complex, integrated design of products, manufacturing processes and manufacturing systems. The concept of digital factory is based on three elements [18]:

- Digital product with its static and dynamic aspects
- Digital design of manufacturing process
- Digital manufacturing with usage of planning data for increasing company processes

Production is designed in digital factory on the basis of pull production. In a pull production system, processes are based on customer demand. The main difference between classic push production systems and pull production systems is that the first one schedules work releases based on demand, and the latter one authorises work releases based on system status. Among other advantages, pull production systems can provide low unit cost, good customer service, and high external quality and flexibility [19].

One of the key factors of the Factory of the Future is autonomous internal logistics. The prevailing, manually controlled, kind of internal logistics which

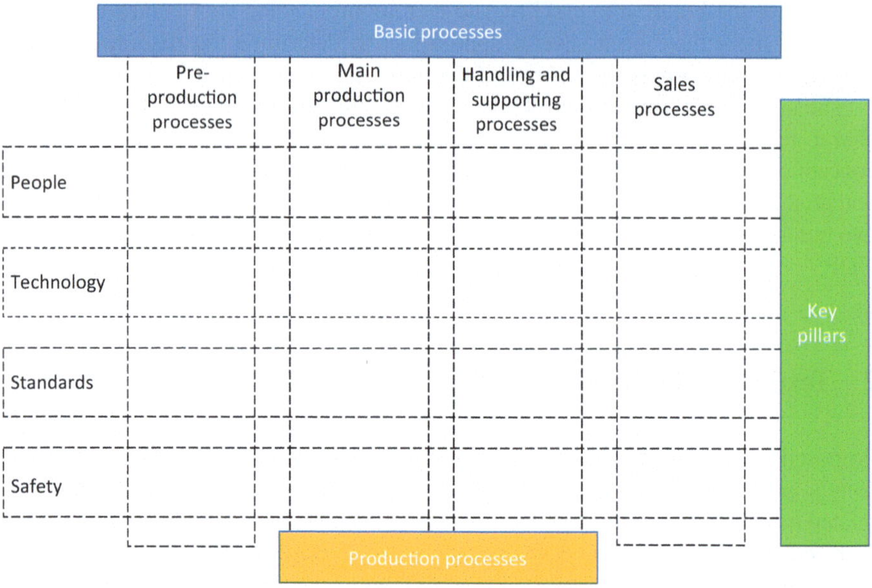

Fig. 1 Linkage of technological pillars and manufacturing processes in factories of the future

provides for handling and transport of material in the production process and also in the preparation and support stages of production creates a high-risk environment with high demands for manual labour. The most commonly used logistics means—handcarts and motorised carts—are usually noisy in service, environmentally unfriendly and required to be operated by experienced and costly operators (drivers). It is estimated that by 2030, up to a half of European factories will be able to operate its own internal logistics with the use of autonomous mobile robotic systems (MRS) [20]. Forklift operators spend most of the shift in the seated position associated with frequent lateral and axial spine rotations. After the 8-h work shift feeling of pain or eventually stiff muscles can be expected in many operators. In the long term, when a spine is loaded repeatedly in the same way, the feeling of pain can turn into a chronic cervical or lumbar back injury. Improvement of ergonomic conditions and therefore elimination of the load on the operator are factors that can be primarily influenced by the producer of the forklift and secondary by the right choice of the type forklift, for the given activity [21]. Along with the development and the application of innovative solutions in production and logistics, interest in modern tools in the ergonomic area does not abate. Thanks to them, in an effort to fit between companies that set out to the way Industry 4.0, it is possible to improve working conditions and the quality level of workplaces [22].

Testing of the readiness of factories was realised in two industrial companies operating in the Slovak Republic. Both factories are involved in global value chains. The evaluation of respondent's answers was realised with usage mathematical-statistical methods. After the evaluation of individual areas, the calculation method within individual areas as well as the overall evaluation was determined.

Assessment (H) in each of the areas (i) is calculated as conjunction of the quotient of the implementation coefficients (K) in concerned area and number of relevant questions in concerned area and maximum point score in concerned area:

$$H_i = \frac{\sum_{j=1}^{n} K_j}{R_i} \cdot M_i \tag{1}$$

where:

H—partial evaluation achieved in individual area

K—implementation coefficient of identified potentials for individual evaluation questions, and can range in values from 0 to 1

R—number of relevant questions in each of the areas that means number of questions which were answered YES or NO by the respondents

M—the maximum number of points that can be earned in evaluated area

i—number of evaluated areas

j—number of responses in individual areas for which the implementation coefficient was determined

n—number of questions in individual areas

Subsequently, the overall rating was carried out. It is referred to as FoF Index—Factories of the Future Index. The proposed overall rating can be simply defined as the sum of the partial evaluations, with mathematical entry given as

$$FoF\ \text{Index} = \sum_{i=1}^{6} H_i = \sum_{i=1}^{6} \frac{\sum_{j=1}^{n} K_j}{R_i} \cdot M_i \tag{2}$$

where:

K—implementation coefficient for individual evaluation questions

R—number of relevant questions in area that means number of questions which were answered exactly yes or no by respondents

M—maximum number of points that can be earned in evaluated area

i—number of evaluated areas; in the proposed model the value is equal to 6

j—number of responses in individual areas by which the implementation coefficient was determined

n—number of questions in individual areas

The model clearly identifies areas in which the company should improve. Practical verification of proposed solution was realised in two factories operating in Slovakia. Considering the fact that both companies allowed the realisations of the model under condition of full anonymity of factories and individual respondents, factories are named as Factory A and Factory B. Factory A is operating long term in Slovakia in the manufacturing industry and currently it belongs to a global concern focused on manufacturing of components for the automotive industry. Factory A is an automotive industry supplier of first tier. Factory B has a long tradition in

Slovakia; for more than 10 years it has been owned by a foreign investor and is part of large global concern. Methodological testing of the readiness was carried out with direct involvement of top management of the factory and involving all necessary employees on different level of management.

3 Experimental Verification

This chapter describes the results from two experimental verification of the proposed methodology that integrate key manufacturing processes and key technology pillars in terms of the Factory of the Future concept. The methodology was verified in two factories from the second sector (manufacturing and industry) of the Slovak Republic.

3.1 Experimental Verification in Factory A

First step of the verification of the methodology was to acquaint with the methodology itself, to provide necessary information on the course and agreement on mutual verification rules. Factory required full anonymity of provided information and of respondents of verification; during the verification process, this was accepted. The first step of the proposed methodology was evaluation of company's strategic documents, in particular vision, mission and strategic dimensions. The aim of this part was to identify whether the factory has the components of the Factories of the Future incorporated in these documents. Based on the evaluation, it was identified that the factory, as part of the global concern, has strategic documents defined on a flat basis for all its manufacturing units (see Fig. 2).

In individual assessment of areas Factory A earned evaluation level between about 60% and 80%. Therefore, the factory is at very good level of implementation of the concept in individual areas, slightly exceeding to excellent level in one area.

The next step after the evaluation of current state was creation of an action plan. To create it, it's necessary to know the areas in which the factory has the highest potential for improvement. Based on the proposed methodology, individual questions were evaluated and potentials for improvements were identified and the number of points that can be earned by factory by eliminating the deficiency.

In terms of the evaluation process, the factory lost overall 32,29 points in a total of 107 questions. Number of questions affecting the overall score is higher than 0,75 point. Based on the evaluation of the assessment results using the proposed methodology, the areas have the greatest impact on the overall score of the factory. Considering this, two action plans were proposed to the management.

Pillar	Maximum	Score
Strategy	10	7,50
People	21	15,25
Technology	28	16,89
Standards	14	11,28
Safety	7	5,13
Culture	20	12,62
FoF index	*100*	*68,66*

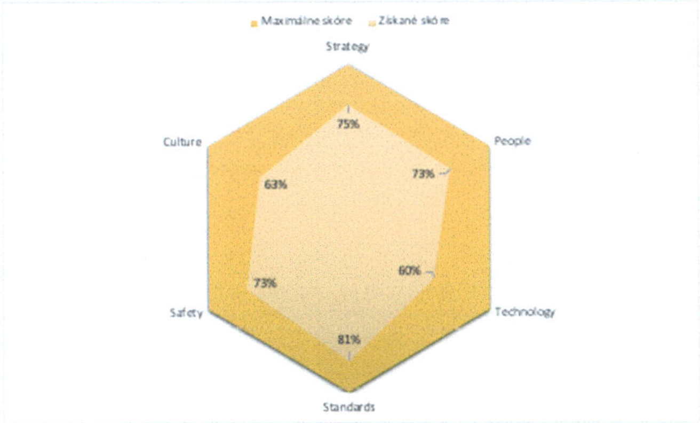

Fig. 2 Readiness of Factory A

3.2 *Experimental Verification in Factory B*

Factory B ranged from almost 61% to almost 90% across each evaluated area. This assessment result, in the context of the global evaluation, means a partial evaluation between the "high level of implementation" and the "excellent level of implementation" of the Factories of the Future. The evaluation in each area depends on the maximum number of points, the number of relevant questions and the individual answers (see Fig. 3).

The overall post-creation report was provided to other members of the team for feedback as well as to top management. Since the evaluations themselves were conducted in the form of a controlled workshop and the recorded inputs were the result of a consensus of the evaluating team, there were no comments from the team members. The results were presented to top management by a team member—manufacturing director. Management has accepted and acknowledged the submitted report. Due to the fact that no comments were made, there was no need to intervene the evaluation report. Based on realised evaluation process, it is necessary to draw up an action plan. To build it, it is necessary to know the areas in which the company has the highest potential for improvement. On the basis of the proposed methodology, individual questions were evaluated and potential for improvement was identified and the number of points that can be earned by a factory by eliminating the lack of implementation. In terms of evaluation, the factory lost in assessment 24,53 points

Pillar	Maximum	Score
Strategy	10	7,75
People	21	16,50
Technology	28	17,00
Standards	14	11,67
Safety	7	6,25
Culture	20	16,30
FoF index	*100*	*75,47*

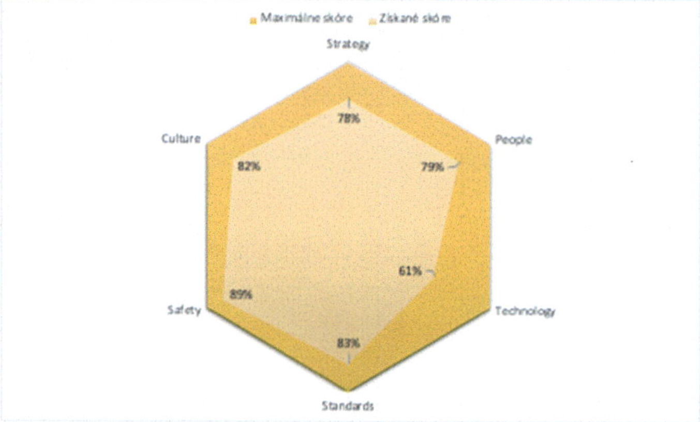

Fig. 3 Readiness of Factory B

in a total of 79 questions. The number of questions affecting the overall score is higher than 0,75 point.

4 Conclusion

A perfect understanding of the current state of the company provides a clear view of the potential areas to be improved. Companies are large and complex entities, and the changes made are directly reflected in a number of areas. On the basis of the audit of preparedness for the new direction of the industry, companies have a real and objective assessment of their state. The prepared assessment of readiness has shown industrial companies in Slovakia a further possible direction of their investments and activities in the company. Regular assessment of readiness enables the company to monitor its progress and also to compare it with competing businesses according to the chosen parameters. Our further research will be based on the creation of a comprehensive benchmarking system where companies will be able to compare on the basis of criteria, such as comparison with competitors, with other companies in the region and so on.

Acknowledgment This work was supported by the Slovak Research and Development Agency under the contract No. APVV-16-0488.

References

1. Brettel, M., Friederichsen, N., Keller, M., & Rosenberg, M. (2014). How virtualization, decentralization and network building change the manufacturing landscape: An industry 4.0 perspective. *International Journal Mechanical, Aerospace, Industrial and Mechatronics Engineering, 8*(1), 37–44.
2. Gregor, M., & Magvaši, P. (2014). Globalizácia a megatrendy. *ProIN, 4*, 67–73.
3. European Commission. (2013). *Factories of the future* (1st ed.). Brussels: European Commission.
4. Associazione Cluster Fabbrica Intelligente. (2015). *Fabrica intelligente* (1st ed.). Bologna: Associazione Cluster Fabbrica Intelligente.
5. Tchoffa, D., Figay, N., Ghodous, P., Exposito, E., Kermad, L., Vosgien, T., & El Mhamedi, A. (2016). Digital factory system for dynamic manufacturing network supporting networked collaborative product development. *Data & Knowledge Engineering, 105*, 130–154.
6. Ferreira, J., Sarraipa, J., Ferro-Beca, M., Agostinho, C., Costa, R., & Jardim-Goncalves, R. (2017). End-to-end manufacturing in factories of the future. *International Journal of Computer Integrated Manufacturing, 30*(1), 121–133.
7. Terkaj, W., Tolio, T., & Urgo, M. (2015). A virtual factory approach for in situ simulation to support production and maintenance planning. *CIRP Annals—Manufacturing Technology, 64*(1), 451–454.
8. Kagermann, H. et al.(2013). Recommendations for implementing the strategic initiative INDUSTRIE 4.0. [online]. Frankfurt/Main: 2013, pp. (82). http://www.acatech.de/fileadmin/user_upload/Baumstruktur_nach_Website/Acatech/root/de/Material_fuer_Sonderseiten/Industrie_4.0/Final_report__Industrie_4.0_accessible.pdf.
9. Cagáňová, D., Čambál, M., & Weidlichová Luptáková, S. (2010). Intercultural management—trend of contemporary globalized world. *Electronics and electrical engineering, 6*(102), 51–54.
10. Čambál, M., Cagáňová, D., & Šujanová, J. (2012). The industrial enterprise performance increase through the competency model application. In J. Surakka (Ed.), *The 4th European Conferences on Intellectual Capital* (Vol. 4). Helsinki: Academic Publishing International.
11. Chen, J., Wang, C. H., Zhao, Z. M., Chen, K., Du, R. Y., & Ahn, G. (2018). Uncovering the face of android ransomware: Characterization and real-time detection. *IEEE Transactions on Information Forensics and Security, 13*(5), 1286–1300.
12. Chen, M., Miao, Y., Hao, Y., & Hwang, K. (2017). Narrow band Internet of things. *IEEE Access, 5*, 20557–20577.
13. Chen, M., Hao, Y., Hu, L., Huang, K., & Lau, V. (2017). Green and mobility-aware caching in 5g networks. *IEEE Transactions on Wireless Communications, 16*(12), 8347–8361.
14. Zhang, L., Luo, Y., Tao, F., Li, B. H., Ren, L., Zhang, X., Guo, H., Cheng, Y., Hu, A., & Liu, Y. (2014). Cloud manufacturing: A new manufacturing paradigm. *Enterprise Information Systems, 8*, 2), 1–2),21.
15. Küsters, D., Praß, N., & Gloy, Y. (2017). Textile learning factory 4.0—Preparing Germany's textile industry for the digital future. *Procedia Manufacturing, 9*, 214–221.
16. Balog, M. (2016). Development factors of cluster organizations in the Slovak Republic. *Ekonomický časopis, 64*(2), 149–168.
17. Vavrík, V., Gregor, M., & Grznár, P. (2017). Computer simulation as a tool for the optimization of logistics using automated guided vehicles. *Procedia Engineering, 192*, 923–928.
18. Bučková, M., Krajčovič, M., & Jerman, B. (2017). Impact of digital factory tools on designing of warehouses. *Journal of Applied Engineering Science, 15*(2), 173–180.

19. Bubeník, P., Horák, F., & Hančinský, V. (2015). Acquiring knowledge needed for pull production system design through data mining methods. *Communications—Scientific letters of the University of Žilina, 17*(3), 78–82.
20. Gregor, T., Krajčovič, M., & Wiecek, D. (2017). Smart connected logistics. *Procedia Engineering, 192*, 265–270.
21. Kramárová, M., Dulina, Ľ., & Čechová, I. (2017). Forklift workers strain of spine at industrial logistics in depending on human work posture. *Procedia Engineering, 192*, 207–212.
22. Gašová, M., Gašo, M., & Štefánik, A. (2017). Advanced industrial tools of ergonomics based on Industry 4.0 concept. *Procedia Engineering, 192*, 219–224.

Development of Employee Extent Commitment in Innovation Processes in Slovak Organisations

Zdenko Stacho, Katarína Stachová, and Dagmar Cagáňová

1 Introduction

The internal potential of an organisation is most significantly influenced by the potential and commitment of its own employees, which can be directly influenced by the individual activities of human resource management [1]. Business support in this area can be directly linked to increased employee performance [2–5], positive perception of the organisation by its employees [4], as well as their satisfaction [6, 7], engagement [8–10], and willingness to cooperate internally within the boundaries of the organisation [11].

To ensure the competitiveness and advancement of an organisation not only in a stable but also predominantly in an unstable environment, its employees must be educated and engaged [9, 10, 12]. It is the dynamically changing environment and the forthcoming fourth industrial revolution that represent a great challenge for businesses, not just in Slovakia, and an opportunity to shift the borders. The most effective and ethical possibility to grasp this opportunity can be the focus of organisations on increasing innovation performance [1].

Creativity as such has been more and more comprehended as a social phenomenon in recent years [13]. As E. Franková states, it primarily results from the fact that the source of all creative ideas is the human mind. A number of empirical studies have confirmed the fact that creative potential is integral to every human, and that every person is, or can be, creative [14–16]. The level of creativity or the level

Z. Stacho · K. Stachová
Institut of Civil Society, University of Ss. Cyril and Methodius in Trnava, Trnava, Slovakia

D. Cagáňová (✉)
Institute of Industrial Engineering and Management, Faculty of Materials Science and Technology in Trnava, Slovak University of Technology in Bratislava, Slovakia
e-mail: dagmar.caganova@stuba.sk

© Springer Nature Switzerland AG 2020

D. Cagáňová, N. Horňáková (eds.), *Mobility Internet of Things 2018*,
EAI/Springer Innovations in Communication and Computing,
https://doi.org/10.1007/978-3-030-30911-4_19

of the development and application of creative potential achieved by individuals primarily depends on themselves and the influence of their environment. Institutions can significantly affect the level and number of their creative employees primarily by implementing suitable methods of recruiting and selecting personnel [17], and by creating environment in the organisation [18, 19].

Suitably set and tested criteria upon recruiting and selecting personnel can help organisations recognise the creative potential of job applicants, and their potential of willingness to commit to their organisation. The given criteria are of importance in the context with an ability and willingness to create and implement innovations for managers as well as specialists or manual workers, as engagement of all employees in innovations is desirable.

Factors such as a leadership style and development of abilities and skills of human capital determine to a large extent a company's innovative direction [20–25]. As Khazanchi et al. [26] remarks, the characteristics of an innovative company include awareness, goal-oriented motivation, as well as a surplus of skills, knowledge, support infrastructure, and a highly participatory culture. According to Herbig and Duphy [27], any creative activity is negatively affected by bureaucracy. On the other hand, innovations are stimulated by free communication, decentralisation, and certain trust between different hierarchical levels, thus enabling them to communicate their ideas and share knowledge within organisational structures [28].

The main objective of the above-mentioned approach (aimed at motivating workers to participate in innovations) is to create an environment where dialogues between managers and other team members are important, and where workers' interests are respected in the matters of decision-making and management [29].

With regard to the fact that the engagement and encouragement of employees should be in the competence of managers, or direct superiors, it is necessary to realise the change that needs to be captured in the context of their competence. While administrative abilities were emphasised in association with traditional managers, the leaders of fundamental transformation changes need to be able to handle new and unfamiliar tasks, and, especially, to persuade their environment to perform them [30]. The key skills of creative managers thus include their abilities to [17, 18, 31]:

- Flexibly react to constant changes in conditions
- Bring creative solutions
- Create the work environment in favour of changes

The creativity and emotional intelligence of managers are among the important sources of success of organisations in the twenty-first century. Creativity is also a crucial input ensuring the quality and success of innovation processes. If an enterprise wants to systematically develop the creative potential of managers, it is only possible through a good selection of managers, their optimal assessment, and systematic education [17]. Creative managers should be not only creative, but also able to accomplish tasks, to inspire people with their vision, to lead people in team work, and to be willing to share credit, thus making their co-workers successful people. Therefore, managers with their creative approach become the key persons in organisations, whom success or failure in the process of change implementation depends on. Managers can turn the entire organisation into an innovation lab, but they need to balance the tension between productive and innovative practices [32].

The creative environment of a company that is in favour of innovation can be characterised by a number of attributes; however, from the perspective of this chapter, we consider the following as of the greatest importance [19, 32–37]:

- Incentive scheme is set so as to positively affect the enhancement of innovation activities; reward is not only associated with performance, but also with innovative behaviour.
- All company employees are supported by management in their commitment in innovations and different improving processes or procedures.
- Managers perceive all employees involved in the innovation process as equal colleagues and they appreciate and support their ideas. That is a reason why employees feel the companionship with their company, and they willingly accept and implement innovations.

The aforementioned clearly indicates that the key to the creation, development, and implementation of innovations in organisations is the presence of a creative manager [17, 18], coordinated interconnection of innovation activities with corporate strategies [38], and comprehensiveness in engaging employees in the innovation processes [32]. The given areas were also analysed by the authors of this chapter.

2 Materials and Methods

Several partial objectives were accomplished in order to achieve the objective of this chapter, particularly including a questionnaire survey, statistical evaluation of the research hypotheses, and evaluation of the overall change of the analysed attributes using fixed-base and chain indices of change.

Research questions were formulated and conditioned by the research objective of this chapter. Research hypotheses were formulated on the basis of set research question, and they are tested by a questionnaire survey and subsequent statistical Pearson test.

Research Question: What was the score achieved by the examined organisations in their activities related to the engagement of all employees in innovation processes? Are there statistically significant differences between the regions of operation of the organisations?

Hypothesis H1: There is a statistically significant relation between the region of operation of organisations and performance of the activities related to the engagement of all employees in innovation processes.

For the needs of this chapter, data obtained from research conducted between 2014 and 2018 were applied, and the top representatives of Slovak institutions were interviewed. Its objective was to uncover the present state of talent management in Slovak organisations. A questionnaire in which the participants in the study responded to 90 questions focused on the issue of formal human resource management in the organisation was used to analyse the application of the principles of talent management. The answers to the questions focused on innovative company's environment were used for the needs of this chapter. The amount of the interviewed institutions was 573–609 every year (depending on the availability of personal contacts of external students, which were used to address the participants in the research), while the response rate of comprehensively completed questionnaires was 60–67%.

In order to define a sufficient research sample, two stratification criteria were determined. The first criterion was the region of operation of the organisation based on the NUTS classification; Slovakia was divided according to the NUTS 2 category, while the structure of the research sample was based on the data provided by the Statistical Office of the Slovak Republic.

The second stratification criterion was a minimum number of employees, determined to 50 employees, thus excluding small enterprises from the research sample on the one hand; however the importance of focusing on a formal human resource management system in the organisations with 50 and more employees was followed on the other hand.

The data provided by the Statistical Office of the Slovak Republic during the monitored period indicated that the number of organisations with 50 and more employees in individual regions was oscillating around similar values, while the regional structure of the organisations with over 50 employees in the given years is provided in Table 1.

Determining an optimal research sample of the given basic group of organisations, confidence level of the research was set at 95%, and confidence interval of the research was set at $H = \pm 0.10$. On the grounds of the given criteria an additional and relevant research sample for individual regions of Slovakia was set in the analysed years (see Table 2).

Table 1 Regional structure of organisations with more than 50 employees (source: data processed according to the Statistical Office of the Slovak Republic [57])

Region—NUTS II	Bratislava region	Western Slovakia	Central Slovakia	Eastern Slovakia
Districts	BA	TT, TN, NR	BB, ZA	KE, PO
Number of organisations 2014	1098	904	644	612
Number of organisations 2015	1105	916	651	613
Number of organisations 2016	1114	923	649	621
Number of organisations 2017	1123	926	654	623
Number of organisations 2018	1125	930	659	626

Table 2 Size of the research sample for individual regions of Slovakia (source: own research)

Region—NUTS II	Bratislava Region	Western Slovakia	Central Slovakia	Eastern Slovakia
Districts	BA	TT, TN, NR	BB, ZA	KE, PO
Number of organisations	1102-1125	904–930	644–659	606–626
Size of research sample	88	87	84	83

The measured values were statistically processed and assessed by calculated chain indices (values changed since the previous year), and fixed-base indices (values changed since the first year).

3 Results

With regard to the aforementioned, we sought to uncover whether the interviewed organisations realised the need to focus on effective commitment of employees in innovation, and whether they were actually dealing with it in practice.

As the comparison of the results for individual years presented in Table 3 indicates, a slight increase in the focus of the organisations on the effective engagement of employees in innovations was recorded each year.

With regard to the aforementioned, we also consider the evaluation of the overall change in the analysed attribute in the monitored period to be necessary. The following table presents the data processed within the performed analysis.

As Table 4 shows, the overall increase of the focus of the interviewed organisations on effective engagement of employees in innovation compared to the first year of the monitored period was recorded in each of the following analysed years. It can be concluded on the basis of this outcome that the share of the organisations focusing on the activities related to the effective engagement of employees in innovations increased in the monitored period.

Given that one of the basic prerequisites for a company to be innovative is the ability of its managers to attract and inspire their subordinates to changes, the authors examined whether the managers created sufficient space for employee involvement in innovation processes, and whether this is a priority in terms of the managers' purposeful guidance of employee innovation behaviour towards organisational strategy (both in the form of rewarding and in the form of correct leadership towards knowledge sharing) [29].

In relation to the researched attribute regarding the level of employee involvement in innovations, the authors also tried to find out whether employees are perceived as a priority source of innovation incentives, or whether all staff categories are involved in innovations (not only managers and various specialists, but also executives or administrative staff) [29].

Table 3 Chain index of the companies focusing on effective commitment of employees in innovation (source: own research)

Focus of organisations on effective commitment of employees in innovation	Share of organisations								
	2014/(*n*)	ci15/14	2015/(*n*)	ci16/15	2016/(*n*)	ci17/16	2017/(*n*)	ci18/17	2018/(*n*)
Company implements the activities related to commitment of all employees in innovation processes	92	1.032	95	1.010	96	1.021	98	1.112	109

Explanatory notes: ci—chain index—values changed since the previous year

Table 4 Fixed-base index of the companies focusing on effective commitment of employees in innovation (source: own research)

Focus of organisations on effective commitment of employees in innovation	Share of organisations								
	2014 (*n*)	bi15/14	2015 (*n*)	bi16/14	2016 (*n*)	bi17/14	2017 (*n*)	bi18/14	2018 (*n*)
Company implements the activities related to commitment of all employees in innovation processes	92	1.032	95	1.043	96	1.065	98	1.185	109

Explanatory notes: bi—fixed-base index—values changed since the first analysed year

Within the analysis we asked the interviewed organisations: **"How do you involve people in the innovation process?"** With regard to the fact that the key personal quality of "innovators" is predominantly the ability to respond flexibly to dynamically changing conditions of the environment as well as the ability to win and enthuse their subordinates and whole working team for changes, we were interested in whether organisations created sufficient room for involvement of them and their teams in the innovation process. The research showed that most of the interviewed organisations, particularly 62.8% ($n = 215$), involved employees in the innovation process only occasionally upon a mistake occurrence in new processes or procedures, and particular innovation had to be approved by management. 28.36% ($n = 97$) of the organisations reported that they involved employees on a regular basis, and they worked on innovations without a common strategically focused coordination (Table 5). Only 26.6% ($n = 91$) of organisations reported that they involved employees in the innovation process on a regular basis with a focus on strategic objective.

The research showed that rewarding for knowledge sharing among employees in the monitored period was at the level of approximately 35%, representing real encouragement of the management of the interviewed organisations to openness and to the creation of the preconditions of creative environment. Employees were motivated to share information in only a minimum number of companies (7–14%) in the monitored period, and almost no change for the better was recorded (Fig. 1).

Table 5 Way of involvement of employees in the innovation process, 2018

How interviewed organisations involve people in the innovation process:	Percentage of companies
Occasionally, upon mistake occurrence	62.8
Regularly without a common strategically focused coordination	23.36
Regularly, in coordination and with a focus on strategic objective	26.6

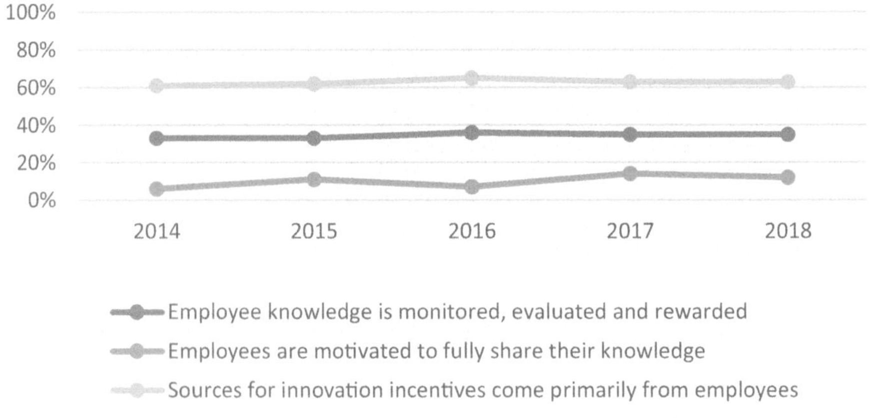

Fig. 1 Encouragement of the openness and initiative of employees in innovations in 2018

Table 6 Sources of impulses of organisations to innovate, 2018

Sources of impulses to innovate	Percentage of companies
Employees	65
Suppliers	22
Wholesale customers	8
Competition (benchmarking)	39
Customers	47
Experts, specialists	40
Own research department	13
Cooperation with universities	4

Within the analysis of the involvement of employees and customers in innovations, we were also finding out: **"From what sources do you most often get impulses to innovate?"** Answers to this question were significant for us not only because of the involvement of employees in innovations but also because the innovative organisation should get impulses to innovations themselves in the greatest possible extent and from the greatest possible number of sources. The research showed a positive result that organisations most often got impulses to innovate from their employees—in 65%. The second most often used source of impulses to innovate was customers; however the share of positively answering organisations, 47%, sounds very negative with regard to their key role in the creation of competitive innovation strategy. Specialists were reported as third, in 40% of organisations; however it did not always concern a person hired by the given company; it was often information obtained from literature, and media such as press, TV, and the Internet. Almost 40% of the organisations used competition and benchmarking to get impulses to innovate. On the contrary, least reported were impulses got from cooperation with universities (Table 6).

The positive finding of the research that the most frequent source of impulses to innovate is employees sounds considerably more negative in association with the finding that only some employee categories are used for this purpose.

In addition to the aforementioned, the objective of the authors was to analyse the relation between the region of operation of the organisation and its focus on the activities related to engaging all employees in innovation processes.

Table 7 presents numbers of the organisations according to the score achieved in the monitored attributes in individual regions for 2018.

The greatest share of the organisations performing the activities related to commitment of all employees in innovation processes was in the region of Eastern Slovakia ($n = 28$), representing 33.73% of all analysed organisations in this region. Overall, the activities related to engaging all employees in innovations in Slovakia were dealt with by ($n = 109$), representing 31.87% of the organisations.

Statistical Pearson parametric correlation test (r) was used to assess individual correlations.

Table 7 Focus of organisations on effective engagement of employees in innovation in individual NUTS II regions (source: own research)

Region—NUTS II	Bratislava region	Western Slovakia	Central Slovakia	Eastern Slovakia	Slovakia
Districts	BA	TT, TN, NR	BB, ZA	KE, PO	Σ
Focus of organisations on effective engagement of employees in innovation					
Company implements the activities related to engagement of all employees in innovation processes	26	28	27	28	**109**

H1: The results of the Pearson correlation test proved that there was no statistically significant relation between the region of operation of organisations and performance of the activities related to the engagement of all employees in innovation processes. The given variable correlates at the level sig. = 0.01 with Pearson correlation coefficient at the level $r = 0.075$; hence this hypothesis was rejected.

4 Discussion

Industry 4.0, which is based on massive digitalisation, increasing intensity of interdependence, and enhancement of the autonomy of both physical and cybernetic technological systems, will have a decisive impact on not only the way how economics will produce and consume, but also the way how people will work. The potential of the conditions of the fourth industrial revolution for the world of work has been the subject of scientific examination and expert discussions, as well as a source of worries about changes in the essence of work, demand for work, and social insecurity. The knowledge of both external and internal environments is the key to defining the competences of the future, which are a commodity on the labour market in terms of Industry 4.0 [39].

Like career management and talent management have been perceived by managers as something only designed for a small group of employees, although the latest studies have indicated the need to regard the two as designed for all employees who are expected to remain in the organisation for more than 5 years, as in such a case, it can be achieved already after 3 years that career management will concern up to 70% of employees in the organisation [40], it can be stated that, unfortunately, engagement of all employees in innovation processes is perceived very similarly by managers. The presented research showed on the basis of both chain and fixed-base indices that although the overall as well as year-on-year increase in the activities of organisations was recorded in this area, as the only one out of the examined areas, absolute quantification indicated negative state. It is due to the fact that the amount of organisations performing the activities related to the engagement of all employees in innovation processes within the analysed sample was less than a third also in the last monitored period. According to the authors, this fact was directly affected by the presence or availability of creative managers in organisations. The research indicated that their number did not change over time; therefore a significant positive change in this area was little probable. It was assumed due to the fact, which was also presented by Horňák [17], Cagáňová et al. [41], Beskovnik and Beskovnik [18], and Hitka et al. [42], that the creative manager was a bearer of innovation activity.

When managing innovation processes, it is necessary to concentrate on people who are considered by many contemporary authors to be the most important resources of companies, i.e. particularly their skills, experience, knowledge, and abilities [30, 43–45]. In order for a company to be able to compete in the current competitive environment, it must have appropriate resources (primarily human resources). On the other hand, the company must also apply such management methods and procedures that will allow it to use their potential in the maximum extent [46, 47].

Employees should therefore be encouraged in their innovation efforts, and motivation can be an appropriate tool. Even the actual manner of communication and collaboration and/or creating a space for self-realisation of employees can be sufficient incentives for them to engage in certain progress [29].

One of the key tasks of managers in supporting the creation and maintaining of innovative environment is to create the work environment in which all employees realise that innovations are necessary for successful business activities, and therefore perceive them positively [34, 47]. The primary role of management is thus to create the environment supporting creative employee behaviour, because the environment in which employees work is among the greatest motivating aspects [18, 48–50]. The survey of a sample of 893 Belgian employees indicated that employment in a family business was positively associated with the commitment of employees in innovations, and that part of these relations could be attributed to their increased perception of support for their organisation and motivation at work [51]. The level, intensity, and targeting of both support and motivation are directly dependent on mangers.

Companies frequently rely especially on individual knowledge of their general managers and top management in the development of innovations. This approach is inefficient, as the knowledge of other employees is not used sufficiently. A selection model of the company Heckman for 305 small enterprises showed that not only the ideas of leaders and managers but also those of non-managing employees contributed to innovation performance [52]. The aforementioned was also confirmed by a study tested on 198 employees of four hospices and palliative care organisations (H&PCOs) for dying cancer patients. Authors found a positive role of knowledge-sharing behaviours in affecting sharers' innovativeness, in terms of propensity and capacity to promote and implement new ideas [53].

On the basis of studies conducted by authors such as, Hajko et al. [54], Tvrdoň and Skokan [55], and Joniaková et al. [56], who focus on analysing disparities between the regions of Slovakia from the perspective of their socio-economic development, the authors of this chapter focused on analysing disparities in the level of using innovation potential by organisations. The objective of the authors was to analyse the relation between the region of operation of the organisation and its focus on the activities related to engaging all employees in innovation processes. On the basis of the results derived in association with the determined hypotheses, statistically significant relation between the determined variables and regions in which the analysed organisations operate can be excluded. This finding is deemed a positive signal for future effects of the development of organisations on the size of disparities between the regions of Slovakia.

5 Conclusion

The authors of this chapter pointed to the level of change in the focus of Slovak organisations on the activities related to engaging all employees in innovation processes between 2014 and 2018 on the basis of partial indicators, such as the performance of the activities related to engaging all employees in innovation processes, coordinated engagement of all employees in innovation processes in the context of corporate strategies, as well as availability of creative managers in organisations. The results were also presented in the context of the outcomes of surveys conducted by authors such as Bammens, Notelaers, Van Gils, Mura, Lettieri, Radaelli, and Spiller, and the level of the results obtained in Slovak organisations was evaluated on their basis. On the basis of the aforementioned, the authors do not regard the present level as sufficient; however the increasing trend, declared by the results of the research, has been evaluated positively. With regard to the fact that the engagement and encouragement of employees in innovation processes are among the key preconditions of achieving effective innovation environment, the authors of the chapter claim that in addition to the aforementioned information and procedures, employees who see their organisation as an attractive employer, as a "great workplace", as an organisation people want to work for, and who are not only able to identify with its objectives but also to see the potential of their advancement are key in ensuring competitiveness and added value in future.

Acknowledgements The chapter is related to Grant Agencies of VSEMvs project IGP 2/2016 Contribution of Selected Attributes of Managerial Work to Creation of Internal Environment Supporting Competitiveness of Companies, project APVV-17-0656 titled Transformation of the Paradigm of Business Management in Industrial Context 4.0, and KEGA project No. 030STU-4/2018 called "E-platform for Improving Collaboration among Universities and Industrial Enterprises in the Area of Education".

References

1. Stachová, K., Papula, J., Stacho, Z., & Kohnová, L. (2019). External partnerships in employee education and development as the key to facing industry 4.0 challenges. *Sustainability, 11*(2), 345.
2. Antonacopoulou, E. P. (2000). Employee development through self-development in three retail banks. *Personnel Review, 29*(4), 491–508.
3. Champathes, M. R. (2006). Coaching for performance improvement: The coach model. *Development and Learning in Organizations: An International Journal, 20*, 17–18.
4. Lee, C. H., & Bruvold, N. T. (2003). Creating value for employees: Investment in employee development. *The International Journal of Human Resource Management., 14*, 981–1000.
5. Nel, P. S., Haasbroek, G. D., Schultz, H. B., Sono, T., & Werner, A. (2004). *Human resources management* (6th ed.). Cape Town: Oxford University Press.
6. Edgar, F., & Geare, A. (2005). HRM practice and employee attitudes: Different measures–different results. *Personnel Review, 34*, 534–549.
7. Georgellis, Y., & Lange, T. (2007). Participation in continuous, on-the-job training and the impact on job satisfaction: Longitudinal evidence from the German labour market. *The International Journal of Human Resource Management, 18*, 969–985.
8. Armstrong, M. (2006). *Jak se stát ješte lepším manažerem*. Praha: Ekopress.
9. Marciano, P. (2013). *Cukr a bič Nefungují - O Motivaci a Zapojení Zaměstnanců* (1st ed.). Prague: Motiv Press.
10. Urbancová, H. (2016). *Talent Management v Organizacích v České Republice: Praktické využití a Přínosy*. Prague: Odborné nakladatelství Vysoké školy Ekonomie a Management.
11. Dvožáková, J. (2007). *Management Lidských Zdrojú*. Prague: C.H. Beck.
12. Kachaňáková, A., Nachtmannová, O., & Joniaková, Z. (2011). *Personálny Manažment. Iura eEdition*. Bratislava: Wolters Kluwer.
13. Watson, E. (2007). Who or what creates? A conceptual framework for social creativity. *Human Resource Development Review, 6*(4), 419–441.
14. Florida, R. (2004). *The rise of the creative class and how is transforming work, leisure, community and everyday life* (1st ed.). New York: Basic Books.
15. Franková, E. (2011). *Kreativita a inovace v organizaci*. Praha: Grada Publishing.
16. Philipps, A. (2010). Drawing breath: Creative elements and their exile from higher education. *Arts and Humanities in Higher Education, 9*, 42–53.
17. Hornak, F. (2009). Development of managers' creative potential. In *Annals of DAAAM and Proceedings of the International DAAAM Symposium* (pp. 1427–1428). Vienna: DAAAM International.
18. Beskovnik, B., & Beskovnik, E. (2017). Transformation of business environment: The introduction of creative environment in Slovenian logistics companies. *Transformations in Business and Economics, 16*(1), 121–133.
19. Cagáňová, D., Bawa, M., Šujanová, J., & Saniuk, A. (2015). *Innovation in Industrial Enterprises and Intercultural Management* (1st ed., p. 126). Zielona Góra: University of Zielona Góra. ISBN 978-83-933843-4-1.
20. Cagáňová, D., Čambál, M., Šujanová, J., & Woolliscroft, P. (2012). The multiculturality aspects and human capital management within Slovak industrial enterprises. In *The 4th European Conferences on Intellectual Capital. ECIC 2012: Proceedings - Arcada University of Applied Science, Helsinki, Finland, 23–24 April 2012* (pp. 106–117). Academic Publishing International. ISBN 978-1-908272-32-4.
21. Carvaja, S. A., Pérez, M. D., Cabello, R. V., & Espinosa, C. C. (2015). Identifying key factors affecting culture of innovation: A case study of Chilean medium mining sector. *Journal of Technology Management and Innovation, 10*(1), 132–145.
22. Hitka, M., Sedmák, R., Alác, P., & Grladhovic, T. (2005). Establishment of motivation programs for workers in manufacturing companies using cluster analysis. *Drvna Industrija, 56*(1), 21–28.

23. Lorincová, S., Schmidtová, J., & Javorčíková, J. (2016b). Employee job satisfaction in furniture manufacturing companies in the Slovak Republic. *Drvna Industrija, 67*(4), 351–362.
24. Mura, L., Horvath, P. (2015). Some aspects of human resource management. In: *International Multidisciplinary Scientific Conferences on Social Sciences and Arts, Albena, Bulgaria* (pp. 863–870).
25. Tokarčíková, E., & Kucharčíková, A. (2015). Diffusion of innovation: The case of the Slovak mobile communication market. *International Journal of Innovation and Learning, 17*(3), 359–370.
26. Khazanchi, S., Lewis, W. M., & Boyer, K. K. (2007). Innovation-supportive culture: The impact of organizational values on process innovation. *Journal of Operations Management, 25*(4), 871–884.
27. Herbig, P., & Duphy, S. (1998). Culture and innovation. *Cross Cultural Management: An International Journal, 5*(4), 13–21.
28. Willems, M. J. T. (2007). *The influence on social capital and cultural dimensions of innovation*. Maastricht: Universität Maastricht.
29. Stachova, K., Stacho, Z., Blstakova, J., Hlatka, M., & Kapustina, L. M. (2018). Motivation of employees for creativity as a form of support to manage innovation processes in transportation-logistics companies. *Naše More, 65*(4), 180–186.
30. Slávik, Š. (2013). *Strategický manažment*. Bratislava: Sprint 2.
31. Kirchmayer, Z., Remisova, A., Lasakova, A. (2016). Ethical Leadership in the Context of CSR'. In: *28th International Business-Information-Management-Association* (pp. 1144–1150). International Business Information Management Association, Seville.
32. Bogers, M. (2018). Innovating by doing: Promoting on-the-job experimentation through a climate for innovation. *International Journal of Entrepreneurial Venturing, 10*(3), 362–382.
33. Bessant, J. (2003). *High involvement innovation* (1st ed.). Hoboken: Wiley.
34. Hitka, M., Kozubíková, L., & Potkány, M. (2018). Education and gender-based differences in employee motivation. *Journal of Business Economics and Management, 19*(1), 80–95.
35. Jurenka, R., Cagáňová, D., Horňáková, N., & Stareček, A. (2016). Smart city in terms of social innovations and human capital. In *Smart City 360 2016 - 2nd EAI International Summit*. Bratislava: The Design Society.
36. Rogawski, E. T., Liu, J., Platts-Mills, J. A., Maphula, A., & Nyathi, E. (2018). Use of quantitative molecular diagnostic methods to investigate the effect of enteropathogen infections on linear growth in children in low-resource settings: Longitudinal analysis of results from the MAL-ED cohort study. *The Lancet Global HealthLancet Glob. Health, 6*, 1319–1328.
37. Urbancová, H., & Vnoučková, L. (2018). Impact of employee development in agricultural companies on commitment, loyalty and performance. *Acta Universitatis Agriculturae et Silviculturae Mendelianae Brunensis, 66*(3), 803–811.
38. Strobel, M., Tumasjan, A., Spörrle, M., & Welpe, I. M. (2017). Fostering employees' proactive strategic engagement: Individual and contextual antecedents. *Human Resource Management Journal, 27*(1), 113–132.
39. Blštáková, J. (2018). *Modely riadenia ľudských zdrojov v ére priemyslu 4.0 (Habilitation thesis)*. Bratislava: EU FPM.
40. Hroník, F. (2007). *Rozvoj a vzdelávaní pracovníků*. Praha: Grada Publishing.
41. Cagáňová, D., Čambál, M., Stachová, K., Stacho, Z. (2014). The level of the focus of organisations operating in Slovakia on comprehensive approach to education. In: *ECIC 2014: proceedings of the 6th European Conference on Intellectual Capital. 10– - 11 April 2014, Trnava, Slovak Republic*. 1st ed. [s.l.] (pp. 10–14). ISBN 978–1–909507-20-3.
42. Hitka, M., Lorincová, S., Ližbetinová, L., Bartáková, G. P., & Merková, M. (2017). Cluster analysis used as the strategic advantage of human resource management in small and medium-sized enterprises in the wood-processing industry. *BioResources, 12*(4), 7884–7897.
43. Lorincová, S., Schmidtová, J., & Balážová, Ž. (2016a). Perception of the corporate culture by managers and blue collar workers in Slovak wood-processing businesses. *Acta Facultatis Xylologiae, 58*(2), 149–163.
44. Tidd, J., Bessant, J., & Pavitt, K. (2007). *Řízení inovací*. Brno: Computer Press.

45. Urbancová, H., & Hudáková, M. (2015). Employee development in small and medium enterprises in the light of demographic evolution. *Acta Universitatis Agriculturae et Silviculturae Mendelianae Brunensis, 63*(3), 1043–1050.
46. Mikešová, S., & Gásbryšová, M. (2012). Balanced scorecard—Revolution in performance management and options of using this method in terms of transportation. *Logi—Scientific Journal on Transport and Logistics, 3*(2), 117–127.
47. Stacho, Z., Potkány, M., Stachová, K., & Marcineková, K. (2016). The organizational culture as a support of innovation processes' management: A case study. *International Journal for Quality Research, 10*(4), 769–784.
48. Bartakova, P. G., Gubiniova, K., Brtkova, J., & Hitka, M. (2017). Actual trends in the recruitment process at small and medium-sized enterprises with the use of social networking. *Economic Annals-XXI, 164*(3–4), 80–84.
49. Stacho, Z., & Stachová, K. (2017). *Organizácia manažérskej práce*. Bratislava: Wolters Kluwer.
50. Papula, J., Kohnova, L., & Papulova, Z. (2018). Impact of national culture on innovation activities of companies: A case of Germany, Austria, Switzerland and the Czech Republic. *Economic Annals-XXI, 169*(1–2), 26–30.
51. Bammens, Y., Notelaers, G., & Van Gils, A. (2015). Implications of Family Business Employment for Employees' Innovative Work Involvement. *Family Business Review, 28*(2), 123–144.
52. Andries, P., & Czarnitzki, D. (2014). Small firm innovation performance and employee involvement. *Small Business Economics, 43*(1), 21–38.
53. Mura, M., Lettieri, E., Radaelli, G., & Spiller, N. (2013). Promoting professionals' innovative behaviour through knowledge sharing: The moderating role of social capital. *Journal of Knowledge Management, 17*(4), 527–544.
54. Hajko, J., Klátik, P., & Tunega, M. (2011). *Konkurencieschopné regióny 21* (1st ed.). Bratislava: Devin Printing House.
55. Tvrdoň, M., & Skokan, K. (2011). Regional disparities and the ways of their measurement: The case of the Visegrad Four Countries. *Technological and Economic Development of Economy, 17*(3), 501–518.
56. Joniaková, Z., Gálik, R., Blštáková, J., & Tarišková, N. (2016). *Riadenie ľudských zdrojov* (1st ed.). Bratislava: Wolters Kluwer.
57. Statistical Office of the Slovak Republic. (n.d.). Retrieved September 10, 2018, from www.statistics.sk

The Smart Logistics

Richard Jurenka, Dagmar Cagáňová, and Natália Horňáková

1 Introduction

In a modern business environment organizations must respond to new challenges, constant changes, opportunities, and also different limitations. Continuous changes in the business environment and technological improvement currently demand from organizations to introduce smart solutions and smart devices and create innovation. Due to the abovementioned reasons contemporary organizations must be flexible, dynamic, constantly improving, and able to adapt to different circumstances. Nowadays the social environment in this century is very dynamic, volatile, unstable, and unpredictable. However on the other hand, the logistics is today an important economic competitive factor, which can make a decisive contribution to success. Therefore it is very important to pay close attention to logistics and its processes.

Logistics is a very large field, which hides a lot of opportunities for improvement. Organizations through these opportunities could better manage their business and compete more effectively on the market. Many companies particularly focus on the production area that is also very important, because in this area rises the biggest added value in the form of changing the input materials into finished products. However, very often it is forgotten how many activities must be done in order to change input materials to final products in time, in the required quantity, and in the required quality.

R. Jurenka · D. Cagáňová (✉) · N. Horňáková
Institute of Industrial Engineering and Management, Faculty of Materials Science and Technology in Trnava, Slovak University of Technology in Bratislava, Slovakia
e-mail: richard.jurenka@stuba.sk; dagmar.caganova@stuba.sk; natalia.hornakova@stuba.sk

© Springer Nature Switzerland AG 2020

D. Cagáňová, N. Horňáková (eds.), *Mobility Internet of Things 2018*,
EAI/Springer Innovations in Communication and Computing,
https://doi.org/10.1007/978-3-030-30911-4_20

Pre-production phases are crucial for successful production of desired products. After producing final products another logistics process follows that is based on delivering desired products to the customers. Aforementioned operations are carried out by logistics; due to their importance it is necessary to pay extraordinary attention to them.

Supply logistics management is in the hands of corporate departments such as manufacturing, purchasing, and trading. Management of these processes should be cross-sectional, for example in the logistics department or industrial engineering department. Only in that way could logistics processes be systematic and correctly optimized in organizations. Supply logistics management is unsystematic and inefficient and ordering of material items is just intuitive in many Slovak and Czech companies. This characteristic brings for companies many problems in production.

The new trend around the world in industries is not using warehouse production that means the companies should not buy or manufacture for warehouse, but first of all companies should quickly and in necessary quantities secure components for goods that have already ensured sales. Smart logistics together with modern and smart devices has the potential to improve all logistics operations, make carried-out actions easier and streamlined, and ultimately increase profitability and performance of industrial companies.

2 Smart Logistics

The term logistics is derived from the Greek base of the word "logos" that in translation means action. However, systematic attention is given to logistics after the Second World War, because an efficient solution of logistic operations has been attributed a significant share in the victory of Allied troops. As an example we could describe landing in Normandy, which was the most important action from Allies.

Logistics is an interdisciplinary science that deals with the coordination; alignment of the flow materials, semi-products, products, and services; as well as flows of information and finances in terms of customer satisfaction and lowest costs [1].

Logistics is inherently an essential activity for every industrial company. Its main task is to move material or different products from A to B. Logistics try to do this efficiently as it is possible to make. Efficiency in moving materials is possible to achieve by doing it in shortest possible time, at the lowest cost, just in time, and in the right amount and requested quality.

One of the most important things for a functioning logistics system is storage that connects together space with time. Storing is an activity that is connected with the physical movement of the stored material. It is very important to know that stored material does not gain a higher value; on the contrary, it increases the costs that affect the profitability of the products and the achievement of the company's profits [2].

The key question then is this: How much does the company spend on improving warehouse processes and developing internal logistics and how much would be the company willing to spend on improvements on warehouse processes, if there will be given positive predictions?

Logistics does not mean only system planning, but also synchronization, management, implementation, and control of the external and internal material flow. The material flow is closely related to information flow; both have got a common goal, that is, to meet customer requirements. The flow of information is enhanced by smart logistics technologies that enable much more data-efficient processing and evaluation.

Logistics is simply focused on fulfilment of the customer's needs as the fundamental effect and this mission is to be achieved with flexibility, accuracy, and economy efficiency. In hard competition and the developed market economy, there cannot exist a business entity that is not able to satisfy the customer needs by delivering [3]:

- The correct product
- Of the right quality
- In the right amount
- In the right place
- At the right time
- At the right price
- And in the correct packaging

Logistics as a whole unit is mainly a supportive activity that must be performed for receiving, storing, and transporting the goods to the final customer. From several experiences, it is possible to generalize the predication that logistics is a field, where it is possible with the right management and organization system to save a lot of funds. Logistics is without doubt a supportive activity with a high added value.

In the past logistics was considered as not so much important. Logistics was very often underestimated and solved in most cases as the last one. Nowadays the great emphas is given on the increasing of logistics automation levels and storage systems. Companies are trying to automate their processes not for labor savings, but mainly in order to accelerate the speed of individual steeps during performing logistics operations and in order to eliminate occurring errors, for example in picking operations. As a result of increasing level of automation process, there is a growing pressure from customers on delivery of reliable goods and materials to manufacturing plants [4].

Companies are developing and expanding regularly every year. Nowadays it is a common trend that several companies transfer their production from or to abroad and regularly purchase new equipment, technologies, and production technologies. Growing production by 10–20% statutory brings also increased demands on logistics.

Logistics is continuously changing with development and expansion of companies. For example, logistics connects together the factual, spatial, and time differentiation of production and consumption. On the one side, this is leading to cost savings, reducing stock levels, releasing funds, and increasing profitability and efficiency. On the other side, what is more important for a market economy? Exactly logistics is a tool for acquiring and retaining a customer, because it provides a lot of benefits. Companies appealing mandatory rules in logistics have better market performance and gain more competitive ability [5].

However, expanding the business and increasing production bring itself various negative situations. One of them may be that the logistics system is overwhelmed by the operation, the incoming and expedition areas are overloaded with a material waiting to be processed, the racking system is hopelessly filled, and there are considerable reserves in the data processing area. So far, occasionally it happens that in the logistics system appears incomplete piece list, and incomplete or missing data. The next inadequate part in logistics is caused by the missing data that has not been collected yet.

Smart technologies and devices are developed to eliminate the creation of negative logistical phenomena and resolve the abovementioned negative situations. This development in logistics is associated with technological advances and intelligent devices and brings itself overall acceleration in all ongoing processes. Currently, smart transport logistics technologies bring revolution, whether they are courier services, large shipping companies, or smaller businesses [6].

Technology is constantly evolving in the world. Contemporary world does not create space for long-term strategies, as the development of information and communication technologies is still faster. On the other hand, the company has to make decisions as quickly as possible to achieve the highest possible profit.

The emphasis on application of innovation is no longer focused only on manufacturing companies but it practically concerns the whole industry—from suppliers through the manufacturer up to the customer. Many companies are not just concentrating on optimizing production processes, but they understand that logistics must also be optimized. Companies now recognized the necessity to innovate all logistics processes, not only through the automation of transport process. Automation must also be implemented to information processes and into warehouses. Qualified logistics personnel raise the costs of company. Therefore reason for automation is very simple, because it has got the potential to significantly decrease costs of the company and make logistics operations easier [7].

Modern technologies and automated solutions come in companies into applications and create SMART solutions, raise efficiency of the workers, and monitor the movement of goods through the sensors. In the past the information flow was isolated only at the level of enterprises, enterprises and warehouses, warehouses and shops, and shops and customers; nowadays there is an attempt to make a complete connection of this information flow.

Wireless technologies provide an enormous scope for improvement of warehouse logistics processes. Devices using wireless technology are often connected to the Internet and allow users a high degree of mobility that ensures smooth data transfer and information transfer practically anywhere. For example, communication can be improved through the use of smart glasses or devices that cooperate together on the basis of voice requirements. Production instructions can be visually transmitted step by step by using the smart goggles and voice devices can also provide additional information in the ear for better performance of the task. Sellers of these technologies promise to increase productivity by 30%. Furthermore, these devices are able to monitor the current mental state of workers and the level of their fatigue or trace their movements and alert them when they enter dangerous zones [8].

Smart technologies are computer-controlled systems that control fully automated work processes and are able to perform several steps in order to manage the whole process themselves. Overall, smart technologies improve ongoing logistics processes, reduce costs, and improve business performance.

2.1 Logistics 4.0

The term Industry 4.0 nowadays resonates in industry field. However, in the twenty-first century this term cannot be perceived merely as an abstract or theoretical concept. It is the current challenge to implement specific solutions and innovations that will enable manufacturing businesses to quickly improve their technologies and practices.

The topic of the fourth industrial revolution was created about 10 years ago in Germany in order to create a "smart factory." The technology base for this smart factory should be the cybernetic-physical systems and the Internet of Things (IoT). Ten years have passed very fast and these factories already exist.

The phenomenon of contemporary world is the interconnection of the Internet of Things, Internet of Services, Internet of People, and related creation of large volumes of data during communication between machine and machine, human and machine, or human and human. The producing environment is also influenced by the introduction of a number of new technologies such as autonomous robots, big data analysis, computer simulation, virtualization, cloud services, 3D printers, and augmented reality [9].

Industry 4.0 is not just a digitization of industrial production; it is mainly a complex system of changes that is associated with human activities. Industry 4.0 thus transforms production from separate automated units into fully integrated automated and continuously optimized production environments.

The vision of Logistics 4.0 is to automate all processes as much as possible, especially to eliminate the human factor. The key factor in smart factories is a perfect overview of the flow of individual materials and components across manufacturing plant. Nowadays, some automotive industrial companies require 100% traceability of individual products and components. We can predict that automotive industries in the future will also require 100% movement traceability of workers who worked on individual products. In others words, the perfect overview of movement of material and workers will be required. Questions like when the material and workers moved, why they moved, from which point they moved, and what this movement caused will be clear and 100% monitored. This kind of traceability can be achieved only by using perfect information systems, which will provide quality and correct data.

The abovementioned kind of data ("master data") are currently considered as a very crucial issue not only in traceability, but also in all other logistics activities, whether we are talking about warehousing operations, delivering material inputs on production lines and workplaces [4].

The vision of Logistics 4.0 also offers a comprehensive interconnection of trucks, and planning, production, and storage software with online traffic data. This interconnection would allow to make the logistics process fully automated. The necessity of this system is therefore a perfect overview of where and what amount of the components is currently located.

The automotive industry is nowadays the "drafter" of ideas and new technologies of the Logistics 4.0 concept. In this segment, we can already meet the elements of Logistics 4.0 such as automatic supply of parts on assembly lines using automated guided vehicles (AGV), robotic workstations, and fully automatic welding lines.

These and various other intelligent logistics elements can be found mostly in the automotive segment.

2.2 *The Logistics of the Future*

The BMW Group in the further development of its logistics will increasingly rely on innovation, digitization, and sustainability in order to make the logistics as flexible and efficient as possible. BMW Group wants this idea to be achieved by using a fully interconnected supplier network, autonomous transport robots, and movement traceability during the delivering process of new cars to customers. Every day around the whole world must 30 million parts arrive in time and in the requested quality into 31 BMW's factories in order to produce 9700 new vehicles. This cannot be done without innovative logistics ideas [10].

The logistics in the future counts with the continuous innovation at all stages of the logistics process. Innovation should ultimately support sustainable growth. The logistics of the future also counts with connected supply network in order to maximize the clarity of all the necessary data. The clarified data flow will allow accurate information in the future about where specific components are located and whether they will be delivered in time.

BMW Group's logistics director says: "Logistics is the heart of BMW's production chain, and the using of innovative and digital technologies is becoming a key factor in complex logistics processes. We are focusing on solutions that are sustainable and resource-efficient. We are already testing the technologies of the future as a part of our several pilot programs" [10].

Future logistics assumes, among other things, the full deployment of a fleet of autonomous transport vehicles, instruments, and robots. Furthermore, it foresees the use of autonomous trailers, interconnected distribution of the final product, and daily use of smart technologies by employees of industrial companies.

The BMW Group tests a wide range of different technologies and innovations from glasses with augmented reality, through autonomous transport systems to electric trucks. The project manager for innovation and Industry 4.0 of the BMW Group in the field of entrance logistics said: "In our company we have a clear vision in the future and we are already working with new technologies. So far, we have identified the potential for innovation at all stages of the logistics process, from deliveries to factories to shipping new vehicles to dealers and distributors around the world" [11].

In the logistics the way of marking and scanning information throughout the entire production and logistics flow is also very important. Therefore, systems and devices (smart technologies) that will be able to read radio-frequency identification (RFID) tags will play an important role in the future.

RFID is an identification component working in the high-frequency bandwidth. RFID represents improvement of the barcode system. The initiator of development of RFID technology was Walmart as in the case of the barcodes. RFID enables contactless and automatic recognition of various objects [12].

Nowadays, there are many companies that can supply both hardware and software equipment for tagging and capturing the items. These systems are modular in most cases, so they can easily adapt to specific customer requirements. Another advantage of these systems thanks to a specific interface is that they can easily communicate with the ERP system (enterprise resource planning system), which is used in industrial companies. On this basis is created a comprehensive system that enables tracking of the amount and movement of individual components.

The vision of Logistics 4.0 brings with itself significant changes in automation, interconnection, and especially digitization. Intelligent interfaces will be an important factor in logistics success.

Nowadays, without IT systems it is not possible to effectively perform logistics process. Therefore, it is almost certain that smart technologies and devices will significantly affect the performance of each company.

Everything that could be digitized will be digitized in the future; this change is unstoppable. The flow of information between all stakeholders will be optimized, and work processes will be speeded up and streamlined. Digitization basically represents a significant improvement in competitiveness of the industrial companies.

2.3 *Radio-Frequency Identification*

RFID is a technology that exists for decades. This technology was first used during the Second World War for identifying enemy aircraft. Scientific laboratories have begun to work for its wider use in normal life spheres at the end of the twentieth century. Nowadays, we encounter various RFID applications in our everyday lives without being aware of it (for example, in automotive immobilizers, contactless identification cards, attendance systems, security systems). Its principle lies on the transmission of data, namely electronic product code, via radio waves [13].

The term RFID includes many different technologies that use different frequencies, protocols, and languages to communicate.

The RFID tag consists of very small silicon chips attached to the narrow antenna. The reading device transmits radio waves, through which the antenna communicates with the RFID chip and then stores obtained information. RFID tags use radio waves for communication. This is the fundamental difference between RFID and barcodes. Another difference is that the RFID can read multiple tags at the same time and the direct visibility of the tag is not needed. RFID tags can also be rewritable, so the data stored in them can be changed and updated at any time. What is more, the RFID technology allows capturing up to several hundred pieces of products at the same time, which is not possible with barcodes [14].

The main advantages of the RFID technology are the following:

- Direct visibility for reading and writing is not required
- Reducing the error rate
- Improving the flow of goods
- Higher degree of automation
- Digital retrieval of information
- The speed of getting information
- Mobility
- Multi-capture capability
- Resistance and variability of the media

2.4 *Smart Gloves*

Smart gloves or scanning gloves are an industrial solution that enables manufacturing and logistics personnel to work more efficiently and safer. The biggest advantage of this solution lies in the simplicity and nature of its use. The process steps can be documented by “free hands” and the user gets immediate feedback about executed operation [15].

Smart gloves speed up and make easier all work steps for its users. For example smart gloves can show if correct component part was used or if work steps were done correctly. Gloves can also record data as well as a regular scanner. The goods can be registered quickly and comfortably.

In addition, the individual work steps for users are also ergonomic thanks to the smart glove. Data logging or material control is integrated into natural hand movements. The codes can be loaded both vertically and horizontally with smart gloves. The loaded code is possible to confirm by the index finger. Accuracy of performed action is confirmed by acoustics or vibration. This ability prevents scanning of incorrect code [16].

The smart gloves simply optimize logistics work. Smart gloves have got an implemented scanner. The long-term test and easy use of the smart gloves convinced many automotive companies that smart gloves could be an ideal solution for series production.

The automotive company Škoda is intensively preparing for Industry 4.0. The Czech automotive company tends on innovative technologies of the future. They started using smart industrial glove (ProGlove). Deployment of smart gloves is designed to facilitate and optimize the work of their logistics. Logistics manager of ŠKODA AUTO said: "At ŠKODA AUTO we are constantly testing modern technologies in order to optimize the everyday work of our employees. Smart gloves (ProGlove) help our team work faster, more efficiently and with fewer errors" [16].

"Hands-free" scanning is already successfully established and commonly used mainly on the foreign market, in companies like Audi, BMW, John Deere, Mahle, Bosch, Lufthansa, Festo, Loxxess, Redcon, and others.

2.5 *Smart Glasses*

Smart glasses, another advancement in scientific and technological progress, accelerate and facilitate picking process in the warehouse. The warehouse operator displays in the smart glasses what items and how much of them are needed to be picked out from storage shelves and where they are exactly stored.

We meet with two types of smart glasses in the warehouse: augmented reality glasses (Fig. 1) and virtual reality glasses (2).

Augmented Reality Glasses

The warehouse operator with attached augmented reality glasses (Fig. 1) sees its surroundings and also specific information about which goods to pick out from the shelf. Some of the glasses have also built-in barcode reader—the warehouse operator looks at the barcode on the packaging, which is consequently recorded. Warehouse operator has got free hands for removing or storing goods [17].

Fig. 1 Augmented reality glasses [17]

Fig. 2 Virtual reality glasses [17]

Virtual Reality Glasses

When an operator puts on virtual reality glasses (Fig. 2), he or she does not perceive his or her surroundings and sees only digital images or scenes, just like someone who plays computer games.

Virtual reality glasses in logistics serve in the training process of operators. Virtual reality glasses can create a simulation model of the entire warehouse and the operator can be trained for example in controlling of a forklift trolley, in the walking through narrow aisles of warehouse, or in picking out items in very high warehouse shelves. Simulation model of virtual reality glasses can plan and schedule effective logistics operations [17].

The using of smart glasses allows workers particularly free movement during performing their workload. The worker is in addition disburdened from manipulating with various stationary terminals or with the paper-based documentation. Using of the smart glasses reduces also the costs of stationary terminals and office supplies. Various smart glasses also allow a video conferencing that has got the ability to improve the efficiency and speed of communication. It is also possible to reduce time for staff training, cycle time for assembly operations, or stress at work.

2.6 Autonomous Vehicles and Transport Instruments

Autonomous vehicles could become serial production technology in the next 10-year horizon. Recently, autonomous vehicles were only utopia or only part of science fiction novels. Nowadays the development of autonomous vehicles is limited by the legislation. Technically everything is possible and only some system components need to be adjusted. The final consequences from the introduction of autonomous vehicles would be crucial for transport logistics, especially if the entire fleet of trucks ride completely automatically on the roads.

The fundamental idea of this vision is to create autonomous means of transport that will be controlled primarily by stereoscopic cameras, radars, and sensors. At present, vehicles are modified to keep control on the road, keep safe distance from other vehicles, or stay in their lane. Current safety systems in transport vehicle can in dangerous situations warn the driver with a special signal and also with an icon on the dashboard; if the driver does not respond within 5 s, the vehicle will gradually slow down until it completely stops. Ability to stay in your lane is secured through stereoscopic cameras.

Radars in vehicles are used to scan a road about 250 m in advance. The disadvantage of these vehicles is the question of their reliability in different places and under different conditions. The tested vehicles yet work without problems only under ideal conditions. The question is how they will react in the urban environment with pedestrians and cyclists, in rain or snow, or at extreme temperatures [18].

The uncontested advantage of these trucks is the high demand for freight transport. It is primarily expected from autonomous vehicles to reduce the number of human-induced transport accidents, reduce human resource costs, increase energy savings, and provide fast delivery options.

Nowadays electric trucks are driven on the roads in Munich and Leipzig in cooperation with other logistic service providers in order to deliver local supplies. The BMW Group strives to use alternative transport technologies with the most cost-effective way in the long-term perspective [11].

The autonomous driving will play an important role in the logistics sector in the future. Abovementioned project serves to get familiar with the various propulsion technologies and for gaining experience. Testing and application of innovations in cost-effective ways of transport will bring enormous cost savings in logistic process.

The first fleet of the intelligent robots with identification smart transport robots provide logistics support for the components in the factory in Wackersdorf. The unique feature of these robots is that they do not need navigational induction coils integrated in the floor of logistics halls. Smart transport robots move completely independently across the halls. They use batteries from the BMW i3 and they are able to take away containers with weight around 500 kg. Smart transport robots use connection with the nearest wireless transmitter in order to determine their position, direction, and distance from the objects. They are able to identify and respond to critical situations by using special sensors, which enable them to use the same paths as people and other vehicles. Currently more accurate navigation system is developed for smart transport robots that is based on 3D camera. The BMW Group launched this project in cooperation with the Fraunhofer Institute IML in Dortmund [10].

Autonomous means of transport in logistics have not only the ability to transport goods from one place to another but they are also able to combine different processes such as loading or unloading goods. The efficiency of logistics processes in warehouse logistics can be increased by combining individual processes.

The efficiency is not the only advantage of this solution, but another one is also increased workplace safety, whereas frequent injuries occur during the process of transporting the goods. Each of these devices works independently and the navigation in the area is provided by various monitoring devices and lasers. These devices have many security features that prevent collisions or can automatically schedule a new route in case of an obstacle [19].

2.7 Automated Guided Vehicles

Due to the constantly increasing competitive pressure, current trends in logistics are gradually being promoted in most of the industrial enterprises [20]. The use of advanced technology in the field of logistics provides space for improvement of business logistics processes and thus secures competition for the company's logistics capability [21]. The trend in industrial production worldwide is no warehouse production, i.e., not buying or manufacturing to the warehouse but carry quickly and in the necessary quantities only the components or the goods, for which the sales are already secured. The effort is to shorten the storage time of goods in the warehouse and shorten the time of shipment, loading, and unloading of goods [22]. Based on the mentioned reasons, the warehouse environment is one of the main areas and opportunities for implementing new innovative smart solutions. Among the best automated warehouses of today belong the warehouses of the companies as Amazon.com, Walmart, Coca Cola, and IKEA which use automated guided vehicles (well known as AGVs), drones, robots, and other smart automated solutions.

In recent years warehouse logistics has been largely devoted to the development of autonomous means of transport capable of transporting goods of various shapes and sizes [23]. One of the smart/intelligent solutions currently used in warehouses are abovementioned AGVs. The main reason of implementation AGV into system is to integrate the material handling flow and increase the material flow rate [24]. AGVs can be used for a specific task, or form part of a completely automated system in a warehouse. AGVs can carry out material handling operations reliably and effectively. AGVs are fully automatic transport systems and electrically powered unmanned vehicles and are equipped with various smart sensors and load protection devices which enable safely transport of all kinds of products without human intervention within production, logistic, warehouse, and distribution environments. AGVs are primarily driven by computers; in some cases AGVs can also be manually controlled [25].

One of the main advantages of the AGV system is its modularity, which constantly develops and transcends its entire structure. Elements of a modular system such as a vehicle and a peripheral and a control system are just basics [21]. Vehicles are driven by electrical energy and most often driven by a magnetic tape located on the floor. Based on RFID tags, the vehicle can determine the route, speed, stop, or communicate with the control system to manage the autonomous

Table 1 The summary of advantages and disadvantages of AGVs [27, 28]

Advantages of AGVs	Disadvantages of AGVs
1. Reduce labor costs – AGVs replace the human factor – There is only the initial investment	1. Potentially high initial investment
2. Increase the safety – AGVs perform tasks, which are considered as dangerous for humans – AGVs are equipped with smart sensors and devices for any obstacle detection that also increase the safety at the workplace	2. Not suitable for nonrepetitive tasks
3. Increase accuracy and productivity – AGVs remove some of the potential for inaccurate workflows, ultimately reducing waste and increasing output – AGVs are capable of running 24/7 and enable to streamline processes	3. Decrease the flexibility of operations – Sometimes operations require flexibility, such as the ability to jump between tasks
4. Easy to expand – Based on their modular system element	4. Forklift service
5. Compatible with other types of automation	5. Longer implementation time

system [26]. Other commonly used AGV navigation methods are the technologies as network of magnetic points and gyroscope, camera, and laser guidance technology.

The main advantages and disadvantages of AGVs are summarized in Table 1.

3 Conclusion

The logistics is undoubtedly an important tool for increasing the efficiency of an organization. Well-organized logistics processes and using of smart technologies bring cost savings, higher profits, higher competitiveness, and higher business performance.

Logistics as a system planning process must carefully consider the material flow that is closely linked to the information flow. Smart technologies used in logistics in the future will allow more efficient processing and evaluation of the data coming from the information flow.

Smart technologies and devices will bring clarification of information and data in the supply network. Technology development will allow us to gain accurate information about where the material is currently located, when exactly the material will be delivered, and whether the amount of components can be flexibly changed. The information of this type in the event of a delay or some unexpected event will allow an immediate response to the situation. For example, if a truck or van becomes a participant in a traffic accident, the linked information flow together with the supply network automatically recalculates alternative solutions for the remedy and initiates the necessary measures in order to carry out the production or delivery according to the plans.

The deploying of RFID technology has a huge potential. The RFID technology provides more functionality than barcodes, but mass replacement of barcodes with RFID tags will not take place in the next few years, as the cost of RFID technology and their procurement is still quite high compared to the barcode. The barcodes are still more appropriate in order to provide basic identification options at the lowest possible cost. The RFID technology allows more frequent data collection at multiple locations than barcodes. It is probable that the barcodes and the RFID will complement each other in the future.

The smart logistics counts with the continuous innovation in all stages of the logistics process. The Logistics 4.0 presupposes among other things the full deployment of autonomous transport vehicles, towing vehicles and trailers, various devices, carts, and robots. Advanced technologies and smart devices will have the goal to eliminate the creation of negative logistical events and overall acceleration of all ongoing processes.

The automotive industry, as a drafter of intelligent ideas and technologies, supports the vision of Logistics 4.0 by all its forces. In the automotive industry are used features of Logistics 4.0 such as automatic supply of parts on assembly lines using automated guided vehicles, robotic workstations, fully automated welding lines, smart gloves, smart watches, smart goggles as device of an augmented reality, or virtual reality. These and various other intelligent logistics devices can be found in other companies that in the majority are suppliers for the automotive segment.

The smart gloves generally make easier and speed up all work steps. For example, the device shows whether the correct component has been used or whether the individual work steps have been performed correctly. Gloves can also record data as well as a regular scanner. Only one movement with smart gloves is necessary for recording data and no other devices are needed. The goods can be registered quickly and comfortably. Data recording or material control is integrated into natural hand movements.

The smart glasses help logistics employees, for example in the process of picking or storing parts. Smart glasses show what items and how much employees should pick up from storage shelves. The smart glasses can also give employee information regarding where the required components are exactly stored. The smart glasses also allow optical quality control. The built-in system can notice to faultlessness of components in a few seconds.

Another challenge in logistics is the full use of drones in logistics processes. While DHL successfully tests its drones and skyports in Germany, Amazon is currently working in the UK on its vision that the drones will be able to deliver to their customers ordered goods within 30 min to their doors from making the order.

Generally, there is a very difficult situation. It will be very important to prove that automated flying drones are safe enough and at the same time they do not interfere the privacy of people living in the areas over which they will fly. Lots of people are afraid that the drones will record on camera the surrounding environment, even when many companies stipulate that the delivery drones will not contain any cameras but only sensors for safe flight.

Therefore, before the drones become quite common in delivering process, it will be necessary to codify the legislative conditions for the full implementation of the drones. It can be concluded that when the drones will start commonly use in the delivery area, the question of privacy will be a very sensitive and crucial.

The smart logistics has many challenges ahead and even greater potential. The logistics, with the help of scientific and technological progress, will bring important changes in automation, interconnection, and particularly digitization. Nowadays, without IT in logistics, nothing moves. Therefore, it is almost certain that smart technologies and devices will significantly influence the performance of each industrial company and will be a decisive factor in achieving success or failure in the competitiveness of other companies.

Acknowledgements This chapter is part of the KEGA project No. 030STU-4/2018 called "E-platform for Improving Collaboration among Universities and Industrial Enterprises in the Area of Education."

References

1. Královenský, J., et al. (2001). *Postavenie dopravy v logistike*. Žilina: Žilinská univerzita v EDIS – vydavateľstvo ŽU. ISBN 80-7100-888-5.
2. MHI. (2016). Visio Picking Technology. [online]. http://www.mhi.org/ofs/solutions-guide/vision.
3. Štefko, R., Rákoš, J. Logistika a jej význam pre manažment podniku. [online]. https://www.pulib.sk/web/kniznica/elpub/dokument/Kotulic7/subor/stefko.pdf.
4. Sidora. J. (2017). Logistika 4.0. [online]. https://www.ipaslovakia.sk/sk/ipa-slovnik/logistika-4-0.
5. Královenský, J., & Gnap, J. (2009). Postavenie dopravy a zasieľateľstva v logistickom systéme. [online]. https://zlz.sk/o-je-logistika/.
6. Herčko, J., & Dulina, Ľ. (2014). *Produktivita s trvalo inovačnou výkonnosťou ako základňa pre konkurenčnú schopnosť*. dvojmesačník CEIT: ProIN. ISSN 1339-2271.
7. PrePriemysel. (2018). Čo potiahne dopravu a logistiku do budúcnosti?. [online]. https://www.prepriemysel.sk/co-potiahne-dopravu-a-logistiku-do-buducnosti/.
8. Suplychain247. (2016). Wearable Technology in the Warehouse. [online]. http://www.supplychain247.com/article/wearable_technology_in_the_warehouse.
9. Mařík, V., et al. (2015). Národní iniciativa Průmysl 4.0. [online]. http://www.businessinfo.cz/cs/clanky/narodni-iniciativa-prumysl 40 71386.html#!&chapter=3.
10. Trans Service. (2016). Logistika budúcnosti: Od virtuálnej reality po elektrické nákladné vozidlá. [online]. http://www.transservice.sk/new/logistika-buducnosti-virtualnej-reality-elektricke-nakladne-vozidla/.
11. TouchIT. (2016). Logistika budúcnosti: Od virtuálnej reality po elektrické nákladné vozidlá. [online]. https://touchit.sk/logistika-buducnosti-od-virtualnej-reality-po-elektricke-nakladne-vozidla/84069
12. Elogistika. Smart logistika: Logistika budoucnosti. [online]. https://www.elogistika.info/smart-logistika-logistika-budoucnosti/.
13. GS1 Slovakia. Rádiofrekvenčná identifikácia, rádiofrekvenčný (RFID) tag, čítacie zariadenia (snímač), RFID na Slovensku. [online]. https://www.gs1sk.org/rfid.

14. Kodys Slovensko. RFID—Rádio frekvenčná identifikácia. [online]. https://www.kodys.sk/rfid-radio-frekvencna-identifikacia.
15. iGROW NETWORK. Smart gloves: Urobte krok bližšie k industry 4.0 vďaka inovatívnemu nástroju pre podporu optimalizácie činností výroby a logistiky. [online]. http://www.innovation.sk/sk/smartgloves.
16. Agentura Auto Data. Škoda používa v logistike smart rukavicu. [online]. https://www.aad.sk/article.php?sid=6632.
17. Talentica. (2018). Novinky vo výrobnej logistike. [online]. https://www.talentica.sk/novinky-vo-vyrobnej-logistike/
18. Invent. (2016). *Industrial engineering—Toward the smart industry*. Žilina: Žilinská univerzita v EDIS—vydavateľstvo ŽU. ISBN 978-80-554-1223-8.
19. Balyo. (2013). Intralogistics is turning to automation. [online]. http://www.balyo.com/en/NEWS/Intralogistics-is-turning-to-automation-Linde-presents-automation-as-a-focal-point-at-WoMH.
20. Krajčovič, M., et al. (Eds.). (2014). *Priemyselná logistika*. Žilina: EDIS – vydavateľstvo ŽU.
21. Stareček, A., Bachár, M., Horňáková, N., Cagáňová, D., & Makyšová, H. (2018). Trends in automatic logistic systems and logistic market in Slovakia. *Acta Logistica., 5*(1), 7–14.
22. Václavík, I. (2005). Súčasný stav logistiky na Slovensku z viacerých aspektov. [online]. http://style.hnonline.sk/autosalon/189414-sucasny-stav-logistiky-naslovensku-z-viacerych-aspektov.
23. Pechanová, Ľ., Stareček, A., Bachár, M., & Cagáňová, D. (2017). Logistic market in Slovakia and analysis of new trends in logistics. *Výkonnosť podniku, VII*(3). ISSN 1338-435X.
24. Wu, S., Wu, Y., & Chi, C. H. (2017). Development and application analysis of AGVs in modern logistics. *Revista de la Facultad de Ingeniería U.C.V., 32*(5), 380–386.
25. Dematic. (2018). Automatic Guided Vehicles (AGVs). [online]. http://www.egemin-automation.com/en/automation/material-handling-automation_ha-solutions/agv-systems.
26. Kerner, S. (2016). Safe autonomous logistics and transportation vehicles in outdoor areas. [online]. https://www.iml.fraunhoferde/en/fields_of_activity/automationembedded_systems/research/SaLsA.html.
27. Conveyco. (2018). The advantages and disadvantages of automated guided vehicles (AGVs). [online]. https://www.conveyco.com/advantages-disadvantages-automated-guided-vehicles-agvs/.
28. Brigant, P., Horňáková, N., Jurík, L., Cagáňová, D., Hrablik Chovanová, H., & Jemala, M. (2018). Implementation of smart solutions in the in-house logistics. In: *3rd EAI International Conference on Management of Manufacturing Systems (MMS 2018). 6–8 November 2018, Dubrovnik, Croatia*. 1. vyd. Gent: European Alliance for Innovation - ISSN 2593–7642. ISBN 978–1–63190-167-6.

Identifying Points of Interest and Similar Individuals from Raw GPS Data

Thiago Andrade and João Gama

1 Introduction

Smartphones and portable devices have become ubiquitous and a part of everyone's life. Due to the fact of its portability, those devices acting as sensors are perfect to record individuals' traces and life-logging generating vast amounts of data at low costs. The popularity of the location-based social networks (LBSN) has increased and raised curiosity and interest among Internet users who are more and more interacting with the many available services like Instagram, Facebook, Flickr. When dealing with raw data, final users cannot make sense of it without processing and apply techniques to extract meaningful information from its content. Many researchers have made efforts in exploring these data in order to find places, locations, and regions [1, 4, 5, 11, 12], as it represents GPS coordinates. They are a pair of two decimal numbers separated by a comma like the following example: 30.2319, 120.14785.

Finding similar users is one of the most important tasks when dealing with social network services since one of the main issues is recommending similar users to a new user or even locations, products, and events. In this sense, some authors have been working to find and propose methods to address this cause. The approach we used in this paper is based on the users' preferred points.

T. Andrade (✉) · J. Gama
INESC TEC, Porto, Portugal

University of Porto, Porto, Portugal
e-mail: thiago.a.silva@inesctec.pt; jgama@fep.up.pt

© Springer Nature Switzerland AG 2020

D. Cagáňová, N. Horňáková (eds.), *Mobility Internet of Things 2018*,
EAI/Springer Innovations in Communication and Computing,
https://doi.org/10.1007/978-3-030-30911-4_21

In this work we tried to answer the following questions:

- *Is it possible to identify from raw GPS history data the users and community locations?*
- *Among those places, which ones are the points and regions of interest?*
- *Which are the users that share the most common places?*

Here we present an approach to analyzing users' raw data looking for points of interest in a given community. We do this by processing individuals' trajectories in order to find stay points (SP) (locations where the user spends considerable time in a given radius). Then we transform the users' found stay points into individuals' location points (LP) by applying a density-based clustering algorithm. Finally, we perform density-based clustering in all users' location points (LP) of a given community in order to find the points of interest (POIs). Another important item of this paper is the method to find similar users based on those points of interest of the community. All the above-mentioned tasks were evaluated in a real-world dataset.

The following section presents the literature review and the most important related works. The remainder of this paper describes the methodology and the dataset in Sect. 3, in Sect. 4 we discuss the experiments and results obtained. Finally, the conclusions and future work are presented in Sect. 5.

2 Related Work

Many researchers have been proposing methods to extract points of interest and calculate the similarity between users for diverse goals. The recommendation of users is one of the most popular among them. According to Cao et al. [1] and Lee et al. [4], several methods based on density have been proposed in order to discover regions of interest although most of these methods are used to aggregate spatial point objects. Some authors were more interested in the semantic movement trajectories. Li et al. [5] introduced a model that makes use of movement datasets which has trajectories defined as sequences of time-stamped stops and moves between locations. These trajectories are enriched with semantic meanings. From this, they define dense regions, where many different users had stop as regions of interest (ROI). The social matching framework was proposed by Terveen and McDonald [7] with the objective to match people by using their physical locations.

In order to help users on planning a trip to unknown places, Lu et al. [6] presented a framework that makes use of images (e.g., Flickr Images) and textual travelogues to perform recommendations. De Choudhury et al. [2] proposed something similar also making use of Flickr tagged images to automatically suggest and construct travel itineraries. Another approach is proposed by Zheng et al. [11, 12] which uses the Geolife GPS logs to understand the relations of trajectories, users, and the locations where they passed through to support the travelers to plan their trips. The authors make use of machine learning to build a recommender system by mining high ranked locations. Zheng et al. [9] performed a study based on the visiting pattern of users in distinct locations to define a similarity metric which detects similar users and their groups.

3 Methodology

Before entering in details of the methodology, we introduce the definition of a trajectory which will be used along this paper:

Definition 1 Trajectory: A trajectory is a list of ordered GPS points, $T = P0, P1, \ldots, Pn$, where $t0 < t1 < \ldots < tn$ and $i = 0, 1, \ldots, n$.

In this work, we denote a new trajectory every time an individual stop moving for more than 30 min.

The first step is the preprocessing task which includes among other activities, the data cleaning process where we perform outliers and noise removal. With the processed dataset we move to the second step where we process all the users' trajectories in order to find the stay points (SP). After having the set of stay points of the individuals' trajectories, we apply the algorithm to find the location points (LP). Further, after processing all users' location points, we perform the step responsible to extract the points of interest (POIs) in the given community of users by using density-based clustering methods. And the last step of the proposed approach is to calculate the similarity between the individuals of the community based on the common POIs each pair of users have visited.

Following we describe the real-world dataset used to apply the proposed method.

3.1 Dataset

All the activities in this list were conducted over the Geolife dataset. This GPS trajectory dataset was collected in (Microsoft Research Asia) Geolife project by 182 users in a period of over 3 years (from April 2007 to August 2012). A GPS trajectory of this dataset is represented by a sequence of time-stamped points, each of which contains the information of latitude, longitude, and altitude. This dataset contains 17,621 trajectories with a total distance of about 1.2 million kilometers and a total duration of 48,000+ h. These trajectories were recorded by different GPS loggers and GPS phones, and have a variety of sampling rates. 91% of the trajectories are logged in a dense representation, e.g., every 1–5 s or every 5–10 m per point. This dataset recorded a broad range of users' outdoor movements, including not only life routines like go home and go to work but also some entertainments and sports activities, such as shopping, sightseeing, dining, hiking, and cycling [10–12].

As we stated in Sect. 1, this work makes use of clustering algorithms based on the density of the data. Density-based methods require data to be collected at more frequent intervals. As mentioned above, more than 90% of the GPS communication intervals in the dataset are less than 10 s. This interval is adequate for the application of density-based methods to distinguish stationary points and regions of interest.

3.2 Preprocessing

When analyzing sources of information and comparing GPS signals with others, one can say that GSM cell tower has an advantage on being available in indoors, while GPS signals are not. In addition, GPS data often suffer from the so-called signal shadowing, when a given sensor is found inside vehicles, behind trees or buildings. On the other hand, GSM cell tower signals give us a more coarse and imprecise register of the location.

Because of the influence of GPS signal loss and data drift, there are a number of outliers in the trajectory data during the data acquisition. Hence, cleaning tasks need to be performed in order to have more trustworthy data. In the given scenario of locating meaningful places, another relevant role of preprocessing is to avoid peaks to fall into other clusters nearby and may form a new stay point unnecessarily.

Examples of common situations are individuals that suddenly took a very high speed in a very short period of time, which is quite improbable. To remove this type of noise we apply a smoothing filter to each pair of GPS points (p1, p0) of the trajectory (Fig. 1).

3.3 User Stay Points Detection

Stay points are regions where a given user has stayed for a while within a defined radius. The algorithm used to extract the users' stay points from a trajectory is

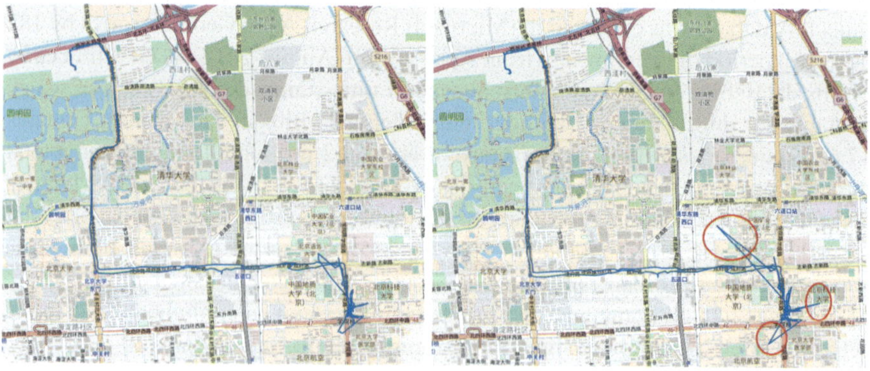

Fig. 1 Data cleaning process removing outliers and spikes in trajectory data. On the right image, we have the red ellipsis highlighting the spikes in the trajectory

a hybrid density and time-based proposal [8]. It follows the logic of calculations between consecutive points where we measure the Haversine distance between two sets of coordinates p1 and p0 in order to find those that are below a distance threshold parameter. While the distance between the successive points remains smaller than the parameter, we keep adding them to a list of candidate points to form the stay point region. Next, the algorithm checks for how long the user stayed in that radius of the distance threshold checking for the second parameter, the time threshold. If the spent time is greater than the parameter value, we add those points to the final set of items which will form the new stay point.

For this experiment, we set the parameters distance threshold as 200 m and the time threshold to 20 min. Hence, individuals having in their trajectories, consecutive points in a region of 200 m radius for more than 20 min are creating new stay points for each similar situation. Having a new set of points we need to calculate their centroid which we denote as the center of mass of all the points in the cluster. To perform this task we calculate the mean of the coordinates of the set of points in the cluster. We can see an example of stay point in Fig. 2.

3.4 User Location Points

A location point is defined as a frequent location visited by an individual. For the purpose of finding points of interest which are described in Sect. 3.5, we need to find the most common locations within all the users in the community individually. In this way, we look for those places a person visits repeatedly in order to form the so-called users' location points.

Location detection techniques are common tasks which make use of density-based methods, this is due because the mechanism of density-based clustering enables it to detect clusters of arbitrary shapes without specifying the number of clusters in the data a priori. It also has a notion of noise and is tolerant of outliers.

In this way, the clustering algorithm we selected to perform this task is DBSCAN (Density-Based Spatial Clustering of Applications with Noise) [3]. Among other advantages, and because of the fuzziness of the trajectory data points logging, this algorithm performs very well with geographic data with arbitrary shape, is easily adaptive to different distance functions, and also is able to detect noise which is

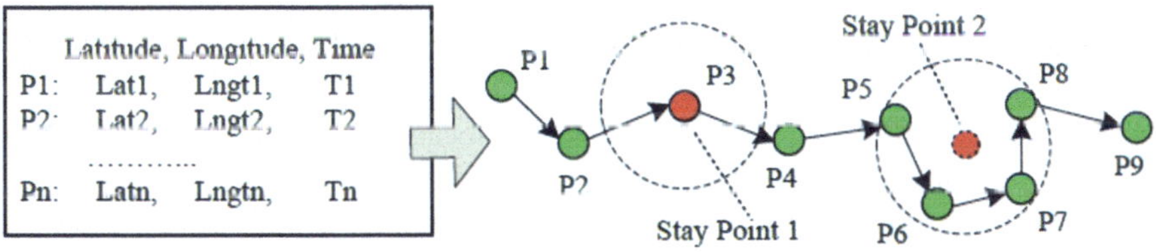

Fig. 2 GPS log and stay points [8]

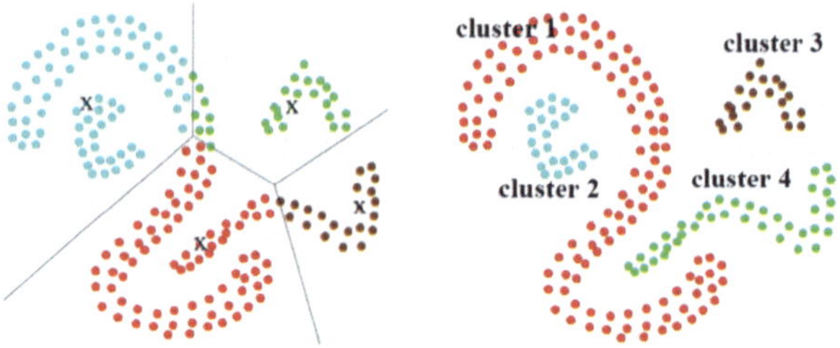

Fig. 3 A comparison between clustering based on centroid distance and density-based cluster

very useful in our case as we do not want points that do not fall close to each other to be part of our points of interest set. In Fig. 3 we can verify a comparison between density based and regular centroid distance cluster.

The parameters used in the algorithms' setup are the eps, related to the distance between the points and the MinPts which are the number of minimum points required to form a cluster. Some techniques were proposed to estimate these parameters, in this work we are following [3] k-dist heuristic to determine their optimal values. After applying the heuristic we ended up with the values 100 m for the eps (distance) and 4 points to the MinPts (minimum points to form a cluster).

3.5 Points of Interest

Aiming to extract the most common locations among all users in a given community, we perform a step of clustering the user location points (LP) extracted in Sect. 3.4, finding then this way, the Points of Interest visited by many individuals. The parameters utilized to perform this task are slightly different than those used to extract the location points as to form points of interest in a community the eps (distance) and MinPts should be increased. The experiments are shown in Sect. 4.

3.6 Users Similarity

By having the set of points of interest of all users in the community, we can apply the method to calculate the users' similarity by verifying the number of points of interest the individuals share in common. The similarity measure utilized in this method is the Jaccard's coefficient described in Eq. 1. This coefficient is denoted by the number of locations shared by two users divided by the size of their locations' union, characterizing the similarity between their sets of locations. In this way, we calculate the users' intersection set which are the locations shared by the two individuals and calculate the union of all locations visited by them. Hence A be the set of locations visited by a user u1 and B the set of locations visited by a user u2, the intersection set $(A \cap B)$ and the union set (A ∪ B), for both users, are, respectively, the set of

Fig. 4 Set of visited POIs of two given users

locations that both individuals have visited and the set of regions visited for at least one of them.

$$Jaccard(x, y) = \frac{|\Gamma(x) \cap \Gamma(y)|}{|\Gamma(x) \cup \Gamma(y)|} \quad (1)$$

In Fig. 4 we have an example of a set of POIs that a pair of users (A and B) have visited. In this particular scenario, the coefficient given by the similarity formula is represented as:

$$|A \cup B| = \begin{bmatrix} 01 & 02 & 03 \\ 04 & 05 & 06 \end{bmatrix} = 6 \; |A \cap B| = \begin{bmatrix} 02 & 07 & 01 \\ 05 & & \end{bmatrix} = 4$$

$$Sim(u1, u2) = \frac{3}{7}$$

4 Experiments and Results

To perform the experiments, a group of 10 individuals representing a community was selected from the whole Geolife dataset in order to evaluate the proposed methods. The subset used to perform the evaluation is shown in Table 1.

4.1 User Stay Points Detection: Results

To perform the search for stay points (SP), we set the algorithm parameters as following: radius distance threshold was set to 200 m and the time threshold was set

Table 1 Example of Geolife dataset

Period	April 2007 to August 2012
Users	10
Number of trajectories	3609
Number of GPS points	1,844,250

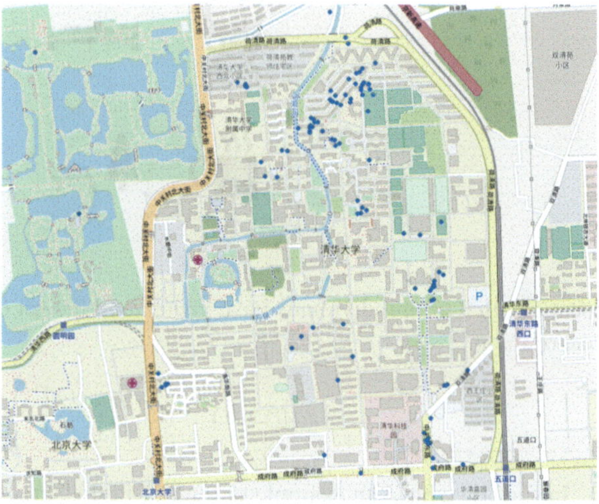

Fig. 5 Example of stay points (SP) found in the subset of GPS raw data by performing the proposed method over the user "000." Each blue dot stands for a distinct stay point

Table 2 Top 10 ranked users according to quantity of stay points detected and number of GPS points

User	Number of trajectories	Number of GPS points	Number of stay points
004	1217	439,397	2671
003	941	485,226	2244
000	458	173,870	2098
002	329	248,217	786
005	164	109,046	457
001	130	108,607	438
007	134	85,531	351
009	109	84,616	314
008	71	77,910	176
006	56	31,830	116
Total	3609	1,844,250	9651

to 20 min. After running the stay points detection method, we were able to identify various points along the city. Figure 5 shows examples of stay points found for the user "000" in the subset.

In Table 2 we summarize the top 10 users ranked by their number of stay points found and the number of GPS trajectory points.

4.2 User Location Points: Results

Following the framework pipeline, the next step was to identify the users' location points (LP). As mentioned in Sect. 3.4, a location point means a person frequently visits the location. In this study, we took care of filtering the locations visited occasionally by the user, as these locations do not represent meaningful places but some odd situations in users' life. To solve this issue we use the location points in order to list the top-k locations. The top-k locations are found by the most visited locations in users' history. When talking about mobility patterns, there are basically two groups of users, those who follow routines, visiting a few locations more frequently and those who use to explore more the region. We perform density-based cluster with a distance threshold (eps) of 100 m and a minimum number required to form a cluster (MinPts) of 4 points, which means an individual that visits the same group of stay points that are located within a shorter radius of 100 m at least 4 times considers that place as a meaningful location. Figure 6 show the user location points found in the experiments.

The set of stay points per user which were converted into location points can be verified in Table 3.

4.3 Points of Interest: Results

The points of interest (POIs) concept is related to small regions where a considerable amount of visits have occurred. To extract the points of interest in the given

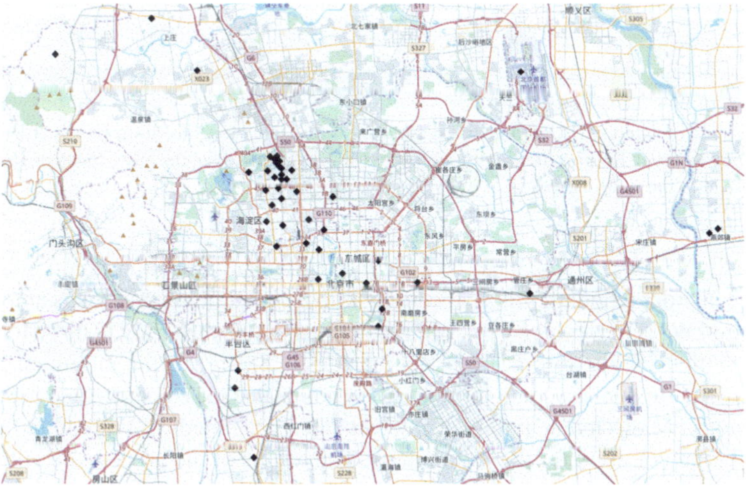

Fig. 6 Location points (LP) identified from the set of stay points (SP) of users in the subset in great Beijing area. Each black dot stands for a distinct location point. The majority of the location points of this given user are concentrated in the northwest area

Table 3 Top 10 users' stay points converted into location points

User	Number of stay points	Number of location points
004	2671	41
003	2244	46
000	2098	38
002	786	23
005	457	5
001	438	18
007	351	8
009	314	4
008	176	7
006	116	2
Total	9651	192

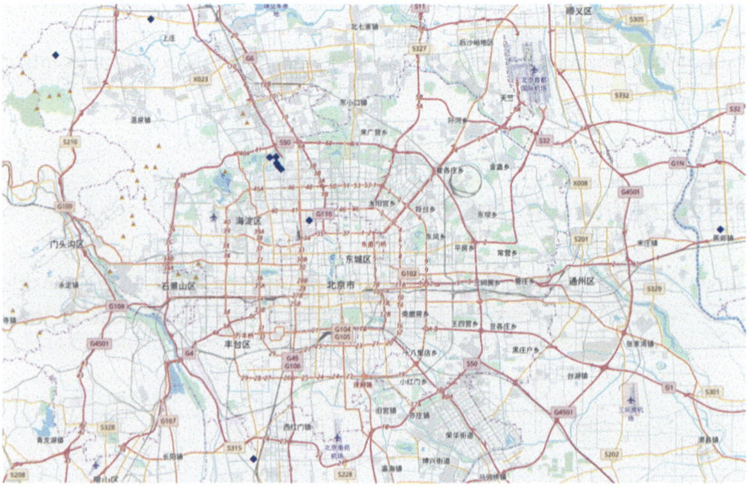

Fig. 7 Points of interest (POIs) identified from the set of all users' location points (LP) of the community in great Beijing area. Although there are POIs all over the city, most of them are concentrated in the northwest area

community of users we need to process the individual's location points found in Sect. 4.2 in order to identify those locations with a higher number of visits among all the dataset. We achieve this by performing density-based clustering in those location points configuring the distance (eps) parameter to 200 m and the minimum points (MinPts) to 4. In this way, only those location points that have been visited in at least 4 occasions are elected as points of interest. This generated 32 POIs found in the selected group of users for this community. The results of the process are in Fig. 7.

4.4 Users Similarity: Results

Regarding the results about the users' similarity it is important to notice that by using individuals' location GPS logs history, we cannot identify the real intention the users have been to a given place as density-based algorithms are supported by a given radius of distance in order to aggregate the closest points. Whereas in the real world, one specific physical location can be related to more than one place, especially when dealing with building with levels such as shopping centers, hospitals, or universities.

The results obtained regarding users' similarities are listed in Table 4. The data is ranked from the most similar users to the less similar ones according to the proposed Jaccard's coefficient mentioned in Sect. 3.6.

In order to analyze the similarity results in a geographical manner, we have Fig. 8 representing the POIs of the two most similar users in the community. One can notice the closeness of their locations having 9 out of 10 locations in common corresponding to 90% of similarity.

The highest the degree, the more likely to be potential friends or have similar location preferences. The results of this simple method can be useful to understand the dynamics and relations between users in a given community. Possible applications of this work are in recommendation tasks for locations, such as shops, parks, restaurants, bars, etc. This also can be used to suggest users with same interests in social network services.

Table 4 Top 5 similar users found in the community according to Jaccard similarity

User A	User B	Similar places	Similarity percentage
000	003	9/10	0.90
000	004	8/9	0.88
003	004	8/10	0.80
005	008	1/3	0.33
003	005	3/10	0.30

Fig. 8 Similarity between a pair of users that are top ranked in the similarity index results. The black dots represent the common POIs shared

5 Conclusions and Future Work

Ubiquitous devices such as smartphones have boosted the interest of researches in location history data such as GPS trajectory records. In this paper, we proposed methods to handle these raw GPS data by applying data mining algorithms to identify meaningful locations. Experimental evaluation was performed in the dataset in order to find meaningful places and similar individuals using the suggested methods. By applying a hybrid method we showed how to extract stay points from users' trajectory data. Furthermore, we were able to find among those extracted stay points those who have more influence in the individuals' routine, here called location points. By analyzing the location points of all the users in a given community we were able to identify the POIs, which are the sets of location points many individuals share in common, helping us to understand the users' behavior. Finally, we showed a metric for finding the most similar users in a given community by applying similarity measures over the extracted POIs. This feature is useful to understand the dynamic between individuals in a given community. The results are quite good when taking into account the precision and the simplicity of the approaches and served as a base to answer the three main questions proposed in Sect. 1.

For future work, the next steps include the development of a method to find the patterns of people visiting and leaving different places at different times in an order (weekly basis, daily basis). Also includes some map matching tasks including external information in order to find the semantic meaning of the individuals' and the collective points of interest or regions of interest.

References

1. Cao, X., Cong, G., & Jensen, C. S. (2010). Mining significant semantic locations from GPS data. *Proceedings of the VLDB Endowment, 3*(1–2), 1009–1020.
2. De Choudhury, M., Feldman, M., Amer-Yahia, S., Golbandi, N., Lempel, R., & Yu, C. (2010). Automatic construction of travel itineraries using social breadcrumbs. In *Proceedings of the 21st ACM Conference on Hypertext and Hypermedia* (pp. 35–44). New York: ACM.
3. Ester, M., Kriegel, H. P., Sander, J., & Xu, X. (1996). A density-based algorithm for discovering clusters in large spatial databases with noise. In *Kdd* (Vol. 96, pp. 226–231).
4. Lee, I., Cai, G., & Lee, K. (2013). Mining points-of-interest association rules from geo-tagged photos. In *2013 46th Hawaii International Conference on System Sciences* (pp. 1580–1588). Piscataway: IEEE.
5. Li, Q., Zheng, Y., Xie, X., Chen, Y., Liu, W., & Ma, W. Y. (2008). Mining user similarity based on location history. In *Proceedings of the 16th ACM SIGSPATIAL International Conference on Advances in Geographic Information Systems* (p. 34). New York: ACM.
6. Lu, X., Wang, C., Yang, J. M., Pang, Y., & Zhang, L. (2010). Photo2trip: Generating travel routes from geo-tagged photos for trip planning. In *Proceedings of the 18th ACM International Conference on Multimedia* (pp. 143–152). New York: ACM.
7. Terveen, L., & McDonald, D. W. (2005). Social matching: A framework and research agenda. *ACM Transactions on Computer-Human Interaction, 12*(3), 401–434.

8. Ye, Y., Zheng, Y., Chen, Y., Feng, J., & Xie, X. (2009). Mining individual life pattern based on location history. In *Tenth International Conference on Mobile Data Management: Systems, Services and Middleware, 2009. MDM'09* (pp. 1–10). Piscataway: IEEE.
9. Zheng, Y., Chen, Y., Xie, X., & Ma, W. Y. (2009). Geolife2.0: A location-based social networking service. In *Tenth International Conference on Mobile Data Management: Systems, Services and Middleware, 2009. MDM'09* (pp. 357–358). Piscataway: IEEE.
10. Zheng, Y., Li, Q., Chen, Y., Xie, X., & Ma, W. Y. (2008). Understanding mobility based on gps data. In *Proceedings of the 10th International Conference on Ubiquitous Computing* (pp. 312–321). New York: ACM.
11. Zheng, Y., Xie, X., & Ma, W. Y. (2010). Geolife: A collaborative social networking service among user, location and trajectory. *IEEE Data Engineering Bulletin, 33*(2), 32–39.
12. Zheng, Y., Zhang, L., Xie, X., & Ma, W. Y. (2009). Mining interesting locations and travel sequences from GPS trajectories. In *Proceedings of the 18th International Conference on World Wide Web* (pp. 791–800). New York: ACM.

Cooperation Models for Employee Education: Analysis on Slovak and Czech Companies

Lucia Kohnová, Ján Papula, and Zuzana Papulová

1 Knowledge Networks and Clusters as Key Source of Knowledge

From a broad perspective, Fayol described the network relationships, but later at the end of the twentieth century, Drucker, Savage, or Noel et al. emphasized the importance of teamwork and network structures in their work. Collaboration with external partners may, however, take a different form. We can distinguish it based on the number of members, the geographical position of the members, the type, or the purpose of the collaboration. Transaction-based partnership cooperation is the most common type of partnership driven by economic factors. This kind of cooperation, however, rarely creates an environment of trust and the form of a relationship that stimulates the exchange of knowledge. On the contrary, there are forms of partnerships as knowledge networks that are designed to promote knowledge creation and sharing. Du Preez [1] has defined knowledge networks as: "a set of people and resources, and relationships between them that enable them to capture, transfer and create knowledge to create value. An integrated knowledge network covers all domains, communities and relationships to promote innovation sustainability, which will continue to support the sustainability of business competitiveness, or its users."

Knowledge networks are also often understood as information networks, expert networks, practice-oriented communities, and strategic alliances, depending on the interest of the members, their amount, or the openness of such a network [2]. For the purposes of this chapter, knowledge networks will be understood as any internally coordinated network—a community of organizations whose main interests are the

L. Kohnová (✉) · J. Papula · Z. Papulová
Department of Strategy and Entrepreneurship, Faculty of Management Comenius University in Bratislava, Bratislava, Slovakia
e-mail: lucia.kohnova@fm.uniba.sk

© Springer Nature Switzerland AG 2020

D. Cagáňová, N. Horňáková (eds.), *Mobility Internet of Things 2018*, EAI/Springer Innovations in Communication and Computing, https://doi.org/10.1007/978-3-030-30911-4_22

promotion and exchange of knowledge. Such networks can also work in the form of online communication between partners, for example through their own e-learning platform or online learning. An optimal form of such a partnership is, for example, business cooperation with universities. The success of the knowledge network, however, depends closely on the relationships and trust of individual members. According to Moron and Taylor, social relationships between network participants can directly excite the learning process [3].

In general, we can distinguish between two types of knowledge networks depending on the formality. Formal knowledge networks are structured networks that are governed by internal guidelines, rules, and agreements that direct their activities. Their primary focus is on knowledge creation, accumulation, and sharing. They can exist within the internal environment of the organization as internal knowledge networks, but also beyond the organization borders. Such networks may include other organizations, sectors, and countries.

Creech and Willard identified the seven principles necessary for the functioning of formal knowledge networks [2]:

1. They must be target oriented.
2. They should be "work networks" (not only for knowledge sharing but also for mutual help in finding solutions).
3. Institutional commitment is essential (not only individuals but also organizations that are interested in network prosperity).
4. They should be built on expertise.
5. They should cover different sectors, regions, or countries.
6. Individual members should support and develop the network's interests.
7. Networks are dependent on communication.

1.1 Cluster

Cluster status in today's economy is still gaining popularity, whether from the point of view of regional or state policies. The OECD sees clusters as an important development tool for countries, like the European Commission, which highlights the positive impact on innovation, growth, competitiveness, empowerment, and information generation.

Michael Porter defined the concept of cluster in 1990 in his book "Competitive Advantage of Nations" as "a geographically close group of interrelated companies, specialized suppliers, service providers and related institutions in a particular field, as well as companies in related fields that compete together but also cooperate, have common features complement each other" [4]. Clusters such as geographical concentration of businesses were described by English economist Alfred Marshall at the beginning of the twentieth century in connection with the concept of industrial districts. He observed the concentration of small businesses in the industries.

Companies with similar profiles have thus benefited both the industry and the entire region [5].

Current definitions of clusters are quite similar [6–8]. Porter, in 1998, broadened his definition and thus included the cluster concept in a broad sense: "Clusters are local concentrations of interrelated companies and institutions in a specific field. Clusters include a group of interconnected industries and other players important for competition. They include, for example, suppliers of specialized inputs such as components, machines and services, and specialized infrastructure providers. Clusters are often expanded downstream to sales channels and customers, and to complementary product manufacturers and companies in the industry, related to knowledge, technology, or common inputs. Many clusters also include government or other institutions—such as University. Normative agencies, research teams or business associations—providing specialized training, education, information, research and technical support" [9].

From the point of view of typology of clusters Spišáková compares Slovak clusters to European clusters based on their life cycle and based on size [10].

1. Distribution based on life cycle—establishment, volume growth, quality growth, maturity, cluster skew [11]

   In Slovakia, 26 clusters were created in 2013. Of these, 13 clusters were created between 2004 and 2009 (up to 7 clusters were created in 2008). Other clusters were established only in 2010. Slovak clusters are younger than European clusters, with the first European clusters formed prior to 1990 (48 clusters).
2. Distribution based on size

In terms of cluster size, similarly, Slovak businesses lag behind European ones. The largest Slovak cluster is the automotive cluster—western Slovakia, Trnava region. In the EU, however, 40% of clusters have 50 or more members compared to zero in Slovakia in the year 2013.

Clusters are a free form of strategic partnership, with the notion of strategic partnership "reflecting the mutual relationship of two or more enterprises that agree to jointly achieve a strategic goal or objectives and to cooperate to achieve this goal" [12]. Examples of free strategic partnerships are alliance agreements, network alliances, and various nonproprietary connections.

The development and support of clusters is also a question of the European Union's strategy. It is mostly communicated through strategic intentions, supportive policies, or memoranda. It is worth mentioning the 2007 Community Guidelines "The Community Strategic Guidelines on Cohesion (CSGs)" (Council of Europe 2006). In 2007, it set up a European Cluster Observatory within the Innova initiative to analyze clusters and cluster policies in the EU. In the same year, the document "Memorandum of European Clusters" was issued to mobilize support for cluster initiatives and policies. EU policy has been gradually expanded to support the achievement of world-class clusters, creating the "European Cluster Policy Group." Mostly, cluster support is included in strategies to support innovation performance and support for SMEs, which are currently also part of the Horizon2020 strategy. Slovak ClusterPoliSEE project, implemented through the Southeast Europe Transnational Co-operation Program, also participates in the cooperation with the European Union on Clusters [13].

However, the Slovak Republic, in terms of cluster experience, is among the countries with the shortest experience (also due to the late adaptation of cluster policies). The first policies emerged in the 2007–2013 strategy. In reality, however, there is still no cluster support organization in Slovakia, so cluster issues include regional development and SME support agencies. In 2010, the Union of Slovak Clusters was founded, the creation of which was initiated by representatives of Slovak clusters. The purpose of this union was to support the phenomenon of clusters in Slovakia, to promote common goals and also the Slovak economy through the regions and their clusters [14].

2 Increasing Importance of Education in Organizations

Nowadays, companies need to flexibly respond to current issues and new development and trends on the market to be able to maintain or to increase their competitive position. Innovation has become companies' key interest in terms of gaining competitiveness and supporting performance. Management of innovation and systematic promotion of innovative activity however require paying attention to factors such as people, knowledge, organizational culture, partnerships, knowledge networks, and advanced information and communication technology.

In this connection, current challenges are especially linked to the changes in external environment [15–19], such as *demographic changes* (e.g., shifts in requested skills and competencies, gap between the education system and practice); *technological changes* that are connected to new technologies (e.g., robotic, robotized production lines, autonomous machineries) and their impact (how technologies will evolve and how and when competitors, suppliers, and customers will respond; how to use technologies to improve the business processes and performance); *legal and ethical constraints* (e.g., constraints imposed by regulators, standard-setting bodies, laws, social mores, and emphasis on business ethics); and *enhanced competition* (new forms of competition, global perspective, innovation, need for higher productivity, and new market development).

To respond to these trends, companies search for proper reactions. To cope with new challenges, it is suggested to focus on these areas [18–21]:

- Permanent innovation and high-quality production and products with the highest possible added value, promoting the most highly qualified human work
- Systematic development of human potential, education and organization development, sharing of information and knowledge
- Involvement of employees in process innovations with supporting culture, communication, and motivation

– Investment in research and related activities and scanning of emerging opportunities (companies should constantly scan, search, and explore across technologies and markets and learn from it)

Orientation on education and development of employees should be among the key strategic priorities. As studies are showing (e.g., [22–24]) investment in education is leading to greater efficiency and organizations can improve and enhance the quality and performance of their employees by providing right education. Also companies can better adapt their processes and workflow to advanced technologies as new types of job design and new forms of working require a different combination of skills [25, 26].

The requirements for education might not be the same with every company as they differ in so many aspects like size, life cycle, dynamicity of the market, level of digitalization and automatization, and other impacts. To ensure effective and efficient process of education, it is important to plan and set up education and training activities to link organization purposes. The steps should involve the following [27]:

1. Set goals that reflect the organization's real needs.
2. Define the clear and measurable expected benefits of education.
3. Choose the appropriate form of education.
4. Provide feedback on the success rate of education.

3 Methodology

The aim of the research was to analyze the level of implementation of the different types of education and cooperation in the educational process between the companies surveyed. The survey was conducted on a sample of 908 companies operating in Slovakia and the Czech Republic. These two countries have been chosen for research purposes based on their common political and cultural history, geographic proximity, and business cooperation and linkages. The Slovak Republic and the Czech Republic are also on similar ranks in innovation scoreboards, with both being moderate innovators.

In the framework of the research analyses, we focused on comparing the behavior of Czech and Slovak companies at the level of the whole sample as well as on the more detailed analyses in terms of size of companies, stage of maturity of companies, and the sector in which they operate. Within the sectors, we focused on the most represented industry sectors in both countries, comparing the level of engagement in education forms and partnerships in the automotive, electrotechnical, mechanical, construction, and pharmaceutical and medical industries. Individual industrial representations in the surveyed countries are shown in Fig. 1. In the group of Slovak companies in selected types of industry, we analyzed 128 companies, and 153 companies in the Czech Republic.

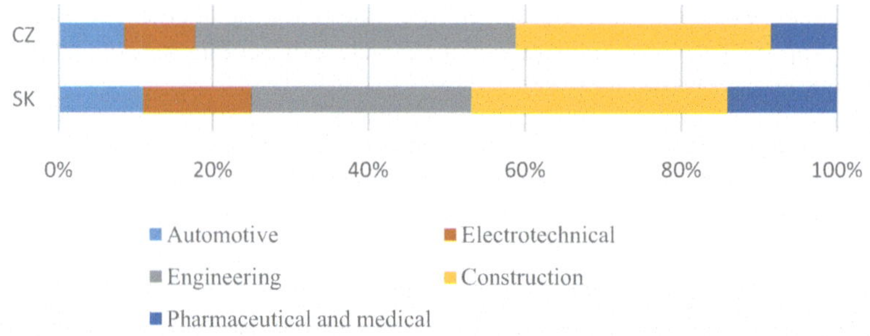

Fig. 1 Distribution of selected industries in research groups of Czech companies and Slovak companies

In terms of research, we have focused on selected forms of education and learning partnerships that companies engage in. The survey was conducted through an electronic questionnaire survey, distributed predominantly between directors and business managers. Respondents answered on the scale of 1–5 (5—all the time, 4—often, 3—sometimes, 2—mostly not, 1—never) to the options of education and learning forms and partnerships based on the level of their engagement:

- Employee training in external organizations
- Training courses, coaching, and development programs
- Staff travel to partner organizations
- Cooperation with research institutions
- Cooperation with secondary schools and universities
- Engagement in knowledge networks, alliances, and joined venture

When analyzing the research questions, we aimed to identify the differences between the surveyed companies, examining the differences in the structure of the companies based on their size, stage of maturity, and the industrial sector. The results were evaluated by means of basic statistics and weighted averages, the significance of the differences being evaluated by the nonparametric chi square test, where the significance level was $p = 0.05$.

4 Research Results

In comparison of the total sample by the country of operations, it can be concluded that Slovak and Czech companies are engaged similarly in research forms of involvement in employee education at a similar level (Fig. 2). The average assessment was where the level of involvement was different between the surveyed countries while the statistically significant difference was in three options (Table 1). Slovak companies are significantly more involved in employee training in external organizations than Czech companies, while Czech companies engage in cooperation

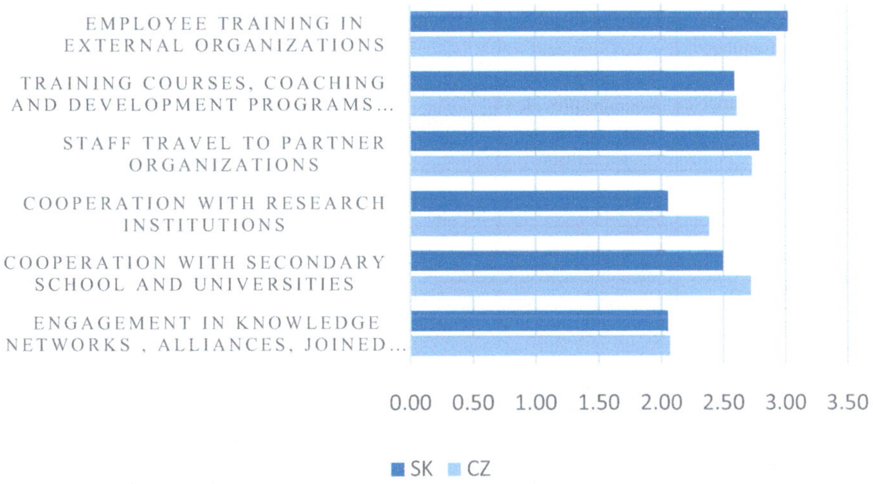

Fig. 2 Comparison of average assessment in selected research questions between Slovak and Czech companies

Table 1 Chi square test results in selected questions

	Employee training in external organizations	Cooperation with research institutions	Cooperation with secondary school and universities
Chi square test, SK-CZ	0,04318935	0,00116184	0,025386416

with research institutions and cooperation with secondary schools and universities significantly more than Slovak companies. Overall, the companies surveyed were the most involved in employee training in external organizations and staff travel to partner organizations and the least engaged in knowledge networks.

When comparing the responses between Slovak and Czech companies surveyed, it can be observed that the distribution of responses in each option is very similar. Figure 3 shows a comparison of median, mode, and response distributions (from left 1–5). The median of responses differed only in the question of cooperation where Czech companies are significantly more engaged than Slovak companies.

From the point of view of the maturity stage of the companies we surveyed, we identified several differences. In the group of Slovak companies, starting companies engaged in different types of education and cooperation in education was less than mature and growing. The highest average rating for Slovak companies was in the group of mature businesses in every question (Fig. 4).

Among the Czech companies surveyed, starting had a lower-than-average rating only in staff travel to partner organizations and cooperation with research institutions (Fig. 5). However, they engaged much more in training courses and engagement in knowledge networks. This can also be observed when comparing to

	Employee training in external organizations	Training courses, coaching and development programs provided by external specialists	Staff travel to partner organizations	Cooperation with research institutions	Cooperation with secundary school and universities	Engagement in knowledge networks , alliances, joined venture
Median						
SK	3	3	3	1	2	2
CZ	3	3	3	2	3	2
Mode						
SK	3	1	1	1	1	1
CZ	3	3	3	1	1	1
Distribution						
SK						
CZ						

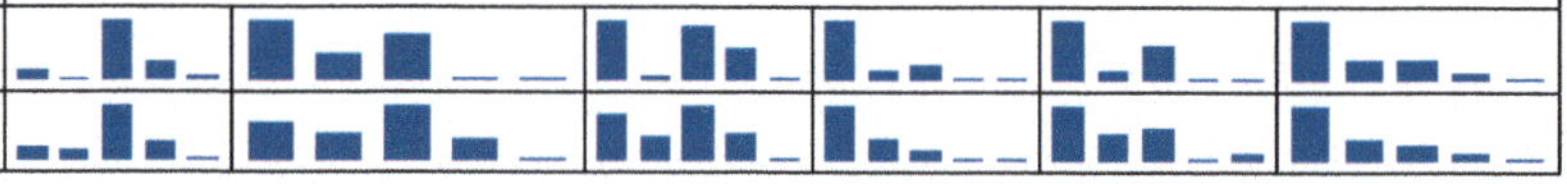

Fig. 3 Comparison of median assessment, mode, and distribution between Slovak and Czech companies

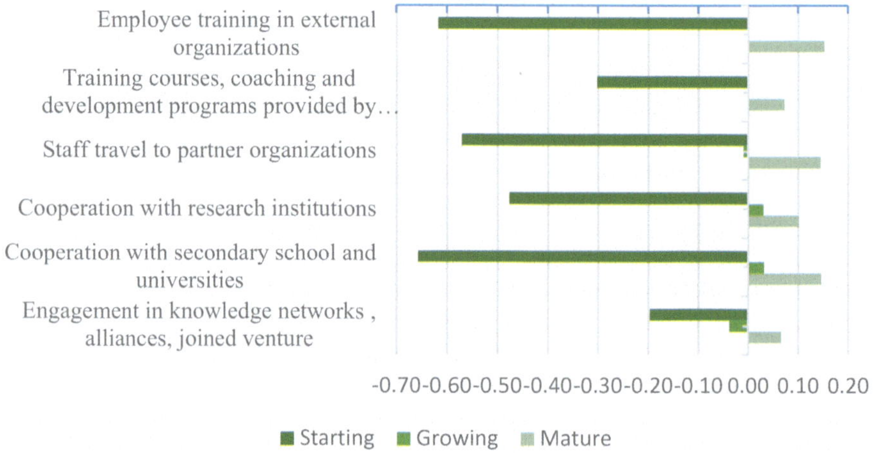

Fig. 4 Comparison of average assessment in the group of Slovak companies based on maturity level

Slovak companies. Among the starting companies in Slovakia, 10% of companies are often or all the time involved in the knowledge networks, compared to 26% of starting companies in the Czech Republic, which was found to be statistically significant. Within the structure of surveyed companies in the Czech Republic, we see that, as well as in the case of starting companies, growing companies have a higher average rating on training courses and engagement in knowledge networks than mature Czech companies.

Based on company size, it is also possible to identify a number of differences between the surveyed countries, but they are not as significant as in the case of

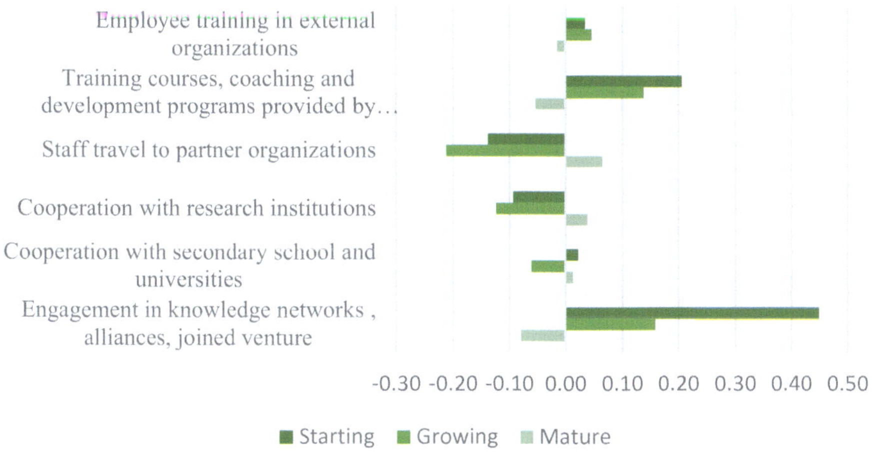

Fig. 5 Comparison of average assessment in the group of Slovak companies based on maturity level

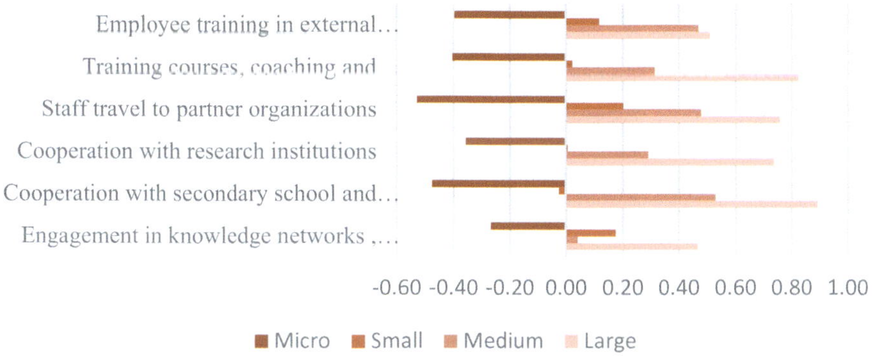

Fig. 6 Comparison of average assessment in the group of Slovak companies based on size

maturity stage of researched companies. From the point of view of the assessment distribution in the group of Slovak companies, all types of educational activities and partnerships of micro-companies were assessed to be below average. Highest over-average assessments can be observed for large companies, followed by medium, except engagement in knowledge networks. On this issue, the second highest over-average assessment is achieved by small Slovak companies (Fig. 6).

Within the Czech companies' structure, micro-companies also achieved the lowest average assessment in all questions (Fig. 7). A markedly different average assessment compared to Slovak companies is observed in cooperation with research institutions where Czech small companies have achieved a significantly higher average assessment than Slovak small companies. By contrast, Czech medium companies had lower average assessment on employee training in external organizations and training courses than Slovak medium companies.

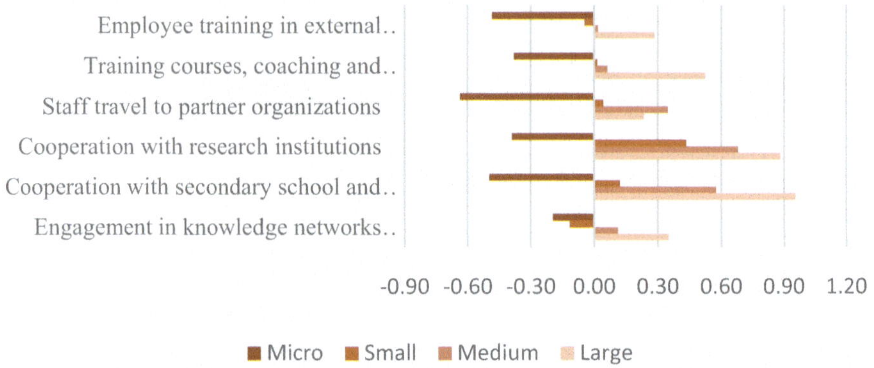

Fig. 7 Comparison of average assessment in the group of Czech companies based on size

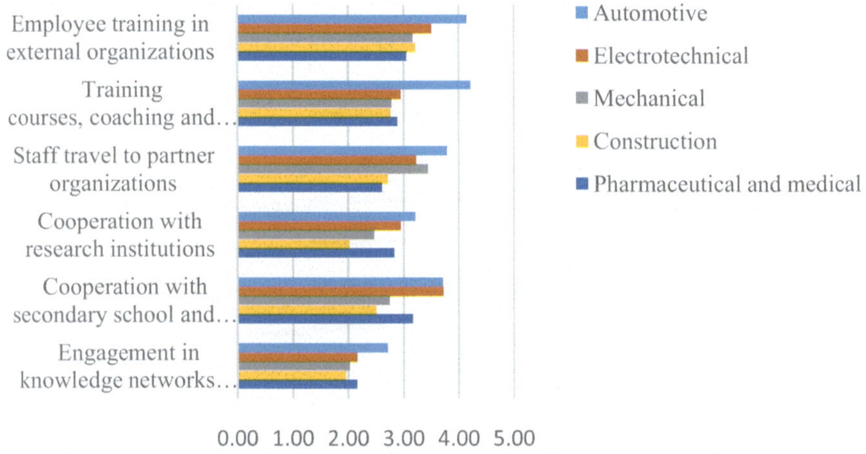

Fig. 8 Comparison of average assessment in the group of Slovak companies in selected industries

From the industry perspective, we looked closer to the five most represented industries. In the group of Slovak companies, the highest average assessment was almost on every question in the automotive industry. The electrotechnical industry has achieved the highest average value in cooperation with secondary schools and universities, the construction industry has generally lower average scores than other sectors surveyed, and the pharmaceutical and medical industry has achieved the highest average rating also in collaboration with secondary schools and universities (Fig. 8).

On the other hand, among the Czech companies, almost all the questions reached the highest average assessment in the pharmaceutical and medical industry (Fig. 9). It was more observable in the question of employee training in external organizations (3,69) and cooperation with secondary schools and universities (3,62), which is significantly higher compared to the Slovak pharmaceutical and medical

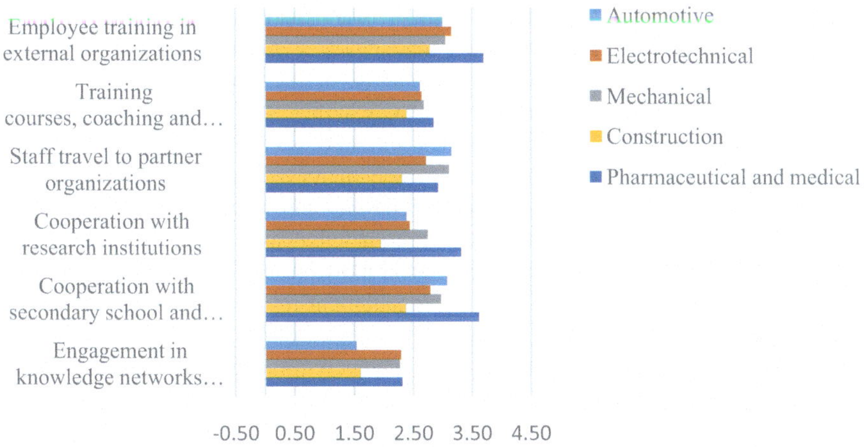

Fig. 9 Comparison of average assessment in the group of Czech companies in selected industries

Table 2 Comparison of median values in the automotive industry in the Slovak Republic and the Czech Republic

	SK	CZ
Employee training in external organizations	4	3
Training courses, coaching, and development programs provided by external specialists	4	3
Staff travel to partner organizations	3,5	3
Cooperation with research institutions	3	2
Cooperation with secondary school and universities	4	3
Engagement in knowledge networks, alliances, joined venture	2,5	1

industry. Low average assessment was achieved by the construction industry in all questions, with the average rating lower than in the case of the Slovak companies from the construction industry. In all questions except of the staff travel to partner organizations, the automotive industry has significantly higher average assessments in Slovakia than in the Czech Republic, also shown by the median numbers in Table 2.

5 Discussion and Conclusion

From the research results we can identify an overall approach to education through cooperation in the Czech Republic and Slovakia. By comparing the results, Czech companies are more active in collaborating with research institutions and schools or universities.

However, in a detailed look at the structure of enterprises, we can find specific differences. Dynamically growing small- and medium-sized enterprises are most dependent on external know-how. The behavior of these companies represents largely progressive approaches in the country, whether it is the traditional use of external lecturers or modern cooperative learning at cluster level.

From the industry perspective, we also see increased cooperative activity in industries that are dominant in the national economy and where there are increased demands for recruiting staff with specific knowledge and skills. They are also sectors that are undergoing major technological innovations.

However, the method and intensity of engagement in cooperative relationships are contingent. It also depends on the maturity and maturity of networks and clusters and their offer of educational activities. It is precisely the degree of cluster activity that initiates subsequent collaboration with universities and research institutions.

References

1. DuPreez, N., Louw, L., & Lutters, E. (2008). In A. Bernard & S. Tichkiewitch (Eds.), *A knowledge network approach supporting the value chain, in methods and tools for effective knowledge life-cycle-management* (pp. 159–168). Berlin: Springer Eds.
2. Creech, W., & Willard, T. (2001). *Strategic intentions: Managing knowledge networks for sustainable development.* Canada: Manitoba.
3. Morone, P., & Taylor, R. (2004). Knowledge diffusion dynamics and network properties of face-to-face interactions. *Journal of Evolutionary Economics, 14*(3), 327–351.
4. Pavelková, D. (2009). a kol. Klastry a jejich vliv na výkonnost firiem. GRADA Publishing a.s.
5. Bialic Davendra, M., et al. (2014). *The clusters phenomenon in the selected central European countries* (pp. 7–8). UK: Cambridge Scholars Publishing.
6. OECD. (1999). *Boostin innovation. The cluster approach.* Paris: OECD Publishing.
7. Mytlka, L., & Farinelli, F. (2000). *Local clusters, innovation systems and sustained competitiveness.* The United Nations University. Discussion Paper Series.
8. Solvell, O., et al. (2003). *The Cluster Initiative Greenbook.* Stockholm: Ivory Tower AB.
9. Porter, M. (1998). *The competitive advantage of nations.* New York: Free Press.
10. Spišáková, E. (2013). Klastre na Slovensku a v Európske Únii z hľadiska počtu členov, veku klastrov a ich právnej formy. *The 16th International Scientific Conference Trends and Innovative Approaches in Business Processes.* www.sjf.tuke.sk/kpiam/TaIPvPP/2013/journal/13_Emilia_Spisakova_KLASTRE.pdf.
11. Otsuka, K. (2006). Cluster-based industrial development: A view from East Asia. *The Japanese Economic Review, 57*(3), 361–374.
12. Smolková, E., & Borovský, J. (2005). Strategické partnerstvá pre malé a stredné podniky. EUROUNION spol. s r.o., ISBN 80–88984-84-X.
13. Duman, P. (2009). a kol. SIEA. Klastre na podporu rozvoja inovácií. Analytická štúdia. www.siea.sk/materials/files/inovacie/slovenske_klastre/Klastre-SIEA.pdf.
14. SBA. (2014). Národný report medzinárodného projektu "Clustrat". www.clusteratr.sk/images/data/file/National%20report_Slovakia.pdf.
15. Caganova, D., Saniuk A., Sujanova, J., Horňáková, N., & Švač, V. (2017). *Managerial skills in industrial praxis.* Zielona Góra.
16. Kohnová, L., Papula, J., Papulová, Z., & Suchoba, M. (2017). Preparation for smart industry, introduction and the comparative study. In *Conference Proceeding SmartCity360.*

17. Pajtinkova Bartakova, G., Gubiniova, K., Brtkova, J., & Hitka, M. (2017). Actual trends in the recruitment process at small and medium-sized enterprises with the use of social networking. *Economic Annals-XXI, 164*(3–4), 80–84.
18. Teece, D. J. (2007). Explicating dynamic capabilities: The nature and microfoundations of (sustainable) enterprise performance. *Strategic Management Journal, 28*(13), 1319–1350.
19. Turek, I. (2006). Globalizácia a jej vplyv na inováciu technického vzdelávania. In *Conference proceeding InEduTech 2006*. Prešove: Prešovská univerzita v.
20. Stacho, Z., Potkány, M., Stachová, K., & Marcineková, K. (2016). The organizational culture as a support of innovation processes' management: A case study. *International Journal for Quality Research, 10*(4), 769–784.
21. Stachová, K., Stacho, Z., & Vicen, V. (2017). Efficient involvement of human resources in innovations through effective communication. *Business: Theory and Practice, 18*, 33–42.
22. Becker, B. E., Huselid, M. A., & Ulrich, D. (2001). *Scorecard: Linking people, strategy and performance*. The HR, Boston, MA: Harvard Business School Press.
23. Rafiei, N., & Davari, F. (2015). The role of human resources management on enhancing the teaching skills of faculty members. *Mater Sociomed, 27*(1), 35–38.
24. Urbancová, H., Stachová, K., & Stacho, Z. (2017). Using of performance appraisal methods in Czech and Slovak organisations. *Quality Innovation Prosperity, 21*(3), 62–77.
25. Bakan, I. (2000). The importance of formal employee education in the world of growing uncertainty. In *Challenges for Business Administrators in the New Millennium* (pp. 341–355).
26. Urbancová, H., Stachová, K., & Stacho, Z. (2015). Methods of recruitment in the Czech and Slovak organizations. *Acta Universitatis Agriculturae et Silviculturae Mendelianae Brunensis, 63*(3), 1051–1060.
27. Štiavnický, P., & Papula, J. (2009). Rozvoj pracovníkov malých a stredných podnikov v období celosvetovej hospodárskej krízy. https://www.efocus.sk/kategoria/manazment-znalosti/clanok/rozvoj-pracovnikov-malych-a-strednych-podnikov-v-obdobi-celosvetovej-hospod.

The Effect of Bankruptcy on the Intangible Asset Valuation

Miriam Šefčíková, Dagmar Cagáňová, and Richard Jurenka

1 Introduction

A company is required to prevent bankruptcy. This condition is characterized by indebtedness, or insolvency. If the company is in danger of being bankrupt, it shall be obliged to take appropriate and proportionate measures to avert this condition without undue delay. The law requires the company—hereinafter referred to as debtor to take steps to satisfy the creditor's claim. True is the fact that the business owners try very quickly to change ownership of valuable assets whose use would be in the future effective. These transactions are done usually at a lower price than would in fact be applied on the market under the same conditions and at the same time, i.e., fair value.

For seven consecutive years (2010–2016, statistics for the year 2017 are not yet available), altogether in Slovakia 10,477 petitions in bankruptcy have been submitted; on average these proposals accounted for less than 1500 proposals each year. Of this total, the bankruptcy proposals made by the debtor were for a total of 6434. The highest number of proposals was made in 2016, when 1612 proposals were received by the competent courts. This quantity represented only 1% increase compared to the previous year. There was a break in 2015 compared to 2014. The negative trend of the proposals submitted was recorded in 2012, when the decrease was 9.4% lower (see Fig. 1).

Under the principle of the market mechanism functioning, the undistorted business competition and responsibility, we could automatically assume that all bankruptcy proposals would be made by debtors themselves. Actually these rep-

M. Šefčíková (✉) · D. Cagáňová · R. Jurenka
Institute of Industrial Engineering and Management, Faculty of Materials Science and Technology in Trnava, Slovak University of Technology in Bratislava, Slovakia
e-mail: miriam.sefcikova@stuba.sk; dagmar.caganova@stuba.sk; richard.jurenka@stuba.sk

© Springer Nature Switzerland AG 2020

D. Cagáňová, N. Horňáková (eds.), *Mobility Internet of Things 2018*, EAI/Springer Innovations in Communication and Computing,
https://doi.org/10.1007/978-3-030-30911-4_23

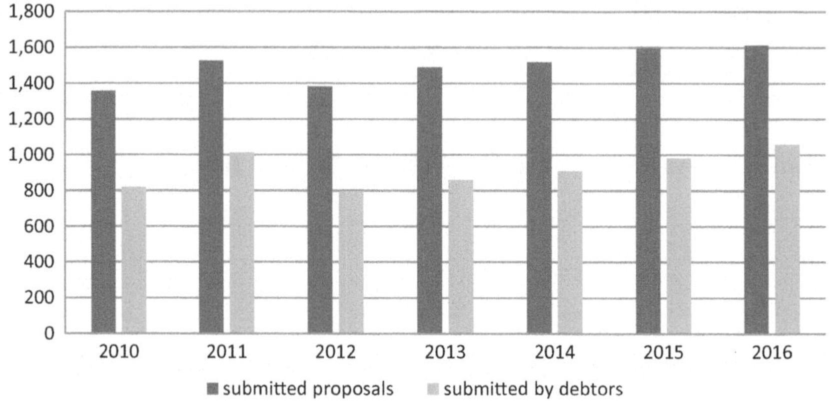

Fig. 1 Statistics of submitted bankruptcy proposals [1]

resent an annual average of only 61.4% of the total proposals submitted. This law degree of debtor responsibility significantly affects the correctness of the business.

A specific part of intangible assets is trademarks. These are regulated by Act No. 506/2009 Coll., which last amendment was effective since the year 2019, that defines that a mark may consist of any sign, in particular words including personal names, drawings, letters, numbers, colors, shape of the goods, or shape of the packaging of goods or sounds, if such marking is eligible to distinguish the goods or services of one person from the goods or services of another person and be expressed in the Register of Trade Marks of the Office of Industrial Property of the Slovak Republic in a way which enables the competent authorities and the public to clearly and precisely determine the subject matter of the protection granted to the trademark owner.

According to Elmore [2], a trademark is an indicative tool for the customer to choose between comparable products with higher quality assurance based on references. On the other hand, the producer represents an increase in goodwill and the growth of its market value. The existence of the trademark promptly raises the question of its value. Reasons for evaluation vary. In the context of expert practice most frequently tasks assigned for the purposes of verifying the value of a trademark are in the sale between related parties before the bankruptcy of the trademark owner, the sale in the execution proceedings, the determination of the price for its verification, or the determination in the usual objective transaction and for the purposes of classifying and registering the mark in accounting. According to Elmore [2], all valuation purposes and tasks can be summarized in three areas as follows:

1. For transactional purposes other than tax compliance
2. For financial accounting purposes
3. For income tax and other tax compliance purposes

This kind of company asset has a time dependency that the older and the more generally known are of higher value. Use of the correct valuation method affects its resulting value. For this reason, careful consideration should be given to judging not only the choice of the correct method but also the setting of relevant parameters within the procedure used.

In accordance with the relevance of the bankruptcy estate created in bankruptcy proceedings for this type of asset, intangible assets are reported that at the beginning of 2019 total assets in Slovakia offered by bankruptcy trustees in the total amount of more than 39.3 mil. EUR, of which only one inventory of assets consists of a combined trademark valued at EUR 50 offered for sale [3].

2 Materials and Methods

The value of intangible assets does not have a permanent persistent. Its value depends on its use, i.e., the company's ability to use it, and market demand. In the case of bankruptcy, and threats of crises, the value of intangible assets may be worthless. Especially with regard to the trademark, this is indisputable especially if it is not registered as a property even if it is protected with registration in the trademark register.

When does the moment occur when the owners of the company (debtor) realize that it is located on the verge of crisis? This relationship is defined by at least the following two situations:

1. By latent costs in the form of lost profits and additional costs arising from the increasing risk
2. By legally set condition

Ad. 1: With regard to the creation of latent costs, these are based on the fact that a company is no longer able to respond to market conditions because of the inadequate composition of assets burdened by disproportionately high commitments and liabilities. Equally, assets are not liquid and the liquidity of the company whether immediate or total is very low. Because of such unused market opportunities, it generates lost profits which represent latent costs for the company.

For financial management, an important indicator is the accounting profit/loss, but it is always adjusted for the latent costs thus generated, so we get the so-called economic profit/loss. This is an indicator of business efficiency that points to whether the company's resources have been used effectively in terms of the costs of capital. If it's negative, that means it's not about capitalizing on the resources of the business, but on the contrary about consumption, which points to the approaching business crisis.

These latent costs are added by extraordinary costs associated with the increase in risk where the creditor is convinced of the growth of business risk with such an entity. These costs include the cost of increasing the creditor's guarantee, bank charges for various guarantee instruments, fees for lawyers and experts, increasing the cost of employing debt capital due to declining liquidity, and so on. Increasing motivational bonuses are also a significant element for skilled workers to prevent their leaving. We also list the so-called management costs that consist of balancing the price between a lower or zero purchase price of employee shares or bonds so that employees are motivated to become more productive through the psychological effect of their shareholding and participation in the employer profit. These costs directly enter the calculation of the accounting profit before calculating the economic profit.

Ad. 2: The definition of a business crisis in Slovakia has become formalized for its legislative definition in the context of the Commercial Code since the year 2016. This situation was by then perceived theoretically and solved solely by law governing bankruptcy and restructuring. However, practice has created the need to include this phase of company's life into the formal legislative amendment. The Commercial Code provides for the performance of business a number of conditions, among which also includes the implementation of its own responsibility for the entrepreneur. He or she is fully exposed to economic risk and is responsible for breaching his or her obligations arising out of his or her business with all of his or her property.

If the company does not manage this risk, it will enter the degressivity phase. This can be solved by canceling a business without liquidation by takeover of another legal entity in several legal procedures or the bankruptcy. Liquidation itself is excluded in this case because it is a legal process in which the assets exceed the company's liabilities and it can be expected that there will be real prospects of excess liquidation balance for its distribution to the company's shareholders.

Recognition of assets and sources of its coverage is always in accordance with balance rule, where assets are equal to the capital, which is divided into equity capital and liabilities. If the equity capital is negative, then the liabilities exceed the assets.

In the case of insolvency monitoring, liquidity ratios are available, i.e., the ratio between liquid assets and liabilities. This ratio is close to zero at the time of bankruptcy or is directly negative. The theory defines three levels of liquidity, from immediate where only the immediately applicable funds to the calculation are counted in proportion to the outstanding payables, until the total liquidity, where the ratio includes a whole group of diverse current assets, and it is necessary to exclude those that are uncashable. However, these indicators of financial analysis are not used to determine insolvency in relation to bankruptcy proceedings. Its terms are strictly defined by law so that there are no differences in its interpretation. The reason is further evidence of liability in the criminal law.

An enterprise is in insolvency if it is unable to pay at least two financial obligations to more than one creditor for 30 days after the due date. One claim for the assessment of a debtor's ability to pay is all receivables that originally belonged to only one creditor within the 90 days prior to the filing of the bankruptcy petition. For purposes of determining insolvency the inability to pay means one who has more than one creditor, who has more than one financial obligation 30 days overdue, and whose financial obligations 30 days overdue are not covered by its financial assets, which means cash, account receivables, deposit, or other forms of deposit with a bank with a notice period of less than 3 months; monetary claims and debt securities with a maturity of up to 30 days; monetary claims and debt securities over a maturity period of no more than 30 days; and monetary claims and debt securities payable on demand (seeing) if it is reasonable to assume with due care that they will be duly and timely satisfied if their repayment is requested the following day.

The second definition of a bankruptcy condition is an extension that can only be defined by a person who is required to keep accounts and has more than one creditor and the value of his or her obligations exceeds the value of his assets. When determining the value of the liabilities and the value of the asset, it is based on the accounting or value determined by the expert's opinion prior to accounting and takes into account the expected results of the subsequent management of the assets or the expected results of the further operation of the company if it is reasonably foreseeable that it will be possible to continue managing the property or operating the business. The amount of the liabilities does not include the amount of the liabilities that are related to the subordination commitment and the sum of the liabilities that would be satisfied in the bankruptcy as the subordinated receivables.

The debtor is required to prevent bankruptcy. As a result, if there is threat of crises or even a company is in a crisis, it must take appropriate measures to continue its business, but not at the expense of its creditors. The cost of debt capital is generally lower than the cost of equity. Therefore, companies seek to finance their assets and business activities, in particular through the use of debt capital, which entails an increasing risk. This risk also consists of the fact that, if the capital is not repaid on time, the creditor is entitled to request additional claims on his original receivable in the form of interest payments, interest and deferred charges, and cost of recovering the claim.

The insolvency of a debtor is a function of bankruptcy proceedings without any further damage to creditors, including employees, social and health insurance companies, tax authorities, business partners, and others. In bankruptcy, the receivables are executed in two copies, one of them being filed with the competent insolvency court and the second with a bankruptcy administrator who was appointed to this position on the basis of its lists kept by the Ministry of Justice of the Slovak Republic. The bankruptcy procedure is considered to be a mixed procedure which contains elements that are disputed, uncontested, and executed. This is a special type of civil procedure [4].

In relation to responsible business it means that the company (debtor) should prevent its bankruptcy by all possible legal methods available so that the creditors are not affected by the loss of business. To satisfy them all business assets up to the amount of creditors' claims should be available. To avoid the reduction of these assets for the benefit of business owners, the legislation defined in the crisis ban restitution substitutes its own resources so as to fulfill the real purpose of bankruptcy and to satisfy claims of creditors to the greatest extent possible.

It is the duty of each business entity (regardless of its legal form) to monitor and quantify the equity and liabilities in accordance with the law in order to avoid a breach of the law in the matter of timely filing of bankruptcy petitions. The main identifier is the threat of crisis defined by this Act. For the year 2016 the marginal ratio was 100:4, which by the year 2018 was to stabilize at a ratio of 100:8. The obligation of an entity is monitoring this ratio throughout the accounting period, although official recognition of the state of management takes place once a year preparing the financial statements of the last day of the period of a calendar year (except for the economical year).

The financial statements are not consistent with respect to the accounting law that divides entities to big-, small-, and micro-sized entities. Classical accounts of entrepreneurs in the double-entry bookkeeping were taken into account in the survey, which is detailed in scope compared to the financial statements of microentity.

Applying this legal requirement we will use the financial statements enacted in Slovakia (data are reported on lines marked numerically). Entering the real balance sheet, we present the assumption of balance sheet equality of assets and its sources of coverage. There in the statement, the information reported on line no. 01 on the asset side must be equal with reported line no. 79 on the side of the capital. Capital is divided into equity, which we report on line no. 80 and the debt capital on line 101. Then we can proceed to the calculation as follows:

- The premise = line no. 101 acquires the ratio of 100/line no. 80 acquires the minimum ratio of 8
- Resulting = line no. 80 must be $\geq$7.41% of the value stated on line no. 01.

In determining the boundaries of bankruptcy, it is necessary to avoid the accounting procedures and before the above calculations it is necessary to apply a specific approach that may not be accounted for in accordance with the state of the resource asset coverage as follows:

1. Is examined as a recognized contribution made by a member of the statutory body, employee reporting directly to the statutory body, company secretary, director of the branch company, a member of the supervisory board.
2. Examines the recorded performance from the person who has a direct or indirect shareholding representing at least 5% of the company's share capital or voting rights in a company or has the ability to exercise influence over the management of the company, which is comparable with the influence corresponding to this share.
3. Examines the performance as recognized by silent partner.
4. Exploration is recognized as the fulfillment of a loved person referred to in points 1st–3rd.
5. It reviews the performance as recorded by a person acting on behalf of the persons in points 1st–3rd.

These transactions referred to paragraphs 1–5 in accordance with the accounting practices and are recognized in equity as part of the liabilities of the accounts Nr. 249 Other short-term borrowings, Nr. 333 Other liabilities to employees, Nr. 361 Amounts owed to affiliated entity and entities within proportionate participation, Nr. 365 Other liabilities to partners and members, Nr. 471 long-term liabilities to affiliated entity and entities within participating interests, and Nr. 479 Other long-term liabilities (possibly different depending on the selected accounting treatment) must be numerically subtracted from the line 101 and added to the line 80 although they are under the accounts placed on lines 108, 110, 130, 131, and 140, i.e., the summary line 101. In case of the absence of transfer within accounting works, it is necessary to justify this calculation in the notes to the financial statements. It follows that the entity has no obligation to charge to the needs expressed by the ratio of 100:8. This subsequently examines the bankruptcy trustee and in case of detection of an altered discharge the responsibilities' consequences to the person of the debtor. This procedure is called compensation of own resources [5].

2.1 The Approach

Business partners would only be interested in it if the other composition and volume of assets and liabilities were favorable to the potential for the use of that trademark. In the original company (debtor), the high level of liabilities by an improper combination of assets, resources, and intellectual capital contributed to the crisis and subsequent bankruptcy, and the use of the trademark was not enough to settle the growing liabilities. In a new company, the branding value of the trademark would be increased by a new set of marketing tools and an appropriate combination of resources. Motivation of this concept often involves a disputable legal act without adequate compensation, whereby the ownership of such a trademark is relatively cheaply used to draw off the principal source applicable to the relative satisfaction of the debtor's creditors. A legal act without adequate consideration means the free legal action of the debtor or the debtor's legal action on the basis of which the debtor has provided or has undertaken to provide a transaction whose usual price is substantially higher than the usual price of the transaction which he or she has or has obtained on his or her behalf. This action is in the bankruptcy subject to contradiction.

The contradiction of bankruptcy acts is one of the consequences of the bankruptcy application for the debtor's assets. Pursuant to the rules on the legality of legal acts, review all legal acts in the last 5 years and examine whether the creditor has been designated on the basis of his rights for any of the debtor's creditors. As a result, it may be required to repay this conflicting advantage either directly in the form of acquired assets or in the form of financial performance. One indicator that a bankruptcy trustee can use in the context of evaluation is the value-added ratio as the average value within the same industry sector against which it compares the performance of the business and indicates the deviations to investigate the causes.

Table 1 The trend of value added return ratio in %

The year:	2008	2009	2010	2011	2012	2013	2014	2015	2016
Value added return ratio %	11.59	10.68	10.52	10.65	9.55	9.96	13.93	14.66	14.91
Index	**−0.079**		**−0.015**	**0.012**	**−0.103**	**0.043**	**0.399**	**0.052**	**0.017**

The index of value added return ratio was applied to the revenue prediction calculation based on revenue reached in the year 2008 at 2.413.363 EUR

Table 1 shows the trend of this ratio for the years 2008 to 2016 for the business sector to which the method is applied in the following parts of this paper.

The general prerequisites for the counterclaim of legal acts in bankruptcy are, for example, that there has been a shortening of the claim of the creditor who has filed a bankruptcy claim. There is at least one creditor. The fact that a legal act is to stint the satisfaction of any of the receivables is a fundamental and general condition of any subject matter of a disputable legal act, without which the right to oppose the legal act will not arise. Furthermore, the property in question must comply with the assumptions of execution.

The purpose of the counterclaim is to secure the legal protection of the bankrupt creditor before the debtor's legal acts, which lead to a reduction in the debtor's assets and thus to the possibility of the creditor failing to satisfy the debtor's assets. The term "shortening" cannot be understood as any reduction in the debtor's assets, but only such that the creditor's claim is not satisfactory either at all or in its entirety.

To ensure this process, the abridgement needs to be defined in monetary units. Here an expert enters the process, which should determine the objective value of the trademark at the date of transfer of ownership to the trademark. An expert by choosing his or her approach and method determines the general value of the asset, which is the resulting fair value of the asset, and which is the expert estimate of the most likely price of the evaluated asset on the valuation date at that location and the time that it would have to achieve on the market under the terms of the free competition, when buyers and sellers will act with due information and caution and with the assumption that the price is not affected by an undue incentive. This value should be at least a lower market price, which represents the average market expectation for the future.

The intangible asset in question—a trademark—is included in the category of industrial rights for the designation for which the licensed analogue method is legally defined and used. The degree of objectivity of determining the value of an asset depends on the extent of the documentation submitted and the availability of information on the particular market in which the company operates. With the extension of the time period between the date of expert valuation and the time at which the value is to be expressed, this rate of objectification decreases. It is also negatively influenced by marketing market analysis and lacking statutory accounting documentation such as main book and other relevant internal documentation.

2.2 The Method Inputs

The relevant business company Bankruptcy Ltd. on the date of the 12th of May 2010 was declared bankrupt by publication in the Commercial Bulletin of the Slovak Republic. On the basis of the records of the financial statements in the Registry of Financial Statements, it was ascertained that the company had published only one financial statement for the accounting period of 2009 (31 December) before the bankruptcy was declared. These financial statements also include data from the previous accounting period of 2008. These financial statements, comprising only two components, namely the Balance Sheet and the Profit and Loss Account, form the basis for determining the general value of the trademark in this case. The purpose is to quantify the value of the trademark registered in the Trademark Register kept by the Industrial Property Office of the Slovak Republic under the registration number as of the date 30.06.2009. The decisive date is the day on which the contract for the transfer of ownership of the trademark between the debtor (the owner of the trademark) and the related partner—other business entities—has become valid and effective.

On the basis of the extract from the Trademark Register of the Industrial Property Office of the Slovak Republic on this trademark date, it was found that on 10.05.2016 the registration of the trademark was renewed and its legal status is registered and valid with the assumed date of validity of the date 10.05.2026. The trademark owner is a related person of the bankrupt (debtor).

3 Methodology Results

Within market approach the method of licensing is used to determine the general value of trademarks. It is assumed that the actual owner doesn't own trademark and therefore must pay a hypothetical third party for a license to use it. The hypothetical trademark royalty payment is calculated as a market-derived running royalty rate multiplied by the actual owner's projected revenue over the remaining useful life of the trademark [2].

The subject of the methodology is not the objective of methodology determination itself but its parameters. The general formula can be taken as follows:

$$\text{General value of trademark } (V) = \text{RV} \cdot \text{LF} \cdot \text{RF} \cdot \text{PM} \cdot \text{CC} \tag{1}$$

RV—annual production range
LF—license fee
RF—recovery factor
PM—share of intangible assets on production
CC—capitalization coefficient

This method is based on the principle that the value of an intangible asset is equal to the price most likely to be paid on the market for consent to the use of the same or similar solution or for the transfer thereof [6].

Annual production range: This is based on revenue prediction. Schedule future revenue of the company under the principle of sustainable business, i.e., under the "ongoing business" principle, must respect the conditions of the business and this prediction is carried out by the company itself. In this case, it is not possible to rely on the company's terms of reference, as there is no historical company data available, however with the exception of balance sheet and profit and loss statements for the years 2008 and 2009, of which a deeper analysis is not possible, only at the level of the underlying financial ratios. The company's own revenues in 2008 compared to 2007 increased by 534.122 EUR, which represents 28.42% increase.

The transfer of trademark ownership came in the following year. In this context, it is reasonable to assume that the business activities of the company have been diminished in that year. Revenues achieved at 31.12.2009 amounting to 609.073 EUR, i.e., the share of 25.2% of revenue from 2008, cannot be considered as the basis of revenue forecasts for the purpose of determining the general value of a trademark as a continuation of business and use of the trademark. Against this background, it can be assumed that 2008 is in line with the conditions of the sustainable business principle and will therefore be the basis for a further procedure for determining the general value of a trademark.

The basic assumption of property, whether tangible or intangible, is the creation of future economic benefits of the enterprise by its use. In this case, even if the company did not report the assets in the financial statements, its existence is evidenced by entries in the relevant register and records.

As mentioned above, the basis for forecasting future revenue by the period of 2016, i.e., the year in which the trademark should be renewed (10.05.2016), is the revenue generated in the accounting period of the year 2008. In the conditions of assuming the same business environment with the participation of competition and other impacts of the sectorial and macroeconomic environment of the company, information was used of the average values of financial indicators of economic activities in the Slovak Republic for the years from 2008 to 2016, which are published for individual statistical classifications of economic activities SK NACE, in this case specifically for group 475 Retail trade in other household goods in specialized stores with respect to the scope of the trademark—the classification of goods and services.

The value-added return indicator was selected. This indicator reflects the share of value-added in the company's revenues (please see Table 1). Value added simplifies the difference between sales of goods, products and services, and costs necessary to acquire, and the so-called production of own products and services. The indicator selection was aimed at respecting the basic business environment.

License Fee: The amount of royalties directly determines the value of the intangible asset to be valued. The term "license fees" is defined as a priority in the OECD Model Agreement as payments of any kind accepted in return for the use of or the right to use any copyright in literary, artistic, or scientific works including

Nr.	CRITERION	POINTS (LEVEL)					RATING
		1	2 - 3	4 - 6	7 - 8	9 - 10	
1.	visual perception	low	weak	average	good	excellent	8
2.	memorability, vocable	low	weak	average	good	excellent	8
3.	aesthetic appearance	very low	low	good	very good	excellent	8
4.	duration of use	1 year	5 years	7 years	10 years	11 years and more	2
5.	scope of territorial protection	SR	2 – 5 countries	6 – 10 countries	11 – 20 countries	21 countries and more	1
6.	level of protection	unprotected unused	unprotected used	registered	registered unused	registered used	10
7.	quality of products	very low	low	standard	high	excellent	6
8.	share of sales	up to 10%	up to 20%	up to 40%	up to 60%	over 80%	10
9.	% of total sales under the label	up to 5%	up to 10%	up to 15%	up to 20%	over 20%	10
10.	impact on marketability	without impact	weak	good	very good	excellent	5
FINAL RATING							68

Fig. 2 Using of the reduction method

cinematographic films, any patent, trademark, design, plan, secret formula, or procedure or information relating to industrial, commercial, or scientific experience.

Their height is given, based on an empirical approach to market conditions. Authors of significant monographs show different intervals in which royalties are moving, with a lower limit ranging from 0.5% up to 20% of the sales price (the sales). The license fee is also determined as a percentage of the profits made. The license fee is subject to the agreement of the two parties involved in the conclusion of a contract for the use of intangible assets. There are also databases where you can obtain a license fee that is current for the industry, country, and competitive, similar trademarks. These databases are subject to paid services whose cost would not be cost effective in this case. In view of this fact, the methodology for alternative determination of the license fee was proposed using the marketing approach of the so-called reduction method shown in Fig. 2.

The final rating—a total of 68 points—is then adjusted by the degree of awareness. The trademark in question is known mainly in the territory of the capital city where the shop is located and also after the transfer of ownership of the trademark is till located. Of the total scale on the scale of 1 (not yet used), 2 (used in the short term), 3 (used in the long term), 4 (very well known), and 5 (grievous), the second stage was selected on the basis of the survey. Thus, the significant factor has reached a total of 136 points. The maximum factor is then determined in percentage by the fraction of the maximum number of points of 500, i.e., in this case it is 27.2%. Subsequently, a referral license fee is applied within a country available from public sources, for example OECD resources, and if such a license fee is 10% the share of this license fee is used based on the use of the calculated maximal factor of 27.2%, resulting in a royalty amount of 2.72%.

Recovery Ratio: When increasing consumer awareness of the trademark, market value added also increases, which should be taken into account by the recovery ratio of its value. The main feature of the trademark is that the longer the protection and use time is, the higher is the value. The ratio in question is determined only on the basis of complete and relevant information from the accounting and the market in which the trademark operates. As this information is missing due to the nature of the bankruptcy, it can be excluded from the calculation. Similarly, its part is in the revenue prediction based on acquired and statistically processed market information for the years from 2008 to 2016.

Share of Intangible Assets on Production: From the point of view of the scope of business of the bankrupt, this proportion was set at 100%. This methodology is set up in the sense of missing accounting documentation; thereby it is hard to apply another share of the trademark's production.

Capitalization Coefficient: Its role is to express the present value of royalties on the valuation date of intangible assets. This factor must reflect the risk to which the planned sales would be achieved. Since it is not possible to use the weighted arithmetic average of capital because of the absence of company documentation, in this case the alternative cost of capital approach has been chosen. Based on the market analysis, it was found that for the government bond issue no. 208 with maturity date until 31st of December 2015, i.e., the period of the trademark's expiry date, the fixed-term interest rate is of 4.20% p.a. [7]. Given the low risk associated with this financial instrument, an alternative to higher risk was taken into account. Thus the basis for setting the discount rate was that stated as the average interest rate on commercial loans to nonfinancial corporations with a fixation period of over 5 years with a loan amount of up to 1 million EUR of 5.48% p.a. for the March 2009 period [8].

Conclusion: Under the conditions for determining all the parameters needed to calculate the general value of intangible assets—the trademark by applying the capitalization rate—the general value of the mark was fixed at 325.000 EUR.

Authors of significant monographs state that the value of an intangible asset—industrial designation law—should range from 10 to 20% of the net asset value of an enterprise and should not exceed that level [9]. As the value of intangible assets was based on the year 2008, the determination of the net asset value will be based on the balance sheet as at 31.12.2008, amounting to 2.401.156 EUR. Then, the total value of the intangible asset—the trademark—determined here is 13.5% of the net asset value and meets the above criterion.

4 Conclusion

The contribution was accomplished with the purpose of linking practice with theory. In fact, the expertise is regulated by a legislative framework that is not clearly set in this case. In the light of his or her experience and analytical procedures, the expert must choose the appropriate approach to correctly determine the value of the asset. This approach can never be limited as the situations that have arisen in practice will always require a modification of the methods and procedures used in the occurred situation. The field of bankruptcy represents a very wide variation of different problems, which have nothing to do with proper and demonstrable business. For this reason, the bankruptcy trustee himself or herself must be able to identify the need to value the asset that should be readmitted to the merits of bankruptcy in order to monetize the creditor's claims. Given the lack of accounting evidence, the asset valuation process, in particular in the retrospective period, is difficult and depends on the expert's experience of how to solve this situation. This chapter was devoted to this process.

References

1. Ministry of Justice of SR. http://www.justice.gov.sk/. Accessed 7 Sep 2018.
2. Elmore, J. D. *The valuation of trademark-related intangible property*. http://www.willamette.com. Accessed 3 Sep 2018.
3. https://ru.justice.sk/.
4. Ďurica, M. (2010). *Konkurzné právo na Slovensku a v Európskej únii* (2. vydanie ed.). Bratislava: Euro Kodex. ISBN 978-80-89447-31-2.
5. Šefčíková, M., & Noováková, R. (2016). Process-based elimination of the company crisis risk origination. In *Fórum manažéra*. Roč. 12, č. 2, s. 57–62. ISSN 1339-9403.
6. Mařík, M. (1998). *Určování hodnoty firem*. Praha: Ekopress. ISBN: 80-86119-09-2.
7. ARDAL. http://www.ardal.sk. Accessed 3 Sep 2018.
8. NBS. http://www.nbs.sk. Accessed 3 Sep 2018.
9. Haigh, D. (2002). *Oceňování značky a jeho význam*. Management Press.

EDGE4ALL: Edge Computing for Smart City

Ricardo Martins, Henrique Santos, Sérgio Lopes, Aníbal Leite, Álvaro Matos, and Paulo Cunha

1 Introduction

The Internet of Things (IoT) can be seen as a parallel world, in constant evolution, where millions of small devices play an important role in several activities. All together and interacting, they form a huge Internet-like network, so complex that it goes far beyond human mind [16]. Along with so many devices, comes a huge amount of data that needs to be collected and processed according to higher-level smart algorithms requirements, being necessary to ensure high performance, as well as information security [10, 14, 23]. Particularly in the field of smart cities, security and privacy are considered core requirements, to take full advantage of this development [11, 20, 21].

To make this scenario possible and because technological development seems to point to it in an unavoidable way [13], it is important to ensure that there is a solid infrastructure to properly support all requirements. Testing and experimenting solutions in this environment becomes very difficult, seeming very unlikely someone to build such a complex prototype without previously knowing its feasibility. This issue led to the development of some simulator frameworks [3], but most of the time focusing on specific applications or not including all required dimensions, in particular information security and privacy.

R. Martins (✉) · A. Leite · Á. Matos · P. Cunha
DigitalSign, Guimarães, Portugal
e-mail: rmartins@digitalsign.pt; aleite@digitalsign.pt; amatos@digitalsign.pt; pcunha@digitalsign.pt

H. Santos · S. Lopes
Minho University, Guimarães, Portugal
e-mail: hsantos@dsi.uminho.pt; sergio.lopes@dei.uminho.pt

© Springer Nature Switzerland AG 2020

D. Cagáňová, N. Horňáková (eds.), *Mobility Internet of Things 2018*,
EAI/Springer Innovations in Communication and Computing,
https://doi.org/10.1007/978-3-030-30911-4_24

Trying to overcome the above limitation, in this paper we propose an Edge-based architecture aiming to satisfy both the performance, security and privacy requirements while implementing light data structures that support the functionality anticipated for a horizontal Smart City application—as horizontal solution, it will be ready to handle a very large range of typical Smart City applications. In parallel and for testing purposes, a simulator has been also developed, which helped to validate the technical solutions adopted.

This paper is organized into seven sections, where: (1) Sect. 1 describes the context of the whole project; (2) Sect. 2 addresses the state of the art concerning fundamental concepts and related work; (3) Sect. 3 focuses on our Edge requirements; (4) Sect. 4 describes the steps followed and decisions taken to design the proposed architecture; (5) Sect. 5 presents the final architecture, together with a general system view to help understand the information workflow; (6) in Sect. 6 we discuss the results obtained, doing also a comparison with the expected outcomes; and finally (7) Sect. 7 presents some possible improvements as future work.

2 State of the Art

Digital technology is an integral part of the contemporary lifestyle, spreading in an increasingly ubiquitous way over all possible application domains, and following an evolutionary logic, apparently well sustained and without signs of slowing down. At the center of this evolution are two important assumptions: the microelectronic components will maintain the same increasing rate of miniaturization and speed (Moore's Law [17]), and data communication networks will also maintain the same increasing rates, both in terms of speed (Edholm Law [4]) and bandwidth (Nielsen Law [15]). Altogether these observations allow us to foresee the possibility of effectively using microtechnology in an integrated Internet-like network of billions of small devices (the so-called Internet of Things or IoT), for diverse utilizations (including Smart Cities), and reshaping the Digital Economy. Notwithstanding all the benefits of this IoT evolution, some security risk needs to be accounted for and failing to do that may easily compromise all the efforts [1]. Several solutions have already been proposed and in [19] the authors present an interesting survey, where Access Control and Authentication appear as critical security controls.

Another result of this evolution led to Cloud architectures, where thousands or millions of computer nodes are aggregated in clusters or through some form of interconnection techniques to promote resource sharing, leading processing, and storing information to become a commodity, available to everyone, everywhere (over the Internet) provided by Cloud companies for different purposes (e.g., Infrastructure as a Service—IaaS, Platform as a Service—PaaS, and Software as a Service—SaaS) [9].

Given the level of resources and computational power available today in the Cloud, this seems to be the right place to implement the Machine Learning and related algorithms to smartly process the data generated by IoTs. However, some studies showed that having a huge number of connection requests directly to a Cloud-based application can easily exhaust the bandwidth available [6]. As response, two similar concepts emerged: Edge Computing [18] and Fog Computing [2]. In both cases the objective is the same, to pre-process information from several local devices and prepare it to delivery to the Cloud—the main difference is the place where those systems operate: at a network bridge level, in the first case, and at the local network level, in the second case. Given the physical location of this components and their direct relation with IoTs, they are in an excellent location to implement the required security controls.

In the context of this work, IoT is used to support Smart Cities. A Smart City can be defined as "a city that monitors and integrates conditions of all of its critical infrastructures, where self-monitoring and self-response are core mechanisms" [8, p. 1]. But the concept of a smart city is yet emerging, without anyone truly knowing its full potential [5]. Even so, some system architectures have already been proposed and all of them approach it in a layer perspective, frequently defining four layers [12, 22], only with minor differences concerning some operations and the layer where they are performed. In general those layers are: data acquisition; data vitalization; common data and services; and application domain. Data vitalization (or data cleaning) and the common data and services are clearly adapted for an edge-like component. There are also some open-source solutions to support IoT implementations used frequently in Smart City applications. In [7] the authors present a comparison of some of those platforms.

Among all the related projects analyzed, as expected, real-time applications are not contemplated and bandwidth limitations are clear. At the data level the authors identify as main issues the interpretation, the data management, deeply linked to decision support systems, and the security and trust issues. The necessity to overcome these gaps in real applications was at the base of this project definition.

3 Expected Edge's Requirements

After discussing the need to implement an Edge in a Smart City context, it is now important to think about its requirements. In synthesis, the following list presents the most important requirements:

- offer a light-weight application protocol for IoTs to communicate;
- the low-level protocol must allow both encryption and authentication mechanisms (passwords, tokens, and certificates);
- real-time operation, namely concerning the expected limit from number of devices/connections;
- adaptive pre-processing techniques according to information source and type of sensors;

- store in a proper database the high-level oriented information, including the corresponding level of trust;
- offer a platform to "monitor" the IoTs;
- offer a documented web API to allow end-users to properly use Edge's information;
- implement a PKI to manage IoTs' certificates;
- ensure that the whole system is easily scalable.

If all these requirements are properly implemented, then it can be considered that this project contributed to a relevant technological improvement, overcoming one of the main obstacles of the Smart City paradigm—the infrastructure.

4 Implementation

A Smart City system architecture is usually described using a set of layers, each one responsible for handling information or performing some sort of control operations, pertinent to that level. We follow a similar approach, but looking for a more effective way to address the previously identified requirements, we decompose the system view in five stages. The next subsections describe each of those stages, including a brief reflection about available technologies and the justification to choose a particular solution.

4.1 Stage I: Data Receiver and Its Classification

As mentioned, an IoT is a small device, normally viewed as a black box, with relatively low capacities (both in memory and processing). So, it is important to use light-weight communication/application protocols, like MQTT, instead of the common known HTTP protocol. In some controversial scenarios,[1] sending 1024 messages, MQTT showed 100% of messages delivered, while HTTP only delivered 50%. Furthermore, this protocol helps to optimize the IoT device power consumption.

Besides the performance advantages of MQTT, to be eligible for the application some security requirements are needed to be fulfilled. It can work with TLS (encryption, enforcing integrity), and it can work with passwords and CRLs (allowing authentication through tokens and certificates, enforcing confidentiality). So, MQTT appears as a good solution, verifying the requirements that the scenario demands.

[1] http://stephendnicholas.com/posts/power-profiling-mqtt-vs-https.

Concerning the message format and looking again to our Edge general description, one of the security goals is to classify information instead of filtering it. So, the work at this stage is simply tag each received message with its source security level. More specifically, we consider four levels of security:

- level 0—for IoTs that talk via a public channel, neither encrypted nor authenticated;
- level 1—for IoTs that talk over TLS, but not authenticated;
- level 2—for IoTs that talk over TLS, with a password/token for authentication; and
- level 3—for IoTs that talk over TLS, with a certificate for authentication, emitted/recognized by the Edge.

To clarify, if a device sends a message containing a temperature value and a GPS location, it will look like this (we assume always the JSON format for message container, given the versatility and acceptance levels):

{'gps' : [41.4507, −8.2933], 'tmp' : 19.5}

and the Edge will tag the message with the device's security level, becoming:

{'gps' : [41.4507, −8.2933], 'tmp' : 19.5, 'security' : 0}

signaling the device communicates via the public channel.

4.2 Stage II: Temporary Data Storage

At this stage the main concern is to decide how to store efficiently the information. For that purpose, the adopted solution has two main contributions for the global goals: firstly handling buffering in a proper way, and secondly storing logs only while they have meaning.

As the Smart City environment is expected to be very crowded, empowered by the high performance of the MQTT protocol, processing all logs in real time is not expected to be possible, becoming necessary to store them temporarily for later process. For this purpose the possibility to use a database was tested. However, the set of actions—inserting, using, and deleting—for every log (thousands per minute or even second), showed to be overkill/inefficient for the system and unhealthy for the hard disk. A better solution is to store them in RAM memory, for instance, using Redis, a key-value map technology that operates in RAM memory, allowing to specify an auto-delete time for each key inserted. This way, the first contribution is achieved, buffering data in an efficient way.

Concerning the store operation, it can take advantage from the native auto-delete feature that Redis implements. Instead of tagging logs with time-stamps and later analysis to build maps or similar structures, requiring application-level processing and storage, it is much more efficient to keep the logs only while they have real meaning. For example, let us assume that a log says that in a specific position the temperature is 15°. Obviously, it is safe to say that the temperature in that place will remain 15° for some time, say 30 min. So, instead of tagging the log with a time-stamp and in every loop check if the 30 min had passed and delete it, ensuring that the log will stay on Redis for 30 min and auto-delete it after that time is provided automatically by Redis itself.

Another feature this auto-delete operation can promote, with a small data overhead, is device tracking. Every time a device communicates with the Edge, if it desires to be tracked, it sends the next time it will communicate, and in Redis it will be kept a log with the device's ID, for that period of time. When the device communicates again, it will "refresh" the previous log. This way, IDs that are in Redis are considered online devices, while the ones who are not are offline devices.

Finally, and looking for the necessary programmable feature, this auto-delete time is implemented in an administrator-level variable called variable-living time, linked to each type of information (map) implemented.

4.3 Stage III: Low to High Level Converter

This stage is probably the most important one, as it will determine the usefulness of the information and the performance of the Edge. Basically, it will transform the low-level data into high-level information, for later machine learning and big data algorithms.

To perform this transformation, it is crucial to understand the output of the previous stage, namely that at any time instant Redis will only have real-meaning logs. So, for every map-type variable (temperature, noise, air quality, etc.) there is a refreshing loop, with a fixed period, remotely configurable by the administrator. In every iteration of this loop, every single log registered on Redis by each device is used, as well as its security level. To define the trust level information a weight is attributed to each security level: security level 0 worth 1; security level 1 worth 3; security level 2 worth 10; and security level 3 worth 30. These weights are also remotely configurable by the administrator. Finally, the last attribute to consider is the ray of affection. Taking the case of temperature, if a log says that in a specific position the temperature value is 15°, it can be assumed that in a range of 1 km, for example, the temperature is 15°. This ray of affection is also a remotely configurable parameter.

With this structure, it can be defined a data model, where at MongoDB (chosen database due to its flexibility and performance) each variable will be defined as a collection, within it the current time-stamp will be the document key, and the document itself will be defined as a JSON object, where the keys are coordinate pairs and the object is the associated stored data, like the variable's value, its security level, the number of devices used to determine it, and more relevant data. This data model definition is primordial because it allows to query the database to get every variables from a specific location and time-range, resulting in an overlap of information (correlation of different variables). This feature will promote a horizontal solution contrary to the current vertical ones.

Concerning to the history, it is not efficient to store these maps in every loop iteration (every 10 s, for example), due to disk-space limitations. To overcome this and for historical purposes, the maps are stored at a larger time interval (for example, every hour), and in this project this periodicity is called historic interval, also defined by the administrator, for each variable.

4.4 Stage IV: Information Distributor (API)

Concerning the Edge, this is the final stage that implements the interface with the outside, through an API. This stage is basically a kind of web service that receives requests, accesses the MongoDB, and responds with the requested information.

Although MongoDB implements direct HTTP requests, for security reasons it is not desirable to give the end-user this kind of direct access. This way, the end-user can only do strictly what the API implements, avoiding attacks through tricky HTTP constructs. This API provides only HTTP responses for requests like getting a specific map, getting a whole history, get some devices information, etc.

This type of operations allows a client application to pull a map whenever it is needed. But this pull mode is not efficient for real-time continuous applications, where a client typically should opt for a callback type solution. To support this operation a WebSocket connection was implemented. This way, when a client opens the connection, the server saves it and every time a new map is processed, the Edge sends it to every connected client, avoiding the pooling from clients.

In addition to the development of this API, a dashboard was also implemented to test the Edge. It shows the information only in real time through the WebSocket connection, over the map in an intuitive and elegant way. Furthermore, a notification/alert area was implemented, where warnings (concerning fires, for example) may pop-up. To track the "registered" IoTs, a section of this dashboard is designed to see the online and offline devices. As explained before at "Stage I," for each registered device, if its ID is on Redis, it is online; otherwise, it is considered offline.

About this last section, the IoT one, a hard task is yet to be implemented. Although the information concerning variables is fully anonymous, the IoT tracking is not. If a company owns some IoTs and wishes to track them, it is not expected that everyone can see them too. This particular section of the IoTs tracking must later implement some authentication, to ensure that each user can only see their own IoTs.

Concerning to dashboards, one of the most important advantages of using an Edge to store all the different types of information is the possibility to correlate them. Overlapping maps of temperature, air quality, bus routes, trash level in bins, and so on emerges the concept of information orchestration. Having the data available and correlate by map positions and time, it opens doors for data exploration through intelligent algorithms.

4.5 Stage V: Security Distributor

This stage is not in the Edge's workflow (is at the Smart City system level), but it must be considered in the Edge's structure. It is responsible for enforcing authentication mechanisms to the IoT devices. As mentioned above, the Edge considers four security levels, two of which implies authentication (one by token, and one by certificates). Concerning certificates, in this project the PKI implemented uses the ADSS software to manage the digital certificates.

In a typical PKI we have a set of RAs (Registration Authorities), responsible for checking the incoming requests and issuing the certificate to the final user (manual certification). However, for a million IoT devices, this process is not efficient, arising the need to automate it. To do so, the ADSS client (built in Java) was adjusted to make it available as a server. To keep the security level of the process, now a final user must present to the platform a developer-certificate and, if it is valid, the ADSS client will ask the ADSS to issue the IoT-certificate. This way, it is ensured that only who has permission to request IoT-certificates can do so. Additionally it was defined an attribute of the developer-certificate to control how many IoT-certificates can be issued by a specific developer.

Since not all devices are capable of handling a digital certificate, a lighter authentication mechanism, like a token, should be supported. In this project, the token used is the JWT token,[2] and to be generated it requires that the final user presents a developer-certificate (authorization to request tokens). To try to keep the management of the tokens similar to digital certificates (security for the private key and a validity), once the token is generated, it is registered in the MQTT password file. This file is hashed with a server aleatory unknown secret, which means it is irrecoverable, even for any administrator. Furthermore, a token is always revoked after a specified duration and can be revoked manually at any time.

In summary, when a final user wishes to authenticate an IoT device, he must present his developer-certificate to the online platform and, if validated, request IoT-certificates or tokens to flash into the devices. To ensure confidentiality and integrity in all the process, it is implemented an end-to-end solution using the TLS protocol.

[2]https://medium.com/vandium-software/5-easy-steps-to-understanding-json-web-tokens-jwt-1164c0adfcec.

5 Final Architecture

This section aims to get all the above stages, and explain how all of them can be implemented, without compromising the proposed requirements (mainly, the scalability and ease of installation). In a first approach to the architecture, virtual machines and containers were compared. Containers were selected mainly because: (1) the portability, as the whole project occupies about 3 MB, and deployment only requires a simple command line, which quickly configure, install, and start everything; (2) the security level it promotes, easily implementing a private network for the containers; and (3) it is independent of the OS, assuming containerization platform is available for a given OS.

Regarding performance, a particular attention was devoted to the usage of asynchronous programming, instead of synchronous. While synchronous paradigm promotes the use of threads to do parallel actions (like receiving one thousand requests), the asynchronous programming uses an event driven paradigm which, through a callback mechanism, passes control to an application only when there is an event (e.g., new data) requiring that. This programming model fits very well with the type of target application and will allow to save a lot of resources (memory and processing time). The asynchronous paradigm is gracefully supported by python (through the native asyncio library and the event loop construct).

Concerning the administrator interface, it was developed a dedicated module that allows to configure all parameters previously defined, without stopping the Edge. In synthesis, the parameters are: (1) map area for tracking; (2) variable-living time; (3) security weights; (4) ray of affection; (5) loop time; (6) and (7) the historic interval.

The complete Edge's architecture is presented in Fig. 1, where pink boxes represent the clients (both sensors on the left and final users on the right); yellow boxers are containers with the receivers, the MQTT brokers; the brown one is the container responsible for the authentication of the devices; green boxes are the containers running Python asynchronous scripts, where all the algorithms are

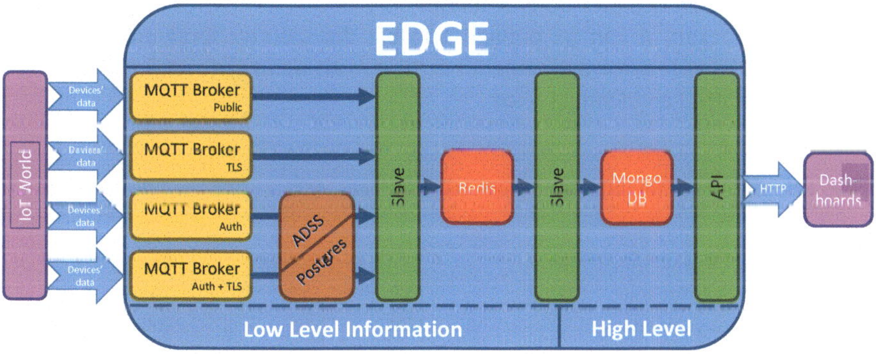

Fig. 1 Edge's architecture

implemented; and the red boxes are the containers that store information, both the logs (temporary one Redis) and the final information (long-term and historical, on MongoDB).

6 Results

The Smart City concept, the related Edge Computing architecture, and IoT paradigm are all relatively new topics, and there are no well-defined metrics to measure their performance, yet. Trying to address the limitation to measure the Edge's performance, an IoT oriented simulator was used, where innumerable devices can be created, each one sending some information to the Edge, following a model similar to what was previous described. To better understand the results, it is important to remember how the information is built: periodically grabbing all the logs from Redis and build a map.

For evaluation purposes, the metric used was the number of devices the Edge can handle, ensuring that the processing time required to build the map does not exceed the defined periodicity.

To perform the test/experiment, the machine used had an Intel Core i5 with 6 cores at 3.8 GHz and 16 GB of RAM at 2100 GHz, and the presets used were: 4 map variables (temperature, air quality, noise, and UV index); each device communicated every 5 s; 80% of them had tracking enabled; random security levels for them (between level 0 and 3); and the periodicity with which a map was built (loop time) was 10 s.

For these presets, four versions of the Edge were tested, each one using a different combination of technological alternatives previously referred, considering protocol (HTTP and MQTT) and the programming model (synchronous and asynchronous). The results obtained can be observed in Table 1. These results helped to validate the hypothesis assumed that MQTT and asynchronous programming would produce an optimal solution, showing a performance improvement of more than 20 times, comparing with the less performer solution.

Also as a result, it can be pointed out that the security level did not affect the results achieved, since assigning all devices with security level 3 (only in last scenario) produced an identical result.

Table 1 Edge's results for different implementation stages

	Presets		Result
	Programming paradigm	Application protocol	Max devices
1	Synchronous	HTTP	400
2	Asynchronous	HTTP	2800
3	Synchronous	MQTT	1100
4	Asynchronous	MQTT	8700

Finally, although theoretically the whole architecture was supposedly to be portable, its demonstration was also important. That was tested when transferring the project to a Data Center belonging to a company supporting the project, making it production ready. The whole architecture was installed very fast and with no limitations, even though it was a different operating system. The whole installation and setup took only about 15 min.

As a final remark, throughout the project some real IoT devices were developed. Over Wi-Fi, a security level 3 was achieved, while over GSM only level 2 was achieved. Up until now, two real devices are installed in two garbage trucks, collecting temperature and air quality logs and sending them to the Edge. A preview of the still under development dashboard-for-reality can be seen through http://edge4all-dev.digitalsign.pt:8000/.

7 Future Work

After concluding this work, it becomes evident that it is a continuous and endless project, as it relies on the creativity and capability to generate more and more useful information. But focusing on the Edge and direct related issues, some evident improvements can be pointed out. For example, a very critical task that needs to be improved is concerned to Stage V, the security distributor. Currently, the final user generates the private key and the CSR needed to ask the Edge to issue the certificate. However, this simple operation violates a fundamental principle of PKI: a private key should be generated and only accessible to the final user. As the final user is the device, and it cannot generate a private key (lack of resources), one possible approach is to introduce a dedicated module that generates the private key, flashing it and the issued certificate into the device. This way, the user would never have access to the private key, and once inside the device, it cannot be recovered.

Another optimization that can be implemented is related to the data traffic consumption and security topic. One of the biggest challenges that the Smart City faces are the GSM communication operators, as the traffic plans are too expensive. For this reason, it is crucial to minimize information transacted. One alternative way to study is to implement security both by certificates and tokens. As the certificates occupy about 1 Kb and the tokens can be smaller (100 bytes or less), the device could use the certificate to request a token, with a daily validity. This way, the certificate would be used once a day, and the token for the regular data transactions.

Finally, another essential operation to be explored is the information correlation. This feature will for sure add value to the whole project, as it marks the difference between common vertical solutions and a horizontal solution as proposed in this document. Define mathematical algorithms to relate different variables, as well as apply some machine learning techniques to different kinds of information, will for sure result in unexpected and still to discover information and knowledge.

Although this "task" is the core of the architecture and the most important final output expected from this project, it is still the most complex one to achieve.

Acknowledgement This work was supported by DigitalSign.

References

1. Adat, V., & Gupta, B. B. (2018, March). Security in Internet of things: Issues, challenges, taxonomy, and architecture. *Telecommunication Systems, 67*(3), 423–441.
2. Bonomi, F., Poledna, S., & Steiner, W. (2017, March). The role of fog computing in the future of the automobile. In *Fog for 5G and IoT* (pp. 189–210). Hoboken, NJ: Wiley. https://doi.org/10.1002/9781119187202.ch8, http://doi.wiley.com/10.1002/9781119187202.ch8.
3. Brambilla, G., Picone, M., Cirani, S., Amoretti, M., & Zanichelli, F. (2014). A simulation platform for large-scale internet of things scenarios in urban environments. In *Proceedings of the First International Conference on IoT in Urban Space. ICST.*
4. Cherry, S. (2004). Edholm's law of bandwidth. *IEEE Spectrum, 41*(7), 58–60
5. Chourabi, H., Nam, T., Walker, S., Gil-Garcia, J. R., Mellouli, S., Nahon, K., et al. (2012, January). Understanding smart cities: An integrative framework. In *Proceedings of the Annual Hawaii International Conference on System Sciences* (pp. 2289–2297). Piscataway: IEEE.
6. Díaz-del Río, F., Salmerón-García, J., & Sevillano, J. L. (2016). Extending Amdahl's law for the cloud computing era. *Computer, 49*(2), 14–22. https://doi.org/10.1109/MC.2016.49
7. Guth, J., Breitenbucher, U., Falkenthal, M., Leymann, F., & Reinfurt, L. (2016, November). Comparison of IoT platform architectures: A field study based on a reference architecture. In *2016 Cloudification of the Internet of Things (CIoT)* (pp. 1–6). Piscataway: IEEE.
8. Hall, R. E., Bowerman, B., Braverman, J., Taylor, J., Todosow, H., & Von Wimmersperg, U. (2000). The vision of a smart city. Tech. rep., Brookhaven National Lab., Upton, NY.
9. Hennessy, J., & Patterson, D. (2011). *Computer architecture: A quantitative approach.* San Francisco: Morgan Kaufmann.
10. Hwang, Y. H. (2015). IoT security & privacy. In: *Proceedings of the 1st ACM Workshop on IoT Privacy, Trust, and Security - IoTPTS '15* (pp. 1–1).
11. Ijaz, S., Ali, M., Khan, A., & Ahmed, M. (2016). Smart Cities: A survey on security concerns. *International Journal of Advanced Computer Science and Applications, 7*(2), 612–625.
12. Jin, J., Gubbi, J., Marusic, S., & Palaniswami, M. (2014, April). An information framework for creating a smart city through Internet of Things. *IEEE Internet of Things Journal, 1*(2), 112–121.
13. Li, S., Xu, L.D., & Zhao, S. (2015, April). The internet of things: A survey. *Information Systems Frontiers, 17*(2), 243–259.
14. Medwed, M. (2016). IoT security challenges and ways forward. *In Proceedings of the 6th International Workshop on Trustworthy Embedded Devices - TrustED '16* (pp. 55–55).
15. Nielsen, J. (1998). Nielsen's law of internet bandwidth. Online at http://www.useit.com/alertbox/980405.html
16. Rathore, M. M., Ahmad, A., Paul, A., & Rho, S. (2016, June). Urban planning and building smart cities based on the Internet of Things using Big Data analytics. *Computer Networks, 101*, 63–80.
17. Schaller, R. (1997, June). Moore's law: Past, present and future. *IEEE Spectrum, 34*(6), 52–59.
18. Shi, W., Cao, J., Zhang, Q., Li, Y., & Xu, L. (2016, October). Edge computing: Vision and challenges. *IEEE Internet of Things Journal, 3*(5), 637–646.
19. Sicari, S., Rizzardi, A., Grieco, L., & Coen-Porisini, A. (2015, January). Security, privacy and trust in Internet of Things: The road ahead. *Computer Networks, 76*, 146–164.

20. Weber, R. H. (2010). Internet of Things – New security and privacy challenges. *Computer Law and Security Review, 26*(1), 23–30.
21. Weber, R. H. (2015). Internet of things: Privacy issues revisited. *Computer Law and Security Review, 31*(5), 618–627.
22. Yin, C., Xiong, Z., Chen, H., Wang, J., Cooper, D., & David, B. (2015, October). A literature survey on smart cities. *Science China Information Sciences, 58*(10), 1–18.
23. Zhou, J., Cao, Z., Dong, X., & Vasilakos, A. V. (2017). Security and privacy for cloud-based IoT: Challenges. *IEEE Communications Magazine, 55*(1), 26–33.

Simulator for Smart City Environment

Ricardo Martins, Henrique Santos, Aníbal Leite, Álvaro Matos, and Paulo Cunha

1 Introduction

Smart Cities (SC) result from the conjugation of several developments, including technological, societal, and governmental, which altogether are looking for ways to promote quality of life and more efficient ways to manage the rising complexity of urban spaces, making them more liveable and resilient. This transformation is much more a process then a concept or a new knowledge area and depending on the approaching perspective it is possible to find other very similar concepts, like Digital City or Intelligent City (from the technological perspective), Knowledge City or Creative City (from the social perspective), Smart Community or Smart Growth (from an institutional perspective) [10]. This work has a more technical approach and adopts a SC definition near to the one provided by BSI (British Standard Institute): the "effective integration of physical, digital and human systems in the built environment to deliver sustainable, prosperous and inclusive future for its citizens" [6].

Concerning technology, there are some important platforms/framework available, frequently for other purposes, but contributing significantly for SC consubstantiation, namely: Cyber-Physical Systems, Internet of Things, Big Data, and Cloud Computing. In [11] the authors present an impressive survey over a large set of such technologies and conclude pointing as an important challenge (in addition to others)

R. Martins (✉) · A. Leite · Á. Matos · P. Cunha
DigitalSign, Guimarães, Portugal
e-mail: rmartins@digitalsign.pt; aleite@digitalsign.pt; amatos@digitalsign.pt; pcunha@digitalsign.pt

H. Santos
Minho University, Guimarães, Portugal
e-mail: hsantos@dsi.uminho.pt

© Springer Nature Switzerland AG 2020

D. Cagáňová, N. Horňáková (eds.), *Mobility Internet of Things 2018*,
EAI/Springer Innovations in Communication and Computing,
https://doi.org/10.1007/978-3-030-30911-4_25

the lack of testbeds to experiment Smart City solutions. Simulators fit into this class of tools and are as important to Smart Cities as it is hard to imagine someone testing a solution in a real-world scenario where errors and failures can have a huge negative impact.

Under that assumption and pushed by the development of another project related to the development of a Smart City architecture—called EDGE4ALL—it was started, in parallel, the design and implementation of a Smart City simulator, aimed to test and validate it. Besides that obvious function, and having in mind also the importance of the human factor in the acceptance of so disruptive technologies, the simulator can be used, if designed properly, as an education tool to familiarize and train citizens with the paradigm change involved on the way they can interact with information, in the Smart City environment. So, this tool can provide a relevant contribution to improve the outcome of some ongoing learning initiatives, like those using MOOCs (Massive Open Online Courses) [5].

This paper is organized into seven main sections, where Sect. 1 describes the context of the whole project, as well as the advantages of such a simulator; Sect. 2 discusses some related projects; Sect. 3 describes the expected features of the simulator; Sect. 4 describes the implementation steps; Sect. 5 presents the final architecture, in the context of a system view, to help understand better the information workflow; Sect. 6 discusses the results obtained, comparing with the initial requirements; and finally Sect. 7 includes some reflection on how to improve the simulator.

2 State of the Art

Nowadays and despite the relative novelty of the Smart City paradigm, it is possible to find several research works and development efforts in this application domain, all over the world and involving a large number of cities [10]. As examples we can point smart parking management systems, which have been implemented in different ways and areas, showing a considerable impact [7]. Another notable example is that of intelligent traffic lights, potentially improving several aspects of traffic management in cities [4, 9]. Notwithstanding the importance of all those solutions, they do not align with the definition of a Smart City, concerning the integration and global analysis of all data related to the city, aiming to augmenting the intelligent capacity of the urban space—those solutions can be described as vertical solutions, candidate to be part of a Smart City.

Giving the integrative and broader nature of Smart City architectures, or frameworks, it is clear the level of complexity and diversity of solutions, making it a very challenging project [2]. Under these circumstances it is desirable to test and verify solutions before committing to project deployments. In this endeavor, some Smart City related simulators were proposed, but only a few have been implemented. An example described in [1] shows a framework that simulates a smart traffic lights network. Another one where a vision of a smart city for exploring "structural criticalities, system vulnerabilities, restoration algorithms, and overall control of the power system" [8] was implemented.

In [3] the authors present a survey of several simulators. Most of them focus on the simulation of IoT devices, with focus on the communication layer. Some of them focus the Edge/Fog level but still more oriented to network level operations. A very small number of solutions exist for code testing and application level testing, using (more or less) real-world scenarios and focused on functionality, usability, and consistence. However, there is a clear lack of tools for testing non-functional aspects like security, privacy, regulatory demands—only one project under development of OWASP (Open Web Application Security Project) seems to be oriented for such objective.

As mentioned before, this work was motivated by an ongoing project aiming to develop an Edge based architecture for SC (EDGE4ALL). This component acts as a general data collector and data preparing unit, with security and privacy objectives, whose test demands the generation of data from very different sources (CPS), as well as application consuming operations, controlling the level of security and trust. None of the studied simulators addresses the core requirement: the possibility to deal with a large diversity of virtual/external devices and events, where different security models can be tested. The simulator under development in this project addresses this limitation, which is believed to be an important idea to explore in a smart city simulator.

3 Expected Simulator's Outcomes

To implement a Smart City simulator, it is now important to discuss its requirements. The main goal of the project is to provide a solid and useful workbench capable to deploy a simulated Smart City architecture and feed it with millions of messages from thousands of devices while giving users the perception of what is going on, adding some metrics to evaluate it. So, the most relevant requirements are:

- define a city model as simple as possible, but also flexible enough to support the largest set of possible architectures;
- more than simulate, it is desirable to emulate devices. Each device must have its own properties, open its communication channel, send its message, and close the channel, so that the architecture sees it as a real device, as much as possible;
- each device must be capable of georeferencing relative to a common axis system, and be parameterized to allow to simulate the largest possible set of devices;
- simulated devices should exhibit some visual properties to distinguish them (e.g., using colors to associate to their security or trust level, or different shapes to represent several kind of entities – wearables, cars, phones, ambulances, etc.);
- to improve interaction it is desirable to allow users to have some control over devices, namely those they create;
- simulate the environment from where the devices will capture "physical" values, treating it as variables. Temperature, noise, and air quality are examples of

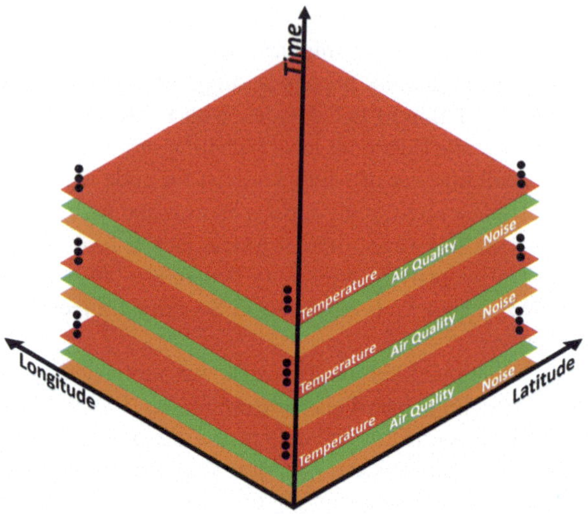

Fig. 1 Representation of the generated map variables

such variables. This simulated environment should be controlled through some interface (possible, a web page) allowing to inject hazardous values linked to external events (like fires, accidents, etc.);

- the simulated environment module should allow anyone anywhere to be able to implement a new model of physical quantity, and all the devices will "read" values from it (this topic is particularly important because it allows the study of new and more precise models and behaviors); and
- all the devices, city map, and simulated services must run online and be as generic as possible, promoting flexibility.

One of the first decisions with a big impact is the simulated city model. Without ignoring other similar projects, the need for this simulator comes from an ongoing project aiming at the development of an Edge based architecture for a Smart City, which will use this simulator as a validation and demonstrator tool. The defined city model is based on a map-like structure, using georeferencing and time as key indexing dimensions. So, we can see this city model as a stack of maps, each one for a different physical variable—see Fig. 1. Although this decision was somehow conditioned by the referred Edge project, the defined data structure corresponds closely to that assumed in any other Smart City projects.

4 Implementation

This section intends to explain in detail the steps followed to build the global simulator architecture. To help understand it, the next three subsections separate the main engines responsible for the whole workflow of the simulator.

4.1 Simulation Engine: City Map Simulator

This simulator module is responsible for displaying the devices, as well as some information about them. As it is intended to run online, this part of the simulator is simply a single web page. In it, it is possible to see a physical map, designed in SVG (Scalable Vector Graphics), because it is absolutely adaptable to any screen resolution. The physical map itself is made up of bi-directional roads, intersections, and roundabouts, for mobile devices, and "open spaces" for the static ones. This physical map represents a GPS-view of the "city," meaning that the devices can pass over another (each device is simulated in an isolated way, and no physical interactions are implemented).

Concerning the development, it was used native JavaScript to implement the logic of the simulator, and initially also to control the devices movement, by manipulating the CSS properties of each device (translations, rotations, and color changes). As this movement control was found to be hard and involves a lot of maths (becoming computationally heavy), a light library for that purpose was adopted, GSAP. This library only manipulates CSS properties, and it stands out for the next reasons: (1) its performance, being incredibly fast and smooth even with slow processors, as some smartphones have; (2) allows to animate any HTML object; (3) compatible with almost any browser, older or newer, and many libraries (JQuery, React, Vue, among others); (4) allows to scale, rotate, move, and smoothly change properties of any object independently; (5) implements as many timelines as desired, allowing to define a set of different animations for each device; and (6) allows to schedule animations and add triggers to it, allowing to set callbacks at the beginning, middle, and end of animations. The adoption of this library increased the web page performance drastically, as well as the attractiveness, which is primordial to better user impact.

About the communication with the devices, it was adopted the MQTT protocol due to its lightweight properties, as compared to the standard HTTP (with MQTT each message can be up to thirty times smaller). Originally, each device sent a message with its ID, color, shape, and speed to be instantiated, and then it would send the ID and routes for the engine to make it travel. This solution was not acceptable because if someone joins the page after a device creation, the user would never see it. To overcome this, every message of each device is self-sufficient, meaning it has the information to instantiate the device. When a message arrives, the engine checks if that specific ID is already assigned to an HTML object and makes it move if so, or initialize it otherwise. This way, when a new user joins the simulator (web page), all the devices will be created and it is safe to state that every user will see the same devices (true online simulator).

Also aiming the user interaction, and since the GSAP library allows it, some interactive features were implemented in order to help the user find its device, such as jump, change speed, change color, and change the shape.

Finally, because the security and privacy are major challenges in IoT field, it was implemented a way to see these properties. So, each device has a visual colorful crown that shows the security level (or rust level) of the device. Furthermore, about privacy, if a device desires to be completely anonymous, there is a flag that will show the device all white, meaning it did not share any private information. These two feature are enabled, but in order to make sense in a Smart City architecture it needs to be set up in a different way, because a device security level and a privacy desire are characteristics defined externally, not by the device itself—meaning that these two flags are received not from the devices, but from some Smart City architecture's component.

4.2 Devices Simulation

After understanding the Smart City simulation engine, it is time to discuss what a device can be. Before this explanation, it is important to note that a device is not exactly simulated, but rather emulated. This difference relies on whether the behavior of the device is "apparent" (simulated) or whether it actually exists (emulated). This feature is particularly important because it makes the difference between a reliable and unreliable benchmark tool for a smart city architecture. It should not be evident to see whether the information comes from a simulator or real devices.

Diving into this module's architecture, it is all built in Python, due to the ease and flexibility of the language, as well as to the vast collection of available libraries. In synthesis, a device is an object (class) that has an initialization, and three loops (methods). At initialization, the device generates an ID, chooses a shape, color, and speed, loads the available roads from the city simulator, and initiates some timers and the three loops. As for the three loops, their purpose and workflow is detailed below, in the next section.

To promote interaction with the users, besides the devices previously described, it was also developed a web page to create them. The user can create a device, choose one of the eight available shapes, change its color, adjust the moving speed, link it to different variables from the available services, and a press button to make the device jump. Furthermore, this page can be accessed by http or https, implying two different security levels (http linked devices are red/public and https linked devices, with encrypted communication, are orange).

Device Behavior This loop is responsible to animate the device in the city map, and its most important role is to decide the device trajectory and detect its position, simulating a GPS sensor. Regarding the trajectory, it is important to understand how the map was built. Every road has one ID (that starts with "P," followed by a number) and has two directions, coded by a "p" tag if it follows the up or right directions and "l" tag for down and left directions. So, for example, a given road that is horizontal is identified by, for example, "P10p" when referring to right direction and by "P10l" when referring to left direction. About the road junctions and roundabouts, its nomenclature is simply the concatenation of the origin road with the destination road, removing the unnecessary second "P." For example, if a device is at "P11p" and arrives to a roundabout, it can follow right ("P15p"), up ("P12p"), or left ("P8l") road, choosing the paths "P11p15p," "P11p12p," or "P11p8l," respectively. For a single roundabout with four exists, there are sixteen different paths, depending on the origin road and destination road desired.

For a device to choose its trajectory, there is a python dictionary that has the available choices, stored in a list, for a specific road. For the above example, the dictionary has an entry with '11p': ['15p', '12p', '8l']. Assuming that the device knows which road it is on, consulting this dictionary it chooses randomly an available road. As already mentioned, the communication with the simulator is via MQTT, so this loop has its own communication channel, and in every iteration, in JSON format, it is sent a payload as follows:

{'id' : '123', 'speed' : 50, 'fig' : 2, 'route' : 'P11p15p'}

This way, the map will instantiate the device (if it is a new one), and animate the device through "P11p15p" and at the end of the roundabout, through the road "P15p." For the map and the device to synchronize, another thing the device has to know is the length of each road. For example, if a desired path has 250 pixels and the device will travel at 50 pixels per second, then it will need to send another route request in $250/50 = 5$ s. While the device waits that the map object travels the path, it goes to "sleep," concerning process status.

The second role of this loop is to offer a "GPS-like" sensor. For this purpose, there is another dictionary that has the starting and ending pixel coordinates of every path. This way, whenever it is requested (and since it is known the speed and for how long the device is in a specific road), it is mathematically possible to do a projection of the device in the path, determining its coordinates.

Generate Data This loop is responsible to promote a reality-like system. As this simulator tries to offer services like noise and air quality, and these quantities correspond to physical characteristics generated by external elements linked to the devices (e.g., cars or persons), each device should also contribute for those maps/services. This way, this loop has a fixed period (defined at instantiation of the device), and in every iteration the device will send (through MQTT) to the pertinent simulated services (explained in Services Simulation subsection) the intended values alongside with its GPS position—except for static devices, for which no GPS information is needed and, by default, it is generated a random value around a reference one, defined during initialization.

Read and Send Data Finally, this loop is the more important one for the smart city architecture. This loop is the common "/* Write your code here */," responsible to establish a communication, performing authentication, read variables, send logs, and end the communication. Here, the user has even freedom to choose the application protocol to use to communicate with the simulator, as well as the URL. This simulator does not forbid any kind of communication or destination.

While still under development, this loop was tested with HTTP requests, MQTT requests, and even with a direct TCP communication. In every scenario, TLS was tested, as well as authentication by passwords, tokens, and x509 certificates. Concerning communication, this loop has absolutely no restrictions.

About the simulated data to send, there are two possible scenarios: (1) values generated by the device itself and depending on its own characteristics (like noise), which may be randomly generated; and (2) and global environmental values, like temperature or humidity, which are provided by simulated services (detailed in the next subsection), responsible for keeping a map with values distribution, which are provided to sensors, when requested, for linked GPS position (this way, two devices at the same position will for sure send the same values to the Smart City system).

4.3 Services Simulation

This last module is responsible for simulating global environment physical characteristics, as described above. Once again, this module must be as generic as possible, concerning the way it receives and sends signals, so that every device can interact with it. To receive information from the devices (like noise), it uses the MQTT protocol, and to send information to devices it uses HTTP or MQTT (both acceptable). In the first case the devices must send a coordinates pair and a value, and the service will save it. In the second case the device sends a request to a specific service and a GPS position, and the service will respond with the value for that position.

As all of this simulator aims to run online, the services module is not an exception. It is even designed to allow services to run in different computers anywhere, allowing a device to ask for a temperature value to a URL, and for noise value to another one. This way, it is possible to anyone to implement a simulated services for a Smart City, exploring different ways and even new mathematical models to simulate physical characteristics. Again, for the server side code, there is already a template done, with the "/* Write your code here */" section, where the users can implement the maths and logics behind the service, to easily integration with the simulator.

At these stage, the following list summarizes the available services, and describes the respective models:

- temperature—service assuming the global average temperature varies from 5 to 25° (from day to night) according to a sine function, always with two focus (one +5 and other −5°) rotating clockwise. The "day to night" time can be set to 1 min or to 1 h. This service has only one input, from a web page, to generate an external event (a fire) at a mouse click position. The fire takes about half a day to be extinguished. OpenCV is used to smooth the resulting map image and to generate an imperceptible (to the human eye) noise, to simulate the differences between sensors;

- noise—service that only has content when there are devices in the map. Each noise log only "lives" for 1 s, and it represents a small focus on the map. When a device is moving, it can be seen its track by the noise it produces. Besides the input of the devices, through a web page it is also possible to create accidents by clicking on the map, being a larger focus, also disappearing after 1 s. This service also uses OpenCV for the same explained reason;
- air quality—similar to noise service, with the exception that the focus is larger and takes 10 s to disappear, as air pollution takes time to dissolve into the air. While noise receives logs from mobile and static devices, this service only responds to the mobile ones.
- ultra violet radiation—the simple service, because it does not receive any external information. At the center, it just describes a negative quadratic function that varies from 0 to 100, which gets attenuated at the borders of the map, simulating the presence of forests, where the radiation does not reach the ground;
- water conduct—this service aims to simulate the water conducts of a city and then change the pollution and rain level, originating floods at the sewage lids. For this, a "tree" system was developed, where each node has an input (for the rain) and each path has a max caudal that must be sized to handle the rain from the upstream lid and the caudal from the previous path. Then, each path has a pollution level that will prevent the max caudal to be reached, originating floods at the upstream lids. As this is a "tree-like" system, any change in any node or path has a recursive effect on the rest of the tree. To control the rain level and the paths pollution, there is a web page where these variables are controlled by scrolling bars.

To promote the reality of these services, some interactions were implemented, like raining that speeds up the fire extinguish, lowers the temperature and the ultra violet radiation decreases, or fires that increases the temperature and the violet radiation.

5 Final Architecture

After describing the operation of individual modules, now it will be summarized the global workflow of the system. Figure 2 tries to schematize the process. The arrows with red borders represent communications with no operator intervention, while the ones with green borders represent the flow of information under direct user control. As can be seen, a device communicates with several services, distinct ones, and can send the same information to one or more (again, distinct) architectures.

For the whole system to perform smoothly, and handle as many devices as possible, two specific details were implemented. The first one, concerning to devices visualization, is about the map itself. Although the library used (GSAP) has already high performance, there is always a finite number of devices that can be displayed. This could be a limitation to test architectures, because despite the simulator

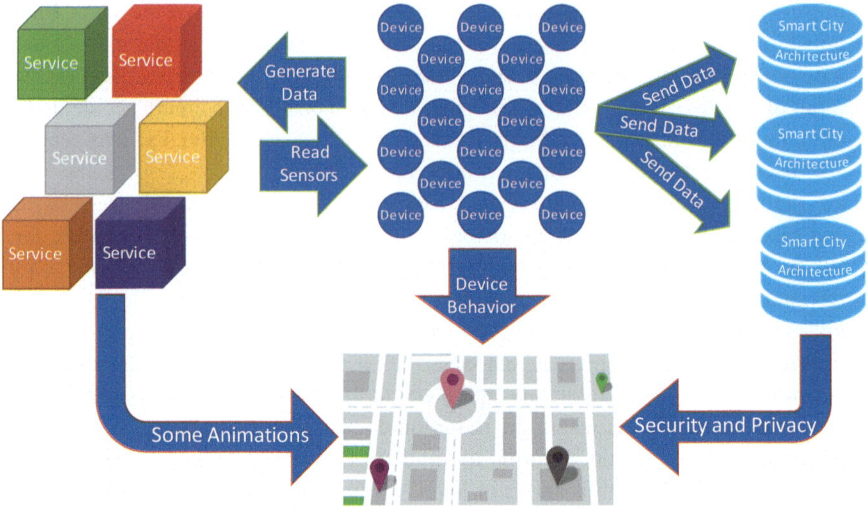

Fig. 2 Global workflow of the simulator

limitations, maybe the architecture could handle much more devices than the ones visible on the map. To overcome this, the initialization of the device receives one more flag ("display") that will prevent the device from sending information to the simulated map. This way, the device will still exist, will generate and read data from the services, will send the information to the architecture, but will not be displayed, overcoming the above limitation.

The second problem is related to threads. Natively, python works synchronously, and implements threads for concurrent jobs. In this particular use case, each device launches three independent loops (3 threads), and each loop does about one or two requests synchronously. This means that every loop would "block" the stack waiting for a response, meaning that other devices would also wait, which is unacceptable. The common/evident solution is also to create a new thread for each request. This way, other devices could perform their requests concurrently. However, if each device launches three threads, and each thread launches two more, one hundred devices would create about nine hundred threads. This huge amount of threads is already too high for such a number of devices, because a regular computer has about two thousand threads limit. To overcome this limitation, asynchronous programming revealed to be a much more suitable paradigm to follow, especially when dealing with a lot of concurrent low-computational tasks. This paradigm implies that a python file executes all in a single thread, but implements an "event loop," which can be imagine as a vertical time ruler, where tasks are hung, and later returned. This way, when a device has to perform a request, he "await" (as called in Python 3) that the answerer arrives. An "await" prefix means that a device does not need "attention" for a while, and the "executor" can give attention to other coroutines. When the answer arrives, the "executor" returns to where it left. This implementation applied to the whole architecture showed incredible improvements, discussed next in the Results section.

6 Results

Concerning results, although there are not yet any metric to measure simulators' performance, some conclusions can be made. Regarding performance, the map simulator itself can show about one thousand mobile devices and three thousand static devices concurrently (for a core i5 8th generation based desktop computer with 16 GB of RAM, on Chrome browser), and about four hundred mobile and one thousand static devices in an Android phone. To measure the simulator true performance, the "display devices" flag was set to off, and about seven thousand devices (mobile or not) was the limit reached.

To see the global impact when applied to a specific Smart City architecture, the EDGE4ALL architecture was used. In that platform, it was possible to see the services' maps recreated, warnings about the external events (fires), the device tracking, and even the feedback from the architecture to colorize the security level of each device at the simulator's map.

This simulator runs currently online, and can be accessed through the LabSecIoT site[1] or via a hub,[2] where the map, devices, services, and even the architecture output are more accessible. For those who access it via the hub, the "External URL" must be the selected one. Furthermore and for demonstrations purposes, the simulation activity is prepared to be projected on the ground, using a special screen, with very interesting pedagogical and training results, in the field of Smart Cities.

7 Future Work

About future work and the next steps of LabSecIoT, one of the evident tasks is to implement more services, aiming to become a reliable enough Smart City simulator, with all that comes with it. Some of the services that are waiting to be implemented are smart parking, smart lamps (from the street), track trash level in rubbish bins, and a people counter.

Another complement to this simulator, that is already under development, is the use of a 3D camera to simulate devices according to their color and height. If a Kinect is pointing to the previously mentioned projected map, placing a yellow cylinder with 10 cm of height could represent a street lamp for instances. With these

[1] labseciot.dsi.uminho.pt.

[2] labseciot.dsi.uminho.pt/hub.

implementations, the combination of different shapes, colors, and heights allows to build a much more attractive 3D map and implement even more services.

About the services, this framework can be used to test new mathematical modules to shape services or even to test the existent ones. For example, while the temperature service is currently described as a sine, maybe a much more precise model could be used. Or about noise, that simply mathematically adds the incoming noises. Since this project aims to be generic and massive online, these two properties leave room for a complex and innovative idea.

Acknowledgement This work was supported by DigitalSign.

References

1. Barba, C. T., Mateos, M. A., Soto, P. R., Mezher, A. M., & Igartua, M. A. (2012, June) Smart city for VANETs using warning messages, traffic statistics and intelligent traffic lights. In: *2012 IEEE Intelligent Vehicles Symposium* (pp. 902–907). Piscataway: IEEE. https://doi.org/10.1109/IVS.2012.6232229
2. Cisco. (2014). *The internet of things reference model*, White Paper. http://cdn.iotwf.com/resources/71/IoT_Reference_Model_White_Paper_June_4_2014.pdf
3. Dias, J. P., Couto, F., Paiva, A. C., & Ferreira, H. S. (2018, April) A brief overview of existing tools for testing the internet-of-things. In: *2018 IEEE International Conference on Software Testing, Verification and Validation Workshops (ICSTW)* (pp. 104–109). Piscataway: IEEE.
4. Gradinescu, V., Gorgorin, C., Diaconescu, R., Cristea, V., & Iftode, L. (2007). Adaptive traffic lights using car-to-car communication. In: *IEEE Vehicular Technology Conference* (pp. 21–25).
5. Hudson, L., Kortuem, G., Wolff, A., & Law, P. (2016). Smart cities MOOC: Teaching citizens how to co-create smart cities. In: *4th International Conference on ICT for Sustainability (ICT4S)*, 29 Aug–1 Sep 2016, Amsterdam, The Netherlands.
6. Institution, B.B.S. (2014). Pas 181: 2014: *Smart city framework—Guide to establishing strategies for smart cities and communities* (p. 52). London: BSI Standards Publication.
7. Litman, T. A. (2006). *Parking management best practices*. New York: Routledge.
8. Lugaric, L., Krajcar, S., & Simic, Z. (2010, October). Smart city—Platform for emergent phenomena power system testbed simulator. In: *IEEE PES Innovative Smart Grid Technologies Conference Europe, ISGT Europe* (pp. 1–7). Piscataway: IEEE.
9. Maslekar, N., Boussedjra, M., Mouzna, J., & Labiod, H. (2011). VANET based adaptive traffic signal control. In: *2011 IEEE 73rd Vehicular Technology Conference (VTC Spring)* (pp. 1–5).
10. Nam, T., & Pardo, T. A. (2011). Conceptualizing smart city with dimensions of technology, people, and institutions. In: *Proceedings of the 12th Annual International Digital Government Research Conference on Digital Government Innovation in Challenging times-dg.o '11* (p. 282).
11. Santana, E. F. Z., Chaves, A. P., Gerosa, M. A., Kon, F., & Milojicic, D. S. (2017, November). Software platforms for smart cities. *ACM Computing Surveys, 50*(6), 1–37

Evaluation of Parking Policy in Bratislava in the Context of Economic Growth and Automotive Transport Development in Slovakia

Julius Golej and Daniela Spirkova

1 Introduction

The continually growing level of automobilization brings increasing demands on space in Slovakia, especially in the most densely populated agglomeration around the capital of Bratislava. It becomes, especially in the inner-city environment, the most important article. It brings new requirements for city urbanization, professionalism to address traffic problems, and high demands on the quality of the environment. One of the most severe current problems closely related to the city space in Bratislava is the traffic situation, namely the issue of static traffic. The current capacity requirements do not meet the design requirements of the 1970s, i.e., before car ownership began to rise sharply. Cities search for solutions to deal with these problems, with measures such as parking regulations and pricing of public parking places [1]. The current equipment of the static transport fails to meet the current needs of the living population. Parking policies are potent instruments for demand management: a city, or an employer, influences the drivers' behavior through the parking policy. According to Anderson and de Palma [2], parking policies in cities typically combine pricing measures and adapting the number of parking places. Underpricing of on-street parking, for example, can elicit the reaction of drivers to cruise for on-street parking which can lead to an increase of congestion. However, communication capacities, in particular the node points of service and collecting communications, and access to large-capacity civilian facilities, as well as conflicts between car traffic and increasingly preferred cycling and walking movements, are also a problem. Its solution is very demanding as the condition is given by the overall poor urban concept which does not allow

J. Golej (✉) · D. Spirkova
Institute of Management, Slovak University of Technology in Bratislava, Bratislava, Slovakia
e-mail: julius.golej@stuba.sk; daniela.spirkova@stuba.sk

© Springer Nature Switzerland AG 2020
D. Cagáňová, N. Horňáková (eds.), *Mobility Internet of Things 2018*,
EAI/Springer Innovations in Communication and Computing,
https://doi.org/10.1007/978-3-030-30911-4_26

the separation of traffic areas with car transport from the pedestrian and cycling movement, even in the position of recreational movement or in the position of traffic servicing. Common parking policies dealing with parking space consist of developing parking facilities outside of the high-density urban centers, i.e., park and ride facilities, introducing parking regimes (restrictions), and handling parking planning within land-use policies [3].

Parking policy has been analyzed in numerous papers. Dulla [4] notes that a systemic and coordinated solution to the situation of parking of vehicles must be based on the conditions of the functional area envisaged and laid down in the land-use planning documentation and the related documents. The goal of parking policy is to regulate transport to improve the public space and the environment. In residential areas, it is essential to provide areas and space for a long-time parking of car users.

According to Young [5], transport and land-use policies are often seen as the main drivers towards achieving urban sustainability. Parking is often seen as having an active impact on transport and land-use quality; however, unless all the policies work towards the same objectives, parking policy will not support the other policies in achieving broader economic, social, and environmental aims.

McShane and Meyer [6] identify six broad categories of goals which may underpin most urban parking policies. These are healthy economic climate; efficient use of transport and land resources; ease of mobility and accessibility; equity of resource distribution; improvement of environmental quality; and enhanced amenity and cultural attractiveness. These urban goals could provide a useful framework for identifying how parking policies can influence urban outcomes.

Litman [7] broadens the discussion of parking policy, putting forward the view that more focus on parking management could increase the utilization of land and transport in urban areas. He focuses on parking supply and price as mechanisms for ensuring that parking provision is appropriate and that its utilization is maximized.

The number of motor vehicles, mainly passenger cars, is steadily increasing in Bratislava and makes the situation of urban mobility and traffic servicing difficult, which is conditioned by sufficient space for stopping and parking of vehicles. Bratislava is the only city in Slovakia that has no established (applied) parking policy, which would serve to the improvement in parking conditions and traffic service as well as to the reduction of the number of unfavorable traffic.

In order to analyze and evaluate the situation, it is necessary to know the input data concerning dynamics; intensity; dynamic transport organization; capacity and management of static transport; occupancy and use of parking areas; socio-ecological, demographic, and urban data related to population; and employment opportunities. Parking facilities tend to use a lot of valuable urban space. More efficient use can be obtained through shared and/or multiple use of parking areas. In the case of shared use, the parking space is available to different users for a certain time period at different times of the day (such as in public parking).

Multiple use consists of using the space for other purposes when the need for parking space is low. The number of parking spaces in a city can be used as an indicator of infrastructure-based accessibility by car [1].

Because parking policy and strategies in city need to be embedded in a whole urban transport policy, data on parking availability and parking prices should:

Differentiate different target groups (residents, employees, visitors)
Be accompanied by data on public transport supply (incl. new mobility services when possible) and prices
Be accompanied by physical environment/demography/land use/built environment information

These data are necessary both for the improvement of urban mobility and for requirements for the development of the territory and its spatial arrangement. Therefore, the significant problem in the area of comprehensive transport solutions in Bratislava is the fact that there is no sufficient database of traffic data and the city does not have enough details about the current state of loading of its urban road network. The city of Bratislava lacks regular traffic surveys and their results, which could identify disproportions for the current situation and its predictive development [8].

Even though Slovakia is still one of the countries with a low level of automobilization compared to the European Union (Fig. 1), its increase in the coming years will bring much higher demands on the space that needs to be reduced at the expense of housing, services, or greenery.

It should be noted that the imperfection and often the inexperience of the current solutions can be seen virtually in every city, in every larger residential area in Slovakia.

The capital city of Bratislava, as the largest city in the Slovak Republic, has the biggest problems in the long run not only with the daily traffic situation but also with the static traffic in the city [9]. Therefore, further in this chapter we focus on the problem areas, which significantly affect its static traffic.

2 Materials and Methods

The materials, which focus on the problems of static transport, are used and analyzed in the chapter. The statistical data processed in this chapter was obtained from the EUROSTAT database, the Statistical Office of the Slovak Republic, the database of the Ministry of the Interior of the Slovak Republic, the Ministry of Transport and Construction of the Slovak Republic, the Big Data statistics, and Market Locator population analytics. The main condition for the selection of statistics was to ensure that the data available for a certain period is uninterrupted.

Other relevant documents on parking policy problematic in Bratislava, which are analyzed in detail in this chapter include the following:

- The methodology of design capacities for lay-by and parking areas is fully covered in the Slovak Technical Standard STN 736110: Local Roads Design 2004, updated with amendments (STN 736110 / Z1) from 2014 [10, 11]. These

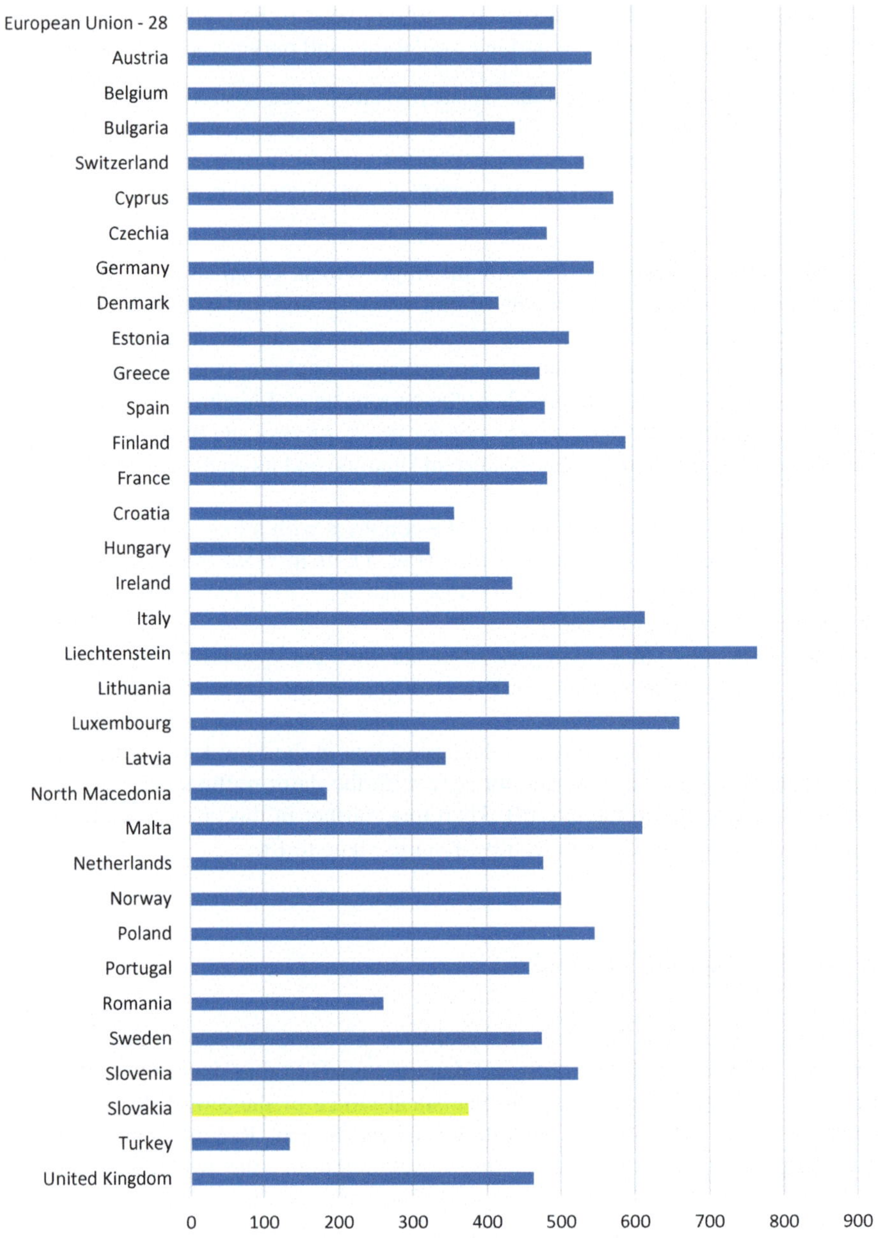

Fig. 1 Number of passenger cars per 1000 inhabitants in 2015 in EU [7]

standards replaced the original construction standards from 1977 and 1986, according to which most of the lay-by and parking areas were designed, and they exist in this form to this day.
- Act no. 135/1961 Coll. on roads, which also deals with parking spaces and parking [12].
- Basic Rules of Parking Policy of the Capital City Bratislava [13]: The document speaks about proposed parking rules associated with the gradual and necessary change of traffic behavior of inhabitants of Bratislava. In this process, the capital takes the example of the modern metropolitan cities of Europe and the world, which by the transition from the individual car transport to other types of urban transport provide a better quality of life to its inhabitants, a better use of public space, cleaner air, or a healthier lifestyle (Municipality of the Capital City of the Slovak Republic, Bratislava, 2015).
- The Land Use Master Plan for Transport of the Slovak Republic, Bratislava [14]: The parking policy of the capital city of the Slovak Republic, Bratislava, is based on conditions where the primary step is a unified system of organization and operation of parking with the possibility of local system operation by city districts (The Land Use Master Plan for Transport of Bratislava, 2015, Annex 3.3.4.3.b, 2015).
- Generally binding regulation of the capital city of the Slovak Republic Bratislava no. 12/2016 of 8th December 2016 on temporary parking of motor vehicles on defined sections of local roads in the territory of the capital city of the Slovak Republic, Bratislava, the amount of payment for temporary parking of motor vehicles, the payment method, and the proof of its payment regulates the conditions of temporary parking on defined sections of local roads in the territory of the capital city of the Slovak Republic, Bratislava, and determines the method of ensuring the operation of parking spaces, the minimum and maximum amount of payment for temporary parking, the method of payment, and the proof of its payment [15].
- Master Plan of the Capital City of the Slovak Republic, Bratislava [16].

3 The Problem Areas, Which Significantly Affect the Static Traffic

The following problem determinants, which negatively influence the static transport in Bratislava, might be considered:

1. Development projects and increasing density in the area
2. Migration flows to Bratislava—labor market mobility
3. Automobilization rate
4. Non-existing parking policy in Bratislava, etc.

3.1 Development Projects and Increasing Density in the Area

The territorial development of the capital of the Slovak Republic, Bratislava, is inherently linked to an increase in transport relations. These relations, in particular in the category of individual car transport, represent increased demands on the capacity of the road network, the possible congestion of which results in an unwanted effect in the daily life of the city as well as in a negative evaluation of the attractiveness of the development projects by their users. The transport-capacity assessment of the impacts of investment projects is the subject of the "Methodology" (2014), which is elaborated for the purposes of a single assessment of the impacts of large investments on the territory of the capital of the Slovak Republic, Bratislava, manifested in the form of both road traffic congestion and high demands on parking places.

In the Central European countries, since the early 1990s, unfavorable trends in the division of transport work dominated, which resulted from the increase in road transport to the detriment of more environmentally friendly modes of transport. One of such countries is also the Slovak Republic, where public transport performance dropped by 32.7% from 1995 to 2003, but on the other hand the individual motorized transport performance increased by 40.3%, which represents an increase in the number of individual cars by 381,694. This development reinforces the negative impact of transport on the environment. Transport produces mainly emissions that pollute the air, higher levels of noise, soil contamination, and use of raw materials [17].

In the interest of sustainability of the transport situation, larger development projects should automatically be subject to the traffic-capacity assessment. The procedures that are used in the authorization process of an investment plan in relation to its impact on the transport-capacity sustainability of the territory are mostly presented in a nonuniform manner, varying in scope and reflecting the subjective approach of their processors.

The objective of assessing the impact of large investment projects on the transport situation in built-up areas (not only new developments in developing areas but also projects that densify stabilized residential areas) is to demonstrate the functionality of the concerned road infrastructure after the projects have been put into operation. At the same time, the fundamental relationship between the subject of a particular project and its transport service requirements in a territory must be transparently reflected. In order to ensure the sustainable development of the city, it is, therefore, necessary to monitor the investment project in its preparatory phase in such a way that its final form guarantees the availability of transport. An important condition in the preparation of large investment projects should be the traffic-engineering analysis of impacts on the functionality of the road infrastructure. The system for assessing transport-capacity impacts should summarize the principles that will commit larger developers to cooperate actively in addressing the traffic situation changed by their construction projects, which should support the sustainability of the transport situation of large cities.

Usually, a construction investment, the primary static transport requirements of which exceed 300 parking lots, is considered as a large investment project. The capacity of a road network is directly dependent on the capacity of crossroads in it. With regard to the fact that a concerned territory is usually represented by several investment projects proposed on a particular area, the goal should be to set the same rules of traffic regulation for each project in a concerned area. The assessment parameters are mainly dependent on the capacity of the affected road infrastructure, i.e., the size of its reserve for a new transport potential. For this reason, it can be assumed that the degree of possible regulation of investment projects will vary in different cities. The traffic load is the number of vehicles on the road network, where it is necessary to know their routing. The value of a transport potential of a territory is dependent on its functional use; it means that there should be seen the difference in whether a territory is used with a prevalence of housing or with a prevalence of administration and business. It should also be borne in mind that the objective or more specifically the source of the transport relation (journey) is a parking space, which means that the number of parking spaces for new investment projects must always comply with the provisions in obligations of a developer to build appropriate—standard number of—parking spaces. However, the reality is different in many cases [18] (see the model example Petržalka).

The design of residential areas should be processed according to applicable standards. In many cases, the reality is completely different—not only the difference between a lay-by and car park is considered, but also the development of mobilization. Transport should be considered as a progressive and significant phenomenon, which should be considered in the preparation of projects for decades to come.

From the perspective of the quality of the environment in which the inhabitants live, the more serious is the fact that whoever does not find a lay-by place or a place on the road is standing on the sidewalk or greenery. A pedestrian becomes a runner through obstacles; a cyclist becomes a stuntman, and a mother with a stroller becomes an undesirable element. Whom are the pavements for? On the one hand, professional terminology says that they are for pedestrians, and on the other hand legislators say that they are also for cars. They argue that "traffic would collapse in Bratislava if it was forbidden to stand on the pavement."

The required free space of width of 1.5 m for pedestrians is a mockery of not only the cultural way of life but also even more the expertness. In traffic analyses, we operate with the capacity of lanes for motor vehicles. On the other hand, however, it should be said that not only in Bratislava but in most places in Slovakia, a strip for cyclists is an unknown term. Wherever a progressive designer designed it, the vehicles are parked. A cyclist has moved to the pavement, which he/she uses along with pedestrians, strollers, wheelchairs, and skaters. Due to these problems in the Slovak cities there is prevailing individual passenger transport. The increase of its share in the division of transport work is becoming more striking. As a result, there is a sharp increase of requirements for the capacity of the urban road network as well as the ever-increasing lack of parking and lay-by areas [19].

3.2 Migration Flows to Bratislava: Labor Market Mobility

The most attractive area to which migration flows are currently directed is the Bratislava region, namely the capital city Bratislava, the wealthiest region of Slovakia. GDP per capita in purchasing power parity is 2.5 times higher than Slovakia's average, and the Bratislava region is also among the ten richest regions in the EU. At the same time, it has the highest job offer and provides a wide range of opportunities for study. This fact is also a cause of high migration flows to Bratislava (or the Bratislava region), whether in the form of daily and weekly mobility or moving. The main migratory flows between the regions are shown in Fig. 2. These are the average annual flows (number of migrations) involving more than 500 changes of permanent residence. It should be stressed out that Bratislava has a special position in the migration system of Slovakia (see Fig. 2—we have obtained data available until 2009; new data are currently being processed).

More than 130,000 people commute daily to Bratislava for work, education, or entertainment [21]. This was confirmed by big data and Market Locator's population analytics data from mobile phones located in the capital [22]. According to the data from the Statistical Office of the SR in 2015, 422,932 inhabitants live in the capital city. Due to the change in lifestyle, this standard and static figure gives a distorted view.

People from near as well as wider surroundings commute daily to Bratislava. In many cases, the average distance is 100 km. As shown by the data from the Statistical Office of the Slovak Republic in 2013, in Bratislava, almost 250,000 more people spend the night than the number of people with permanent residence [22]. Research on an anonymous sample of more than 2.7 million of Slovaks and the movement of SIM cards allowed to distinguish the number of people sleeping in the territory of the capital while not being inhabitants of Bratislava [22]. Table 1

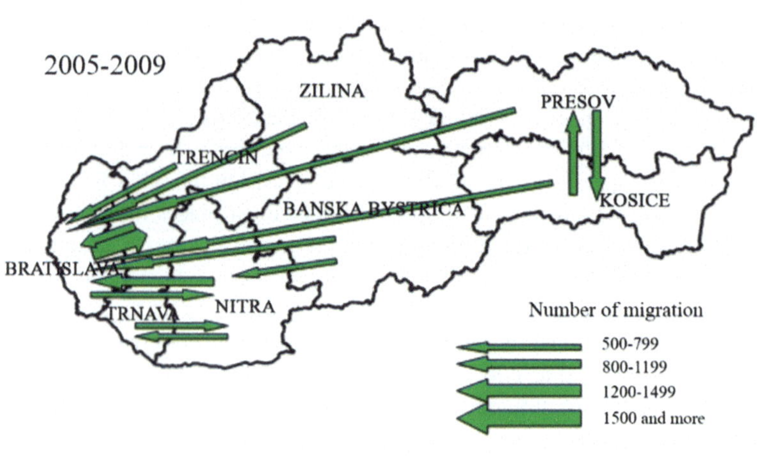

Fig. 2 The main migration flows between regions of Slovakia (number of transfer) [20]

Table 1 Comparison of data from Statistical Office of the Slovak Republic with data from Market Locator

City districts in Bratislava	Statistial office of the Slovak Republic	Market Locator
I	38,823	69,774
II	111,051	188,544
III	62,546	115,824
IV	93,948	143,178
V	111,021	148,944
Bratislava total	417,389	666,264

Source: processed with the data [21, 22]

shows the comparison of data from the Statistical Office of the Slovak Republic and Market Locator research.

From the perspective of the city of Bratislava, in the year 2015, there were 629 passenger cars per 1000 inhabitants (422,932 inhabitants were registered in Bratislava as of December 31, 2015) [21]. With an area of 3,724,448 m^2, these vehicles occupied 0.98% of the city space, representing 367 km^2. The numbers are at first sight negligible but remember that it is only the placement of passenger cars without the right for a road.

Based on the statistical extrapolation of real data, the most substantial part of more than 130,000 people who commute to work in the capital each day consists of workers, then there are students, and the smallest group consists of retirees. The biggest daily onslaught is mainly in business and shopping centers in the city center and the wider center. During a regular business day, there are about 10,000 people (regularly occurring, not in one time). It is a conservative estimate of the daily migration of people to Bratislava, as in the above figures those who come to the capital as foreign tourists are not included.

3.3 *Automobilization Rate*

Bratislava, the capital city of Slovakia, is divided into 5 national administrative districts and 17 borough city parts (see Fig. 3).

These boroughs vary in size and population, from the smallest Lamač and least populated Čunovo to the largest Podunajské Biskupice and most populated Petržalka. It is the largest city in the Slovak Republic. In 2015, 1000 people in Slovakia accounted for 375 passenger cars, which represents 2,031,571 passenger cars per 5,243,801 inhabitants [24, 25]. If we calculate that the average space required for one vehicle is 14 m^2 (7×2), then 28,484,036 m^2 is required for the placement of these vehicles on parking spaces, which is 28.4 km^2 of space.

Automobilization rate is reflected by the ratio of the number of inhabitants per one passenger car or the number of passenger cars per 1000 inhabitants. At the high rate of motorization, the sustainability of the actual transport systems is significantly

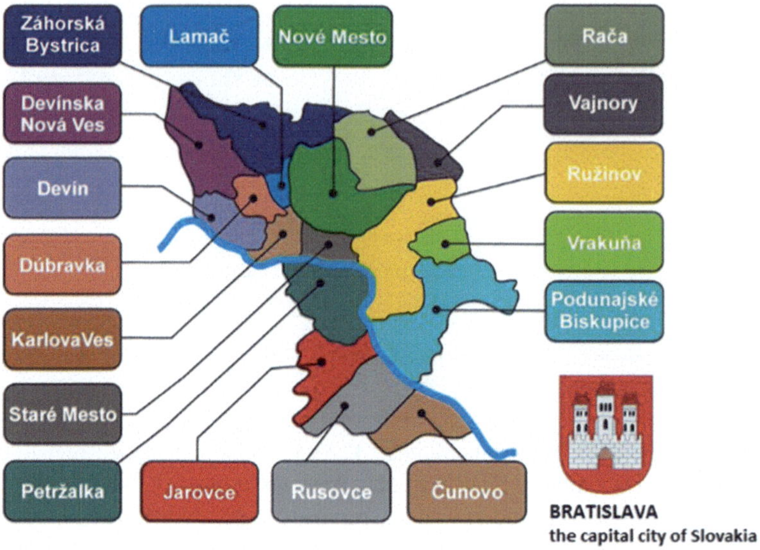

Fig. 3 City districts in Bratislava [23]

affected, and thus there are gradual changes in the final distribution of the transport work between individual car transport and public transport. The increasing number of cars and their use, particularly in densely populated areas, have a positive impact on the development of the economy and employment, and affect the way people live, but on the other hand it has a negative impact on the environment of towns, their roads, and parking.

The European Union countries show a significant share of transport development and a strong transport system on the economic growth of the country, thanks to the effective provision of transport requirements for working-age residents.

As it can be seen from Fig. 1, Slovakia, compared to the EU countries, still belongs among the countries with a lower motorization rate. On the other hand, it can be said that in the last 20 years the number of passenger cars in Slovakia has increased by almost 100% (Table 2).

At present, a passenger car is a dominant type of transport for reaching the citizen's destination, and it is often considered, especially in postcommunist countries, to be a symbol of freedom and a social status. The development of motorization and automobilism in Slovakia has been gradually developing. With a gradual increase, Slovakia is approaching an average level of motorization of the EU. Already at the value of 3.5 people/passenger car, problems with the road safety, its fluency, and especially parking are starting. The capital of the Slovak Republic has a very high motorization rate compared to the rest of the country. As shown by the materials of the Ministry of Interior of the Slovak republic and Statistical Office of the Slovak Republic (2017), the rate of automobilization is currently 2.56 (passengers/passenger car) (Table 3).

Table 2 Development of number of passenger cars in Bratislava

Year	Population* persons	Passenger car number	Automobilization rate	Number of passenger cars/1000 inhabitants
2003	425,533	155,362	2.74	365.10
2004	425,155	165,075	2.58	388.27
2005	425,459	161,713	2.63	380.09
2006	426,091	181,852	2.34	426.79
2007	426,927	190,696	2.24	446.67
2008	428,791	201,177	2.13	469.17
2009	431,061	203,831	2.11	472.86
2010	432,801	214,029	2.02	494.52
2011	413,192	225,092	1.84	544.76
2012	414,391	235,947	1.76	569.38
2013	416,489	243,038	1.71	583.54
2014	418,534	248,156	1.69	592.92
2015	422,932	266,032	1.59	629.02
2016	425,923	279,302	1.54	655.76

Source: processed with the materials [21, 26]
*population with permanent residence in Bratislava

Table 3 Development of number of passenger cars in the Slovak Republic

Year	Population* persons	Passenger car number	Automobilization rate	Number of passenger cars/1000 inhabitants
1995	5,363,676	1,015,794	5.28	189.38
2000	5,400,679	1,274,244	4.24	235.94
2005	5,387,285	1,303,704	4.13	242.00
2010	5,431,024	1,669,065	3.25	307.32
2015	5,423,801	2,034,574	2.67	375.12
2016	5,430,798	2,121,774	2.56	390.69

Source: processed with the materials [21, 26]
*population with permanent residence in Bratislava

3.4 *Non-existing Parking Policy in Bratislava*

Automotive transport is one of the most significant problems in Bratislava, namely the sharp decline in public transport (PT) and the increase in individual car traffic (ICT), the depletion of some roads, and the lack of parking spaces. The issue of parking policy in Bratislava is extensive. In connection with this issue, it is mostly about the lack of parking spaces, but the city notably lacks the order and rules. On the other hand, the city's offer to stop and park the vehicle has a significant impact on the mobility of the city, territory, space, and its serviceability. The capital city of the Slovak Republic has become the only regional city that still has no parking policy implemented; that is, it does not have uniform rules and parking regulations.

Bratislava "excels" with its parking problems even in Europe, where it becomes one of the few without rules and regulations. The solution of parking and parking policy issues and the responsibility for them are in the competence of the local government. Local authorities administer each city district. Most roads of the third and fourth class are in the administration of a city district.

On the territory of the Slovak capital, there are about 150,000 parking places available on the roads of first to fourth categories. There are more than 210,000 cars registered in Bratislava, and another 100,000 come each working day from the surrounding towns, cities, or regions of Slovakia [27].

In the current situation, users park their vehicles in all available areas. However, these areas are primarily not intended for parking and lay-by of vehicles, but they have, for example, the function of greenery and walkways. At present, it is permissible to park on a walkway outside the marked parking area if pedestrians, strollers, and wheelchairs have a free space of 1.5 m. However, street parking does not satisfy all demands placed on it, and this distance is not always met, which often prevents the safe movement of people for whom the walkway is primarily designated. At present, parking in streets is entirely free or cheaper than in parking garages in Bratislava.

4 Impact of GDP Growth in Slovakia on the Growth in the Number of Passenger Cars

The current development of automobilism in Slovakia, as well as the comparison with the development in Central European countries, suggests that the further development of automobilism will depend mainly on the stable economic situation in the Slovak Republic without political and economic shocks. It will ensure a stable living standard for the population comparable to the countries where the motorization rate has been saturated.

Based on these experiences, the future development of individual automobilism in the Slovak Republic can be expected in the dimensions which indicate that the growth of the number of passenger cars and the associated use of them will reach the saturation level of around 2.27 inhabitants per a passenger car around the year 2030 (Table 4). According to Faith [24], a steady increase in GDP levels, a rise in real population incomes, a steady development of population spending, and a gradual stabilization of passenger car prices are envisaged.

For the prognosis of individual motorization, we have considered:

- A moderate annual GDP growth of about 1.2%
- A maximum target level of saturation of owned passenger cars in the ratio of 441 passenger cars per 1000 inhabitants
- An increase in passenger car prices by an average of 2.8% per year
- An increase in incomes of inhabitants on average by about 3% per year

Table 4 Forecast of automobilization development in the Slovak Republic

Year	Population* persons	Passenger car number	Automobilization rate	Number of passenger cars/1000 inhabitants
2020	5,416,585	1,998,740	2.71	368.77
2025	5,430,887	2,198,740	2.47	375.12
2030	5,450,838	2,401,250	2.27	390.69

Source: processed with the materials [24]
*population with permanent residence in Bratislava

Fig. 4 Development of passenger cars in Slovakia [25]

  - The development of the ratio of between a passenger car price and an average monthly salary, which would decrease by 1.4% annually, and in the end of the year the average passenger car price would reach 29 times the average monthly salary.
  - The development of the total annual costs of car operation (fuel, insurance, spare parts and accessories, as well as repair and servicing costs) that would not significantly increase.

- The development of cash incomes, which will develop proportionally in line with the development in inflation and the raising of income levels in EU countries.
- A slight increase of the life span of passenger cars and no significant administrative interference with the ownership of a passenger car (tightening of technical and emission controls, tax burdens, motorway and expressway charges, parking).

The growth in the number of passenger cars (Fig. 4) in Slovakia was supported by GDP growth (Fig. 5). Continuous GDP growth was interrupted by a slight decline in 2008 due to the impact of the global economic crisis. The decline in the number of cars in 2004 was due to the adoption of new legislation aimed at modifying the conditions for the operation of vehicles in road traffic—specifically in the field of vehicle registration.

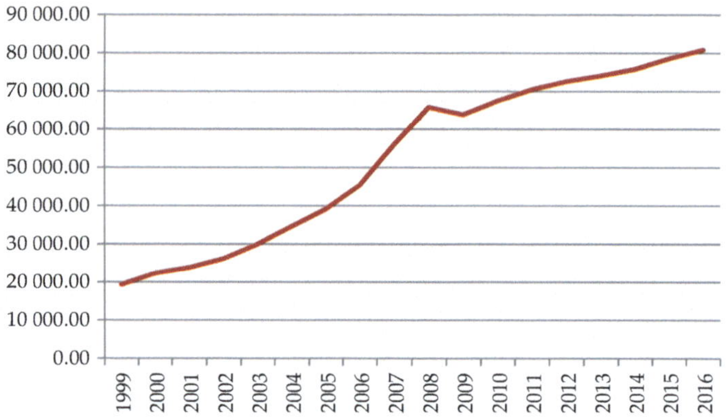

Fig. 5 Development of GDP in Slovakia [25]

Based on these assumptions of the motorization forecast, it is possible to talk about a realistic variant. The realistic variant is a variant for which we expect to have about 2.4 million passenger cars in Slovakia in the year 2030, which is considered a saturation level. This level represents an increase in passenger cars by 76% compared to the current level of mobilization, which can be considered as a realistic estimate of future developments. This number of passenger cars should guarantee a level of motorization with 2.27 inhabitants per one passenger car, which is 441.49 passenger cars per 1000 inhabitants. It would mean that regarding the population growth and the number of household members, a single car would come to each household.

We can state that based on the current figures on the number of passenger cars shown in Table 2, the development was much more dynamic than forecasted by the experts mentioned above. Specifically, in 2016, the number of passenger cars in the Slovak Republic reached 2,121,774 compared to the forecast for 2020 (see Table 4), where the estimated volume is 1,998,740 passenger cars.

5 Park and Ride System: The Suggested Locations for Bratislava

An increase of the motorization rate significantly affects the growth dynamics of population mobility. This leads to crucial solutions to maintain the current level of public transport and how to effectively influence the development of the use of passenger cars when traveling to work. There are a number of possibilities to reduce the growth in the use of passenger cars, but not all solutions can be implemented without difficulty and in the foreseeable future so that there are no road congestion and overloading of some sections of roads.

For the future, it is assumed that the regulation of the use of passenger cars, especially on everyday journeys to work, will have to be reduced by construction of park and ride system (P&R), even at the expense of substantial charging for the use of cars, especially in the cities. This will at least partially result in the transfer of passengers to use public transport or to travel on a bicycle as well as on foot when commuting to work.

One possible solution for improving the situation in the field of static traffic in Bratislava could be the realization of the park and ride concept on the basis of the elaborated design study [28] and its implementation in the city's parking policy. The authors assume that changes will be carried out in the master plan of the capital city of Bratislava [16], processing of the project documentation of specific sites, and subsequently construction of several parking facilities. In the future, they could fulfill the required function of "placing a car aside" and continuing the journey by means of public transport.

Park and ride facilities are places where passengers transfer from passenger cars to public transport vehicles (railways, light urban railways, buses). Although park and ride systems have been developing for many years in a number of cities, there are still many communities in which a much smaller significance is accorded to such systems. P&R is generally associated with the notion of multimodal transport, which is the use of two or more modes to form a complete trip between its origin and destination [29]. The P&R concept however is somewhat narrower in definition than this as it is a tool that provides an interchange facility for transfer specifically between private and public modes [30]. As the construction of R&R facilities is financially demanding, these systems should be planned in a rational manner [31].

According to authors [31], the European experience greatly varies as to the planning and evaluation of park & ride systems. Individual countries do not have much in common with respect to the concepts applied so far, except, of course, for the main principle: park your passenger car and resume travel by public transport. There are several possible classifications of P&R facilities. One of them is based on the mode of public transport, so we differentiate:

- P&R facilities near railway transport systems
- P&R facilities near bus transport systems
- P&R facilities near combined rail and bus transport systems

According to [32] P&R sites are typically found at the edge of urban areas. This enables the relative benefits of both private and public transport to be utilized. The flexibility benefits of private transport mean that P&R can be accessed by passengers from dispersed origins such as low-density suburban areas. The use of public transport to access high-demand destinations such as urban economic centers provides efficiency benefits and can offer time savings to users (in terms of both journey time and search time for car parking) while removing traffic from the urban area. In terms of the public transport mode for which access is provided by P&R, there are a number of variations. In Europe, P&R on both light and heavy rail systems has been used, for example in Germany, France, and the Czech Republic [33].

The suggested locations for the P&R facilities for Bratislava are divided according to their relationship to the built-up area of the city and their actual use. The localities are divided into areas already used for the P&R parking system, sites designated for this purpose also according to the Master Plan and the Land Use Master Plan for Transport, external sites located on the outskirts of the built-up area of the city, and internal sites bound to the main routes formed by the highways D1 and D2, main city radials, and the route of the external traffic semicircle [28].

Currently prepared concepts are oriented towards redirecting passengers coming to Bratislava from individual car transport to the railway. In connection with this solution, an optimal network of railway stations and stops will be sought to ensure the best possible accessibility for passengers. To reduce the burden on the existing road network, it is proposed to create conditions for the transition between individual car and rail transport in the form of "park and ride" capture car parks. Capture car parks are designed mostly outside the city to capture passengers from adjacent municipalities and offer them a quick and comfortable alternative to individual car transport. The prepared concepts, apart from the use of the railways to transport visitors coming to Bratislava from surrounding towns and municipalities, are considering the use of railway lines in the territory of Bratislava as well as other public transport systems allowing fast access to the city center independently of the traffic situation on the roads [29]. In the proposed locations (Patrónka, Devínska Nová Ves zastávka, Lamačská brána (Bory), Trnávka, Ružinov, Vrakuňa, and Mladá garda) it is proposed to create integrated passenger transport terminals (IPTT) linking several public transport systems with railways as the main transport system creating an efficient and flexible system for fast city-wide traffic.

6 Discussion

Parking, connected with the penetrative development of individual car traffic in inner-city transport, is becoming a major problem for the capital. The current, largely unadjusted development of these areas is causing problems in overcrowding of road traffic, day-to-day traffic congestions, and acute shortage of parking spaces and often spontaneous parking, which directly affects the loss of public space for the pedestrians and greenery and leads to deterioration of the air quality. Failure to solve the problem would undoubtedly lead to a further reduction in environmental sustainability and degradation of the quality of life [15].

The survey of the Land Use Master Plan for Transport from 2015 [14] shows that the total distribution of transport work of addressed people living in Bratislava is as follows: 39.6% individual car transport, 32.6% public transport, 26.2% walking, and 1.6% cycling. Other modes of transport (e.g., taxi, plane) were excluded from this analysis due to low frequency. It has been found that the dominant type of all journeys related to work is individual car transport. In case of these journeys, respondents said that they in 52.8% cases used individual car transport, in 35.4%

public transport, and only in 11.8% a "soft type," i.e., walking or cycling. The inhabitants of Bratislava prefer traveling by car to public transport.

Another research was carried out as the night research of static transport with automatic transport counters in May, June, September, and October 2014 within the time range of 11 PM–4 AM. 97,899 cars were recorded with a unique vehicle registration number, of which 78,980 were from Bratislava and 18,919 were from other parts of the Slovak Republic and from abroad, and 9686 were illegally parked vehicles (mainly in the central part). It was also found that 80% of cars were parked in public areas, 18% on private land, and 2% in paid garage houses.

7 Conclusion

The continuously increasing number of passenger cars dramatically aggravates the mobility of cities and their traffic, which is subject to sufficient space to stop and park cars. Cities, built over many centuries, cannot be capable of enduring the current or future growth of car traffic without implementing the concept of coordinated city development and the specific guarantees of mobility in it.

According to the authors, the absence of a uniform parking policy in the capital city f Bratislava, failure to respect the standards in the area of construction of a specified number of parking spaces in case of development projects, and high migration flows into Bratislava (or the Bratislava region) in the form of daily or weekly mobility as well as moving are problematic areas, which significantly affect Bratislava's static transport.

Opportunity to solve parking problems, as seen by the authors, is in defining areas potentially suitable for the location of capture car parks—the park and ride system (P&R system), which more than exceeds the estimated parking capacity of 29,063 parking spaces required for the city of Bratislava.

The creation of capture car parks (P&R system) can become a stimulus for the development of adjacent areas and contribute to the creation of a new alternative for parking of passenger cars, which will at least partly help to reduce the deficit of static areas in the territory of the city.

Acknowledgments This chapter was supported by Slovak Research and Development Agency under the grants VEGA No. 1/0604/18 Economic Aspects of Sustainable Construction within Intelligent Urban Structures.

References

1. Marsden, G. (2006). The evidence base for parking policies—A review. *Transport Policy: Special Issue on Parking, 13*(6).
2. Anderson, S., & de Palma, A. (2007). Parking in the city. *Regional Science, 86*(4).

3. Steenberghen, T., Pourbaix J., Moulin, A., Bamps A., & Keijers, C. (2013). *Study on harmonised collection of European data and statistics in the field of urban transport and mobility*. Final report.
4. Dulla, I. Implementation of city parking policy objectives (Realizácia cieľov mestskej parkovacej politiky). http://www.szsi-sk.sk/index.php?option=com_content&view=article&id=70:parkovacia-politika-ciele-a-podmienky-rieenia&catid=36:odborne-clanky&Itemid=104&lang=de.
5. Young, W. (2008). A view of parking policy in an Australian City. In S. Ison & T. Rye (Eds.), *The implementation and effectiveness of transport demand management measures* (Vol. 8). Hampshire: Ashgate Publishing Limited.
6. McShane, M., & Meyer, M. (1982). Parking policy and urban goals: Linking strategy to needs. *Transportation, 11*.
7. Eurostat. Passenger cars in the EU. http://ec.europa.eu/eurostat/statistics-explained/index.php/Passenger_cars_in_the_EU.
8. Methodology of traffic capacitive impact assessment of investment projects. (2014). The Annex to Decision of the Mayor of the capital city of the Slovak Republic, Bratislava.
9. Spirkova, D., Golej, J., Panik, M. (2014). The issue of urban static traffic on selected examples in Bratislava in the context of economic sustainability. *Proceedings of the Second international conference on traffic and transport engineering. Belgrade, Serbia.*
10. STN 73 6110 Planning of local communications. (2004). Slovak Institute for Standardisation (p. 65).
11. STN 73 6110/Z1 Planning of local communications. (2014). Slovak Institute for Standardisation (p. 2).
12. Act no. 135/1961 Coll. on roads. https://www.slov-lex.sk/pravne-predpisy/SK/ZZ/1961/135/vyhlasene_znenie.html. Accessed 20 Nov 2017.
13. Basic Rules of Parking Policy of the Capital City Bratislava. http://www.bratislava.sk/. Accessed 10 Nov 2017.
14. Land Use Master plan for Transport, Bratislava. http://www.bratislava.sk/uzemny-generel-dopravy/ds-11007012. Accessed 10 Nov 2017.
15. Generally binding regulation of the capital city of the Slovak Republic Bratislava no. 12/2016 of 8th December 2016 on temporary parking of motor vehicles on defined sections of local roads in the territory of the capital city of the Slovak Republic Bratislava, the amount of payment for temporary parking of motor vehicles, the payment method and the proof of its payment. http://bratislava.sk/vseobecne-zavazne-nariadenie-hlavneho-mesta-slovenskej-republiky-bratislavy-c-12-2016/d-11052856/p1=11049947. Accessed 10 Nov 2017.
16. Master plan of Bratislava City. http://www.bratislava.sk/uzemny-plan-hlavneho-mesta-sr-bratislavy-zmeny-a-doplnky-02/d-11031298. Accessed 25 Oct 2017.
17. Kožák, Š. *Externalities as a complementary form of investment evaluation (Externality, ako doplnková forma hodnotenia investícií).* http://fpedas.uniza.sk/dopravaaspoje/2005/1/kozak.pdf.
18. Spirkova, D., Ivanicka, K., & Zubkova, M. (2013). *Economic aspects of sustainable development of built environment in Europe*. Franfurt am Mein: MCDP International UG.
19. Čelko, J. *Cities and static transport (Mestská a statická doprava).* http://www.uzemneplany.sk/zakon/mesta-a-staticka-doprava.
20. INFOSTAT, Slovak Republic. http://www.infostat.sk/web2015/en/.
21. Statistial Office of the Slovak Republic. https://slovak.statistics.sk/.
22. Market Locator. http://www.marketlocator.sk/.
23. Bratislava: City Districts. http://www.bratislava.sk/.
24. Faith, P. *Trend development, automobilization and its impact on the distribution of transport work. (Trend rozvoja, automobilizácia a jeho vplyv na deľbu prepravnej práce).* http://www.svetdopravy.sk/trend-rozvoja-automobilizacia-a-jeho-vplyv-na-delbu-prepravnej-prace/.
25. Ministry of Transport and Construction of the Slovak Republic. http://www.telecom.gov.sk/index/index.php?ids=261&lang=en.

26. Ministry of Interior of the Slovak Republic. https://www.minv.sk/?celkovy-pocet-evidovanych-vozidiel-v-sr.
27. Energetic Center Bratislava. http://www.e-centrum.sk/.
28. Chudík, M., Adamczak, M., Polonský, F., & Kajdi, K. (2017). *Search study for the possibility of realization of capture car-parks and parking houses in Bratislava. (Vyhľadávacia štúdia možností realizácie záchytných parkovísk a parkovacích domov v Bratislave)*. Bratislava: The Institute of Spatial Planning.
29. Krygsman, S., & Dijst, M. Multimodal trips in the Netherlands: Microlevel individual attributes and residential context. *Transportation Research Record, 1753*, 11–19.
30. Spillar, R. J. (1997). *Park-and-ride planning and design guidelines*. New York: Parsons Brinckerhoff.
31. Krasić, D., & Lanović, Z. (2013). Park & ride facility planning. *Građevinar, 2*, 111–121.
32. Meek, S. (2008). Park and ride. In S. Ison & T. Rye (Eds.), *The implementation and effectiveness of transport demand management measures* (Vol. 9, pp. 165–188). Hampshire: Ashgate Publishing Limited. ISBN 978-0-7548-4953-3.
33. Topp, H. H. (1991). Parking policies in large cities in Germany. *Transportation, 18*, 3–21.

Rediscovering Urban Smart Communities

Mauro Romanelli

1 Introduction

Cities of the future should emerge as urban platforms and communities that ensure high quality of services and promote social interaction in order to support knowledge sources for continuous innovation that relies on the use of information and communication technologies (ICTs) to support the life and business within sustainable urban ecosystems.

As smart communities, cities contribute to promoting urban growth and sustaining value creation processes within society, improving the quality of life and employing the information technology in order to achieve successful issues driving processes of innovation, knowledge creation and service quality [1–5]. Sustaining social, cultural and economic development and advancements relies more and more on revitalizing economic and productive growth in urban and regional areas, on cities that strengthen the knowledge transfer, following a multilevel governance and stakeholder involvement [6, 7]. Cities should rethink how to plan and implement a satisfying urban development model in order to proceed towards urban sustainability or contribute to sustainable development of urban communities [8, 9].

Today, smart cities have to promote future and urban development of the city as a community that lives in knowledge-based, global economies and open society. Sustaining smart growth relies on designing the city as a community where local government, industry, businesses, education and research centres and people cooperate by the use of information technology to support change and drive innovation in a significant way [10, 11]. This study aims at identifying

M. Romanelli (✉)
University of Naples Parthenope, Naples, Italy
e-mail: mauro.romanelli@uniparthenope.it

© Springer Nature Switzerland AG 2020

D. Cagáňová, N. Horňáková (eds.), *Mobility Internet of Things 2018*,
EAI/Springer Innovations in Communication and Computing,
https://doi.org/10.1007/978-3-030-30911-4_27

how cities as urban smart communities can develop a pathway for sustainability promoting a smart approach to urban development and growth as a source in order to develop smartness as a vision, culture and value for community governance. This study relies on literature review and analysis about the concepts of smart city and smart community as means to promote smartness in service management, urban mentality and governance as sources to build the city of the future as an entity that continuously develops innovation for change. As following a smart approach to urban development, cities tend to design a pathway for sustainability and learn how to build the organizational processes and patterns that drive cities to act and evolve as smart communities.

2 Rethinking Cities as Smart Communities

Within knowledge-based economies and democratic societies, cities have to promote a sustainable development building smart communities. Cities of tomorrow will contribute to developing smart growth and improving the competitiveness and innovation sources within the urban system and knowledge-based economies [6, 12, 13].

Developing a smart approach to governance and economy by sustaining smart growth urban and communities helps the quality of life for people within urban areas [14]. Sustaining smart growth relies on cities that develop the urban community as a smart community that helps promote economic development and job growth and sustain high and increased quality of life [15]. As smart communities, cities of tomorrow should use the potential of information technology to interact with citizens to solve problems, sharing data, information and knowledge [16], to create digital platforms for advanced services and to support business and facilitate public life [17]. A smart community relies on using ICTs to develop economic and social growth, support the education of citizens, plan smart initiatives and embrace a clear vision that enables citizen engagement [18]. In an information age, smart communities rely on ICTs in order to promote social and economic development to improve the quality of life within the community. Smart communities go online to connect local governments, schools, businesses, citizens, and health and social services that tend to work together in order to transform the community in a significant way to create specific services to address local objectives and to help advance collective skills and capacities to revitalize local economies and to reduce the uncertainty emerging from the acceptance and implementation of ICTs. Building smart communities relies on the active involvement and coalition of business, education, government and individual citizens [10, 11, 15]. Government, industry, business, citizenry, research centres and universities develop forms of cooperation and collaboration within cities as smart communities [19–21].

3 How Cities Design a Sustainable Pathway Within Urban Communities

Designing a pathway for sustainability within urban spaces and communities relies on cities that select a smart city approach as the strategic vision and orientation for driving sustainable urban development in order to support smart governance and promote a culture of smartness. Sustainable cities as smart urban communities tend to construct an ecosystem in order to create social, economic, cultural and public value by developing a smart city approach in order to drive urban growth and development, sustaining smart governance and promoting a culture of smartness by facing problems of urban development and growth.

3.1 Following a Smart City Approach to Drive Urban Growth and Development

Smart cities contribute to creating social and economic value for innovation and growth in urban and regional areas leading communities to develop benefits and improve the quality of life within society. Cities tend to design a smart approach to drive urban and sustainable urban growth by using information technology to make economy, governance, people, mobility, environment and living smart [22].

Smart cities proceed towards sustainability achieving long-term results and successful issues [4], employing strategically information technology to ensure high quality of life because of the connections between productivity, growth and human capital, and industries that use information technology for production [23]. Smart cities attract and retain young and skilled people that are searching higher education, with better employability being expected to have a high quality of life. Smart cities employ information technology, systems, services and capabilities in an organic network [24], investing in human and social capital, and developing new communication infrastructure (transports and technology) promoting sustainable economic growth and participatory governance to ensure a high quality of life. Technology, organization, land, government and policy enable managerial capability creation and innovation to drive institutional urban problems attracting smart and knowledge-oriented workforce [1–3]. Building smart cities relies on cities that design service platforms which enable citizens to be connected to the governing bodies and offer knowledge and experience in designing service infrastructure and utilities within urban environment [17].

3.2 *Redefining the Relationship Between City Government and Citizens*

ICTs contribute to promoting new models of governance value-driven to transform organizational and institutional arrangements engaging people in smart city initiatives and community [25, 26]. Cities should adopt a user-oriented approach in order to meet the needs of different user groups by designing e-services [27].

Redefining tasks and competences of municipal government helps cities to be smart and proceed towards a sustainable growth also promoting civic engagement as an issue of smart city approach [28, 29]. As following an evolutionary pathway, city governments tend to interact with citizens as consumers of rights and users of personalized services coherently with administration for people (e-administration mode), involving citizens as agents of policy formulation and implementation in city governance (e-governance mode) and enabling citizens to learn (learning city) [30]. City governments should promote transparency, high quality of services and citizen engagement [31], ensuring efficiency and effectiveness of internal organization, working with public and private organizations (e-partnering) and strengthening the participation of citizens in government trust-based relationships (e-democracy) [32].

3.3 *Promoting a Culture of Smartness to Develop Knowledge Sources and Creation*

As considering smart approach as a necessary stage to drive cities to proceed towards a pathway for sustainability, it is necessary to promote a culture of smartness as immaterial infrastructure that enables people, local governments, businesses and other stakeholders to assume values and beliefs coherently with an orientation to continuous change that relies on developing knowledge sources and creation. Cities should learn how to transform knowledge resources and infrastructure in local development [33].

Smart city is a new paradigm for driving urban development that relies on sustaining knowledge sources and cooperation between private and public organizations [34]. As sustaining innovation and encouraging public-private partnerships, cities tend to create digital platforms to facilitate the business and life developing a culture of smartness for fostering sustainable urban development [18, 19]. As sustaining a smart approach, cities are drivers of cultural change by promoting a smart mindset and developing a continuous dialogue with civil society and supporting civic participation in order to involve people and represent collective interests [35]. As embracing a triple-helix model for driving urban growth and innovation, cities tend to stress the role of the civil society as critical stakeholder that contributes to innovation processes and systems sustaining the action of and the cooperation between industries, government and university [20, 21].

As planning initiatives for driving urban development, cities contribute to promoting triple-helix models in order to benefit managerial and organizational processes within companies. As means that enables private and public companies to develop processes of value co-creation within urban communities and spaces, the diffusion of a culture of smartness should help to support the creation and development of evolving corporate cultures and knowledge management processes in enterprises [36] and contribute to strengthening knowledge assets as the key drivers of competitive advantage [37] and stressing the multicultural aspects of human capital management [38]. As promoting a smart city vision, cities contribute to introducing a smart mentality in urban policies and development projects that should contribute to shaping the city of the future [39].

4 Discussion

Cities of tomorrow should involve different stakeholders in order to develop the city as smart community that relies on efficient public institutions, businesses, research and education centres and people that use knowledge and information sharing for innovation, sustaining democratic and participatory cooperation as a source to design sustainable ecosystems. As engines of economic and social growth, cities tend to strengthen the human capital and develop knowledge sources to support a continuous organizational and individual learning and to engender knowledge sources in order to promote value creation. As developing a smart city approach, cities should plan initiatives to involve city government, people, associations and businesses to construct that sense of community that enables social and public value creation within urban ecosystem. Smart cities tend to identify a pathway for sustainability. Cities aiming at promoting urban sustainability should continuously develop a strategic view innovation-oriented in long-terms to support a smart city approach and reinforce the city as smart community future-oriented. Sustainable urban communities are driving change, continuous development and innovation, designing long-term policies. Sustainable cities should promote trust-based relationships with citizens encouraging people to participate in policy choices to build smart governance, rediscovering both urban identity and citizenship as values that drive collective action and help to support innovation processes. Cities of tomorrow should develop the city as a community where city government, people, business, research and education centres and institutions cooperate to create public value by strengthening social and economic growth and defining policies for constructing future development perspectives. As living in the city, people, businesses and organizations should collaborate in order to design a smart view for urban development that enables both efficient and effective service management and creation of a smart culture as means for continuous orientation to innovation and change within urban ecosystem. The users of city services tend to become urban stakeholders that play an active role in the urban development policies of the city. As urban communities, cities should develop smart growth ensuring services

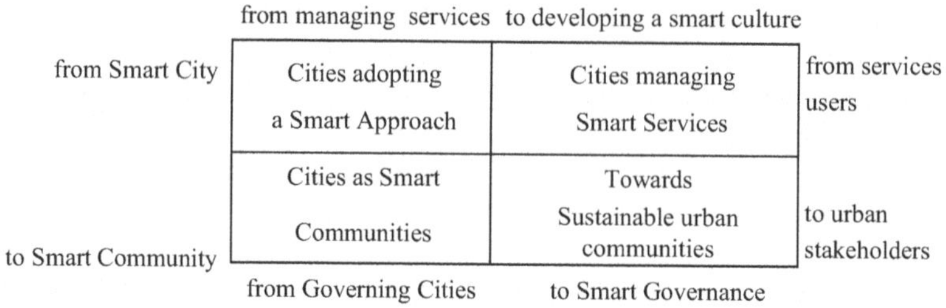

Fig. 1 Designing smart cities as communities oriented towards sustainability

and involving all the stakeholders to play a proactive role in the urban value co-creation process making the city a smart community to rediscover the benefits of smart governance as shown in Fig. 1.

5 Conclusions

Cities should plan the future of urban growth encouraging private-public partnerships to develop knowledge creation and innovation, to ensure high quality of life for people in urban areas. As smart communities, cities should design new models of social and economic development for value creation structuring roles and mechanisms of governance to drive the city as a learning community that evolves by continuous search for knowledge sources to support innovation systems looking at the future and change. Sustainable and inclusive cities will construct the city as a community that believes in cooperation and collaboration for urban growth opportunities. Future research perspectives imply to further investigate how the urban stakeholders and actors (local government, business, industry, civil society, people, citizens, education and research centres) are working together to promote a pathway for sustainable development and growth in small- and medium-sized cities in Italy.

References

1. Nam, T., & Pardo, T. A. (2011). Smart city as urban innovation with dimensions of technology, people and institutions. In *Proceedings of the 5th International Conference on Theory and Practice of Electronic Governance* (pp. 185–194). New York: ACM.
2. Nam, T., & Pardo, T. A. (2011). Conceptualizing smart city with dimensions of technology, people and institutions. In *Proceedings of the 12th Annual International Digital Government Research Conference: Digital Government Innovation in Challenging Times* (pp. 282–291). New York: ACM.
3. Dameri, R. (2013). Searching for Smart City definition: A comprehensive proposal. *International Journal of Computer & Technology, 11*(4), 2544–2551.

4. Czarniawska, B. (2002). Remembering while forgetting: The role of automorphism in city management in Warsaw. *Public Administration, 62*(2), 163–173.
5. Paskaleva, K. A. (2011). The smart city: A nexus for open innovation? *Intelligent Buildings International, 3*(3), 153–171.
6. European Commission. (2012, July 10). Smart cities and communities—European Innovation Partnership, C (2012) 4701 final, Brussels.
7. European Commission. (2017, November 20). Report from the Commission to the Council on the Urban Agenda for the EU, COM(2017) 657 final.
8. Satterthwaite, D. (1997). Sustainable cities or cities that contribute to sustainable development? *Urban Studies, 34*(10), 1667–1691.
9. Haughton, G. (1997). Developing sustainable urban development models. *Cities, 14*(4), 189–195.
10. Eger, J. M. (1997). Cities of the future. The role of smart and sustainable communities, (pp. 114–121).
11. Lindskog, H. (2004). Smart communities initiatives. In: *Proceedings of the 3rd ISOneWorld Conference* (16).
12. Begg, I. (1999). Cities and competitiveness. *Urban Studies, 36*(5–6), 795–809.
13. Eger, J. M. (2009). Smart growth, smart cities, and the crisis at the pump. A worldwide phenomenon. *I-Ways Journal of E-Government Policy and Regulation, 32*(1), 47–53.
14. European Union. (2011). Cities of tomorrow. Challenges, visions, ways forward.
15. Eger, J. M. (2005). Smart communities, universities, and globalization: Educating the workforce for tomorrow's economy. *Metropolitan Universities, 16*(4), 28–38.
16. Chourabi, H., Gil-Garcia, J. Ramon, Pardo, T. A., Nam, T., Mellouli, S., Scholl, H. J., Walker, S., & Nahon, K. (2012). Understanding smart cities: An integrative framework. In *45th Hawaii International Conference on System Sciences*.
17. Anttiroiko, A.-V., Valkam, P., & Bailey, S. J. (2014). Smart cities in the new service economy: Building platforms for smart services. *AI&Society, 29*(3), 323–334.
18. Coe, A., Paquet, G., & Roy, J. (2001). E-governance and smart communities: A social learning challenge. *Social Science Computer Review, 19*(1), 80–93.
19. Deakin, M. (2014). Smart cities: State-of-the-art and governance challenge. *Triple Helix, 1*(7), 1–16.
20. Lombardi, P., Giordano, S., Farouh, H., & Yousef, W. (2012). Modelling the smart city performance. *Innovation: The European Journal of Social Science Research, 25*, 137–149.
21. Leydesdorff, L., & Deakin, M. (2011). The triple-helix model of smart cities: A neo-evolutionary perspective. *Journal of Urban Technology, 18*(2), 53–63.
22. Giffinger, R., Fertner, C., Kramar, H., Kalasek, R., Pichler-Milanović, N., & Meijers, E. (2007). *Smart cities: Ranking of European medium-sized cities*. Vienna: Centre of Regional Science (SRF), Vienna University of Technology. http://www.smart-cities.eu/download/smart_cities_final_report.pdf.
23. Shapiro, J. M. (2006). Smart cities: Quality of life, productivity and the growth. Effects of human capital. *The Review of Economics and Statistics, 88*(2), 324–335.
24. Albino, V., Berardi, U., & Dangelico, R. M. (2015). Smart cities: Definitions, dimensions, performance, and initiatives. *Journal of Urban Technology, 22*(1), 3–21.
25. Meijer, A., & Bolìvar, M. P. R. (2015). Governing the smart city: A review of the literature on smart urban governance. *International Review of Administrative Science, 82*(2), 392–408.
26. Ferro, E., Caroleo, B., Leo, M., Osella, M., & Pautasso, E. (2013). The role of ICT in smart cities governance. In *Proceedings of 13th International Conference for E-democracy and Open Government* (pp. 133–145). Krems an der Donau: Donau-Universität Krems.
27. Ho, A. T.-K. (2002). Reinventing local governments and the E-government initiative. *Public Administration Review, 62*(4), 434–443.
28. Auby, J.-B., & De Gregorio, V. (2015). Le smart cities in Francia. *Le Istituzioni del Federalismo, 4*, 975–993.
29. Portney, K. (2005). Civic engagement and sustainable cities in the United States. *Public Administration Review, 65*, 579–591.

30. Michel, H. (2005). E-administration, e-government, e-governance and the learning city: A typology of citizenship management using ICTs. *The Electronic Journal of e-Government, 3*(4), 213–218.
31. Sandoval-Almazan, R., & Gil-Garcia, J. R. (2012). Are government internet portals evolving towards more interaction, participation, and collaboration? Revisiting the rhetoric of e-government among municipalities. *Government Information Quarterly, 29*, S72–S81.
32. Carrizales, T. (2008). Functions of E-government. A study of municipal practices. *State and Local Government Review, 40*(1), 12–26.
33. Knight, R. V. (1995). Knowledge-based development: Policy and planning implications for cities. *Urban Studies, 35*, 225–260.
34. Kunzmann, K. R. (2014). Smart cities: A new paradigm of urban development. *Crios, 4*(1), 9–20.
35. Rondinelli, P. (2015). Opinion of the European and Social Committee on Smart cities as drivers of development of a new European industrial policy.
36. Čambàl, M., & Cagàňovà, D. (2010). Corporate culture influence on effective initialization and application of knowledge management in enterprises. In *Proceedings of 11th European Conference on Knowledge Management (ECKM 2010): Universidade Lusíada de Vila Nova de Famalicao, Portugal, 2–3 September 2010* (pp. 176–181). Sonning Common: Academic Publishing International.
37. Čambàl, M., Cagàňovà, D., & Šujanovà, J. (2012). The industrial enterprise performance increase through the competency model application. In *The 4th European Conferences on Intellectual Capital. ECIC 2012: Proceedings—Arcada University of Applied Science, Helsinki, Finland, 23–24 April 2012* (pp. 118–126). Sonning Common: Academic Publishing International.
38. Cagàňovà, D., Čambàl, M., Šujanovà, J., & Woolliscroft, P. (2012). The multiculturality aspects and human capital management within Slovak industrial enterprises. In *The 4th European Conferences on Intellectual Capital. ECIC 2012: Proceedings—Arcada University of Applied Science, Helsinki, Finland, 23–24 April 2012* (pp. 106–117). Sonning Common: Academic Publishing International.
39. Vanolo, A. (2014). Smartmentality: The smart city as disciplinary strategy. *Urban Studies, 51*(5), 883–898.

Moving Towards Network State

Marcel Ševela

1 Introduction

The end of politic and economic bipolarities in 1990s, rising global benefits of globalization, and fast-introducing information and communication technologies in everyday life shall constitute the ideal conditions for long-run growth at least for better prepared developed economies. Unfortunately, the reality is not so perfect and substantial obstacles appeared almost immediately. The society and its elements were not prepared for almost revolutionary impact of incoming digital economy. Moreover, the digital economy in globalized world demonstrated its fragility from the very beginning. The shifts in economic and thus politic powers bestirred the well-established order we were used to for decades since post-WWII era, and lit the anti-globalization and anti-technology movements.

The accelerating globalization uncovered a lot of severe consequences of strong unregulated competition resulting into more unequal society, stronger exploration of nature and society, and finally generally weakened role of state as sovereign ruler over at least the formal part of institutional framework above all. These tendencies are also reflected in radically changing environment for employment and the whole society [1]. The pressure to competitiveness in pervasive while the transnational corporations interconnected into the global networks influence all aspects of human life including dissemination of omnipresent information and technology. Work and leisure time almost lost their boundaries in 24/7 uninterrupted life.

The chapter aims to summarize the driving trends in economics in digital and knowledge-based era that result in the need for changes in economic state

M. Ševela (✉)
Mendel University in Brno, Brno, Czech Republic
e-mail: sevela@mendelu.cz

© Springer Nature Switzerland AG 2020

D. Cagáňová, N. Horňáková (eds.), *Mobility Internet of Things 2018*,
EAI/Springer Innovations in Communication and Computing,
https://doi.org/10.1007/978-3-030-30911-4_28

governance. The bureaucratic and network approaches are mutually compared to demonstrate the advantages and disadvantages of each of them. Finally, the possible variants of network governance are outlined. The need for change in governance style is more and more obvious; the need stems mainly from gradational malfunction of traditional governance style in highly flexible, distributed, and efficient communication technology-based economy. The communication technology seems to be the critical trigger. Are the seeds of required communication changes already visible as well as the changes in behavior of market players? These changes could be analyzed through the ICT penetration in more complex activities of market players—the households and enterprises.

2 Bureaucratic vs. Network State

The above mentioned requires the significant transformation of economic and social control at national and international levels while the pre-digital economy institutions are able to keep pace no more. From early 1990s the different approaches to transformation of economic institutional framework started to mutually compete. Taking into consideration the three most important aspects, the bureaucracy and network-based governance systems could be distinguished [2].

In bureaucratic system the state does not resign to its sovereign and central role in managing and protecting the national economy and society in quite dynamic environment. On the other hand, in network approach the state becomes rather coordinator and supporter of decentralized fragmented and gridded networks. New ICT could open the connections among the networks not only at local but also at global level, and incorporates them into differentiated production and consumption networks of post-Fordism style. Thus the state shall abandon its leading and governing role in society and pay attention only to institutional framework.

Summing it up, the bureaucratic approach is an adaptation managed by "enlightened" central government while the network system relies on the initiative from bottom, from individual entrepreneurs and social groups. Japan, South Korea, other newly industrialized countries of Asia, and partly China are the best representatives of bureaucratic approach, with Ireland, Israel, Ireland, etc. being examples of the latter one [3, 4]. The approaches of the United States and European Union cannot be easily classified; they involve important aspects of both approaches in different proportions.

Although since 1980s the ICT sector began to grow rapidly at national and international levels, it unambiguously demonstrated its importance for GDP since 1990s. Rapidly growing contribution of ICT to GDPs was the driving force for GDP growth in majority of developed economies over the whole decade. The Y2K panic significantly contributed to exaggerated growth in stock prices of technological companies, that started in 1996, inflating further the bubble of technological companies.

Table 1 Comparison of developmental bureaucratic and network states

	Bureaucratic state	Network state
Collective development strategy	Building national champions by managing dependency • Strategic use of protectionism • Industrial subsidies • Domestic banking system	Building global regions by mediating global connections • Local networks around global capital • Taking local innovation networks global
Institutions of embedded autonomy	Coherence of state bureaucracy • Embedded • Internal accountability • Tightly coupled	Flexibility of state structure • Multiply embedded • External accountability • Loosely coupled
Dilemmas of global post-Fordist development	• Internationalization of capital and politics of national control • Rigidity of state bureaucracy	• Internationalization of society and politics of inequality • Fragmentation of state

Source: [5]

The bubble burst in March 2001 uncovered the overvalued stocks of tech companies caused by unreal expectations of Internet growth, a profitability of investments in the ICT sector. The so-called dot com crisis exposed the need for all-economy transformation and for discussion about the new society governance paradigm. The new development was embedded in unsuitable old framework not reacting adequately.

It was not the nature of computer and the ITC sector, but the new ways of communication and information exchange in all other sectors of the economy that are connected with the need to transform the coordinating mechanisms in the economies. The production of computers and use of Internet by military and academics in 1960s and 1970s cannot be held as a critical point. The old neoclassical framework seemed to be suitable with these new products till their worldwide spread resulted into radical change in technology of communication (Table 1).

The commercialization of Internet during 1990s changed fully the situation and exposed the need to transform the economic and noneconomic framework. Already during the 1970s the United States and European developed countries lost their competitiveness in production of electronic and computer industry in favor of Japan and some newly industrialized Asian countries. They showed unprecedented transition from economies copying the technologies to economies developing their own advanced technologies during about two decades of purposeful investment.

These Asian economies demonstrated the success of bureaucratically managed state and economy in coping with new technology. The alliances of state and huge national corporations (keiretsu) enable the substantial and permanent investment in research and development, human capital, production capacity building, etc. [2, 6]. Foregoing with low wage level, modest attitude to life, and enormous labor effort resulted in their international price competitiveness, and then later transformed to more enduring non-price competitiveness. The bureaucratic state proved as an ideal system to catch up the train and took control over high-technology sector worldwide.

3 Digital Economy and Information Society

But what was ideal for mastering the mass production in Fordist style seemed to fail in transforming the economy to economy based on information exchange. In late 1980s and mainly 1990s these economies got stuck due to their economic and social inflexibility with respect to turbulently changing technologic and economic environment versus their rigid societies. The system profiting for decades from central planning or indicative planning failed similarly like centrally planned economies of socialistic type in early 1990s.

Not only the impossibility to adapt in the flexible way but also the exhaustion of incentives for development and the will of consumers to diminish the asserted austerity and not to limit themselves in consumption disrupted for centuries fixed rules of traditional behavior that created the competitive advantage [7]. Thus these countries lost their growth drive and fell into the stagnation being endangered by more flexible systems.

The search for next successful systems paid detailed attention to independent local economies based on high technology, mainly the ICT industry of Silicon Valley or Israel kibbutz type. These decentralized and democratic managed groups demonstrated their viability under the new conditions and moreover also ability to compete internationally. It is the return back to decentralized market economy of early capitalistic era, where the small entities cooperate mutually through the competitive market rather than the fixed relations drawn upon some sophisticated centralized plan [5, 8]. The self-initiative became the most important growth factor again.

The development in 1990s was not smooth also in other developed countries. They recorded a few economic downswings, among them the financial crisis in 1998 and 2008 being the most severe. They were also forced to adapt and test new approaches. The logical step was to profit from the experience of local entities and independent networks. These local networks inspired the development of new strategies at national level. Moreover, the local networks could be capitalized as building elements to form wider network grids [9]. The initiative of such state independent entities and renewed market forces are gaining in power. There should be no artificial political borders to building such a network grip; only the economic incentives shall be decisive.

It seems more and more obvious that the cooperation centered on these successful localities can lead to better results than the organization of economic activities on national principle. The successful region becomes the center point—gravity centrum attracting others and thus bundling up together other entities, thus making the gravity centrum stronger [10]. In this process the national differences can be easily overcome in favor of natural regionally based cooperation [11].

Differences usually follow from institutions embedded in society; they are results of long-run development path similar to accumulated society heritage. Ascending the ladder in international competitiveness will not be the result of one-shot reform, but of gradual transformation of the economy-related formal and informal institutions to prepare the "mycelium" for local success followed by whole grids of networks [12] spread through the whole global economy.

Highly productive and sustainable economy is about thinking smarter and creating novel and innovating ideas to generate new and successful revenue models. Smart means producing improved products and services at competitive prices. To meet this aim, the government is supposed to implement a range of policies to make the economy more dynamic and adaptable [13].

4 Network State Governance Styles

The historical predetermination, business environment, development path specifying current institutional framework, examples of good practice, etc. are headed towards formulation of three possible paths of future network economy [2, 8]. Each variant is based on quite unique attitude towards capital/market economy.

The neoliberal stems from equal individual capabilities and possibilities, while the second conservative relies on driving role of elite class. The third variant does not rely on specific role of individual or social group but profits from mutual sharing of knowledge, possibilities, infrastructure, etc. and finally outcomes. The basic characteristics of each approach are summarized in Table 2. These attitudes are then reflected in appropriate governance of the economy and also structure of the whole state (Table 2).

The neoliberal approach was already tested with success in the United States, and the conservative in European continent. Both resulted in competiveness primacy in some historical era, but not in digital society. Moreover, the social inequality and exclusion resulting from these approaches invoke some corrections and social regulation [14]. The social democratic style has not been already tested for a

Table 2 Variants of network state

	Neoliberal	Conservative	Social democratic
Work and inequality	High reward to private investment in human capital	High reward to unequal public investment	Socializing knowledge and technical communities
Risk and security	Individualized portfolio of benefits	Two-tier welfare state	Flexibility within universalism
Governance	Free markets, strong state	Solidarity without equality, uneven extent of social partnership	Decentralized institutions within encompassing social partnership

Source: [5], shortened by author

reasonable time period. Its characteristics including possible social inclusiveness are promising. The critical question is whether the development stage of technology and social knowledge will keep it viable also from economic and long-run growth point of view.

The socializing of knowledge among the communities, decentralization of institution including their self-governance, and general social system flexibility are extremely information intensive. It not only exchanges huge amounts of data, but moreover the whole process of collecting, verifying, processing, storing, etc. of all data shall be automatic and cost effective. It also required the user of this system to adapt this new process and significantly change their market behavior.

5 E-activities of Individuals

Nowadays, the access of households and enterprises to modern information and communication technologies is decisive no more, but their ability and intention to use these ICT instruments for running their daily market and administrative activities are critical. In the past the Internet was only the tool for pure and simple gaining of the information and/or spending of the leisure time. It shall to become the tool for everyday arranging the administration and pursuing of all economic activities.

The use of ICT in this way and the spread of this use in the economy and society are the objectives of regular surveys not only in EU. The EU approach to this matter provides the systematic and comprehensive data from Digital Single Market process monitoring the EU digital economy and society in 2016–2021 [15].

The households and individuals' behavior can be described using two aggregate aspects. Firstly, it is the use of ICT for shopping online as the substitute for going to shop. The share of online shopping has risen significantly during the last decade, but there is still enough scope for further growth. Taking the data of 2018, only 60% of EU individuals in average shopped online in the last 12 months. This share of online shopping spreads from very low circa 20% for Bulgaria and Romania and reaches above 80% for Denmark and the United Kingdom. Concentrating more on higher frequency online shoppers in last 3 months, above six purchases during this period were done only by 44% of individuals in the United Kingdom, 31% in Denmark, and 22% in Germany and the Netherlands. The other countries accounted for far below 20%.

The second aspect—the electronic communication with public authorities and administration—demonstrates entirely different results. The outcome is highly influenced by the grade of e-government in each country. Denmark still keeps the pace and takes the front place, while the United Kingdom loses evidently.

Detailed results of household's e-activity penetration are presented in Table 3 describing the percentage shares of individuals with respect to the whole population. A few activities were selected to illustrate the behavior in each EU member country. The following indicators were used upon data availability for 2018:

Table 3 E-activities of individuals in 2018 (percentage shares)

	Online purchase: in the last 12 months	Six times or more online purchases in the last 3 months	Interaction with public authorities	Obtaining information from public	Downloading official forms	Submitting completed forms
European Union	60	16	52	44	31	34
Belgium	61	10	56	46	31	37
Bulgaria	21	1	22	17	9	9
Czechia	59	6	53	50	26	26
Denmark	84	31	92	90	46	73
Germany	77	22	57	56	35	19
Estonia	61	14	79	69	48	71
Ireland	59	18	54	42	35	49
Greece	36	7	50	47	29	24
Spain	53	10	57	49	39	41
France	67	13	71	46	37	59
Croatia	35	6	36	35	20	16
Italy	36	4	24	20	17	15
Cyprus	32	3	42	39	27	26
Latvia	45	5	66	53	15	50
Lithuania	43	6	51	46	28	41
Hungary	41	5	53	48	38	37
Malta	54	16	46	42	33	23
The Netherlands	80	22	82	77	55	59
Austria	60	16	66	56	38	45
Poland	48	9	35	24	22	25
Portugal	37	3	42	36	21	30
Romania	20	1	9	7	5	4
Slovenia	51	6	54	49	28	19
Slovakia	59	8	51	46	24	16
Finland	70	14	83	78	67	65
Sweden	78	19	83	75	49	74
The United Kingdom	83	44	59	47	36	45

Source: Eurostat

- Last online purchase: in the 12 months
- Frequency of online purchases in the last 3 months: six times or more
- Interaction with public authorities (last 12 months)
- Obtaining information from public authorities' web sites (last 12 months)
- Downloading official forms (last 12 months)
- Submitting completed forms (last 12 months)

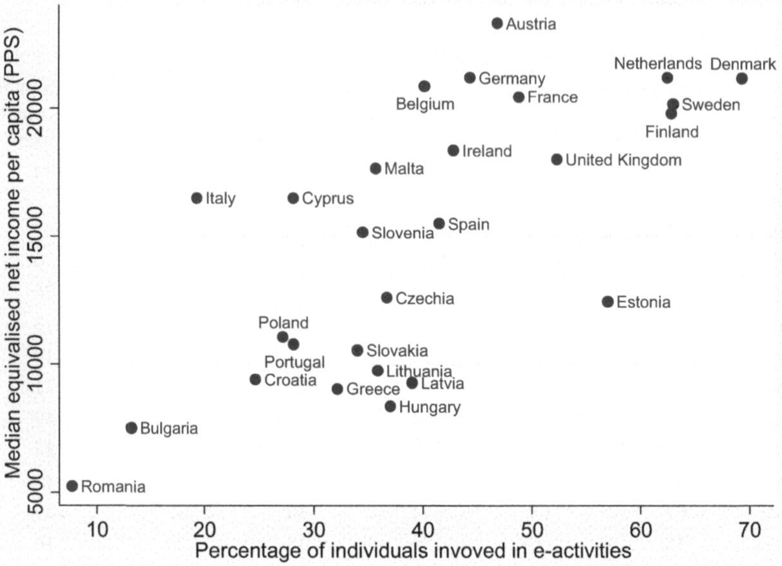

Fig. 1 E-activities and net income (in EUR, PPS, 2017) of individuals. Source: Eurostat

Although the group of six indicators is only the rough picture, it gives us the report about the individuals' behavior in EU member countries. The involvement in e-activities is highly correlated with the net income level of individuals in Euros (PPS, 2017) as presented in Fig. 1. The general involvement in e-activities by individuals is calculated by taking the simple average of the abovementioned six indicators.

The countries are aligned to rising line creating a few clusters. Romania and Bulgaria create the cluster with the worst performance, followed from significant distance by the most numerous cluster of former transition economies and Portugal and Spain. Estonia is a specific outlier because it has comparable economic level with former transition countries but almost double involvement in e-activities mainly due to highly developed e-government. Similarly, Italy is an outlier, but in negative direction.

6 ICT Use in Enterprises

Introduction of modern ICT in enterprises was held as a competitive advantage one or two decades ago; nowadays it is a vitally important element of any business. The simple use of ICT to present the final product or communicate with partner via e-mail is not sufficient; more and more attention is paid to complex systems managing the whole businesses. In addition, the market research is more and more precise and automated through the continual data collection and processing of the so-called big data [16, 17].

The enterprises are involved in these two activities in different degrees; there are also significant differences among the state. To describe the involvement of enterprises, six indicators were selected from the annual Eurostat Model Questionnaire of ICT [15]. The selected indicators strive to cover the use of electronic managerial systems and level of big data processing. The following indicators were selected according to available data for year 2018:

- Enterprises having Enterprise Resource Planning software package to share information between different functional areas
- Enterprises using customer relationship management to capture, store, and make available client information to other business functions
- Enterprises buying cloud computing services used over the Internet
- Enterprises analyzing big data from any data source
- Enterprises analyzing own big data from enterprise's smart devices or sensors
- Enterprises analyzing big data generated from social media

Only the enterprises with more than ten persons employed and businesses outside the financial sector were taken into account. The relative shares presenting the percentage shares of enterprises running the specified activity are presented in Table 4.

Focusing on the usage of managerial systems, the differences among the members are not so significant compared with big data processing. There is a huge cluster of countries having more than 40% of enterprises relying on electronic managerial system like ERP or CMR or both (Denmark, Spain, Lithuania, the Netherlands, Austria, Portugal), with Belgium being the leader with 54% share. The results in CRM are slightly weaker.

The big data processing is governed by a four-member group of countries with shares 20% or above (Belgium, Ireland, Malta, the Netherlands). The exploitation of data collected from smart devices and social media is still very low, but the great increase is awaited.

Using the information derived from electronic managerial systems and/or from big data shall be reflected in a more efficient use of resource; thus it shall increase the productivity of factors of production. This inclination is obvious from Fig. 2, where the scatterplot of labor productivity index (EU28 = 00) is presented. The productivity data corresponds to the year 2017. The explanatory variable is a simple average of the abovementioned six indicators.

The majority of countries follow the rising regression line, Ireland being the outlier demonstrating high labor productivity due to its specific industry structure focusing on high-productive ICT sector and also partly profiting from off-shoring. The majority of countries are crowed in bottom-left corner indicating low performance, while Belgium, the Netherlands, Denmark, and Finland profit from high usage of network for business and its productivity.

Table 4 ICT use in enterprises in 2018 (percentage shares)

	Using ERP software for internal management	Using CRM to communicate with clients	Using cloud computing services	Analyzing big data	Analyzing big data from smart devices or sensors	Analyzing big data from social media
European Union	34	32	26	12	4	6
Belgium	54	42	40	20	7	9
Bulgaria	23	18	8	7	2	2
Czechia	28	18	26	8	4	3
Denmark	40	36	56	14	5	8
Germany	38	46	22	15	3	7
Estonia	28	23	34	11	4	6
Ireland	28	31	45	20	8	13
Greece	37	18	13	13	2	7
Spain	46	34	22	11	4	5
France	38	27	19	16	4	7
Croatia	26	19	31	10	4	5
Italy	37	29	23	7	3	3
Cyprus	35	42	27	5	2	3
Latvia	25	15	15	8	2	3
Lithuania	47	33	23	14	6	8
Hungary	14	13	18	6	2	3
Malta	29	24	37	24	8	15
The Netherlands	48	46	48	22	10	12
Austria	40	43	23	6	2	3
Poland	26	23	11	8	2	2
Portugal	40	24	25	13	5	7
Romania	17	13	10	11	3	5
Slovenia	30	25	26	10	7	3
Slovakia	31	22	21	9	3	3
Finland	39	37	65	19	8	9
Sweden	31	34	57	10	4	5
The United Kingdom	19	31	42	:	:	:

Source: Eurostat

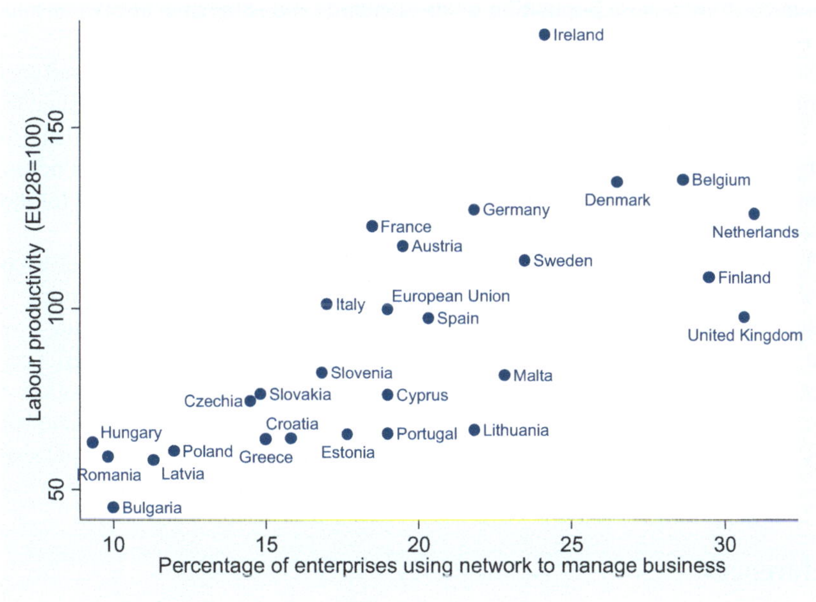

Fig. 2 ICT use and labor productivity index for nonfinancial enterprises with ten and more employees. Source: Eurostat

7 Conclusion

The above-described shares of individuals and enterprises using ICT for their consumption, administration, and business activities indicate the significant difference among the member states. On the other hand, there is already important penetration of ITC in some countries. These countries could take the lead in approaching to network state and network economy governance.

Possible attitudes and their variants could be mutually compared using various sets of partial criteria, but the total effectiveness of the system shall be decisive. There is no exact method to measure the total effectiveness of system performance from the viewpoint of general population. Limiting the evaluating method only to economic growth or productivity is too restrictive while it omits the structural changes in economy. Moreover, the network economy is associated with significant social changes and shall take into account also the noneconomic impacts of economic growth that are viewed as more and more jeopardizing.

Being inspired by mutual competition of two prevailing governance styles in the era of bipolarity, the more effective system cannot be unambiguously determined from the early stages. Although the centrally planned economies seemed to be more effective in scarcity conditions during the postwar reconstruction, in the long run the market-based approaches proved definitively their superiority. The short-run quantitative results usually do not reflect the long-run performance and quality evaluated from general population living standards and subjective utility consumed for more generations.

Similarly, the current competition among the neoliberal, conservative, and social democratic variants of governance of network economy and thus network state may not be evaluated upon the simple quantitative measures using incomplete data from early stages. Each approach already demonstrated its viability in specific conditions (Silicon Valley, Irish tiger, Japan innovative circles, etc.), but it is too early for more general conclusion.

The long-run performance resulting in sustainable higher living standards considering the positive and negative impact to other systems like global policy and global environment cannot be simply anticipated from the limited set of data. It will take a couple of decades to accumulate enough data and subjective experience to assess the individual variants thoroughly and pass the clear recommendation. Of course no pattern is easily transferable into the different institutional framework; thus the final recommendation is never the simple manual for guaranteed success.

References

1. Rodrigues, M. J. (2003). *European policies for a knowledge economy*. Cheltenham: Edward Elgar Publishing.
2. Undvall, B., & Rodrigues, M. J. (2002). *The new knowledge economy in Europe: A strategy for international competitiveness and social cohesion*. Cheltenham: Edward Elgar.
3. Hämäläinen, T. L. (2003). *National competitiveness and economic growth: The changing determinants of economic performance in the world economy*. Cheltenham: Edward Elgar.
4. Barro, R. J. (1998). *Determinants of economic growth: A cross-country empirical study*. Cambridge: MIT Press.
5. Riain, S. O. (2004). *The politics of high-tech growth: Developmental network states in the global economy*. Cambridge: Cambridge University Press.
6. Todaro, M. P., & Smith, S. C. (2011). *Economic development* (11th ed.). Harlow: Pearson Education Limited.
7. Hess, P. N. (2013). *Economic growth and sustainable development*. London: Routledge.
8. Esping-Andersen, G. (1999). *The social foundations of postindustrial economies*. Oxford: Oxford University Press.
9. Jarvenpaa, S., & Wernick, A. (2011). Exploring paradoxical tensions in open innovation networks. *European Journal of Innovative Management, 14*(4), 521–548.
10. Almirall, E., & Casadesus-Masanell, R. (2010). Open versus closed innovation: A model of discovery and divergence. *Academic Management Review, 35*(1), 27–47.
11. Beniger, J. (2009). *The control revolution: Technological and economic origins of the information society*. Cambridge: Harvard University Press.
12. Christou, G., & Simpson, S. (2007). *The new electronic marketplace: European governance strategies in a globalising economy*. Cheltenham: Edward Elgar Publishing.
13. Bawa, M., Caganova, D., Szilva, I., & Spirkova, D. (2016). Importance of internet of things and big data in building smart city and what would be its challenges. *SMART CITY 360 Book Series: Lecture Notes of the Institute for Computer Sciences Social Informatics and Telecommunications Engineering, 166*, 605–616.
14. Boehnert, J. (2018). Anthropocene economics and design: Heterodox economics for design transitions. *The Journal of Design, Economics and Innovation, 4*(4), 355–374.

15. Eurostat. Science, technology and digital society database [online]. Retrieved from https://ec.europa.eu/eurostat/data/database
16. Knieps, G. (2009). Internet of thinks, big data and economics of networked vehicles. *Telecommunication Policy, 43*, 171–181.
17. Larson, J. F., & Park, J. (2018). From developmental to network state: Government restructuring and ICT-led innovation in Korea. *Telecommunication Policy, 38*, 344–359.

Are Home-Based Businesses New Trends in Working Environment for Youth and Senior Entrepreneurs in Europe?

Anna Pilkova and Marian Holienka

1 Introduction

Entrepreneurship is a backbone of all economies. While its significance in general has already been recognized both by academia and policy makers, there are still specific areas which haven't attracted enough attention yet. Home-based businesses and inclusive entrepreneurship are two of them. The importance of home-based businesses has been growing worldwide. For example, according to OECD/European Commission [1], as many as 58% of Australian businesses are home based, almost two-thirds of enterprises in Austria are solo enterprises, and approximately half of them operate from home; almost 60% of businesses with no employees and 24% of small businesses with employees in the United Kingdom are based at home, and finally in the USA more than half of non-farming businesses are based at home. According to GEM 2014 data, among Europe's new and established entrepreneurs, almost 45% operate their business activities primarily from home (own calculation).

Home-based businesses have a long tradition in some sectors of economies, like arts, crafts, and farming. However, the rapid development of new technologies and particularly ICT (third industrial revolution) have significantly contributed to the development of creative and innovative industries, access to new markets, and possibility to run a business virtually from anywhere if there is a relevant space, Internet access, and the skills to know how to use them. Development of home-based business has also contributed to the creation of the new model of business incubator—home-based operations combined with co-working spaces.

A. Pilkova (✉) · M. Holienka
Faculty of Management, Department of Strategy and Entrepreneurship, Comenius University in Bratislava, Bratislava, Slovakia
e-mail: anna.pilkova@fm.uniba.sk; marian.holienka@fm.uniba.sk

© Springer Nature Switzerland AG 2020

D. Cagáňová, N. Horňáková (eds.), *Mobility Internet of Things 2018*,
EAI/Springer Innovations in Communication and Computing,
https://doi.org/10.1007/978-3-030-30911-4_29

Many entrepreneurs start up and remain in these incubators, but some leave the home and continue to grow in terms of employment and revenue, which is of great social and economic significance [1]. According to the latest research findings, home-based businesses can be seen as a vehicle providing a pathway to improved work-life balance and quality of life [2], opportunities for women and marginalized groups to develop their entrepreneurial activities, or contribution to develop alternative, more sustainable growth models [3–5]. However, resources necessary for starting and developing a home-based business are not equally available to all, and thus some groups of people may be underrepresented, such as the young, seniors, immigrants, and disabled [1]. In Europe there is remarkably little known about home-based businesses and, to our knowledge, no comprehensive study on home-based businesses of senior and youth exists.

This exploratory study has an ambition to contribute to close this gap and its main objective is to identify key characteristics and differences of senior and youth home-based entrepreneurs and their businesses in Europe. In addressing this objective, the chapter focuses on three questions. First: are senior and youth home-based entrepreneurs and their businesses different? Second: to which extent are senior and youth home-based and nonhome-based entrepreneurs different? Third: is home-based entrepreneurship the future trend in working environment which deserves attention of inclusive entrepreneurship policies?

2 Literature Review

Research on home-based business (HBB) has been initiated in 1990s. Since that time, a number of definition and home-based business typologies have been presented in literature. However, formulation of a common and universally accepted definition of home-based businesses has been affected by several aspects. For example, definitions lacked differentiating between home-based business activities and a home-based business [6], or between home-based business as an enterprise and home-based entrepreneur as person [7]. On the other side, various definitions of home-based businesses have also some common elements, such as emphasizing home as the main location. For example, Wynarczyk and Graham [8] define a home-based business as "a business that uses the family residential property as a substitute for commercial premises for the conduct of business," while Mason et al. [9] emphasize both employment status of a person and location of a business and define HBB as "any business entity engaged in selling products or services into the market operated by a self-employed person, with or without employees, that uses residential property as a base." Nevertheless, considering only the physical location (in this case, it is a home location) itself is not enough to separate home-based entrepreneurs and those who work from home. Accounting for this aspect, Salazar [10] defines home-based business entrepreneurs as those that are both self-employed and self-managed, which helps to separate home-based business owner-managers from those who are in employment status and work from home (e.g., employees using teleworking) as well as from self-employed individuals who are not self-managed (e.g., contracted labor). In contrast to this very specific understanding of HBBs, OECD/EC ([1], p. 6) offer a rather broad definition of home-based business: "Home-based businesses are private enterprises that have no commercial business premises but use the home of the owner as premises or base for the business activity. Most often the residential address of the owner is the business address. Home-based businesses can be registered (incorporated) or unregistered (unincorporated), and comprise the self-employed and owner managers of companies. Business activities may occur outside the owner's home. Home-based business owners may travel to provide their services or use formal and informal meeting places for their business activities. The home can be used as premises or base also when the business has employees." This extended definition, contrary to Salazar's [10] perspective, includes also self-employed individuals who are not necessarily self-managed. As for the typical features of home-based businesses, they are often described as small- or micro-sized business activities typically not oriented on growth (e.g., [11]). However, at the same time, some studies (e.g., [12]) characterize HBBs as the business segment with the fastest growth and a significant potential to stimulate local and regional economic development. In this respect, combination of HBB and inclusive entrepreneurship policies (supporting less represented segments of population such as women, seniors, youth, migrants, and disabled in entrepreneurship) could significantly contribute to growth by activating underutilized economic resources [1]. However, so far, a majority of research studies that addressed underrepresented segments focused on the gender topic within HBB micro-business research [13]. With this respect, several studies suggest that HBB is especially popular among women (e.g., [14]). Many researchers assumed that a typical HBB owner is a female, for whom working from home is important due to family duties as well as due to lack of other employment options [15]. Contrary to that, numerous subsequent studies (e.g., [9, 16]) challenged the gendered assumption about HBB owners, providing the findings that HBBs are in fact operated merely by males than by females. The only article based on GEM data and dealing with gender issue and HBBs is the work by Thompson et al. [17]. The authors explored attributes of self-employed females in the United Kingdom who manage their businesses from home. Mason et al. [9] also studied characteristics (including age) of home-based businesses on the UK population. They suggest that HBB owners in the United Kingdom are not significantly older than other business owners: 7% of them were younger than 35 years, 54% were aged 35–54 years, and 39% of them were 55+ years old, including 8% of 65+-year-old HBB owners. They also found that only a small proportion of individuals are using HBBs as a means for working beyond the official retirement age. But due to ageing of population and lifestyle of young generation, we can expect that these segments are good candidates for growth of home-based entrepreneurship in the future.

As we explained above, research studies on HBB seniors and youth are missing. Therefore, it is worthwhile to have a look at general studies on these cohorts. Here, one of the recent age-related trends in the field of entrepreneurship is the so-called intergenerational entrepreneurship. Successful application of this concept shall be based on a thorough research on different age cohorts of entrepreneurs with special focus on similar patterns and differences, and consequent implementation of the findings into business and policy-making practice [18].

One stream of the senior entrepreneurship research is concerned with constraints and drivers of entrepreneurial activity of "third-age" individuals (this cohort is usually defined as 50+ years old or 55–64 years old). Here, one of the most frequently considered barriers of entrepreneurship is the age itself [19]. Also, considerable attention of research on this age cohort is paid to entrepreneurship motivations, or more precisely the typical lack of motivation to enter the entrepreneurial path. Senior population usually exhibits lower entrepreneurial propensity when compared to overall adult working-age population. This pattern is explained by several mechanisms, one of them being time allocation preferences [20]. On the other hand, the most frequently cited determinants of senior entrepreneurship are especially high levels of human, social, and financial capital (particularly when compared to younger population). While the accumulated human capital in terms of previous professional experience could be considerably valuable senior entrepreneurs' business endeavors [21], its actual impact in fact depends on the relevance of such experience [22]. Furthermore, the human capital is also subject to ageing, and if not being used for a longer time, its impact usually diminishes [19]. As for the social capital, i.e., relevant social and professional networks of an individual, it has been identified as an important factor in all age categories. Especially networks relevant in the particular business environment can help senior (would-be) entrepreneurs in terms of access to advice or potential business partners [23]. Finally, the effects of financial capital on entrepreneurial propensity among senior individuals are rather ambiguous. While an accumulated capital represents a valuable financial resource for a potential new venture, an excess of finance might at the same time act as an inhibitor of entrepreneurial intentions [24].

Young entrepreneur research (from inclusivity perspective) is focused on unemployment of the youth and self-employment and potential further job creation as its solution [25, 26]. However, youth face several constraints when attempting to start a business. One of the most frequently cited barriers for this age cohort is the limited access to finance [27, 28]. Unlike the older individuals who might have accumulated greater financial capital throughout their lives and have also easier access to financial markets, youth are disadvantaged in this respect. Willingness of banks to finance young entrepreneurs is generally rather limited, and number of programs or support tools available to this target group regarding their financing is insufficient [28]. Another important issue with respect to youth entrepreneurship is the educational system and entrepreneurial education in particular [29]. Educational systems and policies can catalyze but also inhibit intentions to start a business and consecutive engagement in entrepreneurship among young participants and graduates.

On the other hand, young individuals have also certain qualities and advantages relevant for entrepreneurship, compared to their older counterparts. For example, younger entrepreneurs are often attributed to have greater capabilities in terms of individual cognitive attributes [27], such as sensitivity to new information and capability to process them [30]. Also, they are often characterized by higher ability to act upon business ideas and turn them into real business activities [31]. Finally, young entrepreneurs exhibit, compared to senior cohorts, higher levels of energy in terms of dynamism and activity [30].

Connecting age and HBBs, Mason et al. [9] in conclusion of their study predicted that importance of home-based businesses as a direction for business activity shall further increase, with demographic trends being the key determinants of this phenomenon. Trends such as population ageing, increasing longevity, and improved health prospects will result into increased number of postretirement individuals wishing to continue with their economic activity or having a necessity (especially finance driven) to do so. According to the Institute for Future/Intuit [32, 33] the number of older-age individuals running businesses from home in semiretirement age is expected to increase. Furthermore, the young Generation Y (also labeled as the digital generation) is expected to be highly entrepreneurial, resulting into higher probability of business creation efforts compared to previous generations, and the digital technology that this generation is focusing on will encourage them to run their businesses from home. These scenarios fit to the OECD/EC's [1] conclusions: home-based businesses need to be recognized as a distinct type of entrepreneurship. All in all, new perspectives of home-based business related to these segments are ahead of us. Therefore, they deserve deeper attention of empirical research to explore their personal characteristics and businesses and find relevant patterns of their behavior.

3 Data and Methodology

In order to achieve the objectives of our exploratory study, we have employed the following materials and methods in our analysis.

3.1 Sample

Our analyses utilize the Global Entrepreneurship Monitor (GEM) Adult Population Survey data extracted from adult working-age individuals in 2014, which had been to the date the most recent annual cycle in which the survey included a question on home-based character of a business. GEM is the world's largest academic study on entrepreneurship that analyzes entrepreneurial attitudes, activity, and aspirations of individuals deriving representative samples from examined economies [34]. Our sample comprised of data from European countries (the 2014 dataset included 28 countries from Europe) with 85,581 adult population respondents. In this sample we have identified 8380 owner-managers of new and established enterprises, i.e., individuals who own and manage businesses that had been generating income for at least 3 months (nascent entrepreneurs—individuals who are actively involved in starting their own business but have not generated income from it for more than

3 months were not included in our analysis). Among them 3730 entrepreneurs indicated home-based character of their business. In line with our focus on youth and senior entrepreneurs, we further worked with subsamples of 759 youth home-based entrepreneurs (aged 18–34) and 937 of their senior counterparts (aged 55+).

3.2 Measure and Variables

The classification of entrepreneurs as "home-based" as well as their attributes and characteristics of their businesses examined in our analysis were operationalized using the standard GEM variables. We considered as "entrepreneurs" those respondents who indicated that they were currently attempting to start a new business which had already been generating income for at least 3 months and up to 42 months (new entrepreneurs), together with individuals owning and managing a business which had been generating income for more than 42 months (established entrepreneurs). Then, those entrepreneurs who indicated that their business is/will be primarily trading or operating from home were classified as "home-based." Consequently, their business activities were considered as "home-based" businesses.

As for the personal/individual attributes of entrepreneurs, we examined their entrepreneurial self-confidence (belief in having knowledge, skill, and experience required to start a new business), fear of failure (indicating that fear of failure would prevent them from starting a business), opportunity perception (perception of good opportunities for starting a business in the next 6 months), and entrepreneurial social capital (personally knowing early-stage entrepreneurs). Also, we investigated contextual aspects through respondents' perception of social attitudes towards entrepreneurship, as they indicated whether they think that people in their country consider starting a new business a desirable career choice, successful new entrepreneurs enjoy a high level of status and respect, or public media and/or Internet often show stories about successful new businesses.

Considering the character of their businesses, we investigated the prevailing motive for involvement in the current business activity (purely opportunity, partly opportunity, or necessity), its sectoral orientation, innovativeness (bringing products new to at least some customers to markets with few or no competitors), usage of new technologies, size of the business in terms of number of jobs (not counting owners), high job growth expectations (expected job growth of at least ten persons and at least 50% in 5 years), export intensity (proportion of customers living outside the country), and online trading (usage of Internet to sell products or services).

3.3 Analysis

Our exploratory analysis was based on examining and comparing the attributes of home-based entrepreneurs and their businesses within youth and senior subpopulations. For each of the investigated personal/individual, contextual, and business-related characteristics, we looked for the patterns in their occurrence among youth and senior home-based entrepreneurs and businesses, and we searched for significant differences between the two employing chi-square tests. Also, the second dimension of our exploratory analysis was in replicating the investigation on nonhome-based youth and senior entrepreneurs and their businesses to search for similarities and differences between the two distinctive types of business activity (HBB vs. non-HBB) as well as within them (youth vs. senior subpopulations).

4 Results

The first step of our investigation was aimed at identifying representation of home-based entrepreneurs within the different age groups of adult entrepreneur populations (Table 1).

As can be seen from Table 1, the share of home-based entrepreneurs in different age groups shows a slight U-shaped pattern, peaking in the oldest age group (55+) where almost half of the entrepreneurs primarily trade or operate their businesses from home. The second highest proportion of HBBs is found among youth entrepreneurs (aged 18–34). Decomposition of youth age category into subcategories of 18–24 and 25–34 unveils that operating or trading from home is a preferred strategy especially among the younger part of this subpopulation. Result of the chi-square test of independence indicates significant relationship between age category and proportion of home-based entrepreneurs. Thus, we might conclude that the share of HBBs is significantly higher within youth and senior populations, which further justifies our special focus on these two groups through the lens of inclusive entrepreneurship.

The second step of our analysis comprised of investigation of occurrence of selected personal/individual and contextual attributes among youth and senior home-based entrepreneurs, and search for differences between the two age categories (Table 2).

Table 1 Home-based entrepreneurs in Europe in age categories

Attribute	Attribute values	% of home-based entrepreneurs	Chi-square and *p*-value
Age category	18–34	44.3%	14.874
	(Note: 18–24)	(48.3%)	(0.002)
	(Note: 25–34)	(43.2%)	
	35–44	42.8%	
	45–54	44.1%	
	55+	48.5%	
	Adult population total	44.8%	

Table 2 Comparing individual and contextual characteristics of youth and senior home-based entrepreneurs

Attribute	HBB sub-population	Occurrence (%) in HBB subpopulation	Chi-square and *p*-value
Self-confidence	Youth	84.1%	0.867
	Senior	82.3%	(0.352)
Fear of failure	Youth	31.6%	4.582
	Senior	26.7%	(0.032)
Opportunity perception	Youth	54.1%	22.162
	Senior	41.6%	(0.000)
Knowing entrepreneurs	Youth	67.3%	89.966
	Senior	44.0%	(0.000)
E-ship as a good career choice	Youth	59.3%	0.004
	Senior	59.1%	(0.947)
Media attention to entrepreneurship	Youth	56.1%	4.496
	Senior	61.7%	(0.034)
Status of entrepreneurs	Youth	64.4%	0.695
	Senior	62.3%	(0.404)

As can be seen in Table 2, youth and senior home-based entrepreneurs exhibit similar results in three individual attributes, while they significantly differ in four examined individual characteristics. As for the personal/individual characteristics, we see that both youth and senior HBB entrepreneurs show high entrepreneurial self-confidence (more than eight in ten of them believing in having knowledge, skill, and experience required to start a new business) with no difference between the two age groups. However, our findings indicate significant differences between the youth and senior HBB entrepreneurs in having a fear of failure (higher occurrence among the young) as well as in opportunity perception and entrepreneurial social networks (higher frequency among the young HBB individuals). Especially high difference is observed in the last mentioned attribute, as more than two-thirds of young home-based entrepreneurs but not even a half of their senior counterparts indicated personally knowing another early-stage entrepreneur. Thus, entrepreneurship-relevant social networks of senior HBB entrepreneurs seem to be, in general, rather limited.

As for the contextual aspects measured through the perception of social attitudes towards entrepreneurship, we found that both senior and youth HBB entrepreneurs indicate similar occurrence of perceived high social status of entrepreneurs and perceiving of entrepreneurship as a good career choice within society. Thus, it seems that both age groups similarly understand and "translate" the position of entrepreneurship and entrepreneurs in the society. However, on contrary, our findings show significant difference between the age groups of HBB entrepreneurs in perceived high media attention to successful entrepreneurs (to our surprise, seniors show higher perception). We expect that this difference might be caused by

Table 3 Comparing youth and senior home-based businesses

Attribute	Attribute values	% Within youth HBB	% Within senior HBB	Chi-square and p-value
Motive to start-up	Opportunity	54.5%	55.7%	0.799
	Mixed	18.4%	16.7%	(0.671)
	Necessity	27.1%	27.6%	
Sectoral orientation	Extractive	13.4%	21.6%	24.837
	Transforming	28.9%	21.8%	(0.000)
	Business services	28.4%	29.8%	
	Consumer oriented	29.3%	26.7%	
New product market (NPM)	Yes	20.6%	13.3%	15.782
	No	79.4%	86.7%	(0.000)
New technology adoption	Latest	9.5%	3.1%	74.663
	New	19.8%	9.5%	(0.000)
	No new	70.8%	87.4%	
Current number of jobs	No jobs	40.5%	42.9%	5.115
	1–5 jobs	46.0%	46.0%	(0.164)
	6–19 jobs	11.0%	7.9%	
	20+ jobs	2,5%	3.1%	
High job growth expectations	Yes	8.6%	3.0%	25.152
	No	91.4%	97.0%	(0.000)
Export intensity	More than 75%	6.6%	6.6%	18.614
	25–75%	13.3%	8.6%	(0.000)
	Under 25%	38.4%	33.4%	
	None	41.8%	51.4%	
Online trading	Yes	68.9%	53.4%	41.770
	No	31.1%	46.6%	(0.000)

exposure to different media, where a positive picture of business and entrepreneurs is probably more common in the media followed by senior population.

In the third step of our analysis we focused on exploration of business characteristics of senior vs. youth home-based businesses, in order to understand their similarities and distinct features (Table 3).

As shown by our findings in Table 3, senior and youth HBBs significantly differ in six out of total eight examined business characteristics. First, difference was observed in sectoral orientation, with higher focus on extractive sector among senior HBBs and higher share of transforming sector businesses within their youth HBB counterparts. Second, youth HBBs exhibited higher representation of innovative businesses (i.e., bringing products new to some or all customers to markets where few or no competitors are present). Third, youth HBBs are also characterized by higher adoption of new or even latest technologies, while significantly more senior HBBs indicate no usage of new technologies in their business activities. Fourth, significantly more (however still not many) high job growth expectations are found among youth HBBs, compared to their senior-owned counterparts. Fifth, senior and youth HBBs also differ in export intensity. Despite indicating the same share of strongly export-oriented businesses, youth HBBs show higher representation of medium and moderate exporters, while senior HBBs exhibit higher proportion of non-export-oriented businesses. Sixth, there is higher representation of businesses that trade online among youth HBBs compared to senior HBB population. On the contrary, two of the examined attributes showed no difference between youth and senior HBBs, namely the motives for engagement in business activity (where the structure of motives is very similar, and opportunity prevails over necessity) and size of businesses in terms of employee headcount.

The next step of our exploration leads us to analyzing nonhome-based businesses, in order to look for potential specifics related to home-based nature of business. Similar to previous stage of the analysis, we first looked at selected personal/individual and contextual attributes among youth and senior nonhome-based entrepreneurs, and searched for differences between the two age categories (Table 4).

According to our findings (Table 4), youth and senior nonhome-based entrepreneurs differ in four out of the seven examined attributes. In case of the personal/individual attributes, youth HBB entrepreneurs exhibit higher entrepreneurial self-confidence, higher opportunity perception, as well as higher intensity of entrepreneurship-relevant social networks. On the contrary, the difference in occurrence of fear of failure between the two age categories of HBB entrepreneurs was not found to be significant. Concerning the perception of social attitudes towards entrepreneurship, our results show that youth and senior HBB entrepreneurs differ in perception of status attached to successful

Table 4 Comparing youth and senior nonhome-based entrepreneurs

Attribute	Non-HBB subpopulation	Occurrence (%) in non-HBB subpopulation	Chi-square and p-value
Self-confidence	Youth	86.3%	7.712
	Senior	81.6%	(0.005)
Fear of failure	Youth	34.4%	1.997
	Senior	31.4%	(0.158)
Opportunity perception	Youth	46.9%	19.946
	Senior	36.1%	(0.000)
Knowing entrepreneurs	Youth	67.1%	111.828
	Senior	43.2%	(0.000)
E-ship as a good career choice	Youth	58.7%	2.448
	Senior	55.0%	(0.118)
Media attention to entrepreneurship	Youth	51.4%	0.947
	Senior	53.8%	(0.330)
Status of entrepreneurs	Youth	59.9%	7.368
	Senior	53.3%	(0.007)

entrepreneurs in the society, where perception of youth respondents is more optimistic. In case of the remaining two societal attitudes, senior and youth home-based entrepreneurs similarly perceive the media attention devoted to successful businesses and perception of entrepreneurship as a good career choice within society.

Finally, the last part of our analysis was devoted to exploring the selected business characteristics of senior and youth nonhome-based businesses, in order to search for differences between HBBs and non-HBBs, as well as for differences between the two age subpopulations of nonhome-based businesses (Table 5).

As indicated by our findings in Table 5, youth and senior non-HBBs have been found significantly different in five out of the eight examined business characteristics. First, youth non-HBBs indicated higher proportion of new product market efforts compared to their senior counterparts. Second, significant difference

Table 5 Comparing youth and senior nonhome-based businesses

Attribute	Attribute values	% Within youth non-HBB	% Within senior non-HBB	Chi-square and p-value
Motive to start-up	Opportunity	55.9%	57.9%	5.278
	Mixed	20.3%	18.2%	(0.071)
	Necessity	23.8%	24.9%	
Sectoral orientation	Extractive	4.8%	6.8%	5.942
	Transforming	26.3%	26.7%	(0.114)
	Business services	24.7%	21.2%	
	Consumer oriented	44.2%	45.3%	
New product market (NPM)	Yes	17.6%	9.6%	26.452
	No	82.4%	90.4%	(0.000)
New technology adoption	Latest	8.1%	3.7%	40.483
	New	16.8%	9.9%	(0.000)
	No new	75.2%	86.3%	
Current number of jobs	No jobs	30.3%	29.7%	3.565
	1–5 jobs	48.5%	46.0%	(0.312)
	6–19 jobs	13.7%	16.8%	
	20+ jobs	7.5%	7.6%	
High job growth expectations	Yes	8.6%	2.8%	30.627
	No	91.4%	97.2%	(0.000)
Export intensity	More than 75%	7.0%	5.3%	17.346
	25–75%	10.4%	6.7%	(0.001)
	Under 25%	39.1%	36.4%	
	None	43.4%	51.6%	
Online trading	Yes	56.4%	45.7%	22.242
	No	43.6%	54.3%	(0.000)

Table 6 Summary of differences between youth and senior HBBs and non-HBBs

Attribute	HBBs	Non-HBBs
Self-confidence	None	Higher for youth
Fear of failure	Lower for seniors	None
Opportunity perception	Higher for youth	Higher for youth
Knowing entrepreneurs	Higher for youth	Higher for youth
Good career choice	None	Higher for youth
Media attention	Higher for seniors	None
Status of entrepreneurs	None	None
Motive	None	None
Industry	More extractive for seniors, more transforming for youth	None
New product market (NPM)	Higher for youth	Higher for youth
New technology adoption	Higher for youth	Higher for youth
Number of jobs	None	None
High growth expectations	Higher for youth	Higher for youth
Export intensity	Higher for youth	Higher for youth
Online trading	Higher for youth	Higher for youth

was also observed in the adoption of new technology, where youth-owned non-HBBs indicated higher share of latest or new technologies, while senior non-HBBs indicated greater representation of business not using new technologies at all. Third, youth non-HBBs showed significantly higher proportion (however, still rather low) of businesses with high job growth expectations, compared to the senior non-HBB population. Fourth, youth-owned non-HBBs are significantly more export intensive (showing higher percentage in all levels of export intensity), while senior non-HBBs show higher proportion of business without any export activity. Finally, online trading is significantly more present among youth non-HBBs as well. On the contrary, our results indicated no significant difference between youth and senior non-HBBs in the following characteristics: motive for getting involved in current business activity (with opportunity motives prevailing over necessity), sectoral orientation, and business size (in terms of number of jobs).

In order to provide a clear overview of identified similarities and differences between youth and senior entrepreneurs among home-based entrepreneurs as well as among those who don't trade or operate their businesses from home, we constructed a summary table presented below (Table 6).

5 Discussion

Our findings confirm certain intergenerational differences related to youth and senior entrepreneurs in HBB that are generally declared in literature. Among them, we found higher opportunity perception and intensity of entrepreneurship-relevant social networks, higher innovativeness, and growth aspirations of youth than among seniors. However, surprisingly, senior HBB entrepreneurs are similarly self-confident to their youth counterparts, and exhibit even lower fear of failure.

Based on our exploration we are able to derive several main findings and implications for further examination and policy-making.

Contrary to HBBs, in non-HBBs senior entrepreneurs are less confident than their youth counterparts, and also show "only" similar (i.e., not lower) fear of failure. Thus, we might hypothesize that home-based business is a good strategy to overcome concerns about lack in skills, knowledge, and experience (perhaps thanks to easier setting up of and running a business from home) as well as to reduce fear of failure (due to lower potential loss as HBBs usually require lower start-up capital) among senior entrepreneurs.

Also, restrictions linked to operating a business from home don't seem to limit the growth appetite of entrepreneurs and pattern of difference between age groups. The share of high growth-ambitious businesses is almost identical between youth HBBs and non-HBBs, as well as between senior HBBs and non-HBBs. Thus, we hypothesize that adopting home-based operations as strategy to start and run a business does not have a negative influence on growth aspirations. However, at the same time we understand that due to cross-sectional nature of the data we are unable to confirm the conversions of the observed aspirations into actual future growth.

In addition, our findings indicate that in both home-based and nonhome-based businesses, youth entrepreneurs outperform senior entrepreneurs in the same characteristics of their businesses. Thus, we hypothesize that type of business (HBB vs. non-HBB) does not contribute to bridging the gap between youth and senior entrepreneurs in terms of innovativeness, adoption of new technologies, export orientation, or relying on online trading. These seem to be traditional domains of younger generation of entrepreneurs, irrespective of the HBB or non-HBB nature of their businesses.

Finally, our findings on almost identical motive structure between youth and senior entrepreneurs in both HBBs and non-HBBs indicate that both age groups adopt similar behavior in terms of motivation to start-up. Moreover, our results also show only small differences in opportunity-necessity dichotomy between HBBs and non-HBBs in both age groups. Thus, it seems that HBBs are not widely used as escapes from necessity situations as some might expect.

Our approach to exploration of youth and senior home-based businesses also faces several limitations. First, our analysis utilized a pooled sample of individual respondents from 28 countries. Thus, despite bringing a unique and robust insight, some of the findings might require further validation on particular national levels. Also, scope of our analysis was limited to the extent of standard GEM data. This on the one hand enables large-scale analysis of harmonized and high-quality dataset, but at the same time does not allow for deeper or more specific investigation. Thus, our limitations here as well as our key findings should serve as a stepping-stone for further inquiry, as suggested in the next section.

6 Conclusion

This chapter addresses the issues that are so far little known from existing literature and research: investigation of home-based and nonhome-based businesses from perspectives of the European youth and seniors as underrepresented groups of population in entrepreneurship. In general, there are several theoretical reasons as well as myths to believe that entrepreneurial activities of seniors and youth are different. While our findings confirm some of them, the others are left disconfirmed. Based on the results of our exploration, an initial knowledge base has been created regarding senior and youth entrepreneurship in the home-based business context that is relevant to develop policy implications in HBB field tailored to both groups of population from inclusive entrepreneurship perspective. In particular, the following findings that emerge from our exploration ask for support and policy implications on European level:

(a) According to our results, in both home-based and nonhome-based businesses youth are stronger in opportunity perception, innovation, high growth expectations, export intensity, and online trading (see results for HBBs and non-HBBs). This confirms theories and myths that seniors are less creative and innovative. On the other side, within the home-based businesses, seniors exhibit comparable self-confidence in entrepreneurial knowledge and skills, lower fear of failure, and positive attitudes towards entrepreneurship from contextual point of view, and are more involved in traditional extractive industries (like agriculture). These findings, except the one related to industry orientation, are different from what is presented in theory. However, if we want to formulate right policies for home-based businesses as more extensive future trends for youth and seniors, we have to combine our findings with drivers of home-based business in future. According to the Institute for Future/Intuit ([35], p. 3), two expected trends in the area of demographics changes will heavily influence home-based businesses: "Digitally savvy kids grow up and change everything" and "Baby boomers gray but they will not slow down." According to the first trend, Generation Y (digital generation) and Generation Z (Internet generation) will work and live according to "on my own time, place and term." Due to that, we can expect further growth of home-based entrepreneurship in this segment. According to the second trend, traditional retirement will change and "unretirement" and "active engagement" describe this trend the best. Seniors will either continue to work full- or part-time at current job or they will opt a new career, where one of the key options is offered by home-based entrepreneurship. However, because of lack in some required skills, particularly in ICT and Internet literacy and usage, senior-owned home-based businesses' growth might be slower than the one of the youth owners.

(b) Following the results of our exploration described in the previous part, policies for inclusive entrepreneurship should be focused particularly on resources which the studied segments miss for growth of home-based businesses and entrepreneurship. Thus, youth might need mainly financial resources, co-working spaces, etc., while for seniors it is inevitable to acquire ICT and Internet skills.

(c) Finally, following our results there is a room for intergenerational cooperation also through home-based entrepreneurship. Both segments own complementary resources which could be used for both sides' benefits and home-based business trend support.

The results of our exploratory study suggest several topics for further research. Among them, we encourage further research based on qualitative surveys of seniors and youth to learn more about their particularly missing resources, expectations, and conditions for intergenerational cooperation in home-based businesses. Also, it is important to better understand lifestyles of both generations and how the concept of home-based business fits them. Finally, based on all findings stated above, we suggest that home-based entrepreneurship is a good basis for the development of inclusive entrepreneurship policies for both youth and seniors as a new trend in working environment.

Acknowledgements This work was supported by the Slovak Research and Development Agency under the contract No. APVV-14-0647.

References

1. OECD/The European Commission. (2017). *Policy Brief on Home-Based Businesses*. Paris: OECD Publishing.
2. Loscocco, K., & Bird, S. R. (2012). Gendered paths: Why women lag behind men in small business success. *Work and Occupations, 39*(2), 183–219.
3. Ekinsmyth, C. (2014). Mothers' business, work/life and the politics of "mumpreneurship". *Gender, Place and Culture, 21*(10), 1230–1248.
4. Holliss, F. (2015). *Beyond live/work: The architecture of home-based work*. Abingdon, Oxford: Routledge.
5. Steenbeek, W., & Schutjens, V. (2014). The willingness to intervene in problematic neighbourhood situations: A comparison of local entrepreneurs and (un)employed residents. *Tijdschrift voor Economische en Sociale Geografie, 105*(3), 349–357.
6. Deschamps, M., Dart, J., & Links, G. (1998). Home-based entrepreneurship in the information age. *Journal of Small Business and Entrepreneurship, 15*(1), 74–96.
7. Wilson, G., Mitchell, H., & Hall, G. (2004). Home-based entrepreneurs. *International Journal of Entrepreneurship and Small Business, 1*(3/4), 249–254.
8. Wynarczyk, P., & Graham, J. (2013). The impact of connectivity technology on home-based business venturing: The case of women in the North East of England. *Local Economy, 28*(5), 451–470.
9. Mason, C. M., Carter, S., & Tagg, S. (2011). Invisible businesses: The characteristics of home-based businesses in the United Kingdom. *Regional Studies, 45*(5), 625–639.
10. Salazar, C. (2001). Building boundaries and negotiating work at home. In: *Proceedings of the 2001 International ACM SIGGROUP Conference on Supporting Group Work*, pp. 162–170. ACM SIGGROUP, Boulder.

11. Allinson, G., Braidford, P., Houston, M., & Ston, I. (2013). Understanding growth in microbusinesses [Online]. Retrieved February 2, 2018, from www.gov.uk/government/uploads/system/uploads/attachment_data/file/207646/bis-13-895-understanding-growth-in-microbusinesses.pdf
12. Walker, E. A., Wang, C., & Redmond, J. (2008). Women and work-life balance: is home-based business ownership the solution? *Equal Opportunities International, 27*(3), 258–275.
13. Anwar, M. N., & Daniel, E. M. (2016). The role of entrepreneur-venture fit in online home-based entrepreneurship: A systematic literature review. *Journal of Enterprising Culture, 24*(6), 419–451.
14. Loscocco, K., & Smith-Hunter, A. (2004). Women home-based business owners: Insights from comparative analyses. *Women in Management Review, 19*(3), 164–173.
15. Walker, E. (2003). Home-based businesses: Setting straight the urban myths. *Small Enterprise Research, 11*(2), 35–48.
16. Dwelly, T., Maguire, K., & Truscott, F. (2005). *Under the radar: tracking and supporting rural home-based business*. Penzance: Live Work Network.
17. Thompson, P., Jones-Evans, D., & Kwong, C. (2009). Women and home-based entrepreneurship evidence from the United Kingdom. *International Small Business Journal, 27*(2), 227–239.
18. Rehak, J., Pilkova, A., Marian, H., & Jančovičová, Z. (2017). Do senior entrepreneurs differ from youth entrepreneurs? Evidence from Global Entrepreneurship Monitor. In: *Proceedings of the 5th International Conference on Innovation Management, Entrepreneurship and Sustainability 2017*, pp. 836–848. VSE, Prague.
19. Hart, M., Anyadike-Danes, M., & Blackburn, R. (2004). *Spatial differences in entrepreneurship: A comparison of prime age and third age cohorts*. Belfast: Economic Research Institute of Northern Ireland.
20. Levesque, M., & Minniti, M. (2006). The effect of aging on entrepreneurial behaviour. *Journal of Business Venturing, 21*(2), 177–194.
21. Botham, R., & Graves, A. (2009). The grey economy: How third age entrepreneurs are contributing to growth. *NESTA Research Report.*
22. Weber, P., & Schaper, M. (2004). Understanding the grey entrepreneur. *Journal of Enterprising Culture, 12*(2), 147–164.
23. De Bruin, A. & Firkin, P. (2001). *Self-Employment and The Older Worker*. Working Paper. Massey University, Auckland.
24. Kilber, E., Thomas, W., Teemu, K., & Robert, B. (2011). (Work) Life after work: Understanding barriers to older entrepreneurship in London. In: *56th Annual ICBS World Conference*. ICSB, Stockholm.
25. Halabisky, D. (2012a). *Entrepreneurial activities in Europe—Senior entrepreneurship. OECD Employment Policy Papers, No. 2*. Paris: OECD Publishing.
26. Halabisky, D. (2012b). *Entrepreneurial activities in Europe—Youth entrepreneurship. OECD Employment Policy Papers, No. 1*. Paris: OECD Publishing.
27. Minola, T., Criaco, G., & Cassia, L. (2014). Are youth really different? New beliefs for old practices in entrepreneurship. *International Journal of Entrepreneurship and Innovation Management, 18*(2/3), 233–259.
28. Minola, T., & Giorgino, M. (2011). External capital for NTBFs: The role of bank and venture capital. *International Journal of Entrepreneurship and Innovation Management, 14*(2), 222–247.
29. Hytti, U., & O'Gorman, C. (2004). What is 'enterprise education'? An analysis of the objectives and methods of enterprise education programmes in four European countries. *Education and Training, 46*(1), 11–23.
30. Parker, S. C. (2006). Learning about the unknown: how fast do entrepreneurs adjust their beliefs? *Journal of Business Venturing, 21*(1), 1–26.
31. Baker, T., & Nelson, R. E. (2005). Creating something from nothing: resource construction through entrepreneurial bricolage. *Administrative Science Quarterly, 50*(3), 329–366.

32. Institute for the Future for Intuit. (2007a). Intuit Future of Small Business Report. First installment: Demographic trends and small businesses [Online]. Retrieved February 14, 2018, from http://http-download.intuit.com/http.intuit/CMO/intuit/futureofsmallbusiness/SR-1037_intuit_SmallBiz_Demog.pdf
33. Institute for the Future for Intuit. (2007b). Intuit Future of Small Business Report. Second installment: Technology trends and small businesses [Online]. Retrieved February 14, 2018, from http://http-download.intuit.com/http.intuit/CMO/intuit/futureofsmallbusiness/SR-1037B_intuit_tech_trends.pdf
34. GERA. (2018). *Global Entrepreneurship Monitor 2017/18 Global Report*. London: Global Entrepreneurship Research Association.
35. Institute for the future/INTUIT (2010). Intuit 2020 report. Twenty trends that will shape the next decade [Online]. Retrieved February 14, 2018, from https://http-download.intuit.com/http.intuit/CMO/intuit/futureofsmallbusiness/intuit_2020_report.pdf

Smart Solutions for Big/Small Cities

Michal Balog, Angelina Iakovets, and Lucia Knapcikova

1 Urban Traffic

1.1 Safety of Urban Movement

Road safety and its continuity are important parameters and the main driving force in the development of innovative logistics solutions for urban supply and the overall quality of social life. In the logistical system of the state the flow of public and passenger traffic, transport access throughout the territory, and quality of the transport infrastructure are the most important factors that attract investors in the given territory and limit the economic efficiency of the entire state. The annual growth of road transport only over the past year shows an increase of 100,000 vehicles in the territory of Slovakia; unfortunately, old cars are practically not liquidated, and so there is a huge traffic overload, especially during rush hours. We can see these statistics from Fig. 1.

The rise of vehicle collisions that occurred on a certain section of the road starts a partial or complete stoppage of traffic which threatens the normal functioning of the city traffic, the normal movement of fire engine and medical transport, the bus line, the movement of personal transport, etc. As a result, the accident not only leads to direct damage of property, but at the same time it creates a risk of material and nonmaterial damage. Community devises methods and various measures to reduce the number of traffic accidents and subsequent overloads by paying special attention to informative-educational activities and also to the system of measures of state influence. Despite efforts to eliminate accidents, which are caused by driver's

M. Balog · A. Iakovets · L. Knapcikova (✉)
Faculty of Manufacturing Technologies with a Seat in Presov, Technical University of Kosice, Presov, Slovak Republic
e-mail: michal.balog@tuke.sk; angelina.iakovets@tuke.sk; lucia.knapcikova@tuke.sk

© Springer Nature Switzerland AG 2020

D. Cagáňová, N. Horňáková (eds.), *Mobility Internet of Things 2018*,
EAI/Springer Innovations in Communication and Computing,
https://doi.org/10.1007/978-3-030-30911-4_30

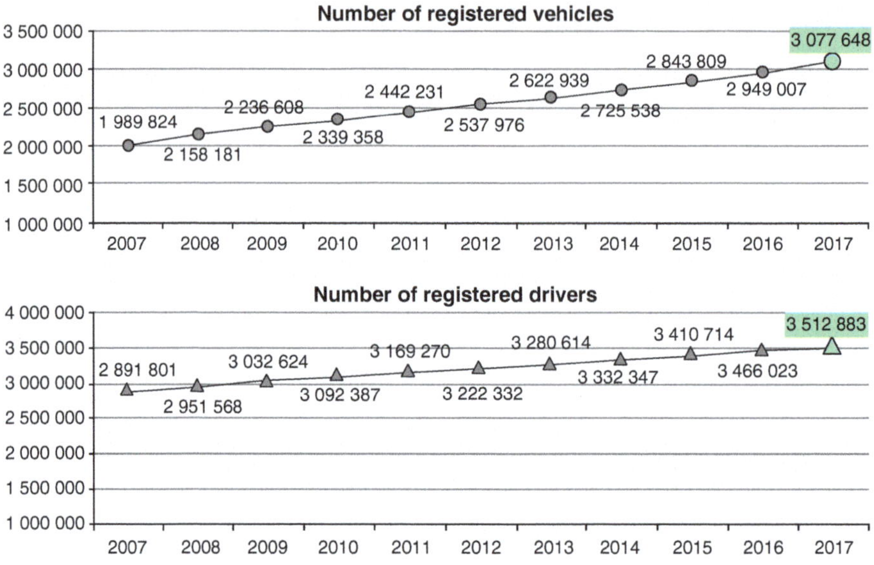

Fig. 1 The number of registered drivers and motor vehicles in the Slovak Republic for the period 2007–2017 [1]

Fig. 2 The number of fatalities on the road in the EU (2017) [2]

fault, the reduction of these accidents is still at an insufficient level. To eliminate the disasters caused by the human factor, information technologies are being involved in the life of the motorists' community in order to help the participants of the urban movement better orient themselves in the urban agglomeration.

Traffic accident statistics show that the riskiest sections of the road are urban areas and rural roads, where the recommended speed is 80 and 50 km h^{-1} (see Fig. 2).

Over the past few years, researchers have carried out statistical calculations that showed which groups of road users, in percentage terms, most often died in fatal accidents (Fig. 3).

The pie chart shows that the majority of the participants in accidents are in cars, and account for 40% of the total road users, and pedestrians—21% of the total road users. The main causes of traffic accidents:

- The quality of the road surface
- Neglect of traffic rules by all road users

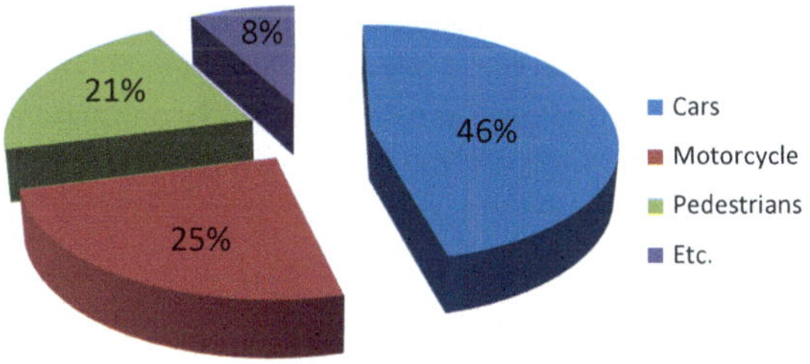

Fig. 3 Mortality from road accidents in the EU by groups of participants in the road movement [2]

Table 1 Comparison of the number of fines awarded by traffic police for individual offenses [3]

Violation/year	2017 ½	2018 ½	Difference
Exceeding the speed limits	64,012	103450	39438
Not giving priority to driving	1916	1540	−376
Alcoholic intoxication	2230	2050	−180
Cyclists for alcoholic intoxication	1345	1012	−333
Bad technical condition	7986	6841	−1145
Unfastened seat belts	9864	7845	−2019
Amount	87,353	122,738	35,385

- Driving under the influence of alcohol, psychotropic substances, drugs, fatigue
- Unreasonably risky behavior and a conscious violation of traffic rules
- Speed limit exceeded
- Loss of vigilance (i.e., distraction for passengers, gadgets, drinks, or food)
- Unpredictable behavior in extreme situations
- Nonobservance of distance
- Violation of the rules of overtaking
- Crossing the place, not designated for pedestrians [4]

1.2 Technical Solutions for Safety Systems in the Automotive Industry

Nowadays, the automotive industry is trying to solve the problems of road safety by developing programs which will signalize the vehicle user about the situation on the road by control unit of the vehicle. Similar programs are mainly focused on reacting to a collision object at a distance no further than 10 m. Together with the automotive industry, this problem is also considered by a group of navigational programs that warn the user about the need to respond to changes in traffic conditions along the chosen route. Despite the fact that there are different types of driving assistance systems, it does not help to reduce the total number of fines paid for traffic violations (Table 1).

According to the table mentioned above, it can be seen that the total number of fines increases, which means that the existing programs cannot completely solve the problem of road accidents. By 2018, cars had active safety systems and the main purpose of the systems is to prevent an emergency. In this situation, the system independently assesses the likely danger and prevents the collision by active intervention in the driving process of the vehicle.

The most popular active security systems are:

- Antilock braking system
- Anti-slip system
- The system of exchange rate stability
- Brake force distribution system
- Emergency braking system
- Pedestrian detection system
- Electronic differential lock [3, 5]

These active safety systems are constructively connected and work closely with the vehicle's braking system and significantly increase its efficiency [3, 6]. In addition to timely alerting of the driver before the onset of possible hazards, the systems also actively intervene in driving and use the braking system and the steering system. Auxiliary active security systems include:

- Parking system
- System of the circular survey
- Adaptive cruise control
- Emergency steering system
- A system for assisting movement along the lane
- A system of assistance in rebuilding
- Night vision system
- Road sign recognition system
- Driver fatigue control system
- Descent assistance system
- Lifting assistance system, etc. [3]

The average position between active and passive security systems is provided by preventive security systems. Depending on the construction of a particular system, the following functions can be implemented:

- Warning the driver of the danger of collision
- Preparation of the braking system for emergency braking
- Activation of individual passive safety devices
- Partial or full automatic braking

Nowadays, automobile security systems are quite widespread and are actively being introduced into cars. Known security systems are the following:

- Pre-sense front, pre-sense front plus, and pre-sense rear by Audi
- Pre-safe and pre-safe brake by Mercedes-Benz
- Collision mitigation braking system, CMBS, by Honda
- City brake control by Fiat

- Collision warning with brake support and forward alert by Ford
- Forward collision mitigation, FCM, by Mitsubishi
- Pre-collision system, PCS, by Toyota
- Front assist and city emergency brake by Volkswagen
- Collision warning with autobrake and city safety by Volvo
- Predictive emergency braking system, PEBS, by Bosch [3]

The principle of operation of these systems can be understood by analyzing the principle of the work of some of them. For example, pre-safe brake from Mercedes-Benz (pre-braking system) uses radar to determine the critical situation. It operates at a speed of 30–200 km h^{-1} and scans 200 m ahead of the car (more parameters in Table 2).

The principle of the system includes the following actions, presented in Table 2.

Collision mitigation braking system from Honda using radar at speeds over 15 km h^{-1} and a distance of up to 100 m fixes moving and standing cars (motorcycles). The operation of the CMBS system is similar to the pre-safe brake system and includes the following (Table 3).

From these two tables, it follows that all systems start responding to the object not earlier than 3 s before the accident [7]. Statistics of the accidents across Europe, in recent years, shows that these security systems are not perfect and do not operate in full capacity of their functions.

Table 2 Safe brake (Mercedes-Benz) [3]

Condition	Action
Obstacle detection (car, person) in the space in front of the car	Time calculation of a probable collision
2.5 s before collision	Three sound signalizing
1.6 s before collision	Partial automatic braking (40% of the maximum braking pressure)
	Seat belt tension
The driver reacted and pressed the brake pedal	Maximum brake pressure
The driver reacted and turned to another lane	Brake pressure reduction
0.6 s before collision, the driver does not respond to warnings	Automatic maximum brake pressure

Table 3 Collision mitigation braking system from Honda [3]

Condition	Action
3 s before collision with an obstacle	The delivery of sound and light signals about the danger
2 s before the collision, the driver does not respond to warnings	Three sharp jerks of a safety belt of the driver
1 s before the collision, the driver does not respond to warnings	Partial automatic braking, seat belt tension
The driver reacted and pressed the brake pedal	Maximum brake pressure

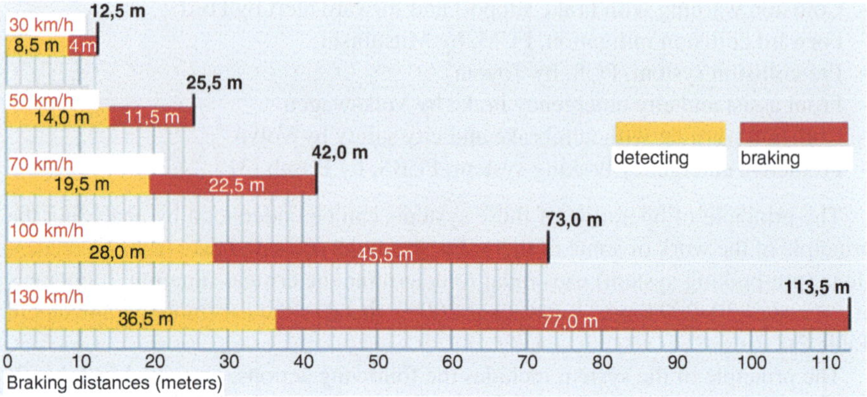

Fig. 4 Braking distances [7]

1.3 Suggested Solutions for Intelligent Automatic Identification Systems Used in Urban Traffic

Investigation of the braking distance at different speeds of the car showed that the braking distance is measured not only at the brake distance from the moment the brake pedal is pressed, but also when the driver detects obstacles on the road (see Fig. 4).

The figure shows that if vehicle goes with the recommended speed in the city km h^{-1}, the stopping distance at standard conditions (high-quality road surface and good weather conditions) will be 25.5 m, and when its speed is 130 km h^{-1} on motorways, the stopping distance will be about 113.5 m. Such conditions require an immediate response from the driver of the vehicle and from the vehicle safety systems. From the analysis of statistical data, it follows that the most endangered road users are pedestrians, cyclists, invalids, and domestic animals. Currently, traffic agency has obliged pedestrians to wear reflective bracelets, which by their lighting warn the driver about the moving pedestrian along a poorly lighted part of the road (Fig. 4).

2 RFID Technology in Urban Traffic

Today, there are a huge number of technologies which help users to read information in many ways; one of such methods is the technology radio-frequency identification (RFID). RFID is a method of automatic identification of objects in which data are read or written down through radio signals in transponders or RFID marks. Any RFID system consists of a reader (reader or interrogator) and a transponder that is also called an RFID tag.

By reading range, RFID systems can be divided into systems:

- Near identification (reading is carried out at a distance of up to 20 cm)
- Identification of medium range (from 20 cm to 5 m)
- Far identification (from 5 to 300 m) [8]

One of the possible solution of a problem is long-range identification is radio-frequency identification (RFID) technology, which allows recording objects equipped with active RF tags at large distances (up to 50 m) from the recording (reading) device. Today, there is a specific proposal for solving road accidents—"The Road Signs Identification System with the help of RFID technology" [6]. The principle of such a system is based on long-range identification, where RFID scanner will scan the code from the road sign and notify the driver by vehicle control system about approaching to the designated road section. The basic principle of proposed warning system through the vehicle control system can be observed in Fig. 5.

There is a possibility to expand this RFID alarming system by applying the opportunity to read RFID tags from pedestrians, cyclists, disabled people, and domestic animals. Behavior of this group of road users is unexpected and that's why it is necessary to follow them. Slovak road administration statistics show that the number of traffic accidents practically does not decrease [9]. The subject of the design of the investigators is to include RFID tag in safety bracelets, which would be

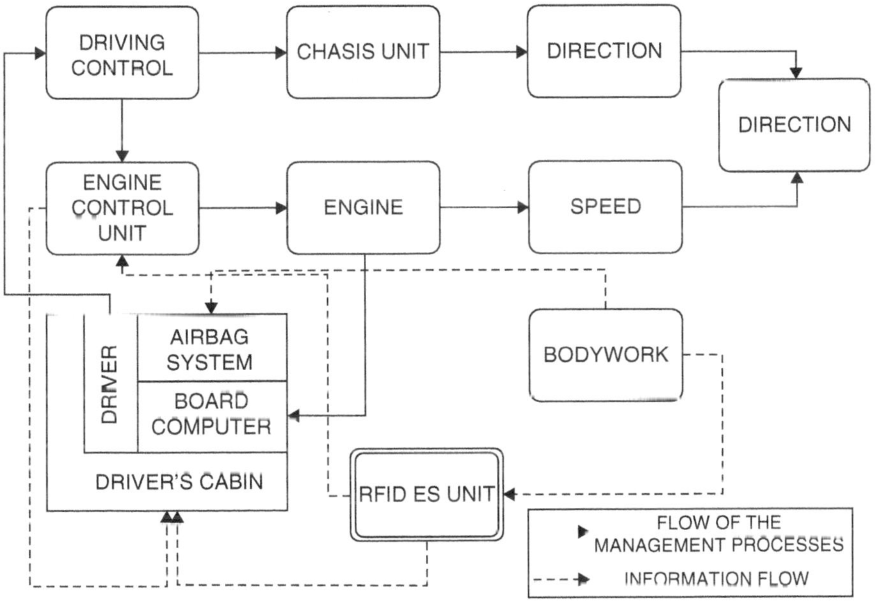

Fig. 5 The current concept of selected elements of a motor vehicle and their activity in interaction with other elements [3]

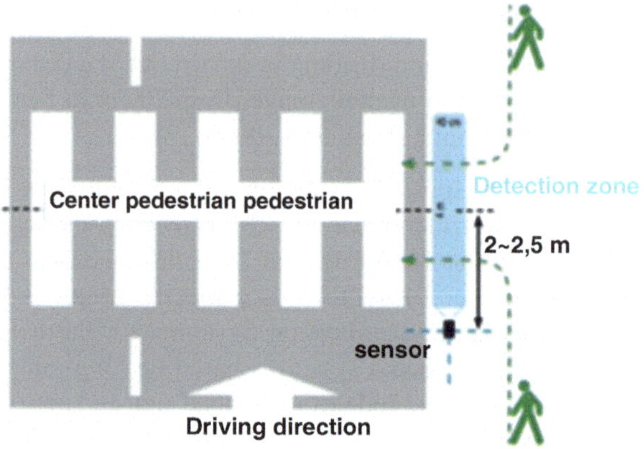

Fig. 6 The detection area of the built-in car security [7]

Fig. 7 Reflective RFID bracelets and accessories [10–12]

legible in every situation and every weather and inform the driver about the possible pedestrian streets. This alarming RFID system is also easy to apply to identification cyclists, animals, and wheelchairs on the road. It is proposed to mount the RFID chip in the reflective bracelet. The bracelet is expected to be effective in any weather conditions also it should be visible for the vehicle system on long-range distances (Fig. 6).

The introduction of a long-range RFID identification system is necessary, since with the proposed scheme and the analysis carried out, it follows that most of the embedded systems react to the obstacle at a distance of up to 3 m. Reflective bracelets (Fig. 7) daily improve through the installation of LEDs in them, changing the shape of the bracelet to other more compact things.

Basically, the bracelets (see Fig. 7) are not visible under clothing or bags and thereby efficiency of the bracelet decreases on the low-light areas of the road. The proposed reflective RFID bracelet, even in the conditions of poor visibility, would give the opportunity to identify a pedestrian, even if the bracelet is in bag or pocket (Fig. 8).

The offered bracelet will be read by the installed receiver in the car and receive the notification code through the car control system to the driver (see Fig. 9).

In the case of installing a powerful receiver antenna to the car system, it will be possible to detect the bracelet at distances of about 300 m (Fig.10) [8].

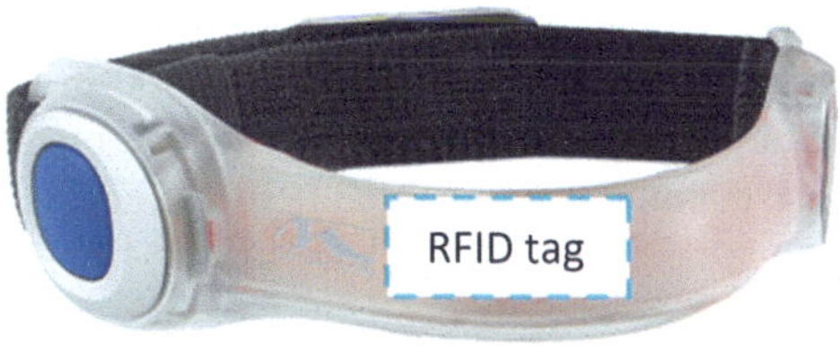

Fig. 8 Reflective RFID bracelet [12]

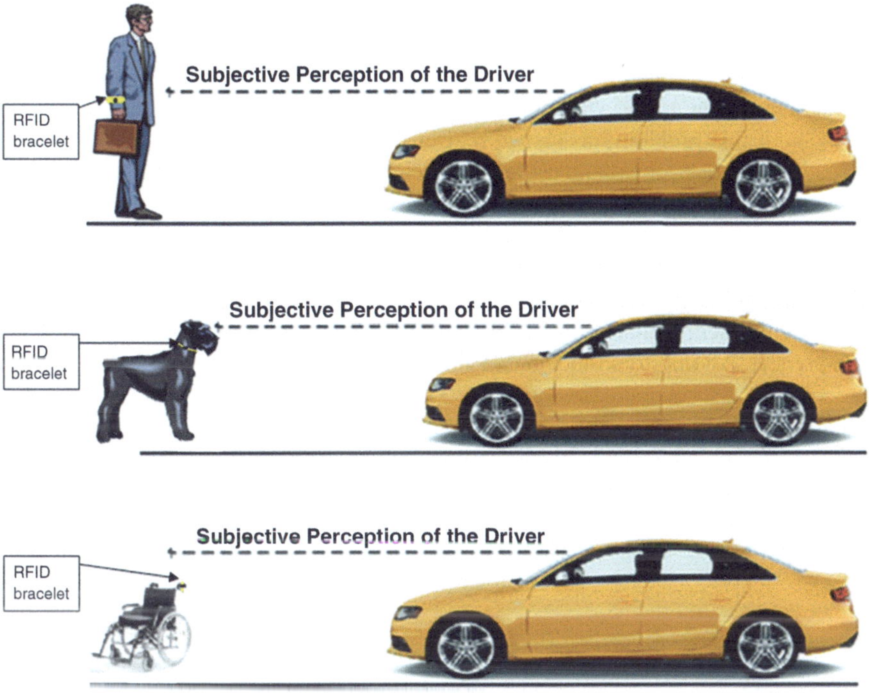

Fig. 9 The principle of reading the reflective RFID bracelet [7]

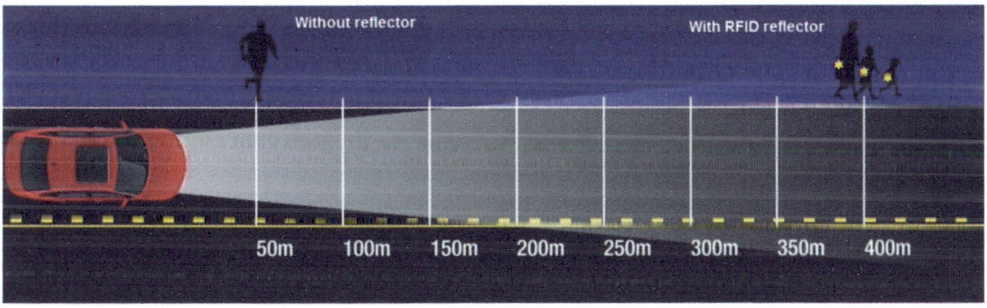

Fig. 10 Model of far identification [13]

The scheme shows that the reflective RFID bracelet will be identified by the system at a distance of more than 300 m under the conditions of using the high beam headlight and a visible reflective bracelet with RFID chip; compliance with these conditions will help to best implement long-range identification.

3 Conclusions

Long-term road safety studies show often recurring as well as new causes of accidents on roads. Despite the fact that Slovakia has the strictest legislation in the field of road regulation, traffic violations remain a frequent cause of accidents on roads. Analysis of this problem has shown that in the present time the most risky group of road users are pedestrians, cyclists, wheelchairs, animals. The relevant ways of preventing traffic accidents still cannot fulfill their function and therefore one of the solutions today is the introduction of reflective RFID bracelets on the market. The RFID chip in the bracelet on moving object over the city communications will send a radio signal through the receiver and will send information to the vehicle control system which will signalize driver by audio or video signal about the approaching to the object. An important fact is that similar alarms will be highly effective for all categories of drivers, including disabled. The advantage of the proposed technology is that active RFID chips will be legible even in bad weather (rain, snow, ice, dust) or if the reflective bracelet is hidden in the clothing or in the bag.

Acknowledgments Research reported in this chapter was supported by EU Structural Funds within the project "Promotion & Enhancement of Centre for Research on Transportation" ITMS code 26220220160.

References

1. Ministry of the Interior of the Slovak Republic. Retrieved September 5, 2018, from https://www.minv.sk/?statisticke-ukazovatele-sluzby-dopravnej-policie
2. Valette, L. (2017). 2016 road safety statistics: What is behind the figures, European Commission Press releases database.
3. Balog, M., Szilagyi, E., & Mindas, M. (2014). Traffic signs in urban logistics with the use of RFID technology. *Smart City 360°, 166*, 584–591.
4. TSN. The main causes of crashes, Retrieved September 5, 2018 from https://ru.tsn.ua/auto/obzory/7-osnovnyh-prichin-dtp-418849.html
5. Bing, G., & Jie, G. (2014). Research of electro-mechanical actuator in electric braking system based on instantaneous torque control with SRM. *The Open Automation and Control Systems Journal, 6*, 519–527.
6. Utility pattern PUV 50009-2016 Identification system for vertical traffic signatures based on RFID technology. Retrieved September 10, 2018, from https://wbr.indprop.gov.sk/WebRegistre/UzitkovyVzor/Detail/50009-2016
7. Systemsauto.ru. Retrieved September 5, 2018, from http://systemsauto.ru/active/preventive_safety_system.html

8. Dudkin, I. A., Fedorov, T. V. (2015). Avtoplas FGBOU VPO, RFID system design, Gosuniversitet—UNPK.
9. Slovak road administration. Retrieved September 10, 2018, from http://www.ssc.sk/sk/novinky/bezpecnost-cestnej-premavky-2018.ssc
10. Reflector. Retrieved September 10, 2018, from http://www.reflexne.sk/reflexne-pasky-na-ruku/
11. Promotion. Retrieved September 10, 2018, from https://www.mpromotion.sk/katalogove-reklamne-predmety/sport-zabava/darceky-pre-deti-hry/reflexne-privesky-na-kluce-mobil
12. Insportline.sk. Retrieved September 10, 2018, from https://www.insportline.sk/reflexne-naramky-a-vesty?page=1-2
13. Kamenaric.si. Retrieved September 10, 2018, from https://kamenaric.si/-avto%20moto/odsevnik_kresnicka.htm

Digital Transformation in the Healthcare Sector: Empirical Evidences of IoHT Benefits and Limits on Chronic Disease Management

Gabriele Palozzi, Daniele Binci, and Irene Schettini

1 Introduction

Digital transformation (DT) allows organisations to radically improve their performance [1] by providing consistent opportunities to disrupt the value creation process [2]. This new conceptualisation is based on technologies such as cloud computing, analytics, social media, mobility and ERP technologies. DT aims to radically improve the reach of organisations by changing business processes and creating new customer and citizen relationships. It is changing the overall society and, progressively, the most important industries such energy, transportation, industrial goods and healthcare.

Particularly, within the healthcare sector, DT of clinical data (medical and health records) [3] consists of the data storage and exchange throughout a combination of big data, cloud computing and business/predictive analytics that leads to vertical and horizontal integration among business units, processes and structure within the organisation [4].

Therefore, hospitals have the possibility of re-thinking service management and delivery [5], by the use of new possibility provided by DT, like IoT applications, and particularly wearable and wireless devices as well as smartphone application [6].

Thus, the remainder of the chapter is organised as follows: Section 2 addresses the research background by providing an overview about benefits and limits of IoT adoption on the healthcare sector (IoHT). Section 3 explains the research problem we addressed through the chapter, by clarifying the research question. Section 4 presents the methodology, while Section 5 highlights the main findings, about benefits and limits of IoHT perceived by practitioners in relation to their real-life

G. Palozzi (✉) · D. Binci · I. Schettini
Department of Management & Law, University of Rome Tor Vergata, Rome, Italy
e-mail: palozzi@economia.uniroma2.it

© Springer Nature Switzerland AG 2020

D. Cagáňová, N. Horňáková (eds.), *Mobility Internet of Things 2018*,
EAI/Springer Innovations in Communication and Computing,
https://doi.org/10.1007/978-3-030-30911-4_31

experience. Section 6 analyses and discusses findings, by underlining the potential impact of IoHT in chronicity management. Finally, Section 7 provides final remarks and suggestions for further researches.

2 IoT and Healthcare: Research Background, Benefits and Limits

Healthcare service is one of the most fertile areas for the IoT implementation [2], due to economic pressures of increasing costs, ageing population and rising of chronic diseases (such as cancer, cardiovascular diseases, diabetes, respiratory failure). Such diseases are crucial matters for healthcare services because they represent one of the first causes of disability, mortality and morbidity [7].

Particularly, IoT, hereafter IoHT (Internet of Healthcare Things), applications and devices, by allowing the connection of anyone, anytime, anywhere, to any service, devices and network [8] can contribute to completely re-define how devices and people (clinicians and patients) interact with each other.

IoHT is also one of the most effective tools for the pressures towards healthcare DT process in terms of quality, cost reduction and patient safety improvement [9], such as lower mortality rates [10–12], higher vaccination rates [13] and higher use of recommended procedures [14–16]. According to current healthcare policy and the "Triple Aim", it allows population service access and universality [17], cost control and patient experience [18].

The introduction of the IoHT can significantly reduce the costs because services can be provided more efficiently, by the scheduling of the usage of the scarce resources on the base of information flow detected by sensors. For example, remote monitoring can cut down unnecessary physical doctors' visits by exploiting the real-time connectivity.

Moreover, digital medical instruments and sensors allow the real-time and continuous access to patients' data by providing better quality information about disease management. This means the possibility to manage accurate data collection, reduce human-based errors and increase the accuracy of treatments with timely health also in terms of hospital stays, by improving patients' experience. This also helps to make predictive actions, optimise drug administration and finally decrease costs.

Over the last years, the steadily ageing population and the increase in chronic illness have dramatically increased the need for professionals, such as nurses and physicians. The scarcity of resources has generated the necessity of creating a new paradigm in order to reduce the physical workload, by fostering interaction between actors involved in the service delivery. This new paradigm should consider an efficient scheduling of resources, also through IoHT, that makes available improved and sustainable healthcare services.

The most important advantages of IoHT [18] are the continuous assistance to patients; the context awareness, as it takes into account both the variables related to the patient conditions and the environmental factors, giving a complete overview of the patients' condition; the learning capacity, a characteristic that is derived from the big amount of data collected by the sensors [19]; the user-friendly approach, since the technologies employed are designed to be easily used without a specific training; the cost-pull logic, due to resource reduction, used only when they are necessary; and the healthcare accessibility, as IoHT allows also the patients living in rural areas to have access to the medical care, in an effective and timely manner.

Some authors [20] have explained which key technology layers refer to IoHT (Table 1) and whose contribution enhances new models for healthcare delivery in terms of service and applications.

Tables 2 and 3, based on [21], show main service and related applications enabled by IoHT.

Literature states that IoHT adoption has some barriers related to human resource issues (i.e. staff skills), legal and ethical factors (i.e. liability, confidentiality, data protection and patient privacy), and culture norms and practices (i.e. doctors' expectations regarding face-to-face patient interaction), along with organisational business strategy [20]. Moreover, such literature often refers to challenges related to healthcare sector implementation; however, there are few empirical evidences about benefits and limits, especially in terms of empirical field studies.

Some authors [8, 18] highlight the main IoHT applications in service delivery. They principally focus on the opportunities and challenges that technologies provide to healthcare organisations, practitioners and patients.

Opportunities refer prevalently to the radical innovation that data digitalisation of healthcare management could enable by enhancing "proximity" between patients and clinicians and their relationship. These regard prevalently the three main following aspects:

1. The clinical area relates to the possibility of remote monitoring vital signs of patients, making them supervised without long interruption and checking also their adherence to clinical follow-up and drug therapy ingestions. This clearly enhances the patients' life quality, in terms of wellness, QALY and life expectancy.
2. The organisational area relates to the impacts of IoHT on the internal process in healthcare delivery. They regard the way in which information and hardware technologies, connected to each other, transmit data able to change the behaviour of both clinicians and patients in their role within healthcare delivery. This leads to change in the necessary activities of health services providing and often the time burden request in patients and/or disease management.
3. The economic area, finally, regards the financial impacts of this new patient and/or disease management triggered by IoHT involvement. If the time burden decreases and patients acquire an active role in their healthcare pathway, the charge of costs is going to decline. Likewise, if the patient's life quality is

Table 1 Key technologies for IoHT layers

Layers	Categories	Goals	Examples
Sensing and identification *Collecting effectively and efficiently a variety of types of personalised health and medical information*	Wearable sensor	Get human health information	– Systolic and diastolic blood pressure – ECG stream – Linear acceleration of movement – Eye movement, etc.
	Ambient sensor	Get information about environment	– Indoor/outdoor temperature – Light on/off state, etc.
Networking *Connecting all devices and allowing personalised health data to be collected, stored, transmitted, shared and aggregated*	Topography	Be a heterogeneous computing grid for collecting enormous amount of vital signs and health data	– "All-in-one" multi-brand device telemedicine platform
	Architecture	Integrate heterogeneous systems or devices. It needs to treat communication sensors, Web services, health application, etc.	Integration between: – Master (professional) – Server (data management and analysis) – Things (connectivity devices)
	Security and privacy	Provide a secure access to e-processing of the data and avoid malicious attacks and misbehaviour	– Cryptographic system – Digital signature system – Protection data
Data processing *Processing a variety of complex health-related data*	Data-driven approaches	Use a large volume of health-related data from different subjects for training general model	Scientific evidence background
	Knowledge-based approach	To represent and transfer knowledge from human expert into computer algorithms to establish computer-aided decision support system	Clinical experience background
	Hybrid	Combine data-driven and knowledge approach to take the advantage of advantages in semantic reasoning with probabilistic models	Machine learning

Source: Own elaboration from Qi et al. [20]

Table 2 IoHT services

Service	Description	Ref.
Ambient-assisted living (AAL)	Implemented for elderly or infirm patients that have to be remotely controlled. Based on NFC and RFID technologies, this platform has been improved through elaborated algorithms that can detect some common problems before they occur.	[22]
Wearable device access (WDA)	Used for many medical applications and for multiparametric checks. All IoT wearable devices have to be integrated ad interconnected.	[23]
Internet of m-Health Things, Iom-HT	Based on the usage of mobile computing connected to sensors that communicate through mobile networks (indeed, it is Internet based).	[24]
Community healthcare (CH)	Useful to create communities around hospitals, or residential areas. This contributes to create "virtual hospital" and exchange information in an energy-efficient way, with lower costs.	[25]
Indirect emergency healthcare (IEH)	Used in all the situations of danger such as accidents or natural disasters. In all these cases the IoT technologies can be a support for its characteristics of information exchange and record keeping.	[26]
Semantic medical access (SMA)	Useful for data and information exchange. The infrastructure of cloud technology is appropriate for making accessible all the data stored and processed.	[27]
Embedded context prediction (ECP)	Used to predict the future change/evolution of a disease. The ECP develops a framework for all the healthcare applications to allow a timely diagnosis of the future health problems.	[28]
Embedded gateway configurations (EGC)	Used to integrate and connect nodes, Internet and medical equipment. It is a fundamental service for the implementation of the whole structure of a service, mainly for all the mobile applications.	[29]
Adverse drug reaction (ADR)	Useful to try to prevent and solve problems related to the ingestion of a wrong dose or combination of drugs. The PDA allows patients to use the right amount of drug according to their electronic heath record.	[30]
Children health information (CHI)	Developed in order to increase the awareness about different issues concerning children. This service is not important only for helping children with health problems but even for making their habits healthier (such as the nutritional habits or the prevention in general).	[31]

Source: Own elaboration from Islam et al. [21]

improved, the resources for managing those situations in which patient's health status gets worst decrease; it means that potential future costs decrease as well.

Limits, even risks related to IoHT usability, refer to implementation setbacks, usability and technology trustworthiness in the real context, once, for example, they are directly by patients' arms.

These aspects regard prevalently the three main following aspects:

1. The technical limit regards the complexity of technology usage, installation and its intuitiveness-usability. Very often, in fact, IoHT should be directly used also

Table 3 IoHT applications

Application	Description	Ref.
Biomedical parameter monitoring	Possibility to remotely monitor some biomedical parameters of patients, i.e.: – ECG: It monitors the electrical activity of the heart – Temperature: Vital sign used to monitor heat stroke and fever – Oxygen saturation: It measures the percentage of oxygen in the blood – Blood pressure: It measures systolic and diastolic blood pressure – Glucose level: It measures the level of glucose into the blood; particularly important for diabetic management	[32, 33]
Medication management	Suitable for several occasions such as the rehabilitation for hemiplegic patients and the language training for children with autism. The IoT can be the right platform to make it easier to connect the resources and create a smart city for medical rehabilitation.	[34]
Wheelchair management	Intelligent wheelchairs that make the life of people with disability easier. These new chairs are provided of several sensors that not only increase the stability of the wheelchair, but also monitor the health condition of the patient.	[35]
Healthcare solutions using smartphones	Availability of different apps on smartphones has improved our everyday life, making accessing to several services easier. The applications can be divided into diagnostic apps, drug reference apps and literature search apps used for giving information about the treatments. Clinical communication apps have a fundamental role in simplifying the exchange of data between the actors of the IoT network.	[36]

Source: Own elaboration from Islam et al. [21]

by not healthcare professionals (patients and/or their caregivers), and this is not easy for "non-trained" people. Then, the presence of many manufacturing brands makes the information exchange very difficult. Often, different devices, involved in the same healthcare pathway, cannot communicate with each other due to "distinctive" business policy that makes incompatible devices with other data management platform.

2. The limit of effectiveness relates with the lack of strong evidences of IoHT employment efficacy. This limit is mostly due to the fact that these technologies are at the beginning of their life cycle, and not so many trials have been successfully concluded yet. This brings to an impossibility to declare their complete reliability in the real context. Besides, the continuous misclassifying between medical devices (CE stamp, FDA declaration of conformity, etc.) and wellness devices (i.e. smartwatches, smartphone apps) contributes to feed doubt about IoHT affordability within users.
3. The security limit, finally, regards mainly the protection of health and sensible information. These limits refer to the architecture of data mining, the way in which data are collected, registered and stored; the way in which date are sent,

Table 4 Benefits and limits of IoHT: interpretation of background

Benefits		Limits	
Issue		Issue	
Clinical	• Adherence to monitor follow-ups/drug therapy • Improving quality of life	Technical	• Integration of "multi-brand" devices • Not always user friendly • Necessity of patients' training and ability for using
Organisational	• Time burden on hospital staff • Service change	Effectiveness	• No general reliability • Substantial difference between medical devices and wellness apps
Economic	• Staff and hospital cost reduction	Security	• Privacy protection • Data storage • Authentication of users • Low level of encryption

Source: Own elaboration

received and processed; and the way to encrypt and unencrypt them, in order to make them available also for the owner and the one responsible of their management. This is also in the lens of the General Data Protection Regulation (GDPR) in effect since May 25th, 2018.

Accordingly, on the base of feature of challenges and opportunities claimed by these authors, Table 4 summarises the benefits and limits, clustered for managerial issues related to the healthcare delivery.

3 Research Problem

The healthcare sector, differently from other ones like media and retail sectors, has been classified into "*late (DT and IoHT) adopters*" by being just at the beginning [37]. Moreover, in such a scenario, Italy is considered "*still in the early stages*" [38]. In fact, as highlighted by literature, several hospitals do not consider healthcare digitalisation as a priority, by underestimating the benefits associated with the digitalisation [2]. Little is known about how digitalisation impacts hospital performance—particularly, in terms of clinicians' job, patients' experience and cost rationalisation [2]. In order to understand more about this gap, our research question is the following:

> What are the main benefits and limits of IoHT applications within a "*traditional sector*", as Healthcare, in chronic disease management?

Referring, in particular, to the chronic cardiac disease field (e g. heart failure, arrhythmia), the aim of this chapter is to discover how practitioners consider the introduction of IoHT technologies in their clinical practice; which are the positive impacts both for patients and clinicians; and which are the threatening ones.

4 Method

In order to understand benefits and limits of IoHT, we adopted the case study method, an "empirical inquiry that investigates a contemporary phenomenon within its real-life context" ([39]: p. 16) on a pilot project.

This study wants to intercept some information-rich key informants, involved in critical cases of IoHT implementation in remote management of cardiac chronicity. This is coherent with the "extreme case sampling" [40] category, in which well-situated and competent people involved with a specific and particular service are listened to. The pilot case study focuses on the new post-discharge process for HF patients' follow-up [41, 42], designed by the Cardiology Department of the Hospital Policlinico Casilino of Rome, whose goal is the introduction of a new chronic care model (CCM) [43, 44], based on IoHT, finalised to improve the patient's quality of life and to lower the costs for the National Healthcare System (NHS).

The cardiology department, after many years' experience on cardiac markers and patterns [45], has developed a follow-up model for the early detection of heart deficiency symptoms that forecast acute cardiac implication and then future hospitalisations [46]. Accordingly, the pilot case is based on open, semi-structured interviews to key physicians belonging to the department, involved with different roles in the cardiac-chronicity management, in the design of the service and in the choice of technologies used. For this reason, the interviewed was subject with the following position within the organisation:

- The Head of the Cardiology Department
- The Coordinator of eHealth Centre of the Cardiology Department
- General Physician of the Cardiology Department

After each quotation in the Findings section, we show a code (letter and progressive numbers) in parentheses to distinguish the interviewees: D1, D2 ... where D1 represents Doctor 1 and so on. According to the qualitative content method, we firstly set the framework for the analysis by deriving the interview questions from the literature on IoHT implementation (Table 5) into the following main topics:

1. Cultural issues in IoHT implementation
2. IoHT implementation drivers
3. Impact of IoHT on decision-making
4. Main IoHT perceived outcomes

We asked participants to talk about such open topics. Then, we imported to NVivo all the transcripts and coded them for each of the four topics. We used sentences, containing aspects related to IoHT implementation, to analyse the data on the basis of such portions of text. Once the coding process was carried out, we sorted and analysed all the portions of text coded. We also analysed the additional secondary sources that would aid us in clarifying and reconstructing the salient events to develop a comprehensive picture of IoHT implementation project (e.g. internal guidelines of Policlinico Casilino inspired by the Italian Board of

Table 5 Main themes based on empirical findings

	Empirical subthemes	
Main themes	Clinicians/hospital	Patients
Cultural issues	Personal attitude (i.e. doctor's sensitivity) Ageing Past experiences and fear for the unknown	Patient literacy Device usability Ageing
Implementation drivers	Marginal economic aspect Data and knowledge management Importance of economic and learning factor Improvement in quality of work Chronic disease typology	Easy access to care and Improvement of quality of life
Decision-making	More autonomy Improvement in the preventive decision Decision-making process more participated	No impact on the patient Active and aware patients with regard to the illness
Perceived outcomes/setbacks	Fewer errors, efficiency and communication Data management Lower costs and higher quality Risk of workload Impact on job (i.e. job loss)	More assistance Technical complaints Feedback perceptions

Source: Own elaboration

Health and Heart Rhythm Society guidelines [47, 48]). Then, we analysed data by generating a large quantity of codes by assigning labels to sections of text that have some distinct meaning in order to categorise, organise and interpret them.

5 Findings

Related to our interviews we derived around 100 raw concepts as free NVivo nodes (approximately 40 pages). This process was finalised to build and refine categories into more specific concepts (Table 5).

In terms of cultural issues, we asked doctors to explain the main cultural gap related to the adoption of IoHT. From the doctors' point of view, we found age and personal attitudes as main issues. One doctor said: "*these new methods [*IoHT*] are considerably increasing and the use depends on the sensitivity of the doctor, technician and all the healthcare staff*" (D3). A similar result arose from the patients' point of view, where it was highlighted how it mainly depends on educational aspects (patients literacy, ageing). One doctor said: "*The level of schooling for patients [is important], the less they are schooled, the more they are wary of IoHT*" (D2).

Data management as well as job quality seems to be the most relevant factor considered as the IoHT driver. When we asked doctors about implementation drivers of IoHT, we found as main topics knowledge management issues and a better job quality. A doctor said: "*[an important driver] in my opinion is related to the possibility of making diagnosis earlier than the usual method, and have a better result from the clinical point of view, in terms of data patient management*" (D4).

In terms of economic motivation, there is quite consensus about a positive pressure related to IoHT adoption; however only one doctor highlighted and stressed such aspect as a fundamental driver: "*the economic aspect is important as motivating factor*" (D5).

From the patients' point of view, we found that the main driver is related to the easy access to care for the patient, the improvement of its quality of life and the perception of a good service: "*patient feels more protected and followed by someone*" (D4).

All the doctors interviewed agree about the improvement of decision-making process thanks to IoHT support. The main answers were, in fact, related to the possibility to improve decision-making process in terms of autonomy and participation. A doctor said: "*we can have a broadly decisional space, such in the case that a patient who can be followed with these methods becomes susceptible to a diagnosis in a faster and safer clinical way*" (D3).

Some doctors perceived no impact of IoHT on patients' experience, while others see the IoHT applications as an opportunity for considering patients as part of the team. When we asked the same question related to patients "Can you also describe if and how the IoT impacts on the patient's decision-making autonomy?" most of the respondents said about any impacts on patient decision-making process, while other respondents highlighted active and aware patients' collaboration regarding the IoHT application: "*The patient can quickly contact a doctor in order to have an answer. A patient who is followed with such a technology also has the ability to call a number where he is immediately answered about that symptom*" (D2).

All the doctors have positive perceptions of IoHT and don't see it as a threat for their job. About the perceived outcome of IoHT doctors agree about the possibility of a better process quality, related to the possibility of reducing error rate, improving efficiency and communication, lowering costs and ensuring higher quality but also increasing the workload. However, the positive perception depends also on the organisation: One doctor said: "... *It depends, for example, for a doctor that use IoHT and has a good organization, surely it is a very useful tool; for those who do not have a good structure, there can be problems with data management because they have to manage a lot quantity of data without the right capacity and time*" (D1).

Doctors agree also about the fact that IoHT can't have impacts in terms of job loss: "*Technology cannot replace doctors because it is himself who has to visit the patient, observes, treats and feels him/her, not all the symptoms that are seen are those and not all the symptoms have a therapy. Technology is a support*" (D4).

Doctors have positive perceptions about patients' reaction to IoHT. When we asked what the main complaints related to the patients were, it seems that they were related to a lack of technical knowledge of the system and its limits. A doctor said: "*... the most important complaints are when the patient transmits a data and does not receive feedback, or he receives it late; it can happen ... the patient may have the perception that the data transmitted by him at midnight or at three o'clock at night should be immediately seen or managed, this is not possible; so the patient's complaint is when he does not understand the limits of using the system*" (D2).

With their general positive perceptions, all the doctors agree that IoHT technologies' employment can particularly enhance a better management of the following specific cardiac chronicity: (i) heart failure, (ii) arrhythmia (i.e. atrial fibrillation) and (iii) syncope.

6 Discussion

In this chapter we mainly analyse IoHT benefits and limits by trying to shed some light on the perceived impact of IoHT and adding some more interesting results related to chronic disease management (Table 4).

According to literature [49], our study highlights the importance of IoHT on clinical area, and particularly on the possibility of making timely diagnosis by having a better result from the clinical point of view.

In terms of data patient management, IoHT can provide clinicians larger amount of biomedical data by improving patient supervision process throughout the reduction of interruptions and checking also their adherence to clinical follow-up and drug therapy ingestions. The mistake reduction in diagnosis can enhance the patients' life quality (wellness, QALY and life expectancy). Moreover, our results confirm the positive perceptions of IoHT on the patients' health status improvement, quality of life and experience [50], thanks to the easy access to care for the patient.

However, big data don't always represent a strength: in some cases, to manage big data is perceived as a limit as it means new knowledge request and new organisational and personal duties. Especially without a solid organisational structure, big data can represent a threat. From the organisational point of view, about the perceptions of IoHT replacement of doctors' decision-making, doctors agree about the fact that IoHT can't have impacts in terms of job loss, even if there is, sometimes, the perception that with IoHT applications patients can undermine doctors' leadership.

Moreover, technical problems are an important issue: All the doctors have positive perceptions about patients' reaction to IoHT, even if it seems that patients can have difficulties in using system as well as understanding its underlying logics. The ageing factor (both of the clinicians and of the patients), together with the educational level of the patients, seems to be very important in order to understand more about the IoHT adoption rate.

In terms of economic motivation, there is quite consensus about a positive pressure in IoHT implementation, even if only one doctor highlighted and stressed such an aspect as fundamental as a driver.

According to physician interviewees the possibility to exchange medical information with patients enhances the healthcare service perception: this allows clinicians to intervene immediately on symptoms, before health conditions get worst. Machines are able to collect a large amount of biomedical data per each patient; significant variances of biomedical parameters form their "standard" (of the singular patient) trigger automatically an alert (also directly triggerable by patient in case of need) that changes the way to detect health status of a patient. Such a condition is a very important matter in terms of organisational issues because it completely changes the healthcare delivery process. Clearly, the opportunity to remotely manage patients at home has economic benefits: (i) contingent cost reduction for hospital; (ii) potential cost savings due to lower hospitalisation rate expectation; and (iii) reduction of social cost for patient.

Further, an IoHT-oriented healthcare organisation could profitably reduce the number of repeated medical tests, by fostering the medical records and patient's information sharing across hospital wards.

However, managing a large amount of data requests more accuracy from clinical personnel in terms of providing, collecting and recording patients' informed consent. After the introduction of the latest General Data Protection Regulation, in particular, the privacy matter is having a disruptive impact on data management (from the transmission to its storage), which, in same case, could be the reason of braking and congestion for healthcare delivery. In the opinion of interviewees, the privacy management is a very tricky phase that involves prevalently bureaucratic and administrative activities, while patients seem to be untouched by this matter.

What, instead, is very important for patients is the effectiveness and the affordability of IoHT technologies. Interviewees are afraid of some misunderstandings, which could involve patients' interpretation of IoHT services, with high risks for their health. Two main issues are fundamental:

(i) The information exchange between patients and physicians by technology is not a first-aid service; it is a support of chronicity management which could not substitute traditional cares.
(ii) Not all devices are suitable for professional medical cares. For example, a smartwatch for fitness aims could not have the same precision and trustworthy of a medical device (CE marked or FDI approved).

This has to be taken into account when an IoHT strategy is going to be implemented, also in terms of organisational know-how and workload chart design in order to promptly answer to patients' claims. For sure, what organisation has to avoid is running into the phenomena of "excessive" diagnosis [51] triggered by patients' pressure and wrong measurement provided by technologies.

To summarise these discussions, Table 6 contains a recap of benefits and limits of IoHT in cardiac-chronicity management as a result of interviews to the five key informants of the pilot case study.

Table 6 Benefits and limits of IoHT in cardiac-chronicity management: interpretation of interviews

Benefits			Limits		
	Referred to			Referred to	
Issue	Clinicians/hospital	Patients	Issue	Clinicians/hospital	Patients
Clinical	– Larger amount of biomedical data – Reduction of mistakes in diagnosis	– Health status improvement – QALY improvement – Unique healthcare organisation as referee	Technical	– New knowledge requested – New duties for big data management – Technical problems: • Hardware • Platform • Internet	– Difficulty in technology usage – New duties – Technical problems: • Hardware • Platform • Internet
Organisational	– Reduction of mistakes in anamnesis – Quick data availability – Increasing speed of clinical decision – Time burden decreasing – Machine learning – Big data usage – Intra-ward medical record exchange increasing	– Increasing speed of claim reaction	Effectiveness/ affordability	– Organisational know-how adequacy – Operational flow chart adequacy – Danger of “excessive” diagnosis	– Not for first-aid aims – Technology availability ≠ Healthcare providing – Only medical devices are trustworthy
Economic	– Potential refund tariff from NHS – Remote management of patient – Hospitalisation reduction – Intra-ward medical test reduction	– Social cost reduction: • Transfer • Lost income • Office visit ticket	Security	– Personal data management – Update to privacy rules – Update to privacy documents	– Data transmission procedures
Cultural	– Increasing awareness – Increasing competences	– Increasing awareness – Increasing perceived safety	Cultural	– Ageing – Personal attitude	– Ageing – Personal attitude

Source: Own elaboration

7 Conclusion

DT, and particularly IoHT, represents a great opportunity for the healthcare sector. Our study, by highlighting some important issues related to clinicians and patients involved in a digitalised process, adds some specific ideas about the benefits and limits of IoHT applications.

From a theoretical perspective, this chapter contributes mainly to the healthcare digital transformation debate. By focusing on the role of clinicians and patients, it particularly offers new elements to further the ongoing debate about IoHT implementation, its benefits and limits.

Hospital digitalisation through the use of IoHT allows improvements in clinician area, patients' quality of life enhancement and hospital cost rationalisation.

However, such outcomes are not automatic, and they are also related to enabling conditions that could also represent limitation to IoHT implementation (Table 6).

From an empirical perspective, the chapter provides healthcare practitioners with specific IoHT case in order to effectively understand the main benefits and limits related to such implementation, by increasing the awareness of such kinds of projects, both from patients' and doctors' point of view.

Finally, our research presents two main limitations. The first one is related to the employment of a qualitative method, which is less reliable for the interpretive role of researchers and limited extension of data.

The second limitation, connected to the first one, is related to the scale level of the study, being focused on a pilot, even if original, scale level. We believe that further and different insights may emerge during a large-scale-level implementation. However, these limitations lay the foundation to examine in depth the theme of IoHT application related to management of chronic disease, which would deserve to be extensively analysed in order to provide new solutions, which are compatible with the healthcare sustainability enhancement.

References

1. Cap Gemini. (2011). MIT Sloan management. Digital transformation: a roadmap for billion-dollar organizations.
2. Gastaldi, L., & Corso, M. (2012). Smart healthcare digitalization: Using ICT to effectively balance exploration and exploitation within hospitals. *International Journal of Engineering Business Management, 4*, 1–13.
3. Tresp, V., et al. (2016). Going digital: A survey on digitalization and large-scale data analytics in healthcare. *Proceedings of the IEEE, 104*(11), 2180.
4. Bharadwaj, A., El Sawy, A. O., Pavlou, A. P., & Venkatraman, N. (2013). Digital business strategy: Toward a next generation of insights. *MIS Quarterly, 37*(2), 471–482.
5. Mettler, T., & Pinto, R. (2018). Evolutionary paths and influencing factors towards digital maturity: An analysis of the status quo in Swiss hospitals. *Technological Forecasting and Social Change, 133*, 104–117.
6. Berman, S., & Marshall, A. (2014). The next digital transformation: from an individual-centered to an everyone-to-everyone economy. *Strategy & Leadership, 42*(5), 9–17.

7. World Health Organization. (2013). Global action plan for the prevention and control of noncommunicable diseases 2013–2020.
8. Rghioui, A., & Oumnad, A. (2018). Challenges and opportunities of internet of things in healthcare. *International Journal of Electrical and Computer Engineering, 8*(5), 2753–2761.
9. Sherer, S. A. (2014). Advocating for action design research on IT value creation in healthcare. *Journal of the Association for Information Systems, 15*(12), 860–878.
10. Devaraj, S., & Kohli, R. (2000). Information technology payoff in the health-care industry: A longitudinal study. Information technology payoff in the health-care industry: A longitudinal study. *Journal of Management Information Systems, 16*(4), 41–67.
11. Devaraj, S., & Kohli, R. (2003). Performance impacts of information technology: Is actual usage the missing link? *Management Science, 49*(3), 273–289.
12. Amarasingham, R., Plantinga, L., Diener-West, M., Gaskin, D. J., & Powe, N. R. (2009). Clinical information technologies and inpatient outcomes: A multiple hospital study. *Archives of Internal Medicine, 169*(2), 108–114.
13. Dexter, P. R., Perkins, S. M., Maharry, K. S., Jones, K., & McDonald, C. J. (2004). Inpatient computer-based standing orders vs. physician reminders to increase influenza and pneumococcal vaccination rates: A randomized trial. *JAMA : The Journal of the American Medical Association, 292*(19), 2366–2371.
14. Kucher, N. K. S., Quiroz, R., Cooper, J. M., Paterno, M. D., Soukonnikov, B., & Goldhaber, S. Z. (2005). Electronic alerts to prevent venous thromboembolism among hospitalized patients. *The New England Journal of Medicine, 352*, 969–977.
15. Parente, S. T., & McCullough, J. S. (2009). Health information technology and patient safety: Evidence from panel data. *Health Affairs, 28*(2), 357–360.
16. Aron, A., Ketay, S., Hedden, T., Aron, E. N., Markus, H., & Gabrieli, J. E. (2010). Temperament trait of sensory processing sensitivity moderates cultural differences in neural response. *Social Cognitive and Affective Neuroscience, 5*, 219–226.
17. Huberfeld, N. (2014). Medicaid at 50: From exclusion to expansion to universality. *Health Affairs Blog*.
18. Terry, N. P. (2016). Will the internet of things transform healthcare. *Vanderbilt Journal of Entertainment & Technology Law, 19*, 327.
19. Newman, L., Bidargaddi, N., & Schrader, G. (2016). Service providers' experiences of using a telehealth network 12 months after digitisation of a large Australian rural mental health service. *International Journal of Medical Informatics, 94*, 8–20.
20. Qi, J., Yang, P., Min, G., Amft, O., Dong, F., & Xu, L. (2017). Advanced Internet of Things for personalised healthcare systems: A survey. *Pervasive and Mobile Computing, 41*, 132–149.
21. Islam, S. R., Kwak, D., Kabir, M. H., Hossain, M., & Kwak, K. S. (2015). The internet of things for health care: A comprehensive survey. *IEEE Access, 3*, 678–708.
22. Dohr, A., Modre-Opsrian, R., Drobics, M., Hayn, D., & Schreier, G (2010). The Internet of Things for ambient assisted living. In *Proceedings of Seventh International Conference on Information Technology, New Generation (ITNG)* (pp. 804–809).
23. Chung, W.-Y., Lee, Y.-D., & Jung, S.-J. (2008). A wireless sensor network compatible wearable u-healthcare monitoring system using integrated ECG, accelerometer and SpO2. In *Proc. 30th Annu. Int. Conf. IEEE Eng. Med. Biol. Soc. (EMBS)* (pp. 1529–1532).
24. Istepanian, R. S. H., Jovanov, E., & Zhang, Y. T. (2004). Guest editorial introduction to the special section on m-health: Beyond seamless mobility and global wireless health-care connectivity. *IEEE Transactions on Information Technology in Biomedicine, 8*(4), 405–414.
25. You, L., Liu, C., & Tong, S. (2011). Community medical network (CMN): Architecture and implementation. In *Proc. Global Mobile Congr. (GMC)* (pp. 1–6).
26. Liu, J. & Yang, L. (2011). Application of Internet of Things in the community security management. In *Proc. 3rd Int. Conf. Comput. Intell., Commun. Syst. Netw. (CICSyN)* (pp. 314–318).
27. Burgun, A., Botti, G., Fieschi, M., & Le Beux, P (1999). Sharing knowledge in medicine: Semantic and ontologic facets of medical concepts. In *Proc. IEEE Int. Conf. Syst., Man, Cybern. (SMC)* (Vol. 6, pp. 300–305).

28. Sai Kiran, M. P. R., Rajalakshmi, P., & Acharyya, A. (2014). Context predictor based sparse sensing technique and smart transmission architecture for IoT enabled remote health monitoring applications. In *Proc. IEEE Int. Conf. Eng. Med. Biol. Soc. (EMBC)* (pp. 4151–4154).
29. Rasid M. F. A., Muzaaliff Wan Musa W. M. N., Kadir, N. A. A., & Noor, A. M. (2014). Embedded gateway services for Internet of Things applications in ubiquitous healthcare. In *Proc. 2nd Int. Conf. Inf. Commun. Technol. (ICoICT)* (pp. 145–148).
30. Jara, A. J., Belchi, F. J., Alcolea, A. F., Santa, J., Zamora-Izquierdo, M. A., & Gomez-Skarmeta, A. F. (2010). A pharmaceutical intelligent information system to detect allergies and adverse drugs reactions based on Internet of Things. In *Proc. IEEE Int. Conf. Pervasive Comput. Commun. Workshops (PERCOM Workshops)* (pp. 809–812).
31. Vazquez-Briseno, M., Navarro-Cota, C., Nieto-Hipolito, J. I., Jimenez-Garcia, E., & Sanchez-Lopez, J. D (2012). A proposal for using the Internet of Things concept to increase children's health awareness. In *Proc. 22nd Int. Conf. Elect. Commun. Comput. (CONIELECOMP)* (pp.168–172).
32. Drew, B. J., Califf, R. M., Funk, M., Kaufman, E. S., Krucoff, M. W., Laks, M. M., Macfarlane, P. W., Sommargren, C., Swiryn, S., & Van Hare, G. F. (2004). American Heart Association, Councils on Cardiovascular Nursing, Clinical Cardiology, and Cardiovascular Disease in the Young: Practice standards for electrocardiographic monitoring in hospital settings. *Circulation, 110*(17), 2721–2746.
33. Istepanian, R. S. H., Hu, S., Philip, N. Y., & Sungoor, A. (2011). The potential of Internet of m-health Things 'm-IoT' for non-invasive glucose level sensing. In *Proc. IEEE Annu. Int. Conf. Eng. Med. Biol. Soc. (EMBC)* (pp. 5264-5266).
34. Pang, Z., Tian, J., & Chen, Q. (2014). Intelligent packaging and intelligent medicine box for medication management towards the Internet-of-Things. In *Proc. 16th Int. Conf. Adv. Commun. Technol. (ICACT)* (pp. 352–360).
35. Yang, L., Ge, Y., Li, W., Rao, W., & Shen, W. (2014). A home mobile healthcare system for wheelchair users. In *Proc. IEEE Int. Conf. Comput. Supported Cooperat. Work Design (CSCWD)* (pp. 609–614).
36. Mosa, A. S. M., Yoo, I., & Sheets, L. (2012). A systematic review of healthcare applications for smartphones. *BMC Medical Informatics and Decision Making, 12*, 67.
37. Fæste, L., Gumsheimer, T., & Scherer, M. (2015). *How to jump-start a digital transformation.pdf*. Boston Consulting Group.
38. Bughin, J., & Manyika, J. (2012). *Internet matters, essays in digital transformation*. McKinsey: New York.
39. Yin, K. R. (2014). *Case study research: Design and methods* (5th ed.). Thousand Oaks: Sage.
40. Patton, M. Q. (2002). *Qualitative evaluation and research methods*. Thousand Oaks: Sage.
41. Palozzi, G., Binci, D., & Appolloni, A. (2017). e-Health and co-production: Critical drivers for chronic diseases management. In M. A. Pfannstiel & C. Rasche (Eds.), *Service business model innovation in healthcare and hospital management*. Cham, Switzerland: Springer International Publishing.
42. Palozzi, G., Chirico, A., Falivena, C., & Calò, L. (2018). How information availability changes healthcare chronicity management: Findings from a pilot case study. In *International Forum on Knowledge Asset Dynamics*.
43. Kadu, M. K., & Stolee, P. (2015). Facilitators and barriers of implementing the chronic care model in primary care: A systematic review. *BMC Medical Education, 16*(12).
44. Coleman, K., Austin, B. T., Brach, C., & Wagner, E. H. (2009). Evidence on the Chronic Care Model in the new millennium. *Health Affairs (Millwood), 28*(1), 75–85.
45. Calò, L., Giustetto, C., Martino, A., Sciarra, L., Cerrato, N., Marziali, M., et al. (2016). A new electrocardiographic marker of sudden death in Brugada syndrome: The S-wave in lead. *Journal of the American College of Cardiology, 67*(12), 1427–1440.
46. Chaudhry, S. I., Wang, Y., Concato, J., Gill, T. M., & Krumholz, H. M. (2007). Patterns of weight change preceding hospitalization for heart failure. *Circulation, 116*(14), 1549–1554.

47. Italian Board of Health. (2014). *Telemedicina—Linee di indirizzo nazionali.* Accessed 9 Sep 2018.
48. Slotwiner, D., Varma, N., Akar, J. G., Annas, G., Beardsall, M., Fogel, R. I., et al. (2015). HRS Expert Consensus Statement on remote interrogation and monitoring for cardiovascular implantable electronic devices. *Heart Rhythm, 12*(7), 69.
49. Wang, Y., Byrd, T. A., & Kung, L. (2018). Big data analytics: Understanding its capabilities and potential benefits for healthcare organizations. *Technological Forecasting and Social Change, 126*, 3–13.
50. Agarwal, R., Guodong, G., DesRoches, C., & Jha, A. K. (2010). The digital transformation of healthcare: Current status and the road ahead. *Information Systems Research, 21*(4), 796–809.
51. Welch, H. G., Schwartz, L. M., & Wolosin, S. (2013). *Sovradiagnosi. Come gli sforzi per migliorare la salute possono renderci malati.* Roma: Il Pensiero Scientifico.

Cost Analysis of Telemedicine Implementation in the Lens of Healthcare Sustainability: A Review of the Literature

Gabriele Palozzi, Antonio Chirico, and Francesco Gabbrielli

1 Introduction

The dynamic environment of the healthcare sector is characterized by pressure on cost reduction, new approaches towards patients, and drastic technological changes. Public and private healthcare organizations are searching for strategies to decrease the healthcare spending, without compromising access, effectiveness, and safety [1]. In this scenario, in which economic crises, digitalization, and new legislation are fundamental drivers, the adoption of innovative telecommunication technologies should be considered as a successful strategy to deliver healthcare services to patients who are far located from hospitals or that have a limited access to specialty cares.

The term eHealth was coined to cover a broad range of data processing and computer networking applications (including use of the Internet) in healthcare [2]. "Telemedicine is a branch of e-health that uses communications networks for delivery of healthcare services and medical education from one geographical location to another. It is deployed to overcome issues like uneven distribution and shortage of infrastructural and human resources" [3]. More specifically in healthcare delivery process, "the term 'telemedicine' was originally used to denote the provision of medical services across distance (this is the literal meaning of the word). As application of telemedicine widened, the term 'telehealth' was

G. Palozzi (✉) · A. Chirico
Department of Management & Law, University of Rome Tor Vergata, Rome, Italy
e-mail: palozzi@economia.uniroma2.it

F. Gabbrielli
National Center for Telemedicine & New Healthcare Technologies, Italian National Institute of Health ISS, Rome, Italy

© Springer Nature Switzerland AG 2020
D. Cagáňová, N. Horňáková (eds.), *Mobility Internet of Things 2018*,
EAI/Springer Innovations in Communication and Computing,
https://doi.org/10.1007/978-3-030-30911-4_32

introduced to reflect a broader scope of health-related functions such as education and administration" [2].

In the Italian context, telemedicine is defined as "a new healthcare delivery process, based on innovative technologies, provided when patient and professional are not physically in the same place" [4]. Based on ICT infrastructure, "it brings a safe biomedical data & information transmission in the form of texts, sounds, images, or other forms necessary for prevention, diagnosis, treatment and follow-up of patients" [4]. Accordingly, telemedicine can contribute to enhance:

- Equity in access to care
- High-quality services
- Improvement in efficacy, effectiveness, and appropriateness of care
- Expenditure containment

Telemedicine services have become a part of modern healthcare management, since it has demonstrated benefits to healthcare access and quality. Accordingly, apart from the concerns with the specific quality of modern devices and services, two major advantages of telemedicine are seen: (1) benefits for undersupplied population and (2) cost savings [5].

Notwithstanding this, many limitations prevent the national acceptance and usability of telemedicine principles. Most of the critics are concerned with its reimbursement possibilities, and, in turn, with the hypothetical amount of reimbursement rates of particular kind of "bundled" services depending on a multiplicity of factors and activities [6]. Further, Hersh et al. argued that limited information on the efficacy of cost of telemedicine is a critical obstacle for the coverage and reimbursement of these services by the third payers [7].

Clearly, in the lens of healthcare system sustainability, fixing the reimbursement tariff of a service provided by telemedicine infrastructure has to depend on the consumption of resources needed for its provision. In accordance with the diagnosis-related group (DRG) logic, the computation of these resources involved has to take into account a deep cost analysis in order to understand the "all-embracing" amount of financial resources needed.

However, with regard to accountability, computation, and categorization of cost of telemedicine, scarce is the literature background; just few authors have analyzed implementation costs of telemedicine in specialist healthcare service delivery. Actually, in fact, literature background prevalently presents inquiries related with cost analysis on specific telemedicine experiences applied to a specific branch of medicine [6, 8–11].

Just Davlaos et al. [1] have analyzed the economic evaluation of telemedicine, by a comprehensive review of the literature. These authors, in 2009, summarized firstly the main applicable economic methods for telemedicine evaluation as follows:

- *Cost analysis*, which identifies the resources used to deliver a service
- *Cost-effectiveness analysis*, which considers both costs and outcomes of a service delivered

- *Benefit-cost analysis*, which compares resource employment with multiple outcomes obtained by an intervention

Even if the sole cost analysis is a limited method for a comprehensive evaluation, this represents the required starting point for the computation of resource consumption of the specific telemedicine services provided. It should allow users and practitioners to understand generic kinds of cost (and their behavior) involved in healthcare provided by telemedicine infrastructure.

Moreover, Davlos et al. [1] have underlined that costs of telemedicine refer to three main key categories of subjects: (1) client/patient, (2) provider, and (3) other stakeholders. In the opinion of the author, these suffer from both fixed and variable costs, depending on the kind or resource requested. Accordingly, Fig. 1 contains the common cost categories of telemedicine programs, clustered per fixed and variable costs, which have been contrasted with the common outcomes achievable (Fig. 2).

However, apart from the Davlos et al. [1] work, in the last decade no other researches have been conducted about telemedicine costs, in particular with regard to general kind of resources requested to deliver healthcare service when patient and physician are not in the same place, in terms of infrastructure, personnel, and equipment.

Given that, the goal of this work is to provide a literature review on the field of cost analysis of telemedicine in order to:

CLIENT/PATIENT	PROVIDER	OTHER STAKEHOLDERS
	Fixed costs	
Time costs (employment, classroom time, or leisure) Medical costs (out-of-pocket)	Equipment/technology (capital investment) Depreciation Facilities (office space)	Costs to the taxpayer from expanded coverage and payment of telemedicine services in Medicare and Medicaid Costs to private insurers from expanded coverage and payment of telemedicine services and costs to their clients if, as a result, insurance premiums increase Loss of productivity (work absences) for the employer from workers' participation in a program
	Variable costs	
	Maintenance and repairs Telecommunication costs (connections, etc.) Administrative support and supplies Training Wages to technicians Wages to staff Other expenses	
	Other costs: program setup	
	Travel (transportation, accommodation, per diem; travel time) Training Other expenses (promoting the program, etc.)	

Fig. 1 Common cost category of telemedicine program. Source: Davlos et al. [1]

CLIENT/PATIENT	PROVIDER	OTHER STAKEHOLDERS
Medical effectiveness	*Healthcare services and others*	*Other outcomes*
Reduced morbidity Avoided mortality	Reduced length of hospital stay Avoided hospitalizations Avoided hospital readmissions Avoided emergency room visits	Increased productivity of workers (less travel, less illness) Avoided cases of communicable diseases More efficient access to healthcare for special groups (prisoners, etc.)
Employment	Avoided laboratory tests	
Increased earnings	Avoided patient transportation to healthcare facilities Avoided physician office/clinic visits	
Healthcare services and others	Avoided referrals	
Increased access to healthcare Increased health knowledge/ability for self-care Faster/accurate diagnosis and treatment Reduced waiting and/or consultation time Increased medication adherence	Reduced length of consultations Increased medication adherence Increased knowledge transfer among practitioners Increased accuracy and faster diagnosis and treatment Increased patient satisfaction	
Decreased travel	*Decreased travel and/or home visits for staff*	
Increased employment/leisure/classroom time Avoided travel expenditures: transportation, accommodation, and other expenses Decreased risk of job loss: less time away from work for travel	Increased employment time (productivity) Avoided travel expenditures: transportation, accommodation and per diem	

Fig. 2 Common outcomes of telemedicine program. Source: Davlos et al. [1]

- Carry out a worldwide descriptive screening of the state of the art about the cost of telemedicine
- Understand which generic cost items have to be considered in case of implementation of healthcare service provided by telemedicine

Finally, this study aims at providing a checklist of cost items, useful to decision-makers to understand which hypothetical financial resources are necessary in order to activate, deliver, and/or modify a healthcare service provided by telemedicine.

Thus, the remainder of the chapter is organized as follows: Sect. 2 explains the methodology used to examine the literature background; Sect. 3 presents findings about the current state of the art of the literature about costs of telemedicine. Section 4 critically analyzes and discusses findings by underlining the potential employment of this inquiry in real practice. Finally, Section 6 provides final remarks in order to clarify the contribution of this study for future researches.

2 Method

In order to get the main goals of this study a literature review was conducted on the field of cost analysis on telemedicine. Consulted databases were Scopus, Web of Science, and EBSCO—Business Source Complete.

The choice of these three databases was due to reason of completeness of information achievement. Although all the three databases are concerned with managerial issues, they have different features and focuses. In particular (1) Scopus is probably the most complete database on humanistic topics, with reference on law, economics, management, and literature; (2) Web of Science is instead focused on scientific sphere (engineering, math, physics, science), but with an overlook on the managerial problems related to the resource employment; (3) EBSCO—Business Source Complete, finally, is a database exclusively devoted to management, business administration, and economics.

The keywords used for the whole inquiry are contained in Table 1. Clearly, the same following set of keywords are used in the three databases.

Keywords contained in the same column are alternative within them. Papers considered relevant for our research are those containing at least one keyword belonging to each column (1st and 2nd keywords) within Title and/or Abstract.

In particular a twofold research strategy was applied in coherence with the two goals.

Firstly, to carry out a worldwide descriptive screening of the state of the art about the cost of telemedicine, a bibliometric analysis of the literature was conducted. This is a statistic-quantitative method aimed at studying the textual and editorial information [12]. Accordingly, this method represents a systematic, transparent, and replicable literature review process that allows users to make a quantitative screening about the bibliographic state of the art of a topic, by statistically analyzing a set of interrelated keywords and terms within a database.

Table 1 Keywords used for the enquiry

1st Keyword	AND	2nd Keyword
"cost accounting"		"e-health"
or		or
"cost analysis"		"ehealth"
or		or
"cost control"		"telecare"
or		or
"cost information"		"telemedicine"
or		or
"cost driver"		"telehealth"
or		or
"cost drivers"		"telemonitoring"
or		or
"full costing method"		"telepractice"
or		or
"full costing"		"telenursing"
or		
"cost control"		
or		
"cost and cost analysis"		
or		
"economic analysis"		

Source: Own elaboration

This first part of research was conducted on two different databases: Scopus and Web of Science (WoS). Choice of two databases depended on the need for more information in terms of a larger number of available papers.

For each database a different software has been used for the bibliometric analysis; these are, respectively, VOSviewer (Scopus) and Bibliometrix R-Package (Wos). The choice of using two software was in order to optimize information from the two databases on the basis of the best features of inquiry of each software. By best feature we mean kinds of information processed, optimized synthesis and representation of data, and clearness of data viewer.

In particular, Scopus was consulted by VOSviewer in order to understand these dimensions of inquiry: (1) research areas, (2) worldwide collaboration across authors, and (3) link about the term "cost" and other topics. The reference time frame was from the year 1978 to August 28th, 2018.

The database WoS was consulted by Bibliometrix R-Package in order to understand these dimensions of inquiry: (1) papers/year, (2) worldwide collaboration across countries, and (3) more advanced countries in the research field. The reference time frame was from the year 1996 to August 28th, 2018.

Moreover, to understand which generic cost items have to be considered in case of implementation of healthcare service provided by telemedicine, a narrative review of the literature was conducted. As declared, the aim of this method is to obtain as many existing evidence-based studies as possible that are relevant to the research being undertaken, irrespective of their published location or even disciplinary background [13]. The basic principles behind adopting this review are the following [14]:

(i) Equality: The review makes no distinction between the type and nature of peer-reviewed journals, edited in English, since 1977 to August 28th, 2018.
(ii) Transparency: By describing each search string and the rationale behind its selection, the review can be repeated to test its rigor and update its findings [15].
(iii) Accessibility: The reviews are made available outside the specialist, academic community in the form of reports and searchable databases.

The extraction template from EBSCO database is included in Fig. 3.

3 Findings

This section presents findings obtained by the literature review. It attempts to give to the reader a road map to orient himself/herself in a literature research about the theme of cost analysis about telemedicine. Accordingly, this section has been divided into two main streams; first one, related to Scopus and Web of Science databases, statistically describes sample of paper, in order to orient researchers in future background scanning. Second one, related to the EBSCO—Business Source

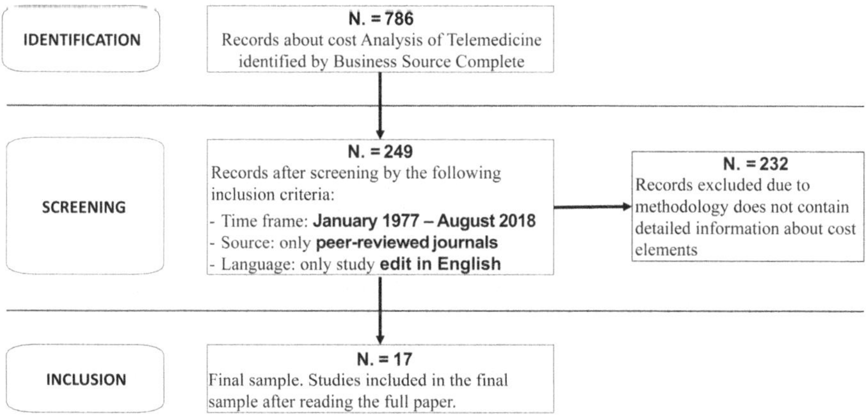

Fig. 3 Extraction template from EBSCO—Business Source Complete database. Source: Authors' illustration

Complete—database, aims at increasing knowledge about the background state of the art about "implementation" and "delivery" costs of telemedicine.

3.1 Statistical Description of Sample

Scopus Database

About the theme of cost analysis of telemedicine, this subsection reports a recap of the main keywords, split per research fields, and most quoted authors in the field. All figures shown were provided by the VOSviewer software.

Figure 4 reports the 30 more recurring keywords on Scopus and their interrelationships. They can be clustered per two main research areas: (1) clinical and (2) healthcare management. To be considered within "the 30 more recurring" a keyword has to be found in the sample at least 49 times. The greater is the keyword circle, the higher is the recurrence of that keyword in the sample.

Figure 5 shows the connection across keywords between the term "cost" and all the others. The sample has been clustered per research fields too.

Figure 6 reports the ten most quoted authors on the topic in Scopus database.

Figure 7 shows the collaboration across the 30 more productive authors world wide. The less quoted author has been cited 133 times. It's easy to note that there no collaborations across authors, except for very small interest groups.

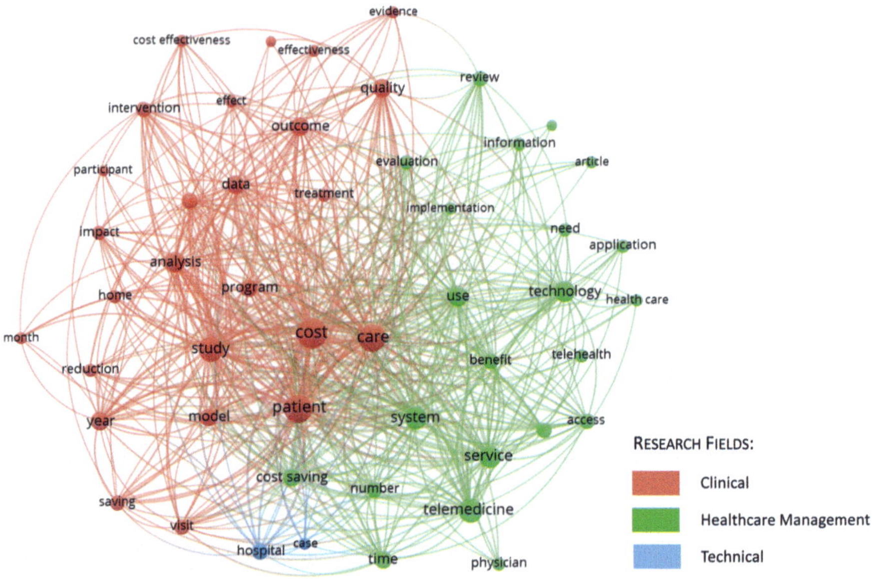

Fig. 4 The 30 more recurring keywords in Scopus database. Source: Authors' illustration provided by VOSviewer software

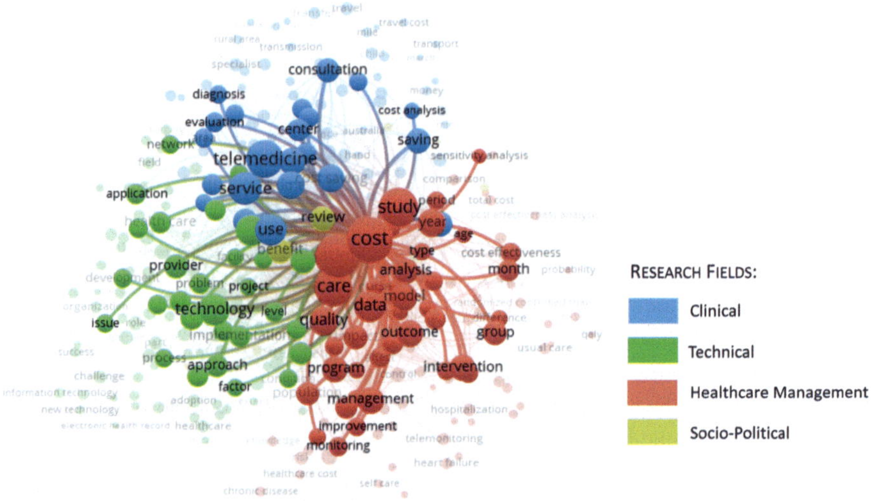

Fig. 5 The connection across keywords between the term "cost" and all the others, in Scopus database. Source: Authors' illustration provided by VOSviewer software

Author	Documents	Citations ∨	Total link strength
ohinmaa a.	8	553	13
hailey d.	5	535	11
roine r.	3	427	9
jerant a.f.	3	348	14
nesbitt t.s.	3	332	11
azari r.	2	303	11
johnston b.	1	280	7
bashshur r.l.	5	256	0
celler b.g.	2	194	2
lovell n.h.	2	194	2

Fig. 6 The ten most quoted authors in Scopus database. Source: Author's illustration provided by VOSviewer software

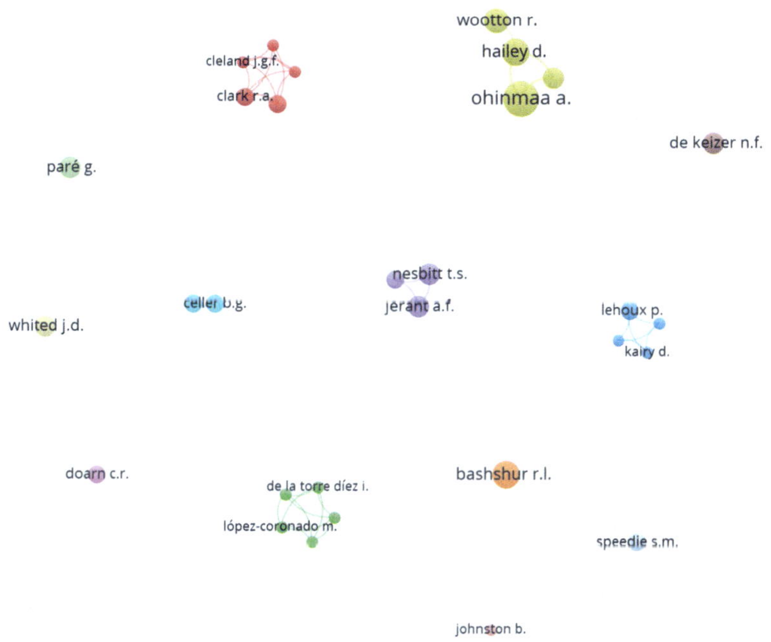

Fig. 7 Collaboration across the 30 more productive authors worldwide in Scopus database. Source: Authors' illustration provided by VOSviewer software

Web of Science Database

About the theme of cost analysis of telemedicine, also this subsection reports a statistical description of sample. All figures shown were provided by the Bibliometrix R-Package software.

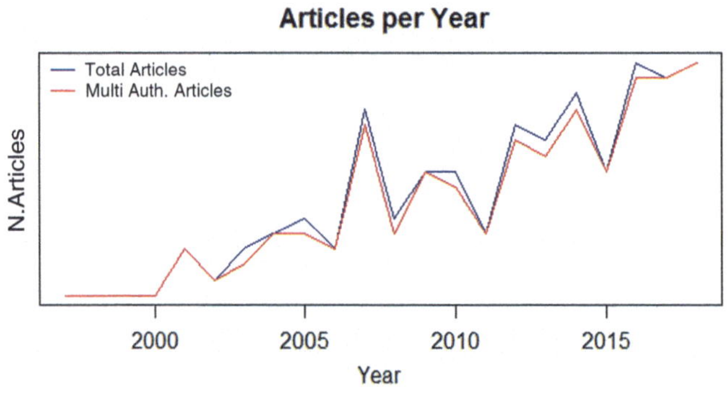

Year	Articles
1997	1
1999	1
2000	1
2001	4
2002	2
2003	4
2004	5
2005	6
2006	4
2007	13
2008	6
2009	9
2010	9
2011	5
2012	12
2013	11
2014	14
2015	9
2016	16
2017	15
2018	16

Fig. 8 Number of paper per year. Source: Authors' illustration provided by Bibliometrix R-Package software

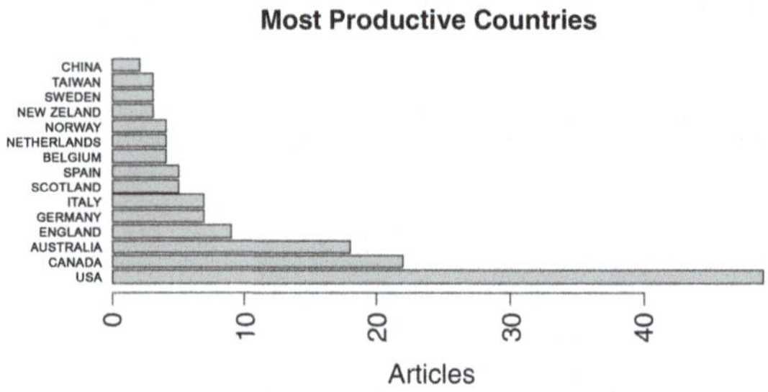

Country	Articles	SCP	MCP
1 USA	49	44	5
2 CANADA	22	21	1
3 AUSTRALIA	18	14	4
4 ENGLAND	9	7	2
5 GERMANY	7	5	2
6 ITALY	7	6	1
7 SCOTLAND	5	2	3
8 SPAIN	5	4	1
9 BELGIUM	4	4	0
10 NETHERLANDS	4	1	3
11 NORWAY	4	4	0
12 NEW ZEALAND	3	3	0
13 SWEDEN	3	3	0
14 TAIWAN	3	2	1
15 CHINA	2	2	0

SCP: Single Country Publications
MCP: Multiple Country Publications

Fig. 9 Scientific production per countries. Source: Authors' illustration provided by Bibliometrix R-Package software

Figure 8 shows the number of paper per year. Graph compares also the number of multi-author papers with the total number.

Figure 9 shows the scientific production per countries. Anglo-Saxon countries (USA, Canada, Australia, England) are those that point more attention to the theme.

Figure 10, instead, reports the worldwide scientific collaboration across countries. The greater the country circle, the higher the level of scientific productivity. Lines show the connection across countries.

Finally, Fig. 11 reports the most productive authors about the theme.

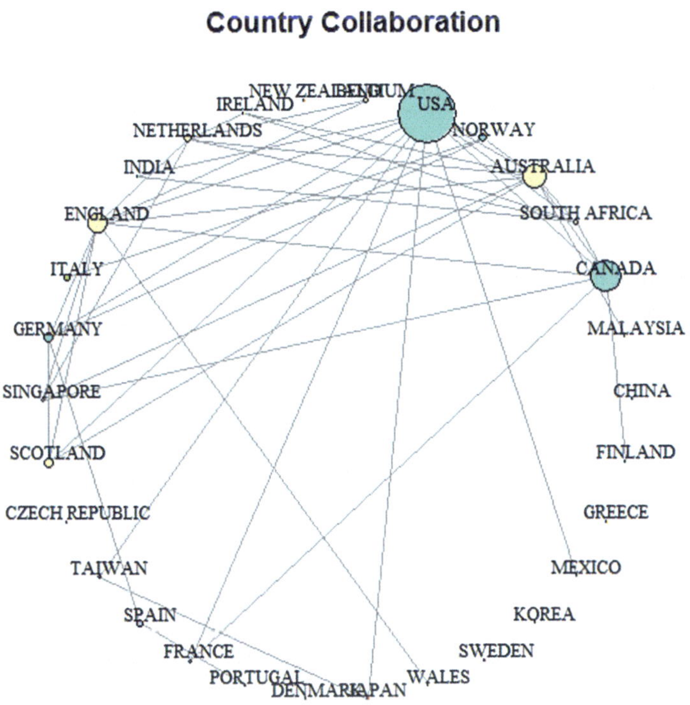

Fig. 10 Scientific collaboration across countries worldwide. Source: Authors' illustration provided by Bibliometrix R-Package software

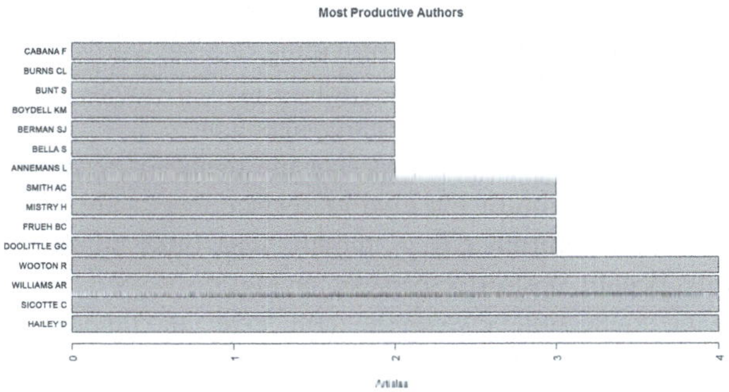

Most Productive Authors

	Authors	Articles
1	HAILEY D	4
2	SICOTTE C	4
3	WILLIAMS AR	4
4	WOOTTON R	4
5	DOOLITTLE G	3
6	FRUEH BC	3
7	MISTRY H	3
8	SMITH AC	3
9	ANNEMANS L	2
10	BELLA S	2
11	BERMAN SJ	2
12	BOYDELL KM	2
13	BUNT S	2
14	BURNS CL	2
15	CABANA F	2

Fig. 11 Most productive authors worldwide. Source: Authors' illustration provided by Bibliometrix R-Package software

3.2 Content Analysis of Sample

EBSCO: Business Source Complete Database

This subsection, which refers to the EBSCO—Business Source Complete database, clusters the literature contributions on the basis of their contents. Papers considered contain at least a reference to "implementation" or "delivery" cost of telemedicine. In different branches of application, these selected authors point out those necessary costs to install a healthcare service provided by telemedicine infrastructure and to manage it in healthcare delivery.

Table 2 contains a recap of all these cost items. For the aim of this chapter, the results of this mini narrative review allow users to generalize all potential resources needed to use telemedicine infrastructure and to deliver a healthcare service based on biomedical information exchange between patients and clinicians.

4 What Findings Tell

The declared aim of this work is to provide a screening of literature background about costs of telemedicine. In this direction both a bibliometric analysis of literature and a narrative review of main contribution were conducted. As shown in the findings section, this chapter has provided a descriptive and a content analysis of the topic in the three important databases which are concerned, directly or indirectly, with the managerial issues related with the cost of telemedicine employment in healthcare.

The first contribution of our findings has been the providing of a keywords road map. Mainly used keywords have been identified. They refer to two different main research fields: (a) technical-clinical and (b) healthcare management.

Reading of the 30 most recurring keywords shows that the theme of resource employment on telemedicine is a function of (1) kind of service (e.g., care, use, service, access); (2) time frame and recurrence (e.g., month, year, time, number); and (3) outcomes (e.g., quality, benefit, cost-effectiveness, savings).

This is confirmed also by the connection between the term "cost" and other keywords. The topic of cost analysis of telemedicine is a matter that links different fields of literature. In particular the theme pertains to clinical, technical, sociopolitical, and healthcare management backgrounds.

Further, this suggests that the theme of resources involved in providing telemedicine service is a very considerable subject, which reflects very different disciplines. Thus, each discipline is fundamental for the understanding of different aspects of the subject of research. Accordingly, analysis of sustainability of a technological innovation in healthcare can't be overlooked from a holistic and multidisciplinary comprehension of phenomena [30–32].

Table 2 Cost elements in telemedicine employment: content analysis of EBSCO—Business Source Complete sample

Papers		Cost elements	
Ref.	Authors	Implementation costs	Delivery costs
[16]	Buchanan (2015)	Equipment Software Setup and maintenance Telegenetics system	Clinic personnel IT specialists
[17]	Burri (2013)	Initial investment Remote monitoring services	Patient management
[18]	Hameed (2017) ***LR***[a]	Telemonitoring™ equipment	TM service TM-based medical supervision Training of personnel Support and consulting
[19]	Kovcs (2017)	Tangible asset: • Rental camera • Technology • Transporting and screening • Staff education	Screening and examination Staff gross salaries Staff education Consumables and medicines
[20]	Kumar (2013) ***LR***[a]	Purchase, install, and maintain software, license, and equipment Technical support Internet service fee Telephony fee Design, construction, or remodeling of space Staff training	Personnel: • Nurse • Physician • Administrative • Support technician
[21]	Marino (2015)[b]	Equipment Device Installation Staff training	Personnel: • Nurse • Dentist • Dental assistant
[22]	Mason (2006)	Equipment (call center)	Personnel (call center)
[23]	Razavi (2016)	Telehealth equipment	Personnel: • Physician • IT • Coordinator
[24]	Roberts (2012)	Technology Staff training IT Scheduling	Personnel (focus on incentive payments for practitioner)
[25]	Rollo (2017)	Materials: • Webcam • Headset • Measuring equipment Telemedicine video platform	Personnel: • Physician • Administrative

(continued)

Table 2 (continued)

Papers		Cost elements	
Ref.	Authors	Implementation costs	Delivery costs
[5]	Rosenberg (2012)	Equipment: • Laptop, printer and other hardware cost • Patient-centered medical home recognition	Personnel: • Administrative • Medical
[26]	Rosenberg (2013)	Depreciation	Personnel Direct material (paper and stationery products)
[10]	Thaker (2013)[c]	Equipment: • Camera • LCD monitor • Wall mount brackets Travel and accommodation Installation Video connectivity	Personnel: Tele-oncology coordinator
[27]	Theodore (2015)[d]	Equipment	Personnel
[28]	Vitaccia (2012)	Device rental Equipment purchase Installation Telecommunication Line charge Maintenance Training/enrolment	Call center Administrative activities Nurse-tutor Specialist
[29]	Williams (2016)	Equipment	Personnel: • Nurse • Administrative
[6]	Zholudev (2017)[e]	NO implementation cost	Personnel

Source: Own elaboration
[a]*LR* Literature review
[b]Costs for consumable parts and clinic sterilizing were mentioned but not computed
[c]Also nursing cost was considered, but it's not directly due to telemedicine
[d]Study analyzes also cost saving for patients, due to telemedicine employment
[e]Study underlines also transportation cost and time saving for patients due to telemedicine employment

From the authorship analysis of the Scopus sample we can note that just few authors contribute to the field with a large amount of papers (only five authors have more than three cited articles). What makes us thinking is the quite total lack of collaboration across authors worldwide. Except small research groups, it seems that evaluations of telemedicine costs are inquiries strongly tied to the specific context. This explains why, except Davlos et al. [1], no other research provides a general framework about telemedicine cost accounting.

Moving to the Web of Science database, this is also confirmed by the limited collaboration in scientific production across countries worldwide. USA, Canada, and Australia are those most prolific countries. Beyond the numerosity of population, the reasons of the interest to the telemedicine theme seem to lie in two main motivations:

1. *The geographical features.* They are extended countries, with rural areas often very far from urban settlements (equipped with hospitals). This makes health policies aimed at "moving" healthcare services to patient/citizen's home necessary.
2. *Tariff.* Reimbursement system of these countries considers just a partial refund for telemedicine service.

Hence, researches on the theme seem to be aimed at fostering both awareness about positive outcomes of remote monitoring services and reexamining reimbursement tariffs.

In the same direction, analysis of the scientific production time frame shows a constant growth of published article about the theme. In our opinion, this shows a worldwide growing inclination of researchers to understand the real sustainability of telemedicine, in the lens of National Health Systems.

Unavoidably, the process of technological innovation adoption in healthcare has to take into account the resources related with their employment. Thus, in this perspective, what matters is the definition of a framework able to list the potential costs related with the telemedicine employment, a recap that allows the user to consider the kind of resources involved.

Thus, the second result of this work is to present a recap of cost items that summarize which kind of resources are involved in providing healthcare services by telemedicine infrastructure. We can cluster these costs within two main groups:

(a) IMPLEMENTATION COSTS: starting investments
(b) DELIVERY COSTS: operational and service

Accordingly, from the 17 authors extracted by the EBSCO—Business Source Complete database, Table 3 shows which cost items have to be considered by healthcare management in approaching with a telemedicine employment feasibility analysis.

5 Conclusion

This dissertation about cost analysis of telemedicine could be considered the starting point to further considerations about a very complex theme, still nebulous and concerning with a multitude of matters.

There is not a clear and homogenous literature background about the topic yet. Moreover, this literature is completely multidisciplinary.

Clearly, this is not a surprising finding; and this is easily justifiable from the interdisciplinarity of the telemedicine theme, which relates with clinical, technical, economic, and organizational matters.

Table 3 Costs of telemedicine employment

Implementation costs	Delivery costs
Investment costs	*Operational and service costs*
Hardware[a]: • Medical devices/home recognition • Office device: – PC – Printer – Office furniture – Etc.	Personnel: • Physician • Nurse • Technical • IT • Call center • Administrative
Software[a]: • License	Staff education: • Continuous training • Upgrade training
IT infrastructure[a]: • Server • Internet • Utility lines • Privacy security	Service hardware: • Depreciation of medical device/home recognition • Rental medical device/home recognition
Call center equipment[a]: • Office device: – PC – Printer	Organization hardware: • Depreciation: – Office device – IT infrastructure – Call center equipment – Office space
Construction—remodeling office space[a]: • Office restorations • New building constructions	Software: • Amortization: – License
Setup costs: • Operative flow chart • Privacy rules	Patient support: • Enrolment • Training • Consulting
Personnel: • New hiring • Training • Privacy	Direct material: • Medical consumables • Office consumables
	Utilities: • Electricity • Gas • Water
	Waste disposal

Source: Own elaboration
[a]Leasing, purchase, or other forms of acquisition (e.g., PPP, service)

However, in our opinion, business, managerial, and economic literatures are still scarce on quite new and "hot" topics, which potentially could have a disruptive effect on healthcare. This strongly limits those analyses focused on the telemedicine sustainability in healthcare delivery process.

Although in its first step of inquiry, this contribution goes exactly in this direction: the chapter summarizes different types of cost involved in providing healthcare services by telemedicine infrastructure. In particular, this work has categorized on a general level those resources needed for both implementation and delivery phases of a telemedicine service.

Accordingly, on the basis of the state of the art of the literature, this work contributes to the design of a synthesis framework about cost analysis of telemedicine.

This framework surely fits with the literature fields of healthcare management, managerial control, and eHealth.

With Davalos et al. [1], this chapter contributes to the systematic understanding and treatment of cost of telemedicine. This could support:

(a) Practitioners and researchers in inquiries about "how much this specific telemedicine service costs"
(b) Practitioners and researchers who want to approach the cost-effectiveness analysis of telemedicine employment

Hence, findings of this study are particularly suitable for supporting empirical inquiries about the theme.

Surely, the beginning inquiry, contained in this work, has some limitations. Undoubtedly, the main one refers to the lack in the content analysis of those cited papers identified by bibliometric examination of Scopus and Web of Science databases.

This limitation, however, lays the foundation for the further and imminent research, which is going to involve the cost-effectiveness of telemedicine. Its aim is to compare costs and benefits (both clinical and economic) achievable by telemedicine potentialities in healthcare delivery process. Accordingly, this could contribute to foster the sustainability of high-performance healthcare system.

References

1. Dávalos, M. E., French, M. T., Burdick, A. E., & Simmons, S. C. (2009). Economic evaluation of telemedicine: Review of the literature and research guidelines for benefit–cost analysis. *Telemedicine Journal and E-Health, 15*(10), 933–948.
2. Fatehi, F., & Wootton, R. (2012). Telemedicine, telehealth or e-health? A bibliometric analysis of the trends in the use of these terms. *Journal of Telemedicine and Telecare, 18*(8), 460–465.
3. Sood, S., Mbarika, V., Jugoo, S., Dookhy, R., Doarn, C. R., Prakash, N., & Merrell, R. C. (2007). What is telemedicine? A collection of 104 peer-reviewed perspectives and theoretical underpinnings. *Telemedicine Journal and E-Health, 13*(5), 573–590.
4. Italian Board of Health. (2014). Telemedicina—Linee di indirizzo nazionali. http://www.salute.gov.it/imgs/C_17_pubblicazioni_2129_allegato.pdf. Accessed 28 Sep 2018.
5. Rosenberg, C. N., Peele, P., Keyser, D., McAnallen, S., & Holder, D. (2012). Results from a patient-centered medical home pilot at UPMC health plan hold lessons for broader adoption of the model. *Health Affairs, 31*(11), 2423–2433.
6. Zholudev, V., Safir, I. J., Painter, M. N., Petros, J. A., Filson, C. P., & Issa, M. M. (2017). Comparative cost analysis: Teleurology vs. conventional face-to-face clinics. *Urology, 113*, 40–44.
7. Hersh, W. R., Hickam, D. H., & Erlichman, M. (2006). The evidence base of telemedicine: Overview of the supplement. *Journal of Telemedicine and Telecare, 12*, 1), 1–1), 2.

8. Elford, D. R., White, H., St John, K., Maddigan, B., Ghandi, M., & Bowering, R. (2001). A prospective satisfaction study and cost analysis of a pilot child telepsychiatry service in Newfoundland. *Journal of Telemedicine and Telecare, 7*(2), 73–81.
9. Jerant, A. F., Azari, R., & Nesbitt, T. S. (2001). Reducing the cost of frequent hospital admissions for congestive heart failure: a randomized trial of a home telecare intervention. *Medical Care, 39*(11), 1234–1245.
10. Thaker, D. A., Monypenny, R., Olver, I., & Sabesan, S. (2013). Cost savings from a telemedicine model of care in northern Queensland, Australia. *The Medical Journal of Australia, 199*(6), 414–417.
11. Palozzi, G., Chirico, A., & Calò, L. (2014). Cost accounting del follow-up annuo per il controllo remoto dei defibrillatori impiantabili. *Mecosan, 90*, 71–98. FrancoAngeli, Milano.
12. Diodato, V., & Brooks, T. A. (1996). Dictionary of bibliometrics. *The Library Quarterly, 66*(1), 103–104.
13. Tranfield, D., Denyer, D., & Smart, P. (2003). Towards a methodology for developing evidence-informed management knowledge by means of systematic review. *British Journal of Management, 14*(3), 207–222.
14. Thorpe, R., Holt, R., Macpherson, A., et al. (2005). Using knowledge within small and medium-sized firms: A systematic review of the evidence. *International Journal of Management Reviews, 7*(4), 257–281.
15. Denyer, D., & Neely, A. (2004). Introduction to special issue: innovation and productivity performance in the UK. *International Journal of Management Reviews, 5/6*, 131–135.
16. Buchanan, A. H., Datta, S. K., Skinner, C. S., Hollowell, G. P., Beresford, H. F., Freeland, T., Rogers, B., Boling, J., Marcom, P. K., & Adams, M. B. (2015). Randomized trial of telegenetics vs. in-person cancer genetic counseling: cost, patient satisfaction and attendance. *Journal of Genetic Counseling, 24*(6), 961–970.
17. Burri, H., Sticherling, C., Wright, D., Makino, K., Smala, A., & Tilden, D. (2013). Cost-consequence analysis of daily continuous remote monitoring of implantable cardiac defibrillator and resynchronization devices in the UK. *Europace, 15*, 1601–1608.
18. Hameed, A. S., Modre-Osprian, R., & Schreier, G. (2017). Identification of cost indicators with significant economic impact on the total treatment costs of chronic heart failure patients—A meta-analysis. *Studies in Health Technology and Informatics, 236*, 161–168.
19. Kovács, G., Somogyvári, Z., Maka, E., & Nagyjánosi, L. (2017). Bedside ROP screening and telemedicine interpretation integrated to a neonatal transport system: Economic aspects and return on investment analysis. *Early Human Development, 106–107*, 1–5.
20. Kumar, G., Falk, D. M., Bonello, R. S., Kahn, J. M., Perencevich, E., & Cram, P. (2013). The costs of critical care telemedicine programs: A systematic review and analysis. *Chest, 143*(1), 19–29.
21. Mariño, R., Tonmukayakul, U., Manton, D., Stranieri, A., & Clarke, K. (2015). Cost-analysis of teledentistry in residential aged care facilities. *Journal of Telemedicine and Telecare, 22*(6), 326–332.
22. Mason, J. M., Young, R. J., New, J. P., Gibson, J. M., Long, A. F., Gambling, T., & Friede, T. (2006). Economic analysis of a telemedicine intervention to improve glycemic control in patients with diabetes mellitus: illustration of a novel analytic method. *Disease Management and Health Outcomes, 14*(6), 377–385.
23. Razavi, H., Copeland, S. P., & Turner, A. W. (2016). Increasing the impact of teleophthalmology in Australia: Analysis of structural and economic drivers in a state service. *The Australian Journal of Rural Health, 25*(1), 45–52.
24. Roberts, L. J., Lamont, E. G., Lim, I., Sabesan, S., & Barrett, C. (2012). Telerheumatology: An idea whose time has come. *Internal Medicine Journal, 42*(10), 1072–1078.
25. Rollo, M. E., Burrows, T., Vincze, L. J., Harvey, J., Collins, C. E., & Hutchesson, M. J. (2018). Cost evaluation of providing evidence-based dietetic services for weight management in adults: In-person versus eHealth delivery. *Nutrition & Dietetics: The Journal of the Dietitians Association of Australia, 75*(1), 35–43.

26. Rosenberg, C., Kroos, K., Rosenberg, B., Hosten, N., & Flessa, S. (2013). Teleradiology from the provider's perspective: Cost analysis for a mid-size university hospital. *European Radiology, 23*(8), 2197–2205.
27. Theodore, B. R., Whittington, J., Towle, C., Tauben, D. J., Endicott-Popovsky, B., Cahana, A., & Doorenbos, A. Z. (2015). Transaction cost analysis of in-clinic versus telehealth consultations for chronic pain: Preliminary evidence for rapid and affordable access to interdisciplinary collaborative consultation. *Pain Medicine, 16*(6), 1045–1056.
28. Vitacca, M., Comini, L., Assoni, G., Fiorenza, D., Gilè, S., Bernocchi, P., & Scalvini, S. (2012). Tele-assistance in patients with amyotrophic lateral sclerosis: Long term activity and costs. *Disability and Rehabilitation. Assistive Technology, 7*(6), 494–500.
29. Williams, C., & Wan, T. T. (2016). A cost analysis of remote monitoring in a heart failure program. *Home Health Care Services Quarterly, 35*(3-4), 112–122.
30. World Health Organization. (2001). *WHO Regional Office for Europe. Institutionalisation of health technology assessment: Report on a WHO meeting*. June 30–July 1, Bonn, Germany.
31. Drummond, M., Tarricone, R., & Torbica, A. (2013). Assessing the added value of health technologies: Reconciling different perspectives. *Value in Health, 16*, 7–13.
32. Sampietro-Colom, L., Lach, K., Cicchetti, A., Kidholm, K., Pasternack, I., Fure, B., et al. (2015). *The AdHopHTA handbook: a handbook of hospital-based Health Technology Assessment (HB-HTA); Public deliverable*. The AdHopHTA Project (FP7/2007-13 grant agreement nr. 305018), Available online: http://www. adhophta.eu/handbook.

How Millennials Will Use Autonomous Vehicles: An Interview Study

Christina Pakusch, Gunnar Stevens, and Dirk Schreiber

1 Introduction

The use of fully autonomous vehicles (AV) will have a disruptive effect on the transport market, as it will fundamentally change current patterns of use, ownership, and business models. With the advent of fully autonomous driving, new business models are also emerging, because the self-driving vehicle can be put on the road in a variety of forms: as a private car, taxi, bus, carsharing vehicle, or a shared taxi [1]. Some experts and researchers predict a positive effect on vehicle ownership and existing growing emissions and congestion problems [2, 3]. In particular, the combination of self-driving technology and mobility as a service concept is being discussed in this context. The result of this combination could be "shared autonomous vehicles" (SAV) [2, 4–6]: fully automated vehicles that consumers do not own privately but use flexibly according to their needs and thus share with others. Since SAV will navigate, collect, and carry passengers autonomously, these innovative services could overcome many disadvantages of traditional taxis and carsharing such as high labor costs or the distance users have to overcome to reach a carsharing vehicle [7]. The success of such mobility concepts, however, depends on user acceptance and its impact on users' mobility behavior.

So far, these are only visions of the future based on a technology that is not yet ready for series production. However, developments are progressing rapidly, with all the major players in the automotive industry and the tech giants in the process

C. Pakusch (✉) · D. Schreiber
Bonn-Rhein-Sieg University of Applied Sciences, St. Augustin, Germany
e-mail: christina.pakusch@h-brs.de; dirk.schreiber@h-brs.de

G. Stevens
University of Siegen, Siegen, Germany
e-mail: gunnar.stevens@uni-siegen.de

© Springer Nature Switzerland AG 2020

D. Cagáňová, N. Horňáková (eds.), *Mobility Internet of Things 2018*,
EAI/Springer Innovations in Communication and Computing,
https://doi.org/10.1007/978-3-030-30911-4_33

of advancing their own developments. This makes it even more important for the transport industry to actively address the possibilities and consequences of AV. This does apply not only to the producers and future operators of SAV, who have an interest in their vehicles being used intensively, but also to the public transport (PT) sector, for which the advent of AV can also be associated with changes and possible risks. Consequently, it is necessary for the industry to address these developments at an early stage, to position accordingly, take advantage of opportunities and counteract risks.

To better understand attitudes and potential effects of the advent of SAV on the mobility behavior of millennial-aged users, we conducted qualitative interviews with a qualitative sample of 25 users. Our findings thus contribute to the ongoing discourse about (S)AV by shedding light on expected adoption and use behavior. The results show that, first, millennials tend to prefer a driverless taxi over a traditional taxi when the cost of a driverless taxi is correspondingly lower and, second, an offer of SAV might have a strong impact on users' mobility behavior; in particular, it will be easier to convince current PT users to adopt SAV than to convince car owners to do so.

The overall structure of the paper takes the form of seven sections, including this introductory section. Section 2 begins by laying out the theoretical background and looks at how (S)AV and their effects are being discussed in scientific literature. Section 3 is concerned with the methodology used for this study. Section 4 presents the findings of the research, focusing on the two key themes. Section 5 analyzes the results of interviews discussing the meaning of the results for the transportation sector. After presenting limitations of our study, the final section gives a brief summary and critique of the findings.

2 Related Work: Shared Autonomous Vehicles

Fully autonomous driving is understood as the autonomous, targeted driving of a vehicle in real traffic without the driver's intervention. An international standard for full automation has been defined to ensure a consistent taxonomy and understanding for on-road motor vehicle automated driving systems [8]. According to that J3016 standard, six levels of automation can be distinguished for road traffic depending on the level of driver intervention and attention required: no automation (0), driver assistance (1), partial automation (2), conditional automation (3), high automation (4), and final stage of full automation (5). Thus, the vehicle is fully automated when the system dynamically and fully autonomously performs all aspects of the driving tasks under all roadway and environmental conditions without the need for a human driver. In this chapter, we refer to the highest level of automation, in line with other studies about AV [4, 9, 10]. We think of vehicles as capturing the driving environment through sensors, communicating between cars and infrastructure independently and navigating without human intervention. Companies such as Audi,

NVIDIA, and NuTonomy have recently announced that these types of driverless cars will be available on the market from 2020 [11–13].

Vehicle automation not only opens up new possibilities for private mobility but also enables new ways of using sharing-based transportation concepts. In that discourse, different terms for these concepts are being used such as Robo-Taxis, autonomous taxis, and autonomous carsharing; in the scientific literature, the term shared autonomous vehicles (SAV) has become established [2–4, 6]. Much of the current literature on SAV pays attention to the adoption of SAV and their impact on transportation systems. Research has recently focused on two different approaches: model-theoretic simulations of SAV fleet on a macro level and empirical user studies.

Simulation studies predict that SAV have the potential to reduce the number of private vehicles by more than 90% [3, 14–16]. Burns et al. [17] simulate an SAV fleet for small-sized, medium-sized, as well as metropolitan cities. They investigate that the average SAV cost per mile is 31% less than the average cost of a privately owned vehicle and calculate that all trips could be executed with a fleet of only 15% of the number of privately owned vehicles. Fagnant and Kockelman [16] and Fagnant et al. [3] conclude that each SAV can replace around 9 or 11 conventional vehicles with reasonable wait times of 1 min or less. Burghout et al. [18] have similar results in their simulation of an SAV fleet. They conclude that it is possible to replace private car commuter trips in a metropolitan area with self-driving on-demand taxis needing less than 5% of the passenger vehicles currently in operation and of the parking places. Spieser et al. [19] simulate an SAV transportation for Singapore and calculate that an effective fleet needs to have one-third of today's private vehicles. The authors generally agree that such sharing concepts offer great economic and ecological potential.

Besides technological, economical, or ethical oriented studies on autonomous driving [4, 10, 20–22], there is a growing number of user acceptance studies. However, most of those studies focus on autonomous cars in general while neglecting the question of ownership as a relevant category. De Winter et al. [23], for instance, conducted a survey finding that only 50% of the respondents had heard of Google's Driverless Car, and 74% of the respondents stated that fully automated driving will reach a 50% market share by 2050. Kyriakidis et al. [24] investigated the public opinion on automated driving showing that over 70% of the respondents agreed that manual driving is enjoyable but around 50% also agreed that fully automated driving would be enjoyable. With regard to autonomous cars in general, a study by Payre et al. [10] reveals that a large majority of the population have a positive attitude and can imagine buying and/or using them. Factors such as age and gender, individual personality, pre-experience with partly autonomous cars, characteristics of the innovation, driving environment, and manufacturer's reputation affect the users' acceptance [25, 26]. Only a few studies have explicitly addressed the willingness to adopt SAV. When asking if people would adopt self-driving technology and in what form, Howard and Dai [27] found that people's willingness to use self-driving cars as taxis was not high, with wealthier people being willing to use a self-driving taxi more often than those with lower income. The study however showed that self-driving taxis still seem more popular than traditional taxis. Krueger et al. [4] identify a young age and multimodal travel patterns to be typical characteristics of potential SAV users. Results of their stated choice analysis showed that service attributes including the given travel time, waiting time, and travel cost are significant determinants of SAV use and dynamic ridesharing acceptance. In a recent study by Moreno et al. [28], 41.5% of the respondents stated that they were willing to use AV as SAV. Menon et al. [5] empirically investigated the willingness of people to relinquish one of their household vehicles if SAV are available. They found that for single-vehicle households, it is the male, well-educated millennials (born in the 1980s and 1990s) who would most likely give up their private vehicle in favor of SAV. The results of Pakusch et al. [7] show that privately owned cars—whether traditional or fully autonomous—will continue to be preferred over shared vehicles in the future.

While quantitative studies to simulate SAV use and hypothetical considerations regarding the influence of AV on mobility behavior have been performed, qualitative studies are lacking. Qualitative methods offer an effective way of shedding light on underlying reasons and aspects why users come to a decision. We therefore contribute to the field of user research in the context of AV by qualitatively unveiling the user's perceptions and attitudes towards SAV and by analyzing potential changes in mobility behavior.

3 Methodology

To address our research question on how millennials will respond to the advent of SAV regarding changes in their mobility behavior, we conducted 25 problem-centered narrative interviews [29]. As a well-established method in empirical social research, qualitative interviews are particularly suitable for exploring the more unknown research fields and the subjective perspectives of users [30]. In our study we focused on the so-called millennials, those born after 1982, a user group that is particularly interesting to consider when exploring attitudes towards technological innovations. To recruit participants, we used a combination of qualitative sampling [31] and snowball sampling [32]. We did not make any demands on previous knowledge of AV; the interviewees took part in the interviews on a voluntary basis and without being compensated. The participants consisted of 8 women and 17 men aged 20–34 years. All participants held a driver license and 19 of them owned a private car. While 15 respondents lived in rural areas and 10 in urban areas, all participants had access to PT and none of them had physical impairments necessitating car use. With those mobility patterns they pretty good represent millennials in Germany (Table 1).

The interview guidelines contained the following topics: We started by asking participants about their current mobility practices and their attitudes to established means of transport. In the course of the conversation, we confronted the participants with the concept of AV and asked them what they had already heard about this topic.

Table 1 Sociodemographic data of study participants

ID	Gender	Age	Job	Place of residence	Driving license	Owning a car	Main travel mode
01	Male	27	Student	Urban	Yes	No	PT
02	Male	22	Student	Rural	Yes	Yes	Car
03	Female	23	Medical assistant	Rural	Yes	Yes	Car
04	Male	23	Marketing manager	Rural	Yes	Yes	Car
05	Male	28	Cutting machine operator	Rural	Yes	Yes	Car
06	Female	26	Geriatric nurse	Rural	Yes	Yes	Car
07	Male	34	Engineer	Urban	Yes	Yes	Car
08	Male	22	Student	Rural	Yes	Yes	Car
09	Female	22	Student	Rural	Yes	Yes	Car
10	Male	21	Student	Rural	Yes	Yes	Car
11	Female	26	Automobile sales person	Rural	Yes	Yes	Car
12	Male	26	Student	Urban	Yes	No	PT
13	Female	34	Architect	Urban	Yes	Yes	Car
14	Male	26	Student	Urban	Yes	Yes	Car
15	Male	21	Student	Rural	Yes	No	PT
16	Male	21	Student	Rural	Yes	Yes	Car
17	Male	22	Industr. Mgmt. assistant	Rural	Yes	Yes	Car
18	Female	20	Apprentice	Rural	Yes	Yes	Car
19	Male	24	Student	Urban	Yes	No	PT
20	Male	22	Assistant tax consultant	Rural	Yes	Yes	Car
21	Female	20	Student	Urban	Yes	No	PT
22	Male	22	Student	Urban	Yes	Yes	Car
23	Female	25	Student	Urban	Yes	No	PT
24	Male	30	Nurse	Rural	Yes	Yes	Car
25	Male	22	Student	Urban	Yes	No	PT

Using a press release of a planned Uber-SAV fleet as an envision stimulus [33], we introduced the topic of SAV to the interviewees, discussing their opinion on AV and SAV as well as the impact of such a service on their mobility behavior. Interviews lasted between 20 and 45 min. All interviews were conducted face to face, were audio-recorded, fully transcribed, and analyzed independently by three authors. The analysis method uses elements of grounded theory [34] and the technique of qualitative content analysis [19]. To organize the data material, deductive and inductive categorization was built with the help of the analysis tool MAXQDA. The formation of the category system was based on the semi-structured interview guideline, following the logic of a deductive category application. The code system was additionally expanded and inductively differentiated by the coders. To do this, we discussed and matched identified categories through an iterative process to ensure consensus and to follow the principle of intersubjective replicability.

4 Findings

First, we give an overview of the situations in which users can imagine using an SAV before we show how users generally think about the question whether to use a traditional taxi or a driverless automated taxi. In the third section, we address the central question of the extent to which the advent of SAV could affect the mobility behavior of users.

4.1 Use Scenarios

Respondents described various reasons and purposes they would use SAV for. In the course of the interviews, we specifically asked them in which situations or for what journeys they would use SAV. In addition, they also casually mentioned routes in the conversations to illustrate other comments. The most frequently mentioned routes include the shorter, every day, and familiar routes to work or home, grocery shopping, and going to the train station or the airport or to transport larger things.

> I'd use it, so if the price is right, I'd use it to shop, go to work, travel longer distances and visit someone, for example, to drive to the airport or something like that. (P1)

As P1's commentary shows for other participants as well, the price plays an important role for the participants. Costs are repeatedly addressed directly or casually as a side condition. Our millennials state that they will only use SAV if they had a significant financial advantage over alternative traditional means of transport. Many car users point out that the costs of an SAV must not exceed the cost of a car and an SAV should be as expensive or cheaper than a traditional taxi. Only for some persons social interaction with the taxi driver outweighs advantages of SAV when costs are left out of the equation:

> … under no circumstances should it be more expensive than a taxi, then you would, of course, prefer the usual taxi. (P18)

The way in which P18 talks about the conditions for using SAV is quite undifferentiated. For her it is clear that only a lower cost would make her prefer a driverless taxi to a traditional taxi. This means that apart from the cost, the traditional taxi generally has advantages over an SAV.

Furthermore, those people who mainly use private cars often refer to the costs of using a car. They state that they will only make use of SAV if the offer will be cheaper than a private car:

> And if [autonomous] carsharing is cheaper than owning your own car, this is the only relevant reason to use [autonomous] carsharing. (P2)

Also, P2 speaks of the costs as *"the only relevant reason to use"* SAV. Just like him, the car-using participants feel their own car to be superior to SAV in all other characteristics. Particularly the permanent availability, the high degree of flexibility, and the comfort are mentioned as special advantages of the own vehicle, which is always ready in front of the door. In summary, it can therefore be said that most respondents are not willing to spend more on the use of SAV than on their current means of transport. In the eyes of most respondents, SAV do not offer a relative advantage over the previously preferred means of transport, which would justify a higher price. Only 4 of the 25 people interviewed were willing to accept a surcharge. Two of them, people who do not currently have an own car and use PT, found higher cost than the current cost of PT reasonable. In a direct comparison of PT and SAV, they see a clear relative advantage:

> ... if I have the opportunity, then I would be ready to pay a little more than the normal bus and train fares for it. After all, there is still a lot of added value to it. [...] The idea is great because you have the advantages of a car and at the same time the advantages of public transport. Time is not completely lost, because you can use it in a different way than when you drive. (P23)

P23 describes that for her, the concept of an SAV combines the advantages of an own car (speed, flexibility, availability, spontaneous use, comfort) with the advantages of PT and thus brings together the best of both variants. As an advantage of PT, she mentions the use of time during the journey. While an SAV is considered to be disadvantageous compared to a car—with the exception of better use of time—the advantages of such a concept clearly exceed those of PT.

To get back to the potential trip purposes, which have been mentioned, it is worth noting that participants so far use their main travel modes for such everyday journeys like grocery shopping or going to work or home. The fact that they can imagine making these journeys with SAV, and not only using them for special occasions, indicates that SAV has some potential to replace their current main travel modes and become a routine travel mode.

In addition to the daily routes, the participants mentioned those situations in which they would currently prefer to take a traditional taxi or would like to take a traditional taxi but are not willing to bear the current taxi costs. Participants describe those situations as situations when they are not able to drive or do not feel fit to drive. In this context, participants often mentioned that SAV could be used well if one was at a party, had drunk alcohol, was exhausted from work, or was in an emergency situation as an ambulance.

> You can go to a party by car and drive back home with the same car being as drunk as you like. (P1)

> ... if you have had a long day and are very tired, then you should not drive by yourself any more, then it's of course good if the car takes over the driving for you. (P25)

Even on routes where users do not feel any pleasure or may even feel stressed by driving, they would prefer to be driven by SAV instead of driving themselves. In addition, the millennials mentioned long and monotonous journeys such as on a motorway or when stuck in a traffic jam as suitable to be carried out by SAV. They see advantages in the fact that they can use the time in the car sensibly and can pursue other activities than driving.

Some participants feel particularly stressed during trips in foreign cities or areas. For those on unknown routes, in very confusing cities and areas where one does not know how to get home, the participants would like to have their control delivered to the SAV computer and then be transported.

> I personally don't like driving in big cities so much, where the roads have multiple lanes (. . .) you get nervous quickly and don't know where you have to go, and it would be quite good if you could rely on the fact that you don't have to drive yourself. (P18)

4.2 Scenario: Traditional Taxi vs. Fully Autonomous Taxi

Our millennials use taxis only very rarely. They recognize the benefits that taxis offer. They appreciate the high comfort and being able to rely on the drivers' knowledge of the area. In addition, those who are PT users appreciate the fact that, compared to buses, taxis do not have long waiting times and that users are picked up at the desired location. Participants who live in rural areas where PT no longer runs regularly at night and on weekends use taxis as an alternative to PT. However, all millennials describe the taxi as a very expensive means of transport that is only used in exceptional cases.

> I only order a taxi now, if there is no other way out, if I can't be taken along by someone else either. (P7)

The high costs of taxis are repeatedly mentioned as the reason why this comfortable means of transport is not used more frequently. The comments in the context of taxi usage show again how very cost sensitive the millennials are when it comes to travel mode choices.

We asked the participants to imagine the following scenario: they were guests at a party, had some drinks, and now want to go home. We then asked them to choose between two means of transport: either they could take a traditional taxi or they could take a fully autonomous taxi that is 30% cheaper than the traditional taxi. Only five people decided to take the traditional one:

> So now, I'd choose the traditional taxi, because I trust people more than the computer when it comes to driving, even if I would have this 30% cost saving. (P17)

Just as P17, those five participants who opted for this variant mentioned safety as a reason. They trust in the familiar variant and do not want to rely on driverless services. As long as there are still taxi drivers, some say, this means that autonomous driving has not yet fully established itself due to safety aspects.

When AV are being discussed in general, many respondents describe job losses as a negative consequence of the increasing automation of the mobility industry. They assess this factor ambivalently: On the one hand, they see it as an economic advantage for the operating taxi companies as they have lower labor costs but, on the other hand, they see the job losses as a clear disadvantage. As soon as the concrete decision situation is at stake, this social aspect takes a back seat. In the present case, the advantage of lower costs clearly outweighed the disadvantage of job losses. This advantage led to the other 20 people opting for the less expensive driverless taxi:

> As a student, I think I would definitely take the cheaper autonomous driving service. Simply because it's cheaper then and yes, the costs, that's what I'm all about. (P22)

With the addition "*As a student*," P22 points to his current situation, which is characterized by the fact that he does not have a large income and therefore keeps his expenses for transportation as low as possible. At the same time, this restriction suggests that his decision is only valid for this phase of his life and may change later if the budget no longer determines his decisions as much as it does now.

For the voters of the traditional taxi, safety was the most important argument for the classic taxi. At the same time, three respondents also named safety as the reason for choosing the SAV:

> ... because I don't always know with the taxi drivers whether they are still roadworthy themselves. Autonomous cars might also be more likely to follow the traffic rules than taxi drivers; driving with less risk. Well, I guess [traditional] taxis aren't very safe. (P14)

For these participants it is not the potential advantages of the SAV that are decisive, but the disadvantages of the current taxi services, which are characterized in part by poor driver behavior.

4.3 Impact on Mobility Behavior

If SAV were available at a high service level and at comparatively low cost as some simulation studies suggest, this availability would have a huge impact on respondents' mobility behavior. Some interviewees thought that, in this case, they did not need their own car and would abandon the private car or would not even acquire one at all. As a reason for this decision, the respondents cited that using SAV would be cheaper than owning and maintaining a private car, and they did not see any disadvantages with a high-service-level SAV compared to their own car.

> If things go well and the car is really quickly available, and you don't have long waiting times, I could imagine selling my own car. (P17)

However, 11 people did not want to give up their own car; 3 of them would still want to drive mainly with their own car but 4 of them can imagine driving less with their private car. People who do not want to do without their own car appreciate the private car being located directly in front of their house so that the car can be used very flexibly in the shortest possible time and without prior booking. In addition, some respondents said that they would miss the driving pleasure if they could not control SAV themselves.

> I still wouldn't want to do without my own car because you sometimes still need it to be able to drive right in front of your door, sometimes it has to be right in front of your front door so that you can start driving straight away … (P22)

> I would probably leave my own car at home more often and fall back on it [SAV]. (P11)

The SAV offer would not only affect car ownership and use but also change the use of other modes of transport. Ten people would reduce or even stop using PT to the benefit of SAV. This change was attributed, above all, to the time savings that would result from using SAV compared to PT. In addition, they would also appreciate the greater flexibility and speed of the new service, as well as the advantage of not having to get to or from the bus stop or station.

> So, I think that if it saves some time and is also financially affordable, I might say I use less public transport. I don't think I'd give up my own car for that. (P8)

Four people stated that they would use SAV like a traditional taxi. One person said that she would walk less. Two respondents indicated that they would undertake additional journeys, thus increasing their overall demand for mobility. Only four people were convinced that they would not change their mobility behavior at all.

5 Discussion

The analysis of the interviews shows that millennials are open to the use of SAV. While it turned out that SAV will mainly compete with traditional taxis and PT, we also noticed that millennials are very cost sensitive in their choice of means of transport. The costs represent the condition that is mentioned as a prerequisite for use in all aspects. Accordingly, most users choose an autonomous taxi when choosing between traditional taxis and cheaper autonomous taxis. The social aspect that the profession of a taxi driver could become obsolete with the advent of autonomous vehicles does not play a role for the millennials in the concrete decision situation. In this respect, it is important to address the consequences of autonomous driving at an early stage, since the profession of professional drivers could at least be altered, if not become superfluous. That is why taxi drivers and taxi organizations, in particular, should draw up a plan as to what kind of right of existence they still have in future and what services and added value they can offer their passengers that go beyond driving.

Further, the analysis suggests that it will generally be easier to convince current PT users to use SAV. Given benefits such as a lower travel time, less transfers, less people to share the vehicle when compared to a bus or train, as well as easier ways of transporting goods support this assumption and stand in line with empirical studies that show how users consider characteristics of private cars, PT, and SAV in comparison to each other [35]. If SAV journeys characterized by a very high level of comfort were financially affordable—as some studies suggest they will be [3, 14–16]—users would prefer to use SAV as a means of transport and might be willing to travel longer distances as they can make good use of the time in the SAV [36]. In contrast, it will be harder to convince private car users and to make them forego private car ownership in favor of SAV. Longer travel times due to detours, less flexibility, as well as users' fear to lose control and driving pleasure reduce the likelihood that drivers will renounce their car. These results are consistent with the expectations of other nonempirical studies [4]. If this would be the case—always remembering that we argue highly speculatively—SAV will result in an increase of road traffic and vehicle-km. Consequently, the efficient PT services will be used less, while the use of private vehicles will increase. Thus, contrary to the predicted and desired effects of SAV to alleviate traffic problems, especially in cities, SAV would contribute to a strengthening of those problems. These implications are of importance for state and environmental policy, PT operators, and planning authorities. The companies that want to offer SAV on the market are interested in achieving the greatest possible success with SAV and generating profits. Accordingly, they will design and offer SAV in such a way that they are as attractive as possible for potential users. In this case, the operators' objectives do not coincide with environmental policy objectives. While the future SAV operators will be interested in an extensive use, it is in the interest of environmental policy to reduce vehicle miles traveled by emission-intensive means of transport. The policy must intervene here by influencing the operators of SAV. To counter this effect, it may therefore be necessary to regulate the SAV market from the very first moment. In addition, environmentally friendly means of transport must be further strengthened, and more effort must be put into making PT more attractive so that in future they are competitive not only against private (autonomous) vehicles but also against SAV.

With its qualitative character, our study does not provide reliable, representative forecasts. Based on the analysis, we suggest to quantitatively survey the questions whether millennials will prefer autonomous taxis to traditional taxis and whether PT users are more likely to adopt SAV than car users are.

6 Limitations

There are two major limitations to the interpretation of the results. First, the sample is small and focuses on a young average age. Since the study was conducted in Germany, the sample certainly has some specific characteristics that should be considered when looking at the results. Nevertheless, this study provides important insights into user's subjective perceptions of SAV. Second, studies such as those relating to future technologies that are not yet on the market generally pose problems: respondents must imagine something that does not yet exist, which they have not yet dealt with, and which they have not yet been able to test. This lack of knowledge and experience inevitably leads to the respondents being influenced by the interviewers' formulations and specifications. On the other hand, without such a priori studies, it would not be possible to anticipate the consequences of the introduction of new technologies.

7 Conclusion

Autonomous vehicles have the potential to address many of today's traffic problems. Experts predict that vehicle ownership will decrease, and cities will be able to provide efficient mobility with up to 90% fewer vehicles. However, the advent of autonomous vehicles could also have negative consequences: the jobs of professional drivers could become obsolete and low-cost SAV could compete with more environmentally friendly modes of transport such as PT. Our analysis shows that the fear of such negative consequences is justified.

Within our study, we have focused on millennials, the generation that grows up with smart technologies and is expected to be the first users of autonomous vehicles. In semi-structured interviews, we confronted 25 millennial-aged people with the concept of SAV to get insights into their subjective perceptions about SAV and to anticipate how the advent of SAV might change their mobility behavior as a result. If SAV can actually deliver on their promises and forecasts, they will have a major impact on people's mobility behavior. The study results suggest that it will generally be easier to convince current PT users to use SAV than to convince private car users to do so, and that ecological and social aspects play a very subordinate role in the mobility decisions of young people. This aspect has already been observed in earlier studies on travel mode choice [25, 37–39]. This study shows how important it is to deal with the effects of new mobility services at an early stage. It is nevertheless necessary to validate these qualitative results in future research to inform transport planners and policymakers. If these forecasts should be confirmed, the policymakers will have to intervene to prevent a worsening of the existing traffic problems. Furthermore, these results show how important it is to put effort into making environmentally friendlier means of transport such as PT more attractive and to expand their infrastructure so that they can compete with new forms of mobility services such as SAV in the future.

Acknowledgements This research was part of the research project "Einfach Teilen" (Easy Sharing) and was funded by the Federal Ministry of Education and Research, grant number 16SV7916.

References

1. Verband Deutscher Verkehrsunternehmen e. V. (VDV). (2015). *Zukunftsszenarien autonomer Fahrzeuge—Chancen und Risiken für Verkehrsunternehmen.*
2. Fagnant, D. J., & Kockelman, K. M. (2014). The travel and environmental implications of shared autonomous vehicles, using agent-based model scenarios. *Transportation Research Part C: Emerging Technologies, 40*, 1–13.
3. Fagnant, D. J., Kockelman, K. M., & Bansal, P. (2015). Operations of shared autonomous vehicle fleet for the Austin, Texas market. *Transportation Research Record: Journal of the Transportation Research Board, 2536*, 98–106.
4. Krueger, R., Rashidi, T. H., & Rose, J. M. (2016). Preferences for shared autonomous vehicles. *Transportation Research Part C: Emerging Technologies, 69*, 343–355.
5. Menon, N., Barbour, N., Yu, Z., Pinjari, A. R., & Mannering, F. (2018). Shared autonomous vehicles and their potential impacts on household vehicle ownership: An exploratory empirical assessment. *International Journal of Sustainable Transportation, 13*, 1–12. https://doi.org/10.1080/15568318.2018.1443178.
6. Pakusch, C., Bossauer, P., Shakoor, M., & Stevens, G. (2016). Using, sharing, and owning smart cars. In *Proceedings of the 13th International Joint Conference on e-Business and Telecommunications (ICETE 2016)*, pp. 19–30. doi: https://doi.org/10.5220/0005960900190030.
7. Pakusch, C., Stevens, G., & Bossauer, P. (2018). Shared autonomous vehicles: Potentials for a sustainable mobility and risks of unintended effects. *EPiC Series in Computing, 52*, 258–269. https://doi.org/10.29007/rg73.
8. SAE International. (2016). Automated driving levels of driving automation are defined in New SAE International Standard J3016.
9. Pakusch, C., Stevens, G., Boden, A., & Bossauer, P. (2018). Unintended effects of autonomous driving: A study on mobility preferences in the future. *Sustainability, 10*(7), 2404.
10. Payre, W., Cestac, J., & Delhomme, P. (2014). Intention to use a fully automated car: Attitudes and a priori acceptability. *Transportation Research Part F: Traffic Psychology and Behaviour, 27*, 252–263.
11. Korosec, K. (2016). Self-driving taxis hit the streets of Singapore. Fortune. http://fortune.com/2016/08/25/self-driving-taxi-singapore/. Accessed 12 Sep 2018.
12. Reuters. (2017). Chipmaker Nvidia's CEO sees fully autonomous cars within 4 years. https://www.reuters.com/article/us-nvidia-ai-chips/chipmaker-nvidias-ceo-sees-fully-autonomous-cars-within-4-years-idUSKBN1CV192. Accessed 12 Sep 2018.
13. Ross, P. E. (2017). CES 2017: Nvidia and Audi say they'll field a level 4 autonomous car in three years. *IEEE Spectrum*. https://spectrum.ieee.org/cars-that-think/transportation/self-driving/nvidia-ceo-announces. Accessed 12 Sep 2018.
14. Bischoff, J., & Maciejewski, M. (2016). Simulation of city-wide replacement of private cars with autonomous taxis in Berlin. *Procedia computer science, 83*, 237–244.
15. Bunghez, C. L. (2015). The future of transportation-autonomous vehicles. *International Journal of Economic Practices and Theories, 5*(5), 447–454.
16. Fagnant, D. J., & Kockelman, K. (2015). Preparing a nation for autonomous vehicles: opportunities, barriers and policy recommendations. *Transportation Research Part A: Policy and Practice, 77*, 167–181.
17. Burns, L. D. (2013). Sustainable mobility: a vision of our transport future. *Nature, 497*(7448), 181–182.
18. Burghout, W., Rigole, P. J., & Andreasson, I. (2015). Impacts of shared autonomous taxis in a metropolitan area. In *Proceedings of the 94th Annual Meeting of the Transportation Research Board*, Washington DC.
19. Spieser, K., Treleaven, K., Zhang, R., Frazzoli, E., Morton, D., & Pavone, M. (2014). Toward a systematic approach to the design and evaluation of automated mobility-on-demand systems: A case study in Singapore. In G. Meyer & S. Beiker (Eds.), *Road Vehicle Automation* (pp. 229–245). Cham: Springer.

20. Sina, N. N., de Winter, J. C. F., Payre, W., Van Arem, B., & Happee, R. (2017). What impressions do users have after a ride in an automated shuttle? An Interview Study | Request PDF.
21. Nordhoff, S. (2014). Mobility 4.0: Are consumers ready to adopt Google's self-driving car? University of Twente. http://essay.utwente.nl/65590/.
22. Nordhoff, S., van Arem, B., Merat, N., Madigan, R., Ruhrort, L., Knie, A., & Happee, R. (2017). User acceptance of driverless shuttles running in an open and mixed traffic environment. In *Proceedings of the 12th ITS European Congress*.
23. De Winter, J. C. F., Kyriakidis, M., Dodou, D., & Happee, R. (2015). Using CrowdFlower to study the relationship between self-reported violations and traffic accidents. *Procedia Manufacturing, 3*, 2518–2525.
24. Kyriakidis, M., Happee, R., & de Winter, J. C. F. (2015). Public opinion on automated driving: Results of an international questionnaire among 5000 respondents. *Transportation Research Part F: Traffic Psychology and Behaviour, 32*, 127–140.
25. Owczarzak, L., & Żak, J. (2015). Design of passenger public transportation solutions based on autonomous vehicles and their multiple criteria comparison with traditional forms of passenger transportation. *Transportation Research Procedia, 10*, 472–482.
26. Rödel, C., Stadler, S., Meschtscherjakov, A., & Tscheligi, M. (2014). Towards autonomous cars: the effect of autonomy levels on acceptance and user experience. In *Proceedings of the 6th International Conference on Automotive User Interfaces and Interactive Vehicular Applications* (pp. 1–8). http://dl.acm.org/citation.cfm?id=2667330.
27. Howard, D., & Dai, D. (2014). Public perceptions of self-driving cars: The case of Berkeley, California. In *Transportation Research Board 93rd Annual Meeting*.
28. Moreno, A. T., Michalski, A., Llorca, C., & Moeckel, R. (2018). Shared autonomous vehicles effect on vehicle-km traveled and average trip duration. *Journal of Advanced Transportation, 2018*, 8969353.
29. Witzel, A., & Reiter, H. (2012). *The problem-centred interview*. Sage.
30. Sanders, E. B.-N., & Stappers, P. J. (2008). Co-creation and the new landscapes of design. *CoDesign, 4*(1), 5–18.
31. Coyne, I. T. (1997). Sampling in qualitative research. Purposeful and theoretical sampling; merging or clear boundaries? *Journal of Advanced Nursing, 26*(3), 623–630.
32. Biernacki, P., & Waldorf, D. (1981). Snowball sampling: Problems and techniques of chain referral sampling. *Sociological Methods & Research, 10*(2), 141–163.
33. Vavoula, G. N., & Sharples, M. (2007). Future technology workshop: A collaborative method for the design of new learning technologies and activities. *International Journal of Computer-Supported Collaborative Learning, 2*(4), 393–419.
34. Glaser, B. G., Strauss, A. L., & Strutzel, E. (1968). The discovery of grounded theory; strategies for qualitative research. *Nursing Research, 17*(4), 364.
35. Pakusch, C., Stevens, G., Bossauer, P., & Weber, T. (2018). The users' perspective on autonomous driving: A comparative analysis of Partworth utilities. In *Proceedings of the 15th International Joint Conference on e-Business and Telecommunications (ICETE 2018)*.
36. Davidson, P., & Spinoulas, A. (2016). Driving alone versus riding together-How shared autonomous vehicles can change the way we drive. *Road & Transport Research: A Journal of Australian and New Zealand Research and Practice, 25*(3), 51.
37. Collins, C. M., & Chambers, S. M. (2005). Psychological and situational influences on commuter-transport-mode choice. *Environment and Behavior, 37*(5), 640–661.
38. Franzen, A. (1997). Umweltbewusstsein und Verkehrsverhalten-Empirische Analysen zur Verkehrsmittelwahl und der Akzeptanz umweltpolitischer Massnahmen.
39. Gardner, B., & Abraham, C. (2007). What drives car use? A grounded theory analysis of commuters' reasons for driving. *Transportation Research Part F: Traffic Psychology and Behaviour, 10*(3), 187–200.

Smart Collaborative Learning Environment for Visually Impaired Children

Ruhiyati Idayu Abu Talib, Predrag K. Nikolic, Mohd Shahrizal Sunar, and Rui Prada

1 Introduction

This paper introduces Smart Collaborative Learning Environment System (SmCLES) designed to support interaction between children of varying sensory abilities to participate in shared narrative experience controlled by user's action. Our main goal is to enable better interaction, understanding, and communication between these groups, so as to overcome barriers present in traditional interactive experiences requiring a consistent set of abilities among the participants. Specifically, in this paper we are focusing on designing a smart collaborative learning system using existing technology that will enable visually impaired children and sighted children to interactively control and observe a communal narrative together, therefore, allowing them an opportunity to play together. As highlighted in several recent works, children with visual abilities do not show full play behavior [7, 11, 28]. Additionally, they spend more time engaging in solitary play interaction or playing only with their parents, which could negatively affect their development of personal and emotional competencies [37]. This situation can arouse feelings of frustration, instead of the self-efficiency and independence that are necessary characteristics for the social competence of healthy children [37].

R. I. A. Talib · M. S. Sunar
Media and Game Innovation Centre of Excellence (MAGICX), Institute of Human Centered Engineering (iHumEn), Universiti Teknologi Malaysia, Johor, Malaysia

School of Computing, Faculty of Engineering, Universiti Teknologi Malaysia, Johor, Malaysia
e-mail: ridayu2@live.utm.my; shahrizal@utm.my
http://magicx.utm.my

P. K. Nikolic (✉)
School of Creativity and Art, ShanghaiTech University, Pudong, Shanghai, China
e-mail: predragnikolic@shanghaitech.edu.cn

R. Prada
INESC-ID and Instituto Superior Técnico, Universidade de Lisboa, Lisboa, Portugal

© Springer Nature Switzerland AG 2020

D. Cagáňová, N. Horňáková (eds.), *Mobility Internet of Things 2018*,
EAI/Springer Innovations in Communication and Computing,
https://doi.org/10.1007/978-3-030-30911-4_34

Social Model Disability [1, 5, 14, 25] views disability as a societal construct and collective issue—caused by the physical environment, inappropriate or inaccessible services and attitudes, and a lack of understanding—rather than one that derives from the health of an individual or medical impairment. Society is made from many individuals, those with disabilities and those without disabilities. But how can we be a society if we never interact with one another?

The environmental factors mentioned above include the built environment, personal and community attitudes, product and technology, and systems, services, and policies. For example, the social model of disability is embedded in Malaysia's Person with Disabilities (PWD) Act 2008 [17] following United Nations Convention on the Rights of Persons with Disabilities [38] in which it recognizes that *disability is an evolving concept and that disability results from the interaction between persons with disabilities and attitudinal and environmental barriers that hinder their full and effective participation in society on an equal basis with persons without disabilities*.

Highlighted here are two key issues regarding environmental barriers that need to be addressed in order for PWD to participate fully in the society:

- Society interaction with disabled person
- Society understanding of disabled person situations

Thirty years after the concept of social model disability was introduced, there are certain barriers a PWD faces in the society that is yet to be challenged, most notably, the hegemony of special education in school, although in further and higher education some disabling barriers have been removed [25].

In addressing the above issues, the authors believe that this novel system is able to contribute to accelerating the social learning curve for visually impaired children, concurrently, it will also have a role in educating sighted children to be more mindful of their peers' physical limitation by improving interactions, understanding, and communication between these groups.

2 Related Work

2.1 Smart Collaborative Learning Environment

The study by the United Nations titled "Childhood disability in Malaysia: A study of knowledge, attitudes, and practice" [2] reported the opinions of Malaysian teachers regarding the educational environment for children with special needs. Those teachers in mainstream education prefer that children with special needs be taught in special classes, or in special schools, due to their lack of capacity and capabilities. However, according to special education teachers, given the chance, these children should be offered an opportunity to study in mainstream schools, to prepare them for future social integration. The same opinion is shared by special education teachers in France [4]. Here it has been understood that children with special needs want to be accepted by their sighted peers and to develop a friendship with them [4]. Unfortunately, due to a lack of common interests and means to interact together, children with special needs have problems in adapting to inclusive environments and interacting with their normal peers when they are placed in a common classroom.

Referring to Piaget's theory regarding cognitive development, humans, and especially children, construct knowledge through interactions with their environments [29–31]. These actions can be described as below:

- The manipulation of an object, for instance, physically moving items via a hand;
- Conversations or discussions regarding certain ideas.

As a result of that action, if the exploration of the object or idea does not match existing schema, such as a block of knowledge or thought structures, a child will experience temporary cognitive disequilibrium, and intuitively they will cognitively record new knowledge to accommodate the new experience.

The Piaget scholar Seymour Papert [27] added to Piaget's theory that this occurrence will happen more robustly if the learner is engaged in building a public, shareable object, such as a robot or a computer program [27]. Papert called this development "Constructionism." It means that an interactive immersive environment for children encourages their new knowledge development with the help of shareable toolkits.

The rapid development of education technology and its use of collaborative space challenges us to view teaching methods and learning spaces differently. Over the span of 20 years, our understanding of learning and the conditions under which it is facilitated have rapidly evolved. It is understood that children learn multiple life skills from observing their environments, using different ranges of cues and imitating gestures. This process is known as *incidental learning* [12, 13]. However, according to this theory, in the case of visually impaired children, a lack of visual cues automatically puts them at a huge disadvantage in terms of learning those skills naturally. On the other hand, many other researchers argue that visually impaired children learn best through the environment, compared to children with other disabilities such as autism [15, 28, 39]. This is because only the visual cue is one sensory not available to them, whereas they still are able to employ other cues (sound, movement) to learn.

Although visually impaired person's perceptions are much more multisensory than that of sighted persons, the interactive multisensory interfaces that are available currently for blind people mostly focus on tactile sensation and auditory perception [10]. Gori et al. [9] implemented a systematic review of assistive technology available for visually impaired individuals. She concluded that there are limited technological devices available for visually impaired children. An analysis of her work shows that there are none who uses technology to create social settings, through which visually impaired children can interact with their peers. The few that have have only focused on the motor-sensory development of visually impaired children.

2.2 Interactive Storytelling as Learning Tools

[21] concluded that storytelling is effective in educating children about moral values, and in encouraging their development of self-esteem. In his paper, [26] established that:

- Storytelling is a rich, interactive process that facilitates imagination, creative thinking, language abilities, and cooperative learning;
- Learners actively construct their own understandings, building on their current knowledge base;
- Social interactions, and working with others on meaningful tasks, help enhance learning;
- Storytelling offers a limitless opportunity for developing more authentic awareness of and respect for children with diverse languages and cultural backgrounds.

Robin (2008) described in his paper that integration of digital storytelling in the classroom is a powerful teaching and learning tool that engages student and teacher alike [34]. Challenges are: how best to use digital storytelling as a meaningful technology-integrated approach in an inclusive classroom environment and create a collaborative space for the students. Therefore, it is important to understand the method of telling a story. No matter the medium, there are and always have been two methods of telling a story: linear and non-linear. Linear means the story moves from A to B, B to C, and so on. Non-linear is a more abstract method of telling a story. A story can jump from A to C, go back to B, and jump to E. Non-linear stories possess a greater freedom of interactivity; they can begin where they want and they can end where they want. Riedl and Bulitko [33] defined interactive narrative as a form of digital interactive experience in which users create or influence a dramatic storyline through their actions. It also means such stories may become more adaptable to the specific needs, interests, and abilities of the learner.

The earliest-known story generators, Novel Writer and Tale-Spin, are examples of emergent-interactive narrative systems [16, 22]. Novel Writer [16], which was programmed in FORTRAN V on a Univac 1108, is a system which generates 2100-word murder mystery stories, complete with semantic deep structures, in less than 19 s each. It is a program that simulates rational behaviors by characters in the created world. According to Meehan [22], in order to create engaging interactive storytelling environment, story plot and character are the two most important elements of a story.

In this paper the authors conceptualized the system as being toys and tools through which users can gather information about each story's character's personal characteristics, such as their shape, texture, voice, and habits. In order to join realities between sighted and visually impaired children, those characteristics have been possible to collect through multiple senses, including touching, hearing, gesturing, and seeing. The same multisensory principles were applied to the SmCLES main board. The purpose of designing the system as a kind of board game with additional interactive and multisensory features has been to help visually impaired children in their learning processes, particularly in developing spatial reasoning, logic and critical thinking [24].

This paper is organized as follows: Section 3 describes the design approach. Sections 4 and 5 describe the process of prototyping the system. We conclude the paper in Sect. 6 with our expectation of the system and future plan.

3 Design Approach

Nardi [23] stated that designing a system has to make sense, from the perspective of its users, in clearly defining the relationships between artifacts and the social groups they belong to. Furthermore, in designing technology for children with special needs, Frauenberger [8] iterated that holistic approach is needed to bridge the perspectives through design and take into account the impact of technology, the support of creativity, the validity of inspiration, and the design of non-digital generative tools to harness children's imagination.

An experiment done by McAlpine [18] concluded that the quality of learning interactions by children with visual impairments is affected both by the children's level of understanding of mind and by limited or non-existent visual information, even when children have acquired some understanding of mind. McElligot and Van Leeuwen [20] discussed in his paper how to design sound tools and toys for blind and visually impaired children by focusing on their abilities rather than compensating for their disabilities. Proffitt [32] observed that human perception amplifies and strengthens by sensory inputs required for perceiving, motivation, and acting.

Thiemes et al. [36] research showed how sense-making of the technology, collaboration, and learning was enabled through an interplay of system design, programming tasks, and social interactions, and how this differed between children with visual impairment and normal children.

Building on their field study and prototype, the following four design guidelines were identified:

- Visual, audio, olfactory, and haptic esthetic qualities, which are beneficial for social and cultural inclusion;
- Multisensory interactions, accommodating different cognitive and perceptual needs, which ease the processes of learning and memorizing [3];
- Non-linear storytelling which reinforces collaboration between visually impaired and sighted children;
- Scenarios should be ludic and reflective, for instance, engaging students in examining what they are learning and relating that to their other experience.

The principal goal of the SmCLES is to develop a platform through which visually impaired children and sighted children can interact and play together. By doing that, we will help develop cognitive, creative, and social skills in visually impaired children, preventing potential psychological problems in their adulthood, due to lack of social competences. The authors implemented a team brainstorming exercise, focusing their approaches by responding to genuine users' needs [19]. This was especially important because the target group was quite extensive, with different sensory and functional abilities. That is the reason why the researchers decided to use Research through Design [41] as a research approach, incorporating multisensory interactive objects for exploring the potential of using mixing experiential realities and shareable content to assist with the early development of social skills while growing cognitive and creative abilities in visually impaired children.

4 Prototyping SmCLES

The prototyping of the system was carried out in three phases. The first phase was an informal discussion between experts with different backgrounds, including designers, software engineers, and game developers. During the sessions, the researchers came up with ideas for the potential design of the multisensory platform itself, which will be enough to offer visually impaired children possibilities to experience interactive non-linear multisensory storytelling, while actively involving them in creating stories, rather than just passively listening to them. The idea was to design a modular multimodal table supported with multisensory disks that were capable of producing a sound and vibration. The system also needs to be able to detect gestures within a progressing storyline context while being triggers for the central computer to generate the next story part, and to connect it with the previous part. Through multisensory disk features, the researchers were in a position to engage multiple senses and interactive modalities, such as touch, sound, gesture, and vibration.

During this phase, the researchers recorded initial probes undertaken by experts, used to test various design concepts for board and multisensory disks, and their meaningful connections with a non-linear storyline (Fig. 1a).

After the probe sessions, it was concluded that a technical solution should be led towards the development of a central computer, or "The Brain", which will maintain multisensory outputs for the board and the disks, rather than designing autonomous devices with all these features integrated. By doing that the researchers extended

Fig. 1 Prototyping In-Visible Island. (**a**) Experts probe on various design concept for board and multisensory disk. (**b**) Probes session to test the idea of non-linear storytelling system on children. (**c**) Structures of non-linear storytelling used in formative study session

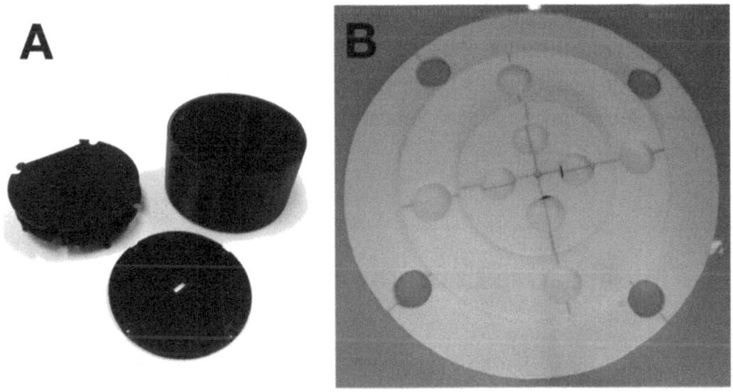

Fig. 2 Low-Fidelity Prototype. (**a**) Circular disk to be fitted with sensor and 3D figure. (**b**) Circular Multimodalities Platform for Storytelling Environment

the platform capabilities to update themselves in the future. *The Brain*, besides controlling sound and vibration, also had a role in controlling the story flow and combining all elements into a meaningful context. Moreover, instead of having only visually impaired children participate in the interactive storytelling, the researchers also included interactions between them and sighted children. The purpose was to create a shareable experience between sighted and visually impaired children, using interactive storytelling as a medium.

In the second phase, based on conclusions from the discussions and the probe sessions, a low-fidelity prototype was designed to initially test user engagement and platform playfulness. The low fidelity design consisted of a table board made out of Styrofoam, and a nonfunctional model of the disk (Fig. 2). The researchers decided to use a round shape for the board design for visually impaired children ease of movement. It is made of three layers (parts) which are in line with the three environments supported by the stories ("forest," "river," and "mountain") with 12 holes as placeholders for disks, which triggered the next story part, and maintained the storyline's context. The idea behind the design of the board with 12 holes is to support four users seated around a table with each will have three different animals character to play and choose from.

In the third phase, the researchers conducted a formative study involving two sighted children aged 5 and 10, facilitated by three researchers and one engineer, to evaluate the board concept and the disks design, using the criteria of playfulness, interactivity, and engagement, in accordance with narrative response read by one of the researchers (Fig. 1b, c). Besides that the researchers wanted to test the suitability of the initially proposed target group of 5–12 year olds. The sessions with children were recorded and analyzed afterward. According to the observations, video analysis, and informal interviews with the two children, the 10-year-old child found the design and the interaction between the parts to be immensely engaging and found the storytelling experience enjoyable. The researchers were aware that a formative study was being done with a small number of participants, and as such it could not be relevant for conclusions related to design decisions and the development of a functional prototype. That is the reason why the researchers combined the collected data with that gathered in the previous session with experts and made design decisions based on research done on the test group of two experts, three researchers, and two users.

5 Smart Collaborative Learning Environment System

SmCLES interactive, multisensory storytelling platform consists of three main parts:

1. **A Central Computational Unit (The Brain)**: The researchers used a Raspberry Pi (Fig. 3a) as the central brain to control the storyline and to send signals to the multisensory disk and the board for audio and vibration modalities (Fig. 3b). A storyline was integrated into the JavaScript Object Notation (JSON) format, and the Java-based application Pickstory was chosen, due to the complexity of the non-linear narrative storyline. The storyline has a direction which complicates the iteration into JSON. To address this issue, the researchers altered its algorithm

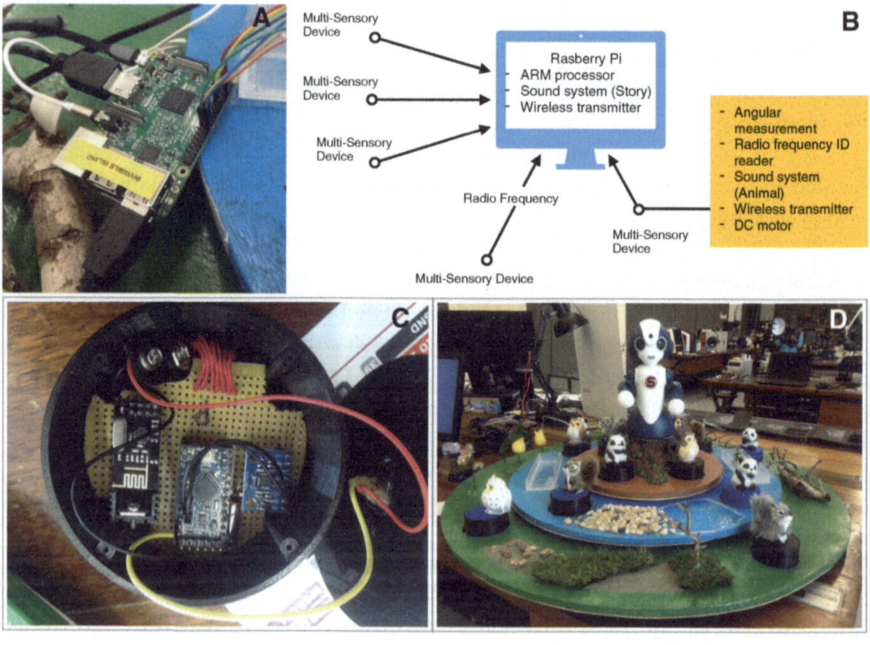

Fig. 3 Prototyping. (**a**) The Brain(Raspberry Pi). (**b**) Illustration of how The Brain works. (**c**) Sensor inside the Multisensory Disk. (**d**) Multi-Modal Interactive Table with three layers of different environment to suit storytelling

to suit the desired story. To execute the program without a monitor, the main program was altered to restart the story, after its branching in JSON ended.

2. **The Multisensory Disk**: To reinforce the tactile and haptic experience, the researchers used 12 tangible playing pieces shaped like disks, which represented characters in the story. The disks were in constant communication with **The Brain**. Based on which character children select, and on which environment they chose out of the "forest," "river," and "mountain," they pick a hole in which to put the disk, The Brain will decide which of the available narrative pieces match the context and will add in the storyline. The story changes according to the players' choices of holes and animal characters.

   To develop the disk, the researchers used Arduino Pro Mini, IMU MPU6050, RFID Reader/Writer, Haptic DC Motor, PCB to assemble all the modules. They also used a Sound Module, Disk Enclosure, and Lithium Polymer Battery—2000 mAh (Fig. 3c). They programmed the disk to have a vibration motor, audio components for generating the sounds of chosen animal characters, and an angular sense. For the angular sense, and to trigger the sound of animals, the MPU 6050 was set to communicate with the Arduino at a 45° angle. When the children move the disk to place it in the "river" environment, the Haptic DC Motor will communicate with Arduino to vibrate, as all the characters are programmed to be "scared of water." By implementing these features, the researchers had the intention to enhance the users' multisensory experiences, increase engagement, and trigger their emotions while creating the story. In the platform's interactive board, the researchers implemented RFID reader for mapping the placement of characters in environment slots, as well as microcontrollers and radio frequencies for realizing the communication between the Brain and the Multisensory Disks. As the stories were developed around animal tales, 3D animals have been placed at the top of each relevant disk, with matching skin and textures, for the benefit of visually impaired children.

3. **The Multimodal Interactive Table**: The board contains three layers of circular movable platforms, which represent the island (Fig. 3d). Each layer consists of a different environment, such as "mountain," "river," and "forest," with different textures and materials being used for easier recognition. For example, pebbles and casing were used to simulate the "river" environment. Each layer has four slots for placing disks. In the board's center, at the top of the mountain, the researchers positioned a robot which played the role of a storyteller, talking to the users.

   The researchers wanted to build a platform that both groups of participants are familiar with, would find easy to understand, and would be able to interact with, without the need for any specific skills development. Evidently, many recent devices designed for visually impaired children have not been widely accepted, as they appear to be hardly acceptable to children [35]. The researchers' intention was also to strengthen positive social interactions between users, through verbal and physical exchanges followed by emotional responses. That was the reason the In-Visible Island platform was designed as a board game. Another aspect the researchers considered when choosing a platform design concept was the ability of such games to teach children valuable social skills, such as communicating verbally, sharing, waiting, taking turns, and enjoying the interaction with others [6]. These accomplishments are found to be important for visually impaired children's development of personal and emotional competencies [40].

As mentioned before, the researchers conceptualized the interactive multisensory disk as both a toy and tool, through which users could gather information by touching, hearing, gesturing, or seeing the story characters, including their shape, texture, voice, and habits. The same multisensory principles have been applied to the main platform board. The purpose of designing the platform as a kind of board game, with additional interactive and multisensory features, has also been to help visually impaired children in the learning process, and with their development of spatial reasoning, logic and critical thinking [35].

6 Conclusions and Future Directions

The principal goal of the SmCLES is to develop a platform through which visually impaired children and sighted children can interact and play together. By doing that we will help develop cognitive, creative, and social skills in visually impaired children, preventing potential psychological problems in their adulthood, due to lack of social competences. The authors implemented a team brainstorming exercise, focusing their approaches by responding to genuine users' needs [19]. This was especially important because the target group was quite extensive, with different sensory and functional abilities.

We designed SmCLES to achieve common multisensory user experience between groups of users with different physical preferences by using tactile and sound stimulations, common elements of realities for both groups, as essential for users' experience design. We assumed that this playful collaboration could lead to positive social interactions between peers and help visually impaired children in social skills development in the early stage as well as improve the social competences of both groups.

The concept of simple non-linear storytelling that we introduce puts the control for story narratives in children's hand. This forced them to interact and communicate with each other in order to achieve desirable storyline and endings. Increased interaction between visually impaired children and sighted children helps the visually impaired children to understand how to work as a team and how to give and receive feedback. As the interaction continued, they will slowly gain self-confidence and increase their self-esteem. Indirectly, sighted children benefited from this interaction as they understood that not everyone in the society is born equal and each individual regardless of any differences has the same feeling of wanting to be accepted. This will make them more empathy towards their differently able peers.

In the future, we are planning to conduct an experiment on SmCLES by having visually impaired children and sighted children play together to evaluate the effectiveness of the system.

References

1. Abberley, P. (1987). The concept of oppression and the development of a social theory of disability. *Disability, Handicap and Society, 2*(1), 5–19.
2. Bedford, K. M. J. (2017, October). *Childhood disability in Malaysia: A study of knowledge, attitudes and practices*. Malaysia: United Nations Children's Fund (UNICEF)
3. Blikstein, P., Sipitakiat, A., Goldstein, J., Wilbert, J., Johnson, M., Vranakis, S., et al. (2016). Project Bloks: designing a development platform for tangible programming for children. *Position paper* (retrieved online), 06–30.
4. Brulé, E., Bailly, G., Brock, A., Valentin, F., Denis, G., & Jouffrais, C. (2016). MapSense: Multi-sensory interactive maps for children living with visual impairments. In: *Proceedings of the 2016 CHI Conference on Human Factors in Computing Systems* (pp. 445–457). New York: ACM.
5. Burchardt, T. (2004). Capabilities and disability: The capabilities framework and the social model of disability. *Disability and Society, 19*(7), 735–751.
6. Caballo, C., & Verdugo, M. Á. (2007). Social skills assessment of children and adolescents with visual impairment: Identifying relevant skills to improve quality of social relationships. *Psychological Reports, 100*(3_suppl), 1101–1106.
7. Fazzi, E., Signorini, S. G., Bova, S., Ondei, P., & Bianchi, P. (2005). Early intervention in visually impaired children. In: *International Congress Series* (Vol. 1282, pp. 117–121). Amsterdam: Elsevier.
8. Frauenberger, C., Good, J., & Keay-Bright, W. (2011). Designing technology for children with special needs: Bridging perspectives through participatory design. *CoDesign, 7*(1), 1–28.
9. Gori, M., Cappagli, G., Tonelli, A., Baud-Bovy, G., & Finocchietti, S. (2016). Devices for visually impaired people: High technological devices with low user acceptance and no adaptability for children. *Neuroscience and Biobehavioral Reviews, 69*, 79–88.
10. Gori, M., Del Viva, M., Sandini, G., & Burr, D. C. (2008) Young children do not integrate visual and haptic form information. *Current Biology, 18*(9), 694–698.
11. Guralnick, M. J., Connor, R. T., Hammond, M., Gottman, J. M., & Kinnish, K. (1996). Immediate effects of mainstreamed settings on the social interactions and social integration of preschool children. *American Journal on Mental Retardation, 100*(4), 359–377.
12. Hart, B., & Risley, T. R. (1975). Incidental teaching of language in the preschool. *Journal of Applied Behavior Analysis, 8*(4), 411–420.
13. Hartup, W. W. (1989). Social relationships and their developmental significance. *American Psychologist, 44*(2), 120
14. Hughes, B., & Paterson, K. (1997). The social model of disability and the disappearing body: Towards a sociology of impairment. *Disability and Society, 12*(3), 325–340.
15. Kekelis, L. S., & Andersen, E. S. (1984). Family communication styles and language development. *Journal of Visual Impairment and Blindness, 78*(2), 54–65.
16. Klein, S., Aeschlimann, J., Balsiger, D., Converse, S. L., Court, C., Foster, M., et al. (1973). *Automatic novel writing: A status report*. Wisconsin: Wisconsin University.
17. Laws of Malaysia act 685 persons with disabilities act 2008.pdf. (2014, July). http://www.ilo.org/dyn/natlex/docs/ELECTRONIC/86297/117930/F139356912/MYS86297.pdf. Accessed 28 August 2018.
18. McAlpine, L. M., & Moore, C. L. (1995). The development of social understanding in children with visual impairments. *Journal of Visual Impairment and Blindness, 89*(4), 349–358.
19. McDonagh, D., & Thomas, J. (2010). Rethinking design thinking: Empathy supporting innovation. *Australasian Medical Journal, 3*(8), 458–464.
20. McElligott, J., & Van Leeuwen, L. (2004). Designing sound tools and toys for blind and visually impaired children. In: *Proceedings of the 2004 Conference on Interaction Design and Children: Building a Community* (pp. 65–72). New York: ACM.
21. Miller, P. J., Cho, G. E., & Bracey, J. R. (2005). Working-class children's experience through the prism of personal storytelling. *Human Development, 48*(3), 115–135.

22. Meehan, J. R. (1977). Tale-spin, an interactive program that writes stories. In: *IJCAI* (Vol. 77, pp. 91–98).
23. Nardi, B. A. (1996). *Context and consciousness: Activity theory and human-computer interaction*. Cambridge: MIT Press.
24. Ogershok, P. R., & Cottrell, S. (2004). The pediatric board game. *Medical Teacher, 26*(6), 514–517.
25. Oliver, M. (2013). The social model of disability: Thirty years on. *Disability and Society, 28*(7), 1024–1026.
26. Palmer, B. C., Harshbarger, S. J., & Koch, C. A. (2001). Storytelling as a constructivist model for developing language and literacy. *Journal of Poetry Therapy, 14*(4), 199–212.
27. Papert, S. (1980). *Mindstorms: Children, computers, and powerful ideas*. New York: Basic Books, Inc.
28. Perez-Pereira, M., & Conti-Ramsden, G. (2013). *Language development and social interaction in blind children*. New York: Psychology Press.
29. Piaget, J. (1965). The stages of the intellectual development of the child. In: *Educational psychology in context: Readings for future teachers* (pp. 98–106).
30. Piaget, J. (2002). *The language and thought of the child* (Vol. 5). New York: Psychology Press.
31. Piaget, J., & Elkind, D. (1968). *Six psychological studies* (Vol. 462). New York: Vintage Books.
32. Proffitt, D. R. (2006). Embodied perception and the economy of action. *Perspectives on Psychological Science, 1*(2), 110–122.
33. Riedl, M. O., & Bulitko, V. (2013). Interactive narrative: An intelligent systems approach. *Ai Magazine, 34*(1), 67–67.
34. Robin, B. R. (2008). Digital storytelling: A powerful technology tool for the 21st century classroom. *Theory Into Practice, 47*(3), 220–228.
35. Rosenfeld, A. (2005). The benefits of board games. *Scholastic Parent and Child, 12*(4), 52–54.
36. Thieme, A., Morrison, C., Villar, N., Grayson, M., Lindley, S. (2017). Enabling collaboration in learning computer programing inclusive of children with vision impairments. In *Proceedings of the 2017 Conference on Designing Interactive Systems* (pp. 739–752). ACM.
37. Troster, H., Hecker, W., & Brambring, M. (1994). Longitudinal study of gross-motor development in blind infants and preschoolers 1. *Early Child Development and Care, 104*(1), 61–78.
38. United Nations convention on the rights of persons with disabilities. http://www.un.org/disabilities/documents/convention/convoptprot-e.pdf. Accessed 28 August 2018.
39. Warren, D. H. (1994). *Blindness and children: An individual differences approach*. Cambridge: Cambridge University Press.
40. Watson, S., Vannini, N., Davis, M., Woods, S., Hall, M., Hall, L., et al. (2007). FearNot! an anti-bullying intervention: Evaluation of an interactive virtual learning environment. In: *Proceedings of AISB'07*
41. Zimmerman, J., & Forlizzi, J. (2014). Research through design in HCI. In: *Ways of knowing in HCI* (pp. 167–189). New York: Springer.

Building an Ecosystem to Support Social Mobility for Person with Disabilities (PWDs): Malaysia Context

Ruhiyati Idayu Abu Talib, Mohd Shahrizal Sunar, Predrag K. Nikolic, and Ruzimi Mohamed

1 Introduction

During medieval times, the words "impairment" and "disable" carry different meaning as not all impaired people were treated as disabled [22]. Prior to Industrial Revolutions, most of the population lives in rural areas where social norms and social status were relatively fixed across intergenerational [28]. At that time, everyone had a certain role in a society and as long as he or she is fulfilling that role, he or she is considered part of a valuable member of society. For example, a dressmaker can be also those that physically impaired or deaf and mute, but as long as she is providing the community with her services, her impairment would not make her any different from somebody else. Industrial Revolutions in the eighteenth century changed the social, structural, and economics landscape of the society and created the barriers which impede person with disabilities attainment of full citizenship [10]. In this context, a person is being disabled not by their impairment but due to the unbearable structural barriers and prejudices of the society around them [11]. It is not a physical limitation that restricts disabled people, but inaccessible infrastructures and negative attitudes. The lack of knowledge

R. I. A. Talib (✉) · M. S. Sunar · R. Mohamed
Media and Game Innovation Centre of Excellence (MAGICX), Institute of Human Centered Engineering (iHumEn), Universiti Teknologi Malaysia, Johor, Malaysia

School of Computing, Faculty of Engineering, Universiti Teknologi Malaysia, Johor, Malaysia
e-mail: ridayu2@live.utm.my; shahrizal@utm.my
http://magicx.utm.my

P. K. Nikolic
School of Creativity and Art, ShanghaiTech University, Pudong, Shanghai, China
e-mail: predragnikolic@shanghaitech.edu.cn

© Springer Nature Switzerland AG 2020
D. Cagáňová, N. Horňáková (eds.), *Mobility Internet of Things 2018*,
EAI/Springer Innovations in Communication and Computing,
https://doi.org/10.1007/978-3-030-30911-4_35

and misconceptions about disability has resulted in stigma against people with disabilities and limited their access to their rights to be a part of society [32].

Social Model Disability [2, 9, 16, 23] views disability as a societal construct and collective issue—caused by the physical environment, inappropriate or inaccessible services and attitudes, and a lack of understanding—rather than one that derives from the health of an individual or medical impairment.

In the Malaysia context, the social model of disability is embedded in Malaysia's Person with Disabilities (hereinafter "PWDs") Act 2008. Both the Policy and the Action Plan on disability were also put into effect together with this legislation. In 2010, Malaysia signed the United Nations Convention on the Rights of Persons with Disabilities (UNCRPD). According to the definition by United Nations Convention on the Rights of Persons with Disabilities [33] in which it recognized that *disability is an evolving concept and that disability results from the interaction between persons with disabilities and attitudinal and environmental barriers that hinder their full and effective participation in society on an equal basis with persons without disabilities*.

In November 2012, Malaysia and together with other members of the United Nations Economic and Social Commission for Asia and The Pacific (hereinafter "UNESCAP") adopted The Incheon Strategy to "Make the Right Real" for PWDs in Asia and the Pacific which comprises 10 goals, 27 targets, and 62 indicators. This is to serve as part of Malaysia's strategy to move forward towards building a stronger policy framework to mainstream persons with disabilities and to ensure their effective participation in the society. Social mobility is the very first goal under Incheon Strategy, which is to reduce poverty and enhance work and employment prospects for PWDs [1].

2 Background Research

Malaysia achieved its Independence in 1957 from British and since then has progressed tremendously in the area of economic, social, and international standing. As a multiracial country, according to World Bank Data [12], Malaysia GDP in 2017 is 257 USD billion, compared to 1.90 USD billion in 1960. Malaysia average GDP is 87.25 USD billion from 1960 until 2016. The poverty rate had been reduced to 0.4% in 2015 from 51.2% in 1957 [13, 17]. Malaysia is on track to become High-Income Country from Upper-Middle-Income Country [6]. Despite all these achievements, disability welfare in Malaysia still has a long way to go. Previous disability research in Malaysia, albeit inadequate, found that disabled people are still subjected to manifold barriers in their daily encounters. Disability is found to be more prevalent among the poor and among women, which they experience significant barriers and discrimination in terms of access to health care, education and training, employment, marriage and motherhood, and violence and abuse [3, 5]. Evidence showed that there is a strong positive correlation between disability and poverty [8, 25, 30] as they were denied these basic accesses to live their life [36].

2.1 Social Mobility

Social mobility is defined as upward or downward movement of social status for individuals, families, and households [15, 26]. From the context of PWDs, their ability to enhance their social status is closely related with their rights to enjoy opportunities in all aspects of life to ensure their full and effective participation in the society and development; equal recognition before the law and access to justice on an equal basis with others; and all forms of discrimination on the basis of disability are eliminated. Hence, it is important to increase public awareness and understanding of the rights of PWDs. Empowering PWDs through economic is very crucial in raising PWDs status and it can only be achieved through the development of infrastructures and creating a supportive ecosystem that will allow independent living in the community for PWDs. Government Plan For Inclusive Development: Eleventh Malaysia Plan 2016–2020 [14] concluded that PWDs should be encouraged, promoted, and supported to become self-employed, develop their entrepreneurship capacity, and own and operate their business as an ideal solution to the above objectives.

2.2 Challenges Regarding the Person with Disabilities

The main challenge faced by the government regarding PWDs issues is collecting comprehensive and integrated data related to the actual number of PWD in Malaysia [4]. WHO estimated that as many as 15% of the total population in Malaysia are PWDs [24]. Based on statistic data of registered PWDs in 2016, there are only 409,269 PWD in Malaysia [20]. Lack of awareness of the rights and facilities available for PWDs and the stigma attached to PWDs is among the many reasons why PWDs are not registered with the authority. Without accurate data, the government will not be able to recognize and identify the disabilities in order to plan and formulate appropriate programs for the development of PWDs [31].

Public and private sectors also lack understanding of the requirement of the laws, regulations, and guidelines pertaining to the needs of persons with disabilities, i.e., to provide PWDs' friendly facilities. Major critique regarding PWD Act of 2008 is that it merely "promotes" the protection of rights instead of actually protecting the rights of PWDs. Many researchers suggested that the statute needs to be amended for the empowerment and progress of the PWDs and to include a penalty clause and to get rid of any shielding provision in order to ensure its effectiveness [3, 18, 29, 34, 35].

Other than that there is also a shortage of qualified teachers, lack of uninformed access to education, and limited availability of respite care services for PWDs. In this paper, first, we will discuss what does it means to be PWDs in Malaysia from social welfare and legislation context. We will analyze text from government reports and multiple works of literature regarding the facilities and initiatives that the government has put in place to assist PWDs. Lastly, we will conclude the paper by recommending future directions to improve social mobility for PWDs in Malaysia.

3 Disability, Legislation, and Social Welfare

3.1 Statistic Data Regarding Person with Disabilities

Before the 2008 Act, there are no accurate statistic data on disability, although there are several statistics that exist based on a sample survey and voluntary registration. Table 1 shows a registered number of PWDs in 2009 before the legislation comes into place [7].

Category of registration of PWDs was changed in January 2010 which now includes people with mental disabilities. Under the new registration category, there are seven categories of people that are known as PWDs [21]. They are those with:

(1) Hearing Disability (hereinafter "DE")
(2) Visually Disabled (hereinafter "BL")
(3) Speech Disability (hereinafter "SD")
(4) Physical Disability (hereinafter "PH")
(5) Learning Disabilities (hereinafter "LD")
(6) Mental Disability (hereinafter "ME")
(7) Multiple Disabilities (hereinafter "MD")

Table 2 shows amended registered number of PWDs before and after the legislation in 2010 [27]. There is a 50% increase from 283,512 in 2009 to 531,932 in 2010. In 2014, changes in the data were due to overlapping data from previous years and unaccounted data for those that have passed away and adjusted the data

Table 1 Registered number of PWDs (2009)

Type	No
Visual	25,757
Hearing	37,201
Physical	92,494
Intellectual	107,164
Cerebral Palsy (CP)	3864
Others	11,211
Total	277,509

Table 2 Registered Number of PWDs (2009–2010)

Type	2009	2010
BL	26,155	27,840
DE	37,729	39,824
SD	0	334
PH	94,331	105,020
LD	109,708	120,109
ME	0	3663
MD	11,521	13,389
Total	283,512	531,962

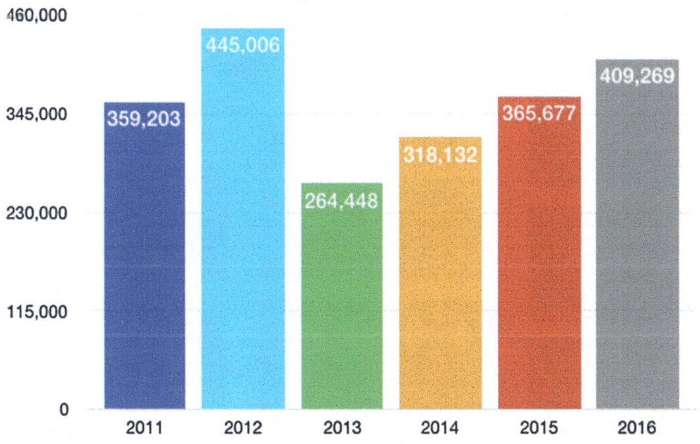

Fig. 1 Number of registered PWD 2011–2016. Source: DSW

accordingly. Figure 1 shows a yearly increase of 10–15% in PWDs that registered with the Department of Social Welfare from 2013 until 2016.

3.2 *Legislation and Social Welfare*

Over the years, there is a paradigm shift in Malaysia in dealing with the issues of PWDs. Before PWDs Act was enacted in 2008, Malaysia's approach in dealing with issues of PWDs was more on a charity based system without specific plan and policy targeted to increase their quality of life and well-being. Figure 2 shows the government agencies and legislation related to PWDs before 2008.

The PWDs Act 2008 is the first comprehensive legislation on disability in Malaysia which reflects the philosophy of the UN CRPD by taking disability as an issue of rights and equality. It is an Act that provides for the registration, protection, rehabilitation, development and well-being of PWDs, the establishment of the National Council for PWDs, and for matters connected therewith. The Act is aimed to ensure that the rights, interests, and welfare of PWDs in the country are protected. The responsibilities of implementing a national plan and policy regarding PWD fall under the jurisdiction of National Council for Persons with Disabilities (hereinafter "NCPWD") chaired by the Ministry of Women, Family and Community Development (hereinafter "MWFCD"). NCPWD recognizes that a collective effort through multi-sectoral and multi-agency collaboration with other relevant agencies was needed to achieve comprehensive and holistic result [19]. Figure 3 shows the government task force related to PWDs.

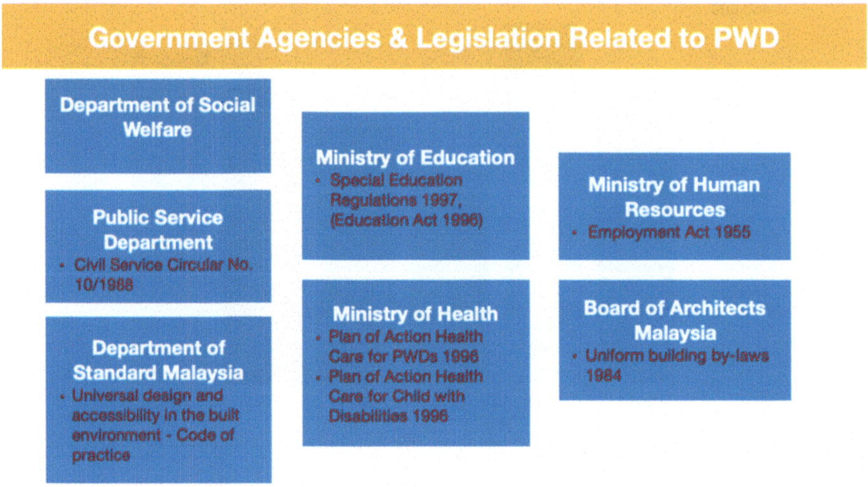

Fig. 2 Government agencies and related PWD legislation. Source: DSW

Fig. 3 National council for persons with disabilities. Source: DSW

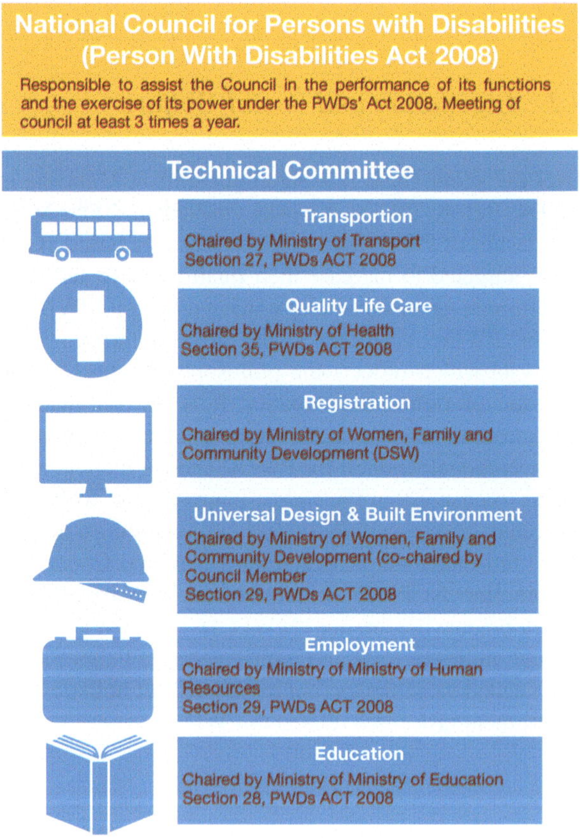

Among the Council's functions (Section 9, the PWDs Act 2008) are:

(1) to monitor the implementation of the national policy and national plan of action related to PWDs;
(2) to make recommendations to the Government on all aspects of PWDs (e.g., support, care, protection, rehabilitation, development, and well-being of PWDs).

The policy and plan for the development of PWDs focus on four targets [7]:

(1) Eradication of discrimination;
(2) Rights and opportunities for participation in the society;
(3) Equal opportunities based on the rights;
(4) Awareness on the rights.

The policy adopts fifteen strategic area, namely:

- Advocacy on rights;
- Accessibility
- Health and medical care
- Rehabilitation
- Education
- Employment
- Social welfare and security
- Environment
- Human resource development
- Social participation
- Research and development
- Housing
- Children with disability
- Women with disability

Below are the list of services for PWDs by both government and private agencies;

- Special aids
- Entrepreneur fund
- Disable workers allowance
- Disability allowance
- Allowance for CBR clients
- Allowance for personal assistants
- Vocational rehabilitation center
- Residential institutions
- Community-based rehabilitation (CBR)
- Sheltered workshop
- Counseling
- Inclusive school
- Allowance for disable students
- Income tax exemption (including parents of PWDs)

- Other tax exemptions (driving license, equipment)
- Special price for public transportation
- Priority in low income housing
- Reduction of medical fee (public hospital/clinics)
- Reduction of fees related to government services, i.e., passport
- Job placement/matching
- Income tax exemption (vehicles)
- Allowance for parents of children with disabilities
- Reduction of fees for utilities, i.e., internet, electricity, cable channel

In its Tenth Malaysia Plan (2011–2015) for the first time government introduced inclusive development as the national agenda. The Tenth Malaysia Plan encapsulates the spirit of Malaysia to create a fair and socially just society with national unity as its ultimate objective. Malaysia launched The Action Plan for Persons with Disabilities (2016–2022) to provide continuity to the existing plan to ensure that the programs keep up with the development and current needs of the group. In this phase, the Government is upscaling the initiatives towards improving the quality of life of persons with disabilities by:

- exploring the provision of new initiatives, i.e., respite care and strengthening independent living programs;
- enforcement through legislation on the rights of PWDs.

4 Conclusion and Future Direction

Malaysia Tenth and Eleventh Plan were anchored under UN CRP and UNESCAP *The Incheon Strategy to "Make the Right Real" for PWD in Asia and the Pacific*. Two of its main goals are:

1. to have community and family-based, inclusive development promoted in order to ensure that all persons with disabilities, irrespective of socioeconomic status, religious affiliation, ethnicity, and location, are able, on an equal basis with others, to contribute to and benefit from development initiatives, particularly poverty reduction programs;
2. PWDs are included in mainstream community life and are supported with life choices equal to those of others, including the option to live independently.

Based on the existing facilities and initiatives provided insofar by the government, there is still plenty of room for improvement, especially in the early intervention programs, in the context of social skills development and entrepreneurship programs for the PWDs. In the future, we will discuss the results of the programs by analyzing the outcome derived from all the initiatives mentioned above and study the effectiveness of the plan and policy to build a fair and socially just society in which all people, with no exception, have the rights, freedom, and capacity to access services and resources to enhance their well-being, and where the most disadvantaged are given extra support to ensure such success.

Acknowledgements Deepest gratitude to Malaysia Technology Development Corporation (MTDC) and Universiti Teknologi Malaysia (UTM) for supporting our ongoing research that will allow us to identify suitable technopreneur skills that PWDs need and the level of their knowledge that will give them the opportunity to participate actively in Industry 4.0 movement and taking advantages of an advancement in technology to change how they are interacting with the society.

This work is supported by Malaysia Technology and Development Corporation (MTDC) and Universiti Teknologi Malaysia (UTM).

References

1. 10 goals to make the right real for persons with disabilities in Asia and the pacific, 2013–2022. https://www.unescap.org/sites/default/files/Incheon%20%20Strategy%20pocket%20version.pdf (Accessed on 10/07/2018).
2. Abberley, P. (1987). The concept of oppression and the development of a social theory of disability. *Disability, Handicap & Society, 2*(1), 5 19.
3. Abdullah, N., Hanafi, H., & Hamdi, N. (2017). The rights of persons with disabilities in Malaysia: The underlying reasons for ineffectiveness of persons with disabilities act 2008. *International Journal for Studies on Children, Women, Elderly And Disabled, 1*.
4. Ahmad, N. A., Kasim, N. M., Mahmud, N. A., Yusof, Y. M., Othman, S., Chan, Y. Y., et al. (2017). Prevalence and determinants of disability among adults in Malaysia: Results from the national health and morbidity survey (NHMS) 2015. *BMC Public Health, 17*(1), 756.
5. Amin, A. S., & Abdullah, S. H.: Malaysian disabled women experiences in employment. *International Journal for Studies on Children, Women, Elderly and Disabled, 1*, 189–194 (online). Available at: http://www.ijcwed.com/wp-content/uploads/2017/01/IJCWED-122.pdf (Accessed 30 Jul. 2018).
6. Authority, M.M.I.D. (April 2017). Malaysia top among upper middle income countries at 28th spot. http://www.mida.gov.my/home/3719/news/malaysia-top-among-upper-middle-income-countries-at-28th-spot/ (Accessed on 10/14/2018).
7. Baseline survey. http://www.jobcoachmalaysia.com/pdf/BaseLineSurveryFinalEnglish24052010.pdf (Accessed on 10/14/2018).
8. Braithwaite, J., & Mont, D. (2009). Disability and poverty: A survey of world bank poverty assessments and implications. *ALTER-European Journal of Disability Research/Revue Européenne de Recherche sur le Handicap, 3*(3), 219–232.
9. Burchardt*, T. (2004). Capabilities and disability: the capabilities framework and the social model of disability. *Disability & Society, 19*(7), 735–751.
10. Cagáñová, D. (2011). *Multiculturality and industrial enterprises*. Anhalt (FH): Hochsch.
11. Caganova, D., Cambal, M., Sujanová, J., & Woolliscroft, P. (2012). The multiculturality aspects and human capital management within Slovak industrial enterprises. In *Proceedings of the 4th European Conferences on Intellectual Capital* (pp. 106–117).
12. Data, W. B. Malaysialdata. https://data.worldbank.org/country/malaysia (Accessed on 10/13/2018).
13. Data, W. B. (2018). Poverty & equity brief, country: Malaysia. http://databank.worldbank.org/data/download/poverty/33EF03BB-9722-4AE2-ABC7-AA2972D68AFE/Global_POVEQ_MYS.pdf (Accessed on 10/13/2018).

14. Economic Planning Unit, Prime Minister's Department. (2016). Eleventh Malaysia plan 2016–2020; Anchoring growth on people. https://mitra.gov.my/wp-content/uploads/sites/2/2019/01/RMKe-11-Book.pdf (Accessed on 05/30/2018).
15. Heckman, J. J., & Mosso, S. (2014). The economics of human development and social mobility. *Annual Review of Economics, 6*(1), 689–733.
16. Hughes, B., & Paterson, K. (1997). The social model of disability and the disappearing body: Towards a sociology of impairment. *Disability & Society, 12*(3), 325–340.
17. Ikemoto, Y. (1985). Income distribution in Malaysia: 1957–80. *The Developing Economies, 23*(4), 347–367.
18. Islam, M. R. (2015). Rights of the people with disabilities and social exclusion in Malaysia. *International Journal of Social Science and Humanity, 5*(2), 171.
19. JICA: 12251765.pdf. (2015). http://open_jicareport.jica.go.jp/pdf/12251765.pdf (Accessed on 07/22/2018).
20. JKM: Buku statistik 2016.pdf. (2016). http://www.jkm.gov.my/jkm/uploads/files/penerbitan/Buku%20statistik%202016.pdf (Accessed on 10/21/2018).
21. JKM: Buku statistik 2016.pdf. (2016). http://www.jkm.gov.my/jkm/uploads/files/penerbitan/Buku20statistik202016.pdf (Accessed on 07/23/2018).
22. Metzler, I. (2006). *Disability in medieval Europe: Thinking about physical impairment in the high middle ages, c. 1100–c. 1400*. New York: Routledge.
23. Oliver, M. (2013). The social model of disability: Thirty years on. *Disability & Society, 28*(7), 1024–1026.
24. Organisation, W. H. (January 2018). Disability and health. http://www.who.int/en/news-room/fact-sheets/detail/disability-and-health (Accessed on 10/20/2018).
25. Sen, A. (2008). The idea of justice. *Journal of Human Development, 9*(3), 331–342.
26. Simandan, D. (2018). Rethinking the health consequences of social class and social mobility. *Social Science & Medicine, 200*, 258–261.
27. Slide 1. https://unstats.un.org/unsd/demographic-social/meetings/2016/bangkok--disability-measurement-and-statistics/Session-6/Malaysia.pdf (Accessed on 10/14/2018).
28. Swain, J., French, S., Barnes, C., & Thomas, C. (2013). *Disabling barriers-enabling environments*. London: Sage.
29. Tah, I. H. B. M. (2013). A need for remedial provision to protect persons with disabilities in Malaysia. In *Proceeding of the Kuala Lumpur International Business, Economics and Law Conference at Kuala Lumpur*, 2–3 Dec 2013 (Vol. 3).
30. The World Bank (1999). Poverty and disability: A survey of the literature. The World Bank.
31. Towards realising the Incheon strategy in Malaysia. https://mind.org.my/wp-content/uploads/2017/09/P1-Reporting-to-UN-on-Current-Status-of-PWD-towards-an-Inclusive-Malaysian-Society-Post-UNCRPD-Ratification-YBhg.-Datuk-Harjeet.pdf (Accessed on 10/21/2018).
32. Unicef-children_with_disability_in_malaysia_2014_lowres.pdf. https://www.unicef.org/malaysia/UNICEF-Children_with_Disability_in_Malaysia_2014_lowres.pdf (Accessed on 10/07/2018).
33. United nations convention on the rights of persons with disabilities. http://www.un.org/disabilities/documents/convention/convoptprot-e.pdf (Accessed on 08/28/2018).
34. Wahab, H. A., & Ayub, Z. A. (2016). Persons with disabilities act 2008: The economic promises for people with disabilities? *International Journal of Economics and Financial Issues, 6*(7S), 313–319.
35. Wahab, H. A., & Ayub, Z. A. (2017). Employment right of persons with disabilities in Malaysia. In *Social Interactions and Networking in Cyber Society* (pp. 217–232). Singapore: Springer.
36. www.who-seajph.org/temp/whosouth-eastasiajpublichealth7284-1713988_044539.pdf. http://www.who-seajph.org/temp/WHOSouth-EastAsiaJPublicHealth7284-1713988_044539.pdf (Accessed on 10/23/2018).

Migration and Diversity Potentials for the Labour Market

Petia Genkova

1 Introduction

The topic of diversity has high priority in Germany, but also in other European countries. The process of globalisation as well as the heterogeneity of European societies is a call to action for politics and organisations. Moreover, executives need to adapt their structures and processes due to demographic changes in diversity and ageing. Europe has an ageing problem. The life expectation of people has doubled in Germany since the nineteenth century. Especially in France, the life expectation is at the highest [1]. As a consequence, a higher number of older people are retiring than younger people entering the job market. This creates a disequilibrium regarding pension funds on the one hand, and the lack of skilled and qualified workers on the other hand.

To persist in a globalised market and stay competitive, European organisations have to select suitable employees, who bring required diversity competences like intercultural, social and digital competences, to address effectively the challenges of the twenty-first century. Intercultural contact at the workplace is no longer a singularity, but a common practice [2]. Yet, this does not exclude the existence of various friction points. People with different cultural heritages are often discriminated in work life and held back from productive participation, although they bear potentials for organisations. It follows that organisations have to proactively support integration measures, especially for elder employees, women and people with an immigration background [3]. The underlying motivation for European organisations

P. Genkova
Osnabrück University of Applied Sciences, Osnabrück, Germany

Springer Heidelberg, Heidelberg, Germany
e-mail: p.genkova@hs-osnabrueck.de; petia@genkova.de; lncs@springer.com

© Springer Nature Switzerland AG 2020
D. Cagáňová, N. Horňáková (eds.), *Mobility Internet of Things 2018*,
EAI/Springer Innovations in Communication and Computing,
https://doi.org/10.1007/978-3-030-30911-4_36

should be flexible modes of operation that are appropriate for modern times. If corporations want to earn the title of "global player", adaptation is needed.

For many large corporations diversity management is seen as indispensable since they have to plan globally and in the long term. And indeed, unforeseen demographic processes such as strong migration flows in the last years to Europe change opportunities and challenges in the labour market. To be resistant against sudden changes in the market, concrete action plans on how to deal with a more diverse European society are therefore of utmost importance.

In order to set a concrete action plan into stone, more than 2600 organisations and institutions signed the Diversity Charter of German companies, which implies that the importance of diversity management has reached a multitude of decision makers from the economy. However, while reviewing the cross section of organisations in Germany, it becomes eminent that approximately two-thirds of corporations do not include active diversity management in their management structures [4]. This poses a problem, considering the fact that employing a diverse workforce is only successful when characteristics and singularities of minorities are taken into consideration and actively managed (e.g. [5, 6]).

2 Diversity Management

Diversity management describes the engagement with individual differences and commonalities in a diverse working force of an organisation [7–10]. Overtly identifiable characteristics like age, religion and cultural values are of interest for a diversity manager, but also assessing more hidden characteristics like intercultural competence [11] and values of equal opportunity and fairness.

Diversity management aims to make use of different strengths coming from diversity in organisations and to promote and cherish the opportunities of diversity in order to change attitudes and behaviours of organisation and working groups [12]. One premise for the valuing treatment of diversity in an organisation is the appreciation of employees and their different individuality [13]. The challenge hereby is to treat all employees equally when it comes to rights and duties within the workforce, but meet their personal demands based on their diverse background individually at the same time [13].

Diversity of employees can affect the productivity within organisation differently. It depends especially on the value and the weight diversity is given. When emphasis is put on social status, the organisation's performance is impacted negatively. If employees bring experiences, e.g. based on their different cultural backgrounds, it positively contributes to the productivity. It follows that promoting the latter example of diversity can contribute to an improvement of work results [14].

New input and perspective change can be brought by people with different cultural backgrounds, which is also appreciated by some native workers [15]. Research on diversity indicates that individuals with pro-diversity beliefs feel comfortable in multicultural working groups and consider diversity as an enrichment. Pro-diversity beliefs enable them to identify themselves with people from the "outgroup" and subsequently feel strong bonds and identification within the group [16]. This accords with the social identity theory by Taijfel and Turner [17]. In their study Wolf and Van Dick [18] demonstrated that people, who see migrants as an asset for society, cherish more interpersonal contact with them and express less racism than those who do not see migrants as enrichment. As Stegman [19] concluded in his meta-analysis—pro-diversity beliefs as well as a positive diversity culture within an organisation create beneficial outcome for groups and individuals by increasing work performance.

Yet there is not a single overarching approach on why and how diversity and its management should be implemented in the company's strategy [20–22]. A variety of different approaches exist on how to deal with that. For instance, the discrimination-and-fairness approach is to be mentioned. In this approach, equal opportunities for every employee are the main goal. This approach even involves firm policies like anti-discrimination laws that should be established in order to reach respective main goal. Another approach that can be followed is the access-and-legitimacy perspective. Hereby, diversity is seen as an opportunity to access global markets and cherished as an added value to the organisation's strategic direction [21]. The last approach highlighted by this chapter is the integration-and-learning perspective. Diversity is seen as a complementary source of new knowledge, creating change in perspective and a learning effect that allows for better problem solution. As one aspect of this approach, managers facilitate diversity on every level and encourage the exchange of different opinions.

Every diversity approach mentioned is successful in its own way. All foster integration at the workplace, but it seems that only the integration-and-learning approach leads to sustainable advantages [21]. Yet, as a common denominator in all three approaches managers have to actively facilitate diversity beliefs in the workforce.

It is to say that mangers take an important role in diversity management. Their decisions have a direct influence on the constitutions of the workforce (e.g. satisfaction, health and performance of employees) [5]. Managers have to balance their moral duties and their work duties. They overtake a role model function on the one hand, whereas they oblige to the aims of the company, which is accompanied by pressure in performance [5]. Especially, decision makers on the highest executional level are of utmost importance to establish diversity management approaches [23]. Decision makers from middle management have to implement the strategy and set an example of a diversity culture within the corporation. Summarising, the highest level of decision-making is responsible for constructing a successful diversity strategy; however without the middle management ensuring its implementation the strategy is useless [24]. That is, by actively introducing employees to the benefits of a multicultural working environment, middle management serves as a speaking tube for a diversity-open company philosophy [25].

Across studies on diversity management failure of implementing respective management efficiently had negative impacts and impaired working conditions for heterogeneous and homogeneous organisations alike.

Summarising diversity can be an advantage in competing on a global market. Yet, it can also be the reason for conflicts and counterproductive behaviour, if not addressed appropriately (e.g. by explicit diversity management). Overall fair personnel selection processes that do not exclude people that deviate from the majority is important for the well-being of the individuals but also for the flexibility and adaptivity of organisations on a diverse labour market [26].

3 Diversity: Migration

Problems and benefits alike accompany the employment of people with migration backgrounds. Especially for the employees themselves, intercultural contact at the workplace can lead to stress for them [27].

The migration flows to Europe, caused by war, environmental catastrophes and other hazards, peaked in 2015. Approximately 4.7 million people have emigrated to the EU states. More than half of those people came from poor and dangerous conditions in third world countries. Germany received the highest amount of immigrants followed by the United Kingdom, France, Spain and Italy. Yet, migration to the European continent is nothing new. Data on the overall migration population in Europe reveal that 35.1 million people were born outside of the Eurozone. Almost 20 million people live in another European state, than their homeland [28]. Among this population, people from Poland, Italy, Romania, Portugal and the United Kingdom make up the biggest part [28]. A shift in population numbers of citizens with different nationalities raises the question of what dynamic relation emerges between sojourners and their new host society.

Effects of migration as the topic of interest for social sciences started in the 1950s. Researchers started to develop a variety of approaches that explain the dynamic relationship between the migrant and his/her environment. For instance, intercultural learning, stress approach, social identity and acculturation are terminologies that were successively invented by migration researchers. They were found to have high applicability in the context of economic effects on migration. The stress approach found is quite prominent in the research of intercultural economics, since short-term and long-term intercultural interactions are accompanied by stress.

Recently, the association between perceived stress and physical and psychological constitution of the migrant was investigated in migration studies [29]. The extent of the experienced burden is high, when the coping behaviour is ineffective or the number of appearing stressors is high [29]. As key factors for perceived stress discrimination, disturbing events and social isolation, e.g. by colleagues or self-inflicted, affect the constitution of a person with a migration background [30]. However, whereas the common idea of discrimination involves blatant racism by individuals, processes within the organisation or institution can prohibit migrants from being employed in the job they are actually qualified for. Subtle disadvantages like this, which are not uttered openly and often are anchored in organisational structures, can be summarised under the term "structural discrimination". An antidote to structural discrimination is structural integration.

Revisiting the social identity theory of Tajfel und Turner [17], structural integration offers the chance to give migrants a sense of belonging to the "in-group" and therefore enhances their motivation to work.

Across organisations types of structural discrimination can be found. Therefore, the chances of success in the labour market are worse for migrants on average. Academic diplomas and achievements from other countries are often not accepted in the new host society; hence migrants often are forced to work below their skills and qualification levels [31]. By looking at research about academic qualifications of migrants, the educational level and the level of perceived integration are inversely related. Apparently, migrants with higher educational backgrounds are more likely to feel rejected by their new environment [32].

This is considered as the glass ceiling effect. Disenchantment after career expectation is not met in the new society leading to a critical and negative evaluation of the host society. Not feeling welcome or accepted by the society is accompanied by stress, and subsequently reinforced by daily contact situations with natives that are misunderstood. Misinterpreting ambiguous cultural norms and values causes at worst conflict and friction points.

The same problem of misunderstand and failed communication during intercultural contract goes for natives. Summarising it poses a challenge for people with and without an immigration background to adjust to the other person's culture and aim for a better understanding [33]. Translating this back to diversity management, managers need to prevent and reduce the possibility of misunderstandings and monitor possible friction points during intercultural communication. Furthermore, they should ease conditions of admission for foreign applicants to fight structural discrimination.

It is the major objective of diversity research to confirm that diversity management has positives on the economic success. The results of the investigation are inconsistent and are highly related to the cultural specific context (e.g. [34]).

However, to find a solution for challenges of migration and integration is getting more and more important for employees and companies in a diverse labour market. One problem is that in fast-changing times, the immediate effects of diversity actions are questioned and fast measurable determinants for "successes" are expected. Yet, integration is a timely process that requires special attention depending on the context.

Since the literature on management structures mainly kept busy with investigating management structures and their effect on the successful use of diversity, context variables have been mainly skipped. By trying to contribute new insights on the topic the research questions of the study therefore are the following. Which context variables are seen as important? Which obstacles prohibit successful implementation diversity strategies? Do executive managers and human resource managers recognise the problem fields?

4 Methodology and Data

The focus of this study lies on the issue human resource management has to face when dealing with diversity. Due to missing manpower, people with migration background and elder people are important to fill in the gaps in organisations. For this purpose, explorative qualitative interviews were conducted. As the main objective, this study wants to reveal important competences in managers that ensure equal opportunities during personnel selection processes. Special attention is given to the diversity component's age and cultural heritage.

Subsequently, 63 telephone interviews with managers ($N = 17$) and human resource managers ($N = 13$) and employees with ($N = 18$) and without an immigration background ($N = 15$) from different organisations were interrogated about equal chances at the workplace, with special focus on chances for people with migration background and elderly employees. In order to answer the research question the following hypotheses were generated:

Explorative hypothesis 1: Managers and employees consider the need for diversity measures as equally important.

Hypothesis 2: Native managers deviate from native employees and employees with migration background regarding the expression of stress level.

Research results on diversity management that have applicability in real life, especially in the context of cultural diversity and age diversity perception in management, are quite rare. As a research method, qualitative interviews were chosen, since the richness of informational material gathered from respective interview makes it possible to paint a realistic picture of challenges diversity brings for executive and human resource management. Furthermore, underlying hints to foster diversity potentials and to avoid obstacles regarding the quality of opportunities for elderly and people with migration background can be extracted. The interviews were conducted with a structured interview guideline.

By asking about diversity beliefs, unheeded aspects of a diverse working force can be revealed and in turn conclusions can be drawn, where the weaknesses of the diversity management lie. Hereby the competences are highlighted, which are of importance for diversity management and to where the application of diversity measurements is needed [35]. Since the method involves a structured interview guideline, priory formulated questions aim at the view of managers and human resource managers concerning the importance of diversity in organisations and their subjective evaluation of lacking skills, e.g. intercultural competences. Furthermore, what impact stress has on native employees and employees with a migration background is investigated. The standardised interview guideline is derived deductively from theory [35]. It was chosen for this method so that the interviews stay within the limits of the research question and data can be compared across interviews [35].

Open-ended and closed questions were used. Here is an example of an open-ended question: "How stressed are employees by a multi-cultural working environment in your company at the moment, in your opinion?" The closed questions were answered on a 5-point Likert scale, e.g. 1 = not at all stressed until and 5 – very stressed. By using both open-ended and closed questions it was possible to receive precise answers but also to give participants the chance to elaborate more.

The structure of the interview was divided into different topics. The first set of questions dealt with problems with diversity as well as company-specific questions like equal opportunities of employees for people with migration background. A different set of questions concerned aspects of cultural stereotypes and prejudices and how they affect personnel selection. The set of questions concerned leadership aimed at revealing accompanied stress. Moreover, it was the objective to determine if employees with an immigration background experience different stress levels than native employees. The last set of questions asked the participants about the importance of competence. In more detail, which competences have the biggest impact on creating equal opportunities during selection processes?

Content analysis was used to dissect and interpret the obtained data [36]. The frequency of reoccurring categories and schemas in each interview was analysed, in order to compare frequencies and mentioned categories across interviews [36]. The program Microsoft Excel was used to transcribe and analyse the anonymised interviews.

The subsample of managers consisted of 17 executive managers and 13 human resource managers. All of them were natives. Seventeen of them were male and 13 were female. On average the managers and human resource managers were $M = 40.83$ years ($N = 29$; SD = 9.30) old.

In the subgroup of employees, 15 employees were natives, whereas 18 employees had a migration background. Twenty were female and 13 were male. On average, the employees with and without an immigration background were $M = 35.48$ years ($N = 33$; SD = 9.99) old.

5 Results

In this section, the results of the empirical study are presented. First, we review the explorative hypothesis: Managers and employees consider the need for diversity measures as equally important.

By looking at the output of our analysis the management subgroup and the employee subgroup considered diversity quotas, e.g. quota of women and migrants, as highly important, which comes as a surprise. Usually, respective guidelines for respective quotas do not exist and are not positively viewed by managers. However, in this study executive mangers argued during the interviews that such quotas are implemented by their company and used as moral guideline for their company's philosophy. Furthermore, they argued in favour for diversity measures such as the establishment of a diversity department, an equal opportunity officer and diversity representative who actively monitor and foster processes that hinder/facilitate equal opportunities.

Diversity measures for gender within their respective companies were implementations that were most frequently mentioned. For cultural diversity measures, executive managers listed language courses, exchange programs and culture trainings that were common practice.

Human resource managers frequently mentioned cultural trainings, membership of the Diversity Charter of German companies and cooperation with the Federal Employment Agency. According to the HR managers "individual intercultural competence trainings are offered for employees who would like to deal with intercultural competences and want to analyse their own intercultural competence". Following their statement, cultural trainings include topics such as "How to recruit internationally? Which shall improve working with cultural differences". Further topics of the trainings are "anti-prejudice [37]" or "emotional competences".

Contrary to those statements native employees and employees with a migration background almost never mentioned any diversity measures implemented in the working processes of the companies. Such discrepancy in statements seems odd and indicates either a lack of diversity management for employees or the unsuccessful application of it. According to the employees most of the diversity measures concern communication like on- and offline platforms, diversity departments and diversity committees. Both subgroups mentioned that some cultural trainings are offered.

The majority of companies from which the sample was recruited signed the Diversity Charter of German companies and have implemented diversity in their company strategy. Therefore it was not surprising that employees stated that the own professional growth and the equal treatment during personnel selection are of importance. Employees mentioned following expectations they have from management: "Anonymous applications maybe. [38] Those, who hold responsible during application processes should undergo intercultural competence training. They have to reflect themselves, if they are able to hire someone, who is different from them. [38] I think that is this question should be answered by human resource man-agers: Am I ready to hire somebody, who is different than I?" Following the statements of employees, also a form of skill analysis can be helpful "[38] which hide the background of people as good as possible. So therefore, only the professional skills, social competence and media competences are assessed without being influenced by the appearance of the person, e.g. gender etc."

All subgroups agreed that openness and interests in other cultures are very important to obtain competence in intercultural contact situations. Social competences, for instance tolerance and self-reflection, were mentioned as being important, followed by the intercultural competence and the cultural knowledge and the cultural awareness. Reviewing the statements so far, the same pattern as in previous results can be observed: native employees and employees with migration background list up more competences important for successful diversity management and differentiate stronger regarding the promotion of intercultural competence, than the manager subgroup.

In contrast, managers thought of facilitating social and intercultural competences as a matter for the future. Currently, they see no relevance in promoting competences for the present. It can be concluded that managers attribute the necessity for diversity measures externally to future scenarios. Interestingly, the management

Table 1 Importance of diversity

Executive managers/human resource managers	Employees with and without an immigration background
Seriousness and urgency of diversity are not clearly recognised Incorrect assessment of the stress of migrants No consideration of the stress level of migrants in the day-to-day management	Seriousness and urgency of diversity are not clearly recognised by employees without an immigration background Good assessment of the stress level of migrants by employees without an immigration background Risk of problematisation of diversity

subgroup stated that risk tolerance constitutes itself a relevant competence for dealing with diversity. In addition, the subgroup of managers stated that emotional stability, intercultural flexibility, social identity, intercultural anti-prejudice as well as sensitisation for cultural differences are important for diversity management. In contrast, employees mentioned openness and empathy, social competence, self-reflection, cultural knowledge and tolerance as key competences.

The quantitative analysis implies that perspective change and experience exchange are the strongest predictors for intercultural decision-making (e.g. [34]). Furthermore, perspective change is a stronger predictor than empathy. Risk tolerance was proven to be a competence, which is not important for diversity management and does not increase intercultural competence. In conclusion the subgroup of managers perceive diversity as a risk yet fail to recognise the stress level of employees with migration background (see Table 1).

Overall, managers and employees consider the relevance of diversity measure as important. However, throughout the interview it became eminent that the subgroup of employees is more involved with the topic of diversity than the subgroup of managers (based on the frequencies in mentioning related concerns). A possible reason for this discrepancy can be the notion that employees are confronted with intercultural contact situations, like with colleagues or customers, than the management subgroup.

Furthermore, the interview material indicates that both groups attribute failed diversity measures and need for improvement to the other subgroup. Yet, the literature on diversity management shows that the effects of diversity management are top-down, indicating the responsibility of manager in that matter [25]. Additionally, the results of recent studies reveal that top-down processes are only successful if diversity-sensitive and intercultural competences are highly represented within the workforce [25].

Hypothesis: Native managers deviate from native employees and employees with migration background regarding the expression of stress level.

Reviewing the results it could be seen that the stress level of employees with an immigration background equals the stress level of native managers and native employees ($T = 0.035$; df 1; 49; $p = 0.972$).

The qualitative analysis confirms this. Prejudices and the impatience of co-workers resulting from language problems as well as differences in work philosophies rooting in intercultural differences impose stress on employees with migration background. Statements by the employees with an immigration background support this analysis: "I have to give more than 120% than my German colleagues. This is one reason why migrants feel more stressed".

Apparently, the demand for diversity management exists. Furthermore it can be concluded that the improvement in competence is attributed to the "outgroup" and not to oneself. This is consistent with prior research.

6 Discussion and Conclusions

From the statements made by managers and employees, it can be derived that diversity seems to be an important topic. However, employees seem to give even more importance to diversity measures than managers. During the interviews, the subgroup of managers frequently mentioned the value of diversity measures for the organisation performance, justice and fairness for employees. Yet, executive managers see these aspects as relevant only for future scenarios and see no actual need in improving them now. It follows that executive managers deny topicality and the actual need for an intervention. They also fail to recognise the stress level of their employees, caused by shifts in diversity. Concerning cultural diversity and creation of equal opportunities for people with immigration backgrounds, they do not see the urgency for change in management. Human resource managers fail to assess the additional stress levels of migrant applicants correctly and do not consider them as problematic in their day-to-day management. In contrast, most employees have recognised the value of diversity. Employees with a migration background consider organisational structures that create equal opportunities as more important than native employees.

However, results show that native employees correctly estimate the stress level of employees with a migration background. In the interviews, employees have already spoken about the risk of problematisation of diversity. Regarding demographic differences, the study has demonstrated that in the western part of Germany, e.g. in the Ruhr area, urgent and drastic diversity measures such as policy change would be redundant, since there a multicultural working environment is a daily routine. Once again it became clear that important diversity competences are intercultural and social competences. Empathy, tolerance and communicative competences, openness, self-reflection and emotional competences as well as cultural knowledge and awareness were also mentioned. In both subgroups, participants advocated for the need of competence trainings and experiences abroad. Such statements seem to be contradictory to the actual results obtained. Both groups did not differ significantly concerning the subjectively assessed social competence. For this competence scores on the self-assessment and the qualitative interview were high.

It is to mention that the sample showed a high diversity in general. Furthermore, it was able to highlight diversity aspects such as age and culture from different perspectives.

Additionally, the interview guideline seems to be robust, since it covered a broad range of questions relevant to diversity management. The combination of closed and open-ended questions supported the robustness and allowed for comparisons between interviews. However, the open questions made it quite difficult to produce 100% comparability. Instead a standardised, quantitative questionnaire could have been used, which would have increased the sample size as well. However, it was the aspiration to receive profound statements and a plenitude of interpretable material. Since the topic itself is not popular in mainstream research, interviews gave us the ability to explore the field. Moreover, the qualitative investigation was used to generate hypotheses and deliver approaches that have not been priory examined and deliver starting points for future research.

While the points mentioned above add to the literature of diversity studies, there are some limitations to mention. The interview questions were partly too superficial. For instance, they did not ask the participants which diversity actions are the most effective for elderly employees. However, covering every aspect of diversity remains impossible, since companies differ in, e.g., size, structure, implementation of a diversity department and other characteristics.

Reviewing regional patterns it became obvious that regional differences exist in Germany concerning the importance of diversity measures. This raises the question if unified diversity measures are applicable across regions. Next, there were differences in responses concerning details given. Therefore, the results had to be compared cautiously to not oversimplify/exaggerate statements. We used the qualitative interview to generate hypotheses for a quantitative questionnaire. By asking more profound questions it was possible to take a stronger focus regarding a feeling of urgency for diversity measures, which might lack comparability.

Also, the sample choice produced some limitations: Only elderly people and people with and without migration background were included in the study. Recruitment criteria could be expanded to other diversity aspects. In order to recruit a more representative sample, managers from different working fields and locations in Germany and other small- and medium-sized and large organisations should be included to cover environmental aspects. The research result shows differences between employees with and without an immigration background with regard to the importance attributed to diversity measures.

The question remains whether the answers were given due to social desirability or deliberate misrepresentation since interviews were used. A deliberate misinterpretation can be excluded because the respondents participated voluntarily and had nothing to expect in terms of any sanctions or other effects by specific statements. Social desirability could be minimised by the anonymisation of the data but this remains quite challenging given the interactive nature of interviews.

Summarising, representatives of corporations should give more attention to diversity measures and register them in company guidelines and manager guidelines. As a premise for that, decision makers should be more sensitised regarding the needs and stress levels of their employees regarding diversity. Most importantly, they should tackle this problem now and not in the future.

References

1. German Federal Statistical Office. (2015). *Jahresbericht 2015*. Retreived from https://www.destatis.de/GPStatistik/servlets/MCRFileNodeServlet/DEHeft_derivate_00021695/Jahresbericht2015.pdf;jsessionid=25D2ACCC3AE3B2F6FF98F4CF11630BE9.
2. Fantini, A. E. (2009). Assessing intercultural competence: Issues and tools. In D. K. Deardorff (Ed.), *The SAGE handbook of intercultural competence* (pp. 456–476). Thousand Oaks, CA: Sage.
3. Fuchs, J., & Dörfler, K. (2005). Projektion des Arbeitsangebots bis 2050. Demografische Effekte sind nicht mehr zu bremsen. Den Betrieben stehen langfristig immer weniger und immer ältere Arbeitskräfte zur Verfügung—Die Zahl der Jungen sinkt dramatisch. *IAB Kurzbericht, 11*, 1–5.
4. KFW Research. (2017). Migranten gründen häufiger und größer: mehr Wochenstunden, mehr Angestellte. Nr. 165, 4. April. Verfügbar unter: https://www.kfw.de/PDF/Download-Center/Konzernthemen/Research/PDF-Dokumente-Fokus-Volkswirtschaft/Fokus-2017/Fokus-Nr.-165-April-2017-Migrantengr%C3%BCnder.pdf [Zugriff am 17 Oct 2017].
5. Franken, S. (2015). *Personal: Diversity management* (6th ed.). Wiesbaden: Gabler.
6. Thomas, A. (2003). *Kulturvergleichende Psychologie (2. überarbeitete Auflage)*. Göttingen: Hogrefe.
7. Becker, M. (2006). Wissenschaftstheoretische Grundlagen des Diversity Management. In M. Becker & A. Seidel (Eds.), *Diversity Management—Unternehmens- und Personalpolitik der Vielfalt (S. 3-48)*. Stuttgart: Poeschel.
8. Bell, S. T., Villado, A. J., Lukasik, M. A., Belau, L., & Briggs, A. L. (2011). Getting specific about demographic diversity variable and team performance relationships: A meta-analysis. *Journal of Management, 37*, 709–743.
9. Charta der Vielfalt & Ernst & Young GmbH. (2016). *Diversity in Deutschland. Studie anlässlich des 10-jährigen Bestehens der Charta der Vielfalt.* Verfügbar unter: https://www.charta-der-vielfalt.de/fileadmin/user_upload/Studien_Publikationen_Charta/STUDIE_DIVERSITY_IN_DEUTSCHLAND_2016-11.pdf [Zugriff am 20 Nov 2017].
10. Krell, G. (2003). Personelle Vielfalt in Organisationen als Herausforderung für die Praxis. In H. Wächter, G. Vedder, & M. Führing (Eds.), *Personelle Vielfalt in Organisationen*. Hampp: München.
11. Milliken, F. J., & Martin, L. L. (1996). Searching for common threads: Understanding the multiple effects of diversity in organizational groups. *Academy of Management Journal, 21*, 402–433.
12. Ashikali, T., & Groeneveld, S. (2015). Diversity management in public organizations and its effect on employees' affective commitment: The role of transformational leadership and the inclusiveness of the organizational culture. *Review of Public Personnel Administration, 35*(2), 146–168.
13. Vedder, G. (2011). Die Grundlagen von Diversity Management. In G. Vedder, E. Göbel, & F. Krause (Eds.), *Fallstudien zum Diversity Management* (pp. 1–17). München: Rainer Hampp.
14. Garib, G. (2013). Diversity is in the eye of the beholder: Diversity perceptions of managers. *The Psychologist-Manager Journal, 16*(1), 18–32.
15. Page, S. E. (2007). *The difference: How the power of diversity creates better groups, firms, schools, and societies*. Princeton: Princeton University Press.
16. Van Dick, R., van Knippenberg, D., Hägele, S., Guillaume, Y. R. F., & Brodbeck, F. C. (2008). Group diversity and group identification: The moderating role of diversity beliefs. *Human Relations, 61*(10), 1463–1492.
17. Tajfel, H., & Turner, J. C. (1979). An integrative theory of intergroup conflict. In W. G. Austin & S. Worchel (Eds.), *The social psychology of intergroup relations* (pp. 33–47). Monterey, CA: Brooks-Cole.
18. Wolf, C., & van Dick, R. (2008). Wennandersnichtschlechterbedeutet-WertschätzungvonVielfaltfördertGleichwertigkeit der Gruppen [When different doesn't

mean worse. Valuing diversity facilitates equality of groups]. In W. Heitmeyer (Ed.), *Deutsche Zustände. Folge 6 [German states. Sequal 7]* (pp. 137–153). Suhrkamp: Frankfurt.
19. Stegmann, S. (2011). *Engaging with diversity of social units—A social identity perspective on diversity in organizations* [online]. Doctoral dissertation. Goethe University, Frankfurt am Main. Retrieved February 6, 2018, from http://nbn-resolving.de/urn/resolver.pl?urn:nbn:de:hebis:30-111582.
20. Ely, R., & Thomas, D. (2001). Cultural diversity at work: The effects of diverse perspectives on work group processes and outcomes. *Administratice Science Quarterly, 46*, 229–273.
21. Van der Zee, K., & Otten, S. (2015). Organizational perspectives on diversity. In S. Otten, K. van Safdar, S., Struthers, W., & van Oudenhoven, J. P. (2009). Acculturation of Iranians in the United States, the United Kingdom, and the Netherlands. A test of the multidimensional individual difference acculturation (MIDA) model. *Journal of Cross-Cultural Psychology, 40*(3), 468–491.
22. Wei, L., & Lau, C. (2012). Effective teamwork at the top: The evidence from China. *International Journal of Human Resource Management, 23*(9), 1853–1870.
23. Wildermuth, C., & Gray, S. (2005). *Diversity training (2. Aufl.)*. Alexandria/Virginia: ASTD Press.
24. Thomas, R. (1990). From affirmative action to affirmative diversity. *Harvard Business Review, 68*(2), 107–117.
25. Dreas, S., & Rastetter, D. (2016). Die Entwicklung von Diversity Kompetenz als Veränderungsprozess. In P. Genkova & T. Ringeisen (Eds.), *Handbuch Diversity Kompetenz. Band 1: Perspektiven und Anwendungsfelder* (pp. 351–370). Wiesbaden: Springer.
26. Kersting, M., & Ott, M. (2016). Diversity-gerechte Personalauswahl. In P. Genkova & T. Ringeisen (Eds.), *Handbuch Diversity Kompetenz. Band 1: Perspektiven und Anwendungs felder* (pp. 679–692). Wiesbaden: Springer.
27. Holmes, T. H., & Rahe, R. H. (1967). The social readjustment rating scale. *Journal of Psychosomatic Research, 11*(2), 213–218.
28. Eurostat. (2017). Statistiken zu Wanderungen und Migrantenbevölkerung. http://ec.europa.eu/eurostat/statistics-explained/index.php/Migration_and_migrant_population_statistics/de. Accessed 6 Feb 2018.
29. Eggerth, D. E., & Flynn, M. A. (2013). Immigration: Implikationen für Stress und Gesundheit. In P. Genkova, T. Ringeisen, & F. T. L. Leong (Eds.), *Handbuch Stress und Kultur. Interkulturelle und kulturvergleichende Perspektiven* (pp. 343–359). Springer Fachmedien: Wiesbaden.
30. Bozorgmehr, K., & Razum, O. (2015). Effect of restricting access to health care on health expenditures among asylum-seekers and refugees: A quasi-experimental study in Germany, 1994–2013. *PloSOne, 10*(7), e0131483. https://doi.org/10.1371/journal.pone.0131483.
31. Hormel, U., & Scherr, A. (2013). *Bildung für die Einwanderungsgesellschaft: Perspektiven der Auseinandersetzung mit struktureller, institutioneller und interaktioneller Diskriminierung*. New York: Springer.
32. De Vroom, T., & Verkuyten, M. (2015). Labour market participation and immigrants' acculturation. In S. Otten, K. van der Zee, & M. B. Brewer (Eds.), *Towards Inclusive Organizations* (pp. 12–28). New York: Psychology Press.
33. Asbrock, F., Christ, O., Duckitt, J., & Sibley, C. G. (2012). Differential effects of intergroup contact for authoritarians and social dominators: A dual process model perspective. *Personality and Social Psychology Bulletin, 38*(4), 477–490.
34. Genkova, P., & Ringeisen, T. (2017). *Handbuch Diversity Kompetenz. Band 1: Perspektiven und Anwendungsfelder*. Wiesbaden: Springer Fachmedien.
35. Mayer, H. O. (2013). *Interview und schriftliche Befragung: Grundlagen und Methoden empirischer Sozialforschung* (6th ed.). München: Oldenbourg Wissenschaftsverlag GmbH.
36. Mayring, P. (2015). *Qualitative Inhaltsanalyse. Grundlagen und Techniken (12. überarbeitete Auflage)*. Weinheim: Beltz.
37. Turner, J. C., Brown, R. J., & Tajfel, H. (1979). Social comparison and group interest in ingroup favouritism. *European Journal of Social Psychology, 9*(2), 187–204.
38. Finkelstein, L. M., Burke, M. J., & Raju, M. S. (1995). Age discrimination in simulated employment contexts: An integrative analysis. *Journal of Applied Psychology, 80*(6), 652–663.

Gender Differences in the Income of Employees Over 50 with Emphasis on Various Sectors of Economy

Jana Stávková, Naďa Hazuchová, Hana Porkertová, and Martina Rašticová

1 Introduction

Gender inequalities may be understood as different treatments for men and women in various areas of human life, which gives structure to an unequal access of women towards money and power. While some differences are difficult to measure, unequal remuneration is one of the spheres which can be quantified. Eliminating the considerable difference between the income of women and men is one of the priorities of the European Union and, as early as 2003, the member states pledged to follow the European Employment Strategy (EES), which aim was to reduce the gender pay gap to a half by 2010 [1]. However, this has not been achieved even by 2018. The latest numbers from Eurostat (for 2016) state the average difference among the European 28 to be 16.2% and the Czech Republic took the uncomplimentary second place among the EU states with the difference of 21.8% [37]. Besides the improvement of conditions for combining work and family life, the issue of equal pay remains one of the main priorities in gender equality.

Pay discrimination does not only concern the gender and feminist movement, women and their families; it has a significant impact on the entire labour market. The existing gap between the wages of women and men causes a decline in the productivity of women's work [2] and consequently a reduction in the gross domestic product [3]. On the other hand, fair remuneration and increasing employment opportunities stimulate economic growth [4]

The issue of equal opportunities when entering the labour market and getting employed as well as the economic impact caused by discrimination gets more complicated with the increasing age of the women and other employees. In the light

J. Stávková · N. Hazuchová · H. Porkertová (✉) · M. Rašticová
Mendel University in Brno, Brno, Czech Republic
e-mail: jana.stavkova@mendelu.cz

© Springer Nature Switzerland AG 2020

D. Cagáňová, N. Horňáková (eds.), *Mobility Internet of Things 2018*,
EAI/Springer Innovations in Communication and Computing,
https://doi.org/10.1007/978-3-030-30911-4_37

of the ageing European population, ageing and ageism are becoming more and more topical [5], especially in the public sector, where the majority of employees are older women [6].

In the text below, we have connected the economic challenges and problems brought by each topic. We have focused on the difference between the remuneration of men and women over 50 considering mainly the public sector and we have looked at how these differences developed regarding the changes brought by the economic crisis. That is why we have chosen the years 2010 and 2015 for comparison.

2 Causes of Unequal Remuneration

When attempting to explain gender pay inequalities in economic circles, the so-called human capital theory prevailed for a long time. It divided the wage gap into adjusted and unadjusted factors. The unadjusted ones were attributed to discrimination. Among the adjusted factors was the unequal investment allegedly made by men and women in their jobs. Women do not have as much work experience, they work for shorter hours, their education is worse and they spend less time on self-education than men. Therefore, they are less productive in their jobs than men, which leads to lower remuneration in the labour market [7].

Nevertheless, even these seemingly objective factors do not lack a gender dimension. Women more often interrupt their career to take care of children, which has a negative impact on their professional advancement, especially in the initial phases of their productive life, when it is important to invest more time and energy in work [8]. Another human capital factor, education, lacks validity. The difference in educational attainments mainly concerns the older generation, which followed different gender paths. Women today are more likely than men to finish secondary school and take up university studies [8]. Czech women are not an exception [9]. However, Vlachová [2] shows that the causality of higher education meaning higher pay is not true in the Czech Republic and men with secondary education earn more than women with a university degree. The wage gap is generally smaller in unskilled professions, where low education is sufficient, and it becomes significantly larger for people with university degrees [10]. Therefore, education explains the gender pay gap only partly, or rather it is insignificant [1]. Other data show that although men and women make the same investments [same education, similar jobs], the gender pay gap remains [11] and even a focus on different sectors does not necessarily mean different investments regarding education, the amount of time spent at work or the overall difficulty of the jobs.

Translating pay inequalities into human capital is not neutral; however, it incorporates the discrimination system elements in an unreflected way. This does not mean that the human capital theory should be completely abandoned. Nevertheless, each of its elements needs to be critically examined and, most importantly, related to a wider gender context [11]. The objective causes cannot be separated from the discriminatory ones in a simple way.

Searching for the causes of pay inequality requires focusing on their interconnection and how they reinforce each other [9]. Various causes are linked by cultural patterns and values carrying stereotypical division into a male and female sphere or aspirations and abilities related to these stereotypes. These patterns are already part of gender socialisation, which develops and appreciates different abilities of male and female pupils and afterwards male and female students often choose different types of school directing them to different professional careers and motivating them to different aspirations [12]. So-called male abilities are related to jobs which offer higher pay and prestige whereas the female ones do not.

In the Czech Republic there is still a deep-rooted idea of a woman as the primary caretaker and a man as the breadwinner, which reproduces gender inequality even further [9]. The idea of women as caretakers is reflected in pay remuneration in the sectors where care is the main job description, and reveals other difficulties related to the human capital theory. Women are considered to be naturally able to take care, i.e. they are not supposed to have to invest so much time and energy in these abilities. Therefore, the wages in these sectors remain low [13]. Gender stereotypes are also reflected in discriminatory practices favouring female and male candidates for traditional posts and at the same time making it more difficult to get the jobs which go against these stereotypes [9]. Moreover, people tend to work in a socially close group, which strengthens the separation of men and women even further [14].

Different directing and ideas of the abilities and aspirations then create and reproduce both horizontal [i.e. different sectors] and vertical [i.e. different posts] gender segregation in the labour market. Jobs where men prevail are better paid than those where there are more women [8, 9, 15]. The relation between the independent and dependent variables related to income is not quite clear. Certain jobs or sectors do not seem to be the independent variable. On the contrary, it is gender. The theory of "devaluation" says that if women start to prevail in a certain type of job, its prestige and remuneration go down [9].

Vertical segregation is primarily connected with the concept of a "glass ceiling", when qualified women do not get into top managerial positions, which increases the overall gender pay gap [8, 16]. The characteristics attributed to women, regardless of whether a particular woman is endowed with them, include empathy, intuition, emotionality, non-competitiveness and lack of desire for managerial positions. They do not overlap with those which are related to traditional managerial achievements and at the same time characterised as male [17]. The pay gap is reflected not only in the importance of the position but also in its character and team composition. Those who lead mainly female teams have lower remuneration and the other way round. Teams consisting of women are usually led by women [15]. Ostroff and Atwater [15] also bring psychological explanations based on the need for trust in the person in charge, which leads to a tendency to exclude people from outside and, on the contrary, favour those who are socially close, which means men in most cases, which maintains homogeneity. If women get into a managerial position after all, they are often under closer scrutiny and criticism and they hold positions which are less promising and more risky. Further advancement is often not guaranteed and the position may even be lost. This phenomenon is called a glass cliff [18].

Besides the glass ceiling and cliff there is another phenomenon described as a glass lift, which is connected with both vertical and horizontal segregation [19]. In branches with a higher representation of women [it is especially the public sector such as education, social care and healthcare] the pay gap is usually bigger than in sectors where men prevail. The reason is that men have a higher chance to gain faster promotion in "female professions" and they hold managerial positions more often than women [18].

The difference in remuneration of men and women is usually smaller in the public sector [8, 20, 21]. The [16] study showed that in most European countries it is between 6% and 16%. Vlachová [2] says that in the private sector there is space for different remunerations based on gender since the employee's wage may be influenced by the employer to a large extent. In the public sector this space is significantly limited by laws determining the salary scale. Moreover, they are usually sectors which do not have so many hierarchical positions [20]; the average income is generally lower and, therefore, there is not such a wide range [22]. However, even the scale given by law does not guarantee equal remuneration because employees are sometimes put into payroll categories based on wrong criteria.

It is, however, still true that both in the private and public sectors there is a pay gap in favour of men, especially in the feminised branches. Like in the private sector, this gap mainly concerns managerial positions, which shows that the glass ceiling and glass lift effects do not elude even the public sector [16]. Another problem is the non-transparency of remuneration, when women [or other disadvantaged groups] do not know how much their colleagues earn [9].

3 Women Over 50 in the Labour Market

Differences in remuneration in the Czech Republic change in the course of the life cycle. The values are the lowest for younger age categories, then the difference gradually grows and at an older age it slightly decreases again. Křížková et al. [9] attribute this mainly to work interruption caused by childcare. The age category of 40–44, when the pay gap is the biggest (it reaches up to 30%), is also a time when the employees reach their peak and this inequality fully reflects the glass ceiling phenomenon. Women do not manage to eliminate the gap at an older age and the gender inequalities accumulate over the years [23]. Moreover, it is characteristic of the generation of women over 50 that the gender emancipation does not concern them so much yet. Their level of education is lower than that of men and more often they follow the traditional paths. Although Czech women, unlike Western Europe and the English-speaking world, where most of the studies come from, spend significantly more time at work and part-time jobs are not so frequent, so-called double burden is common in the older age cohort, which leaves almost all the care of the family and household with the employed women [24].

Lower remuneration is also reflected in the pension rates, which also applies to the Czech environment, where pension rate calculation is highly dependent on previous income [25]. Therefore, older women are at a higher risk of poverty than men [2, 9, 26].

Ageism, i.e. discrimination based on age, is a phenomenon analysed by a lot of authors and together with specialised texts dealing with gender discrimination it forms an extensive network of studies. However, there is very little focus on their interconnection [5]. Current research focusing on linking gender and age follows the so-called intersectional approach which deals with the permeation of various axes of inequalities and oppression. It is not possible to simply add more and more levels of inequalities such as gender, age, class or race. It is necessary to focus on their mutual construction and transformation. Intersection is rather a verb than a noun (Lykke 2010) [27].

Gender and age discriminations strengthen each other and they must be approached together. Women are endangered by ageism more than men, which is given not only by the fact that they are more burdened by taking care of the family [6]. A survey by Duncana and Loretto [28] shows that the age at which a decline in work capacity is expected is lower for women than men. Sometimes, women are considered "old" as early as the age of 40 [5]. Therefore, age cannot be considered an objective category. It is connected with many other attributes such as gender or type of job. Being put in the older age category often means for women a barrier to career advancement or to re-entering the labour market [29]. Jyrkinen and McKie [5] analyse the situation when there is no right age for women to reach the career peak. Younger women deal with taking care of small children and when they exceed the age of 40 or 50, which is usually considered the best age, they encounter age discrimination. The same conclusions are also reached by Itzin and Phillipson (1995) or Loretto and Vickerstaff (2011) [6, 30].

The construction of a woman's age in the labour market, especially at certain posts, is closely related to discrimination based on appearance [lookism]. Especially women working in services are required to meet a certain standard of beauty, elegance and slenderness [31] but the demands on appearance are also common at administrative, especially assistant, posts [32] or in healthcare [33]. Older women are not considered attractive [34] and for women old age is linked to appearance, which is something that men do not have to deal with [6].

"Inappropriate age", which women have to face in the course of their entire career, is also reflected in vertical gender segregation. Jyrkinen and McKie [5] focused in their research on managerial positions in Scotland and Finland from the perspective of gender ageism. Managers' work experience is expected to grow with their increasing age. It could be therefore expected that a more advanced age is not an obstacle but, on the contrary, an advantage. However, this is only true for men, whereas women face age discrimination. Nevertheless, both women and men over 50 occupy fewer top managerial positions than their younger colleagues and especially older women occupy positions at the bottom end of the spectrum [6]. All this is reflected in their remuneration.

4 Methodology

The source of the data used for evaluating income inequalities was the data obtained in the European Union—Statistics on Income and Living Conditions (EU—SILC)—project. The data file in question is made up of individuals who have income from employment and who are 50 years old or over. The basic variable is the annual gross income from their main job greater than 0 in the years 2010 and 2015. In 2010, there were 2706 individuals surveyed at the age of 50 and over. In 2015, the sample group was made up of 2254 individuals. The identification attributes of each individual whose annual gross income from the main job was investigated as the basic variable were sex [male or female], age (50+) and industry. After recalculation using a weight coefficient, the scope of the 50+ file corresponds in both years to 11% of the total population of the Czech Republic and for this proportion of the population generalisation of selection characteristics has been made.

The industries have been categorised based on the NACE classification of economic activities.

Industry (NACE):

1. Agriculture, Forestry and Fishing (Section A)
2. Manufacturing (Section C which contains 24 different productions)
3. Construction (Section F)
4. Wholesale and Retail Trade; Repair of Motor Vehicles and Motorcycles (Section G)
5. Transportation and Storage (Section H)
6. Accommodation and Food Service Activities (Section I)
7. Financial and Insurance Activities (Section K)
8. Public Administration and Defence; Compulsory Social Security (Section O)
9. Education (Section P)
10. Human Health and Social Work Activities (Section Q)
11. Others, which were formed by combining 11 sections with a lower total number of employees (Section B—Mining and Quarrying, D—Electricity Production and Supply, E—Water Supply, J—Information and Communication Activities, L—Real Estate Activities, M—Professional, Scientific and Technical Activities, N—Administrative and Support Service Activities, R—Arts, Entertainment and Recreation, Others: S—Other Activities, T—Activities of Households as Employers, U—Activities of Extraterritorial Organisations and Bodies)
12. Unspecified (0). For SILC data, CZ-NACE says: the 00 group includes the following activities—"Production, trade and services not listed in Appendices 1 and 3 of the Trades Licensing Act". They are small trades not listed in Appendices 1 and 3.

To calculate the difference in the income of men and women in each industry, we used the Gender Pay Gap Index created by the authors. It expresses the ratio of men's and women's income in each sector (expressed as a percentage of the total income of all the employees over 50) to their frequency (expressed as a percentage of the total frequencies of employees over 50) in those sectors. This eliminates the influence of the number of men and women in each industry on the income level and the overall comparison expresses both the difference between men's and women's income and the difference between the sectors. The income-to-frequency ratio higher than 1 in a particular group means that its share in the total gross income is higher than in the total frequency. In other words, it is a rough criterion of whether these individuals fall into the above-average or below-average income group in relation to the data file in question, where the average income would equal 1.

To generalise the results to apply to the entire population of the Czech Republic for comparison of the statistical significance of the differences in gross income from the main job based on gender (when the influence of the industry is eliminated), on industry (when the influence of gender is eliminated) and on factor of joint influence of gender and industry, an unbalanced two-factor scattering analysis was performed. The group of people aged 50+ was scarcely represented in some sectors, being thus deviated by outliers. In both the years in question there were only units of people in this sector and extremely high income, which led to big differences between the 2 years. For this reason, the financial industry is not further analysed.

5 Results

5.1 General Numbers

In 2010, the Czech Republic was in the middle of an economic crisis, which started to subside in 2015. The unemployment rate was at its peak and it was 7.3%. In 2015, it fell to 5% [Czech Statistical Office] [38]. Although an economic crisis usually has a greater impact on the private sector, the public sector, especially the industry, does not remain unaffected [9]. Saving of the public finances enforced by politicians means wage stagnation, which does not respond flexibly to an improvement in the situation. Therefore, fluctuations are expected to be larger in the private sectors than in the public ones. However, an improvement in the overall economic situation does not have to mean a rise in the income of all the employees even in the private sector. Higher employment rates bring growth of groups with low qualification and income, which concerns mainly women. High employment rates for women then increase the overall pay gap [9].

Income inequalities based on gender are evident from the analysis of the group of individuals over 50 surveyed in 2010 and 2015. The income levels and the frequencies of men and women in this age group are shown in Table 1.

The analysis of the income volume based on gender clearly shows that men earn more. Although the total volume of the men's income in 2015 rose by "mere" 13 billion CZK and women's by almost 20 billion CZK compared to 2010, this

Table 1 The number of individuals employed, their income from the main job, by gender, in the 50+ age bracket

	Frequencies of 50+ employees		Gross income from the main job		Gender Pay Gap Index
	In total	Percentage	In total [bn. CZK]	Percentage	
2010					
Male	623,240	54.14	162.11	60.27	1.12
Female	527,952	45.86	106.88	39.73	0.86
In total	1,151,193	100	269.00	100	
2015					
Male	587,391	51.77	175.57	58.18	1.12
Female	547,325	48.23	126.18	41.82	0.87
In total	1,134,716	100	301.74	100	

relatively positive development in favour of women is only given by their increased frequencies. The Gender Pay Gap Index remains unchanged. The income volume per one employee in 2010 reached an annual income of 260,000 CZK for men and almost 58,000 CZK less for women, i.e. 202,000 CZK. In 2015, the annual income per one employee was 298,000 CZK for men and 231,000 CZK for women.

The total number of men and women over 50 employed between 2010 and 2015 fell slightly. We must look for the causes of this phenomenon because the number of people employed in this category was expected to rise as a result of the changes in the pension system made in 2013, which shifted the age limit for retirement. The number of economically active individuals fell by approx. 16,500, i.e. by approx. 1.5%, and further development cannot be accurately estimated. What is noticeable is the increase in the number of employed women [by 3.7%] and, on the contrary, a significant decrease in the number of employed men [by 5.5%]. This difference in frequencies is closely related to the stable Gender Pay Gap Index, which did not change between 2010 and 2015. The rise in the employment rate of women could lead to a bigger pay gap; however, it was balanced by the decrease in the number of employed men. Therefore, we cannot claim that no changes have taken place although the final numbers remain the same. To monitor specific changes, we must focus on the industries and the processes which take place there, which can be seen in Tables 2, 3, 4 and 5.

From the different frequencies of men and women in the labour market it is clear that the numbers of income active men and women are getting closer. The original 8% difference in the numbers of economically active men and women dropped to approx. 3%. This trend may partly be caused by the positive economic development in the Czech Republic bringing lower unemployment rates and thus the possibility for women to succeed in the labour market even with part-time employment, and also the longer lasting economic activity of women if it does not have to be full-time. Considering the fact that the pension level assessed for women is lower in the long term than the pensions received by men, this economic activity balances out the financial deficit. Last but not least, a certain role may be played by legal changes in social politics and availability of social benefits for both the sexes.

Table 2 Frequencies of employed people over 50 in each industry, the proportion of men and women in 2010 and 2015, $n = 1{,}134{,}716$

Sector	2010			2015			Total
	Total frequency	Proportion of men [%]	Proportion of women [%]	Total frequency	Proportion of men [%]	Proportion of women [%]	increase/decrease [%]
Agriculture, forestry and fishing	48,041	78.68	21.32	44,879	67.47	32.53	6.58 ↓
Manufacturing	253,647	67.43	32.57	269,121	65.22	34.78	6.75 ↑
Construction	76,689	84.07	15.93	79,649	90.05	9.95	3.72 ↑
Wholesale and retail trade; repair of motor vehicles and motorcycles	83,032	32.48	67.52	81,327	36.22	63.78	2.05 ↓
Transportation and storage	74,004	73.78	26.22	68,421	70.52	29.48	7.54 ↓
Accommodation and food service activities	21,590	22.13	77.87	15,841	38.39	61.61	26.63 ↓
Financial and insurance activities	–	–	–	–	–	–	–
Public administration and defence; compulsory social security	88,191	42.30	57.70	81,336	40.28	59.72	7.77 ↓
Education	94,205	28.19	71.81	104,133	18.20	81.80	9.53 ↑
Health and social care	73,628	16.47	83.53	92,429	16.14	83.86	11.31 ↑
Others	150,162	64.19	35.81	161,037	63.96	36.04	6.75 ↑
Unspecified [0]	166,681	51.13	48.87	115,713	43.44	56.56	31.58 ↓
In total	1,151,193	*x*	*x*	1,134,716	*x*	*x*	

Table 3 The total volume of the income of employed people over 50 in the industries in 2010 and 2015

Sector	2010			2015			Total increase/ decrease [%]
	Total volume of income [bn. CZK]	Proportion of men's income [%]	Proportion of women's income [%]	Total volume of income [bn. CZK]	Proportion of men's income [%]	Proportion of women's income [%]	
Agriculture, forestry and fishing	10.99	84.30	15.70	11.44	70.99	29.01	3.93 ↑
Manufacturing	59.59	74.62	25.38	73.51	73.73	26.27	18.94 ↑
Construction	20.53	82.66	17.34	22.87	88.59	11.41	10.23 ↑
Wholesale and retail trade; repair of motor vehicles and motorcycles	15.69	38.87	61.13	19.22	51.52	48.48	18.37 ↑
Transportation and storage	19.22	78.10	21.90	20.24	74.25	25.75	5.04 ↑
Accommodation and food service activities	3.25	28.80	71.20	3.62	37.27	62.73	10.22 ↑
Financial and insurance activities	–	–	–	–	–	–	–
Public administration and defence; compulsory social security	25.44	49.60	50.0	25.79	43.63	56.37	1.36 ↑
Education	24.41	32.83	67.7	28.67	23.15	76.85	4.86 ↑
Health and social care	19.07	22.41	77.9	26.18	22.91	77.09	27.26 ↑
Others	41.72	72.27	27.3	46.46	66.41	33.59	10.2 ↑
Unspecified [0]	18.80	61.15	38.5	14.33	54.64	45.36	23.78 ↓

Table 4 An overview of the ratio of income to frequencies and their combination in 2010 for employees over 50

	Total gross income [%]		Frequencies [%]		Gender Pay Gap Index	
2010	Male	Female	Male	Female	Male	Female
Agriculture, forestry and fishing	3.44	0.64	3.28	0.89	1.05	0.72
Manufacturing	16.53	5.62	14.86	7.18	1.11	0.78
Construction	6.31	1.32	5.60	1.06	1.13	1.25
Wholesale and retail trade; repair of motor vehicles and motorcycles	2.27	3.57	2.34	4.87	0.97	0.73
Transportation and storage	5.58	1.57	4.74	1.69	1.18	0.93
Accommodation and food service activities	0.35	0.86	0.42	1.46	0.84	0.59
Financial and insurance activities	–	–	–	–	–	–
Public administration and defence; compulsory social security	4.69	4.77	3.24	4.42	1.45	1.08
Education	2.98	6.09	2.31	5.88	1.29	1.04
Health and social care	1.59	5.50	1.05	5.34	1.51	1.03
Others	11.21	4.30	8.37	4.67	1.34	0.92
Unspecified [0]	4.27	2.72	7.40	7.08	0.58	0.38
In total	60.27	39.73	54.14	45.86	*x*	*x*

5.2 Comparison Between 2010 and 2015

Although the Gender Pay Gap Index eliminates frequencies and only shows differences in income, it is interesting to notice that a stable Gender Pay Gap Index (as was obvious in Table 1) does not mean that there are no considerable changes. This is also reflected in the frequencies which (together with income) are used to subsequently calculate the Index.

The data confirm that the representation of men and women in each industry develops. Some sectors show a change in the number of employees, whether it is a significant decrease (agriculture, forestry and fishing; transportation and storage; accommodation and food service activities; public administration and defence + compulsory social security; unspecified) or an increase (manufacturing; education; health and social care; others). A dramatic decrease may be noticed in the unspecified section, where there are primarily small tradesmen, which indicates adverse conditions for their activities. In the agriculture, forestry and fishing category the overall decline in the number of employees meant an increase in the proportion of women. However, in the accommodation and food service activities section the overall decline led to a decrease in the proportion of women in this sector. In the education category, on the contrary, the total number of employees rose and the increase in the proportion of women is caused by a decrease in the number of

Table 5 An overview of the ratio of income to frequencies and their combination in 2015 for employees over 50

	Total gross income [%]		Frequencies [%]		Gender Pay Gap Index	
2015	Male	Female	Male	Female	Male	Female
Agriculture, forestry and fishing	2.56	1.23	2.67	1.29	0.96	0.96
Manufacturing	18.20	6.17	15.47	8.25	1.18	0.75
Construction	6.76	0.82	6.32	0.70	1.07	1.18
Wholesale and retail trade; repair of motor vehicles and motorcycles	3.34	3.03	2.60	4.57	1.29	0.66
Transportation and storage	5.06	1.65	4.25	1.78	1.19	0.93
Accommodation and food service activities	0.48	0.72	0.54	0.86	0.89	0.84
Financial and insurance activities	–	–	–	–	–	–
Public administration and defence; compulsory social security	3.52	5.03	2.89	4.28	1.22	1.17
Education	2.21	7.29	1.67	7.51	1.32	0.97
Health and social care	2.07	6.61	1.31	6.83	1.57	0.97
Others	10.17	5.23	9.08	5.11	1.12	1.02
Unspecified [0]	2.48	2.27	4.43	5.77	0.56	0.39
In total	58.18	41.82	51.77	48.23	*x*	*x*

men and by a significant increase in the number of women in this industry. The number of male teachers has been falling steadily (Czech Statistical Office), which can also be seen in the older age category [39, 40]. There is always a shortage of teachers [35] and a substantial increase in the number of women means that the strong age group of female teachers who were in the younger age category in 2010 reached the age of over 50. Another industry from the public sector which saw a significant increase in the number of employees [even though with the same proportion of men and women] is health and social care, which remains highly feminised. This data proves that there are many older women working in the public sector in the Czech Republic [6].

Despite a slight decrease in the number of employees over 50 in 2015 compared with 2010, the national economy improved a little, which can be seen in Table 3.

Despite a sharp decrease in the number of employees in the categories of agriculture, forestry and fishing; transportation and storage; and accommodation and food service activities, as Table 2 clearly shows, the volume of income in these industries slightly increased. A small increase can also be seen in public administration and defence + compulsory social security, which also shows a decrease in the number of employees and, moreover, they belong in the public sector. A dramatic increase in income with the number of employees almost unchanged occurred in the wholesale and retail trade, repair of motor vehicles and motorcycles sector. Another considerable increase in income can be seen in the health and social care sector, which is not so big due to an increase in the number of employees, but it still reaches almost 16%. The only category showing a decrease in the total volume of income is unspecified. However, if we consider the decrease in the number of employees, which was close to one-third, we will see that there was an increase in income per one employee as well. The only sector where the average income decreased is education since the increase in the volume of income by almost 5% does not level the almost 10% increase in the number of employees. Considering the fact that the average income of teachers did not decrease, this indicator reflects the growing age of other employees in education such as cooks or cleaners. Still, in general, the sphere of education in the Czech Republic has been underfinanced for a long time and the income in the public sector stagnates even after the subsiding of the economic crisis [9].

By calculating the ratio of proportional income to frequencies we will get the Gender Pay Gap Index. The comparison of this Index for the years 2010 and 2015 will show how the inequality between the remuneration of men and women over 50 developed in each sector of the national economy.

Between 2010 and 2015 the average wage rose; however, not all the sectors showed the same increase. What is interesting at first glance is the same value of the Gender Pay Gap Index in agriculture, forestry and fishing, which is 0.96. The shift from the inequality in 2010 (which was 1.05 for men and 0.72 for women) is a typical example of factors leading to equal income. Table 2 clearly shows that there was a significant decrease of the difference between the numbers of men and women relative to the total frequency in this sector. In industries with a more balanced representation of men and women the gender differences in remuneration are usually less considerable [9]. Moreover, this sector includes the manual, less skilled professions, where the gender pay gap is also usually smaller [10] and thus the Gender Pay Gap Index is even. While a more even ratio of women to men in the sectors may be considered a positive trend, the fact that the more balanced income occurs primarily in the unskilled jobs, where the wages are generally lower [which is also proved by the Gender Pay Gap Index which is lower than 1], is a negative phenomenon.

A similar ratio of men to women (67–33, or 65–35) can also be seen in manufacturing, where the Gender Pay Gap Index went slightly down. While the men's income increased from 1.11 to 1.18 in relation to the average income, for women it decreased from 0.78 to 0.75. The manufacturing sector is also similar to agriculture, forestry and fishing in the fact that there are many unskilled and manual jobs. However, there are considerably more employees (269,121 to 44,879 in 2015) and unlike in agriculture, etc. there was not an overall decrease in the number of employees compared with 2010 but, on the contrary, an increase. This confirms what Křížková et al. [9] say—a higher employment rate leads to an increase in people with low skills and thus a low income—primarily women. The subsiding of the economic crisis and increase in the employment rates did not, therefore, necessarily lead to an increase in incomes.

A similar trend as in the agriculture, etc. sector can be seen in accommodation and food service activities, where the Gender Pay Gap Index remained almost the same for men (0.84 in 2010 and 0.89 in 2015), whereas for women there was a considerable increase in relation to the average income (0.59 in 2010 and 0.84 in 2015). This industry also showed a more balanced gender equality with regard to the ratio of men to women (22–78%, or 38–62%), the total number of employees fell (by as much as almost 27%) and there were mostly little skilled or unskilled jobs. Particularly the profession of a waiter/waitress is characterised by a low basic wage which is increased by tips. However, a low wage base has a negative impact on pension calculation, which is the current problem of people over 50 as it makes them liable to poverty at the retirement age [26].

The only industry where the Gender Pay Gap Index is in favour of women is construction. After 2010 the income in this sector did generally decrease in relation to the average income (1.13–1.07 for men and 1.25–1.18 for women); however, women are still doing relatively well. This difference is mainly given by the structure of positions—in construction, women mostly do administrative work unlike men, who do the manual jobs. The difference in the remuneration of men and women is in fact even greater because there are plenty of labourers working illegally in construction and they are not included in the statistics.

From the three industries falling into the public sector only public administration and defence + compulsory social security showed an overall decrease in the number of employees, and the others showed an increase. This makes the public sector different from the private one, where most of the industries showed an outflow of employees over 50. Despite a slight decrease and the preservation of the ratio of men to women in the public administration and defence + compulsory social security industry, men's and women's income is becoming more equal and comparable (1.45–1.22 for men and 1.08–1.17 for women). This sphere remains the only one in the public sector where men's and women's income is balanced. In education [1.29–1.32 for men and 1.04–0.97 for women] and in health and social care (1.51–1.57 for men and 1.03–0.97 for women) the difference in remuneration grew even larger and remains big. In 2015, women did not even reach the average income. Although the gap in the remuneration of men and women is usually smaller in the public sector [8, 20, 21], which is also clear from the Czech data [2], it is the most considerable from the industries in question for the over-50 age category, which is because they are sectors with a predominance of women, which became even more significant in education. A decrease in the index value for both men and women is primarily caused by the larger increase in the overall average income compared to education and health and social care. The health and social care is strongly segregated by individual positions, when women usually occupy the worst-paid ones [36]. A big difference in remuneration in education shows that older women, unlike older men, do not occupy leading positions and the existence of both the glass ceiling [8, 16] and the glass lift [19] in this sector is difficult to eliminate. Moreover, education remains quite a conservative sphere with a gender-segregated idea of who is endowed with authority and leadership abilities. Arulampalam et al. [16] say that due to the fact that the public sphere is not focused on profit, there may be a stronger tendency towards discriminatory treatment. The private sphere may be more flexible and able to monitor who leads the enterprise to efficiency and financial profit, unlike the public sphere, where the cultural gender patterns are more maintained.

6 Conclusion and Discussion

The purpose of this text was to point out the development of the employment rate and income inequality in the remuneration of men and women over 50 in the different sectors of the Czech national economy between 2010 and 2015. We primarily focused on the situation of women with regard to the public sector, where older women prevail [6]. These are later prone to poverty due to the dependence of the pension rate calculations on previous income [2, 9, 26].

Between 2010 and 2015 significant changes occurred in the Czech economy. After the economic crisis subsided the incomes rose quite considerably, the employment rate for women over 50 increased and the employment rate for men in the same age category decreased, which shows that the numbers of the income of active men and women over 50 got closer. This made the difference in the total volume of income smaller; however, these changes did not mean a decrease in the overall pay gap based on gender and the data researched clearly show that men have a higher income both in the private and in the public sectors.

A closer analysis shows that it is necessary to focus on several factors that are responsible for the gender pay gap. The first one is frequency, which shows whether the sector is masculinised or feminised or whether the ration of men to women is balanced. Our paper has shown the validity of the thesis that a more balanced representation of men and women usually brings smaller gender differences in remuneration [9]. On the other hand, the most considerable differences favouring men are seen in the highly feminised sectors [18]. The reason is particularly the existence of the so-called glass lift, when men in female teams achieve fast career advancement [19].

Another significant factor that we identified is the character for the industry. Sectors with lower incomes, where there are a lot of unskilled and manual jobs, are characterised by smaller differences between the incomes based on gender and our results therefore correspond with other literature (e.g. [10]). A higher employment rate for women over 50 in 2015 compared to 2010 confirms the statement of Křížková et al. [9] that a higher employment rate leads primarily to an increase in employees with low qualification and income and thus even the subsiding of the economic crisis did not result in an increase in the average income of women.

The public sector is a sphere which, to a certain extent, combines both the factors that contribute to a significant pay gap—i.e. they are feminised sectors and although there are not predominantly manual or unskilled jobs they are generally characterised by low incomes. This is particularly evident in education, which was the only industry with a decrease in income in relation to the average income between 2010 and 2015. The differences in income in education are caused not only by the glass lift phenomenon, when men often occupy the leading positions, but also by the manual and less skilled jobs from cleaners to cooks, which are women's domain. Our results show that in the 50+ category in the public sector the differences in income are among the largest from all the industries and therefore it is not true for this category that the pay gap is usually smaller in the public sector than in the private one [2, 8, 20, 21].

The data from the "European Union—Statistics on Income and Living Conditions (EU—SILC)"—project, which was the source for our analysis, enables an international comparison between the European Union countries. A major drawback may be the fact that our analysis of the employment rate and income inequalities for employees over 50 did not consider particular positions. Nevertheless, we have shown that, especially in the feminised sectors, it is possible to see that their division and the leading positions are occupied primarily by men, and we are working on a further analysis that will take particular positions into consideration.

To analyse the differences, we used the Gender Pay Gap Index which is related to the average income. The already widely accepted Gender Pay Gap (GPG) does show the differences in income; however, it does not say anything about the economic power of each sector of the national economy. It is therefore suitable to combine both the indicators and put them into mutual context, which opens up the potential for further research even in international context. At the same time, by comparing the years 2010 and 2015, we have shown that although the total figures expressing an income inequality may be similar, on closer examination based on frequencies we can see that extensive changes occurred in the economy during that time. One number may thus cover a lot of ongoing processes and, for a conclusive analysis considering a number of connections influencing the phenomenon in question, other quantifiable indicators should also be examined.

References

1. Rubery, J., Grimshaw, D., & Figueiredo, H. (2005). How to close the gender pay gap in Europe: Towards the gender mainstreaming of pay policy. *Industrial Relations Journal, 36*(3), 184–213.
2. Vlachová, H. (2014). Genderová platová mezera v soukromém a veřejném sektoru České republiky. *Acta Oeconomica Pragensia, 22*(5), 89–100.
3. Cavalcanti, T., & Tavares, J. (2016). The output cost of gender discrimination: A model-based macroeconomics estimate. *The Economic Journal, 126*(590), 109–134.
4. Stotsky, J. G. (2006). Gender and its relevance to macroeconomic policy: A survey. IMF Working Paper. WP/06/233. http://hussonet.free.fr/imf6233.pdf. Accessed 8 Feb 2018.
5. Jyrkinen, M., & McKie, L. (2012). Gender, age and ageism: Experiences of women managers in Finland and Scotland. *Work, Employment and Society, 26*([1]), 61–77.
6. Loretto, W., & Vickerstaff, S. (2011). The relationship between gender and age. In M. Parry & S. Tyson (Eds.), *Managing and Age Diverse Workforce* (pp. 59–79). Basingstoke: Palgrave Macmillan.
7. Blau, F. D., & Kahn, L. M. (2000). Gender differences in pay. *Journal of Economic Perspectives, 14*(4), 75–99.

8. Jacobsen, J. (2016). Gender wage gap. In N. A. Naples (Ed.), *The Wiley Blackwell encyclopedia of gender and sexuality studies* (1st ed., pp. 1–3). Chicester, UK: John Wiley & Sons, Ltd..
9. Křížková, A., Vohlídalová, M., Pospíšilová, K., & Maříková, H. (2017). *Aktuální rozdíly v odměňování žen a mužů v ČR*. Praha: Ministerstvo práce a sociálních věcí.
10. de la Rica, S., Dolado, J. J., & Llorens, V. (2005). *Ceiling and Floors: Gender Wage Gaps by Education in Spain. Series: IZA Discussion Paper Series 1483*. Bonn: Institute for the Study of Labor (IZA).
11. Lips, H. M. (2013). The gender pay gap: Challenging the rationalizations. Perceived equity, discrimination, and the limits of human capital models. *Sex Roles, 68*(3–4), 169–185.
12. Jarkovská, L. (2014). *Gender před tabulí: Etnografický výzkum genderové reprodukce v každodennosti školní třídy*. Praha: SLON.
13. Palmer, E., & Eveline, J. (2012). Sustaining low pay in aged care work. *Gender, Work & Organization, 19*(3), 254–275.
14. Groshen, E. (1991). The structure of the female/male wage differential: Is it who you are, what you do or where you work? *Journal of Human Resources, 26*, 457–472.
15. Ostroff, C., & Atwater, L. E. (2003). *Does whom you work with matter?: Effects of referent group gender and age composition on managers' compensation*. Washington, D.C.: American Psychological Association.
16. Arulampalam, W., Booth, A. L., & Bryan, M. L. (2007). Is there a glass ceiling over Europe? Exploring the gender pay gap across the wage distribution. *Industrial Laboratory, 60*(2), 163–186.
17. Macarie, F. C. & Moldovan, O. (2012). Gender discrimination in management. Theoretical and empirical perspectives. Babes Bolyai University. http://rtsa.ro/en/files/TRAS_35E_11_MACARIE,%20MOLDOVAN.pdf. Accessed 3 Apr 2018.
18. Ryan, M. K., & Haslam, S. A. (2005). The glass cliff: Evidence that women are over-represented in precarious leadership positions. *British Journal of Management, 16*, 81–90.
19. Williams, C. L. (1992). The glass escalator: Hidden advantages for men in the "female" professions. *Social Problems, 39*(3), 253–267.
20. Bradley, S., Green, C., & Mangan, J. (2015). Gender wage gaps within a public sector: Evidence from personnel data. *The Manchester School, 83*(4), 379–397.
21. Hedija, V. (2014). Gender pay gap in different sectors of Czech economy. In *32nd International Conference Mathematical Methods in Economics 2014* (pp. 275–280). Olomouc: Palacky University.
22. Jurajda, S. (2003). Gender wage gap and segregation in enterprises and the public sector in late transition countries. *Journal of Comparative Economics, 31*, 199–222.
23. Ross, C. E., & Mirowsky, J. (2002). Age and the gender gap in the sense of personal control. *Social Psychology Quarterly, 65*, 125–145.
24. Asklöf, C. (2003). Proč a jak vypracovat plán rovnosti žen a mužů. 1. vyd. Ministerstvo práce a sociálních věcí, Praha.
25. Dudová, R. (2013). Genderové souvislosti reforem důchodového systému v ČR. *Gender, Rovné Příležitosti, Výzkum, 14*(2), 56–67.
26. McLaughlin, D. K., & Jensen, L. (2000). Work history and U.S. elders' transitions into poverty. *The Gerontologist, 40*, 469–479.
27. Lykke, N. (2010). *Feminist studies: A guide to intersectional theory, methodology and writing*. New York: Routledge.
28. Duncan, C., & Loretto, W. (2004). Never the right age? Gender and age-based discrimination in employment. *Gender, Work and Organisation, 11*(1), 95–115.
29. Moore, S. (2009). "No Matter What I Did I Would Still End Up in the Same Position": Age as a factor defining older women's experience of labour market participation. *Work, Employment and Society, 23*(4), 655–671.
30. Itzin, C., & Phillipson, C. (1995). Gendered ageism: A double jeopardy for women in organisations. In C. Itzin & J. Newman (Eds.), *Gender, culture and organizational change: Putting theory into practice*. London: Routledge.
31. Shilling, C. (2003). *The body and social theory*. London: Sage.

32. Handy, J., & Davy, D. (2007). Gendered ageism: Older women's experiences of employment agency practices'. *Asia Pacific Journal of Human Resources, 45*(1), 85–99.
33. Clarke, L. H., & Griffin, M. (2008). Visible and invisible ageing: Beauty work as a response to ageism. *Ageing and Society, 28*(5), 653–674.
34. Gibson, D. (1996). Broken down by age and gender: The problem of old women redefined. *Gender & Society, 10*(4), 433–448.
35. Píšová, M., & Hanušová, S. (2016). Začínající učitelé a drop-out. *Pedagogika, 66*(4), 386–407.
36. Campos-Serna, J., Ronda-Pérez, E., Artazcoz, L., Moen, B. E., & Benavides, F. G. (2013). Gender inequalities in occupational health related to the unequal distribution of working and employment conditions: a systematic review. BioMed Central Ltd. http://www.equityhealthj.com/content/12/1/57. Accessed 21 June 2018.

Statistics:

37. http://ec.europa.eu/eurostat/statistics-explained/index.php/Gender_pay_gap_statistics. Accessed 5 July 2018.
38. https://vdb.czso.cz/vdbvo2/faces/cs/index.jsf?page=vystup-objekt&pvo=ZAM01-B&skupId=426&katalog=30853&pvo=ZAM01-B&str=v467&u=v413__VUZEMI__97__19. Accessed 5 Aug 2018.
39. https://www.czso.cz/documents/10180/34193311/2300421623.pdf/6083b745-1628-4009-a809-ddfaab4c878a?version=1.3. Accessed 5 July 2018.
40. https://www.czso.cz/documents/10180/34193311/2300421612.pdf/40859bfa-0082-4535-96ed-e1ce83fa8801?version=1.3. Accessed 5 Feb 2018.

Analytic Network Process as a Tool of Innovative Management at Manufacturing Technology Selection

Peter Monka, Katarina Monkova, Andrej Andrej, Jozef Tkac, Vladimir Duchek, Milan Edl, and Helena Zidkova

1 Introduction

The analytic network process (ANP) method is one of the multi-criteria decision-making methods that support the user in choosing the best option from the available options. This method presents the problem as a network of criteria, sub-criteria, and alternatives, but it can only be applied if the criteria, sub-criteria, and alternatives are dependent. ANP method was defined in 1996. It is a generalization of analytic hierarchy process (AHP) that was introduced by Thomas Saaty (1980) as an effective tool for dealing with complex decision-making, and may aid the decision maker to set priorities and consequently make the best decision. By reducing complex decisions to a series of pairwise comparisons, and then synthesizing the results, the AHP helps to capture both subjective and objective aspects of a decision. In addition, the AHP incorporates a useful technique for checking the consistency of the decision maker's evaluations, thus reducing the bias in the decision-making process. The advantages of the ANP method are particularly straightforward methodology; a network structure of the decision-making process for greater understanding of the problem; possibility of combining the results with other methods; the ability to include the qualitative decision criteria in the decision process; and also that the method has been verified in real cases [1].

P. Monka · K. Monkova (✉) · A. Andrej · J. Tkac
Technical University of Kosice, Faculty of Manufacturing Technologies with the Seat in Presov, Presov, Slovakia
e-mail: peter.pavol.monka@tuke.sk; katarina.monkova@tuke.sk; andrej.andrej@tuke.sk; jozef.tkac@tuke.sk

V. Duchek · M. Edl · H. Zidkova
Faculty of Mechanical Engineering, University of West Bohemia, Pilsen, Czech Republic
e-mail: duchekv@kto.zcu.cz; edl@kpv.zcu.cz; zidkova@kto.zcu.cz

© Springer Nature Switzerland AG 2020
D. Cagáňová, N. Horňáková (eds.), *Mobility Internet of Things 2018*,
EAI/Springer Innovations in Communication and Computing,
https://doi.org/10.1007/978-3-030-30911-4_38

The main goal of this chapter is to find the most favorable method for the cutting of composite material (namely bone cement). This material is known as polymethyl methacrylate (PMMA). Bone cement is being used for over 50 years, mainly in clinical practice. It features excellent biomechanical properties, long-term stability and tolerance by the human body, and possibility of the application through a thin needle. PMMA is used to repair dentures, in repairs of bone and skull defects in humans, in total hip joint replacement, etc. Currently there is a new type of bone cement, which is characterized by good viscosity. This reduces the risk of cement leakage outside the reserved area during the application [2, 3].

2 Comparison of Progressive Cutting Technologies

Traditional working methods use a "classic" cutting tool and mechanical energy to cut the material. But unconventional methods use other forms of energy in various chemical and physical events. The special features of these advanced methods are defined below [4]:

- Material removal does not depend on the mechanical properties (such as hardness, strength, machinability).
- Cutting force does not work on the site of the separation of particles; the cutting resistance is not formed; therefore, the worked pieces are not deformed.
- Less heat passes into the worked piece during the particle removal, because the cutting of the particles is micro sized with a large number of sites and the frequency of elementary removals is high.
- The entire surface of the worked piece is machined at the same time.
- When during a work the micro- and nano-particles are obtained.
- The size of the workpiece is defined by the energy base equipment (102–103 kW).
- The energy consumption during material removal is greater and removal rate is much lower than in traditional processes.

Comparison and selection of the most appropriate method were implemented from the following cutting technologies. Specifically, authors focused on nine unconventional methods shown in Fig. 1.

3 Defining Defects

When choosing the most appropriate cutting technology in between the abovementioned nine methods, the deficiencies that may arise from fragmentation of bone cement will play a big role. These shortcomings need to be defined and focused on. The selected technology should minimize these flaws to acceptable levels.

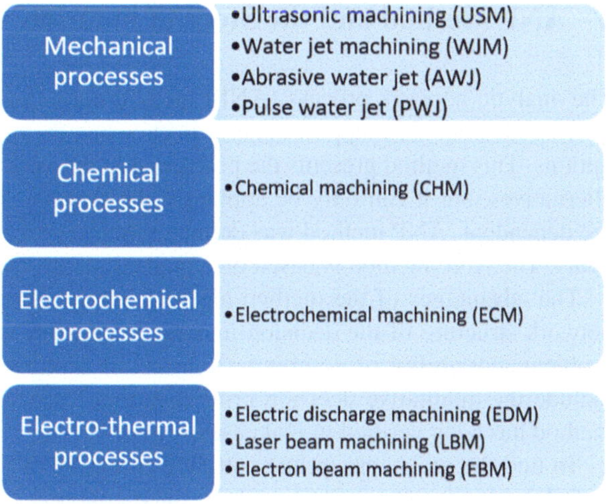

Fig. 1 Comparison of cutting technology

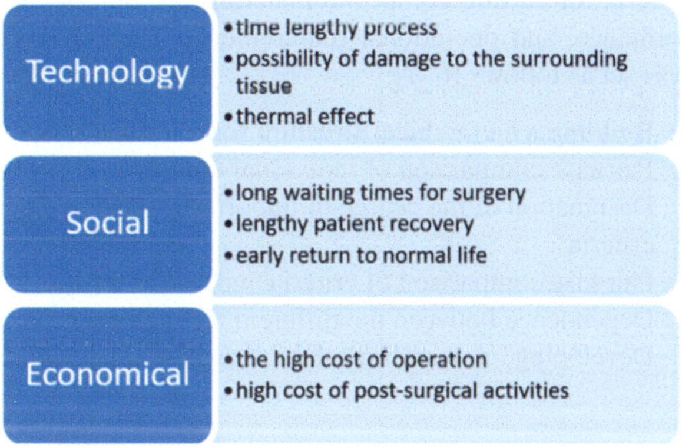

Fig. 2 Defined defects

The deficiencies are broken down into three areas (Fig. 2), namely in the social, technological, and economic fields.

The aim is to find a cutting technology, which minimizes technological shortcomings, such as that the cutting will be done without damaging the surrounding tissue, that there is no heat effect in the actual disintegration of bone cement, and that the overall process will not be time consuming.

Minimizing these shortcomings should result in a shorter surgery; the patient would remain for a shorter time under anesthesia, and thus the risk of infections and other complications would be minimized. Patient's quality of life will depend primarily on reduction of waiting time for surgery; elimination of post-surgery complications; and also a shorter recovery and quicker return to normal life [5].

4 ANP Method and the Steps of Solution

The analytic network process (ANP) method is one of the multi-criteria decision-making methods that support the user in choosing the best option from the available options. This method presents the problem as a network of criteria, sub-criteria, and alternatives, but it can only be applied if the criteria, sub-criteria, and alternatives are dependent. ANP method was defined relatively recently, in 1996 by Thomas L. Saaty. The ANP method is based on a mathematical calculation [6].

The advantages of the method are particularly straightforward methodology; a network structure of the decision-making process for greater understanding of the problem; possibility of combining the results with other methods; the ability to include the qualitative decision criteria in the decision process; and also that the method has been verified in real cases.

To find the most appropriate cutting technology of bone cement, it is necessary to follow the decision criteria, which must contain certain sub-criteria. The main criteria for cutting composite materials are process performance, divisibility, performance, and operational requirements. The sequence of steps of ANP method was set as follows [6, 7]:

- Building a hierarchical structure for the material of bone cement
- Pairwise comparison of individual criteria
- Designation of the degree of importance of individual technologies based on five criteria
- Pairwise comparison of criteria and sub-criteria
- Dependence between the different technologies
- Developing an overall resultant matrix w_p and summary results W_{ANP}

4.1 The Preparation of a Hierarchical Structure

Figure 3 shows in a hierarchical structure individual unconventional cutting technologies with their criteria and sub-criteria. These are ultrasonic machining (USM), water jet machining (WJM), abrasive water jet (AWJ), pulse water jet (PWJ), chemical machining (CHM), electrochemical machining (ECM), electric discharge machining (EDM), laser beam machining (LBM), and electron beam machining (EBM) [8].

4.2 Pairwise Comparison of Individual Criteria

Pairwise comparison of individual criteria, namely *MA* material, *DI* divisibility, *PP* process performance, *EC* economy, and *Trade* operation requirement. In Table 1 the decision maker needs to select the values on the Saaty scale 1–9, where 1 means that

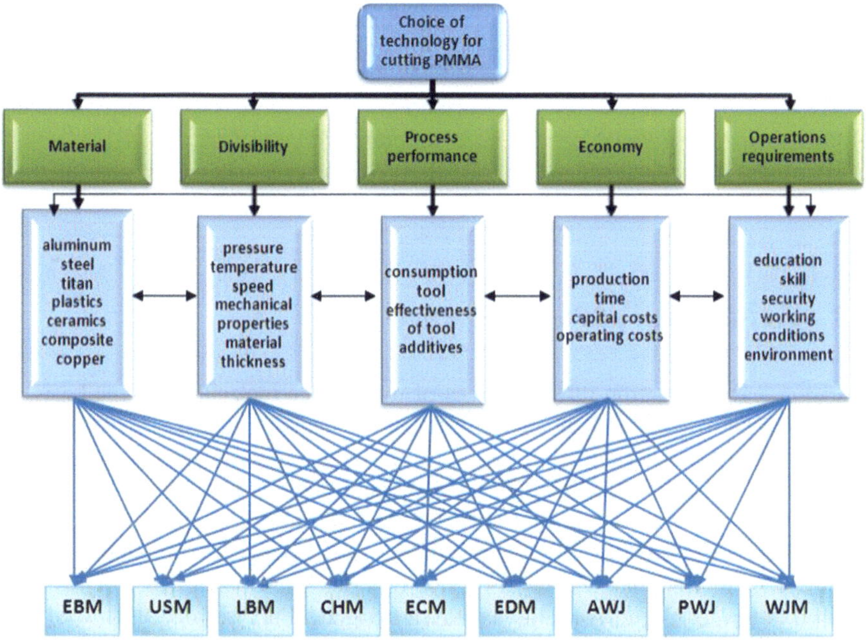

Fig. 3 Requirements for cutting PMMA in a hierarchical structure ANP

Table 1 Paired comparison criteria

	MA	DI	PP	EC	OR	Geomean	w_1
Material (MA)	1	4	7	7	7	4.2410	0.5356
Divisibility (DI)	1/4	1	6	6	6	2.2206	0.2804
Process performance (PP)	1/7	1/6	1	2	2	0.6248	0.0789
Economy (EC)	1/7	1/6	1/2	1	2	0.4735	0.0598
Operation requirements (OR)	1/7	1/6	1/2	1/2	1	0.3589	0.0453
						7.9189	1.0000

the criteria are equally important. Fractions represent the reversed (inverse) value. A geometric mean G ("Geomean") has been calculated by means of the formula (1) [6]

$$G = \sqrt[n]{x_1 \cdot x_2 \ldots x_n} \tag{1}$$

where G is the geometric mean of the line, n is the radical of the conjunction of the values, and $x_1, x_2, \ldots, x_n$ are the values assigned to the criteria.

Weight of each criteria "w_1" has been calculated (2) as the geometric mean of the outcome of each line divided by the sum of the final resultant average [6]:

$$w_{tj} = \frac{x_g}{\sum \left(x_{g1} + x_{g2} \ldots x_{gn}\right)} \quad (2)$$

where w_{ij} is the calculated weight of each criterion in line, x_g is the geometric mean of the matrix row, and $x_{g1}, x_{g2} \ldots x_{gn}$ is the sum of all geometric means.

The input and calculated values of paired comparison criteria are presented in Table 1.

4.3 *Construction of the Importance Stage*

Decision and designation of the degrees of importance took place on the basis of theoretical knowledge and subjective view of the decision maker. The results are presented in the table (Table 2). It should be noted that values in Table 2 could differ according to operator's (technologist's) experience and opinion.

The final values of importance are then recalculated according to formula (2), where the results are the values representing the dependence of importance of unconventional technologies based on five criteria. These values of weights will be used in further calculations in the form of matrices (Fig. 4).

Table 2 Degree of process importance

	MA	DI	PP	EC	OR
A1 (EBM)	7	3	1	3	8
A2 (USM)	9	7	3	4	8
A3 (LBM)	7	3	8	3	3
A4 (CHM)	7	5	3	3	1
A5 (ECM)	3	5	4	1	1
A6 (EDM)	3	3	1	3	8
A7 (AWJ)	9	8	7	4	7
A8 (PWJ)	9	9	9	7	9
A9 (WJM)	9	7	7	4	7

$$w_{21} = \begin{pmatrix} 0{,}1111 \\ 0{,}1429 \\ 0{,}1111 \\ 0{,}1111 \\ 0{,}0476 \\ 0{,}0476 \\ 0{,}1429 \\ 0{,}1429 \\ 0{,}1429 \end{pmatrix} \quad w_{22} = \begin{pmatrix} 0{,}0600 \\ 0{,}1400 \\ 0{,}0600 \\ 0{,}1000 \\ 0{,}1000 \\ 0{,}0600 \\ 0{,}1600 \\ 0{,}1800 \\ 0{,}1400 \end{pmatrix} \quad w_{23} = \begin{pmatrix} 0{,}0233 \\ 0{,}0698 \\ 0{,}1860 \\ 0{,}0698 \\ 0{,}0930 \\ 0{,}0233 \\ 0{,}1628 \\ 0.2093 \\ 0{,}1628 \end{pmatrix} \quad w_{24} = \begin{pmatrix} 0{,}0938 \\ 0{,}1250 \\ 0{,}0938 \\ 0{,}0938 \\ 0{,}0313 \\ 0{,}0938 \\ 0{,}1250 \\ 0{,}2188 \\ 0{,}1250 \end{pmatrix} \quad w_{25} = \begin{pmatrix} 0{,}1538 \\ 0{,}1538 \\ 0{,}0577 \\ 0.0192 \\ 0{,}0192 \\ 0{,}1538 \\ 0.1346 \\ 0{,}1731 \\ 0{,}1346 \end{pmatrix}$$

Fig. 4 The resulting scale matrix

Fig. 5 The total scales of the resulting matrix

$$\begin{pmatrix} 0{,}5089 & 0{,}2865 & 0{,}2014 & 0{,}1562 & 0{,}1852 \\ 0{,}2571 & 0{,}5492 & 0{,}1875 & 0{,}1852 & 0 \\ 0{,}1145 & 0{,}1002 & 0{,}5518 & 0 & 0{,}1562 \\ 0{,}0680 & 0{,}0641 & 0 & 0{,}6586 & 0 \\ 0{,}0515 & 0 & 0{,}0593 & 0 & 0{,}6586 \end{pmatrix} x \begin{pmatrix} 0{,}5356 \\ 0{,}2804 \\ 0{,}0789 \\ 0{,}0598 \\ 0{,}0453 \end{pmatrix} = \begin{pmatrix} 0{,}3865 \\ 0{,}3176 \\ 0{,}1400 \\ 0{,}0938 \\ 0{,}0621 \end{pmatrix}$$

4.4 Pairwise Comparison of Individual Criteria and Sub-criteria

To create a pairwise comparison of individual criteria and sub-criteria is the same as in Sect. 4.2 with the difference that the comparison is not carried in respect of all criteria, but only one. Gradually, the weights of pairwise comparison w_{31} to w_{35} are calculated. These are summarized in a next matrix, which when multiplied by the resulting weights w1 according to formula (3) obtains the value of w_c.

$$w_3 \cdot w_1 = w_c \tag{3}$$

It represents the result of the overall balance between the weights of all possible interdependencies among criteria and sub-criteria weights and the resulting pairwise comparison of the five criteria, according to Fig. 5.

4.5 Dependence Between Different Technologies

We continue by calculating the resulting weights w_{41} to w_{45}. Using them we express the resulting relations between the alternatives of unconventional methods of cutting in respect of each criterion individually. Implementation process is similar to that in Sects. 4.1–4.4, except the fact that dependence is not expressed with regard to the criteria, but with regard to a technology with specific criteria. The resulting weights w_{41} to w_{45} were gradually multiplied by weights in a range of w_{21} to w_{25}, while weights of w_{p1} to w_{p2} were reached. These result matrices are further allocated to the final resultant matrix w_p.

4.6 Developing an Overall Resultant Matrix w_p and Summary Results W_{ANP}

The resulting total matrix w_p obtained from the resulting matrix of weights w_{p1} to w_{p2} has normalized values shown in the following Fig. 6.

The final step is to calculate the matrix W_{ANP} that is expressed by multiplying the derived weights w_p and w_c using formula (4). Figure 7 shows the calculated value.

Fig. 6 The resulting total matrix w_p

$$w_p = \begin{pmatrix} 0{,}0042 & 0{,}0182 & 0{,}0199 & 0{,}1202 & 0{,}0392 \\ 0{,}0245 & 0{,}0547 & 0{,}1137 & 0{,}1445 & 0{,}0638 \\ 0{,}0316 & 0{,}1059 & 0{,}0619 & 0{,}1121 & 0{,}1505 \\ 0{,}0316 & 0{,}0623 & 0{,}1137 & 0{,}1121 & 0{,}0163 \\ 0{,}0188 & 0{,}0710 & 0{,}0885 & 0{,}0292 & 0{,}0163 \\ 0{,}0188 & 0{,}1059 & 0{,}0199 & 0{,}0741 & 0{,}1312 \\ 0{,}0547 & 0{,}1293 & 0{,}1095 & 0{,}0874 & 0{,}1166 \\ 0{,}0668 & 0{,}2244 & 0{,}2268 & 0{,}2161 & 0{,}2690 \\ 0{,}0788 & 0{,}2282 & 0{,}2461 & 0{,}1042 & 0{,}1971 \end{pmatrix}$$

Fig. 7 Expression of matrix W_{ANP}

$$\begin{pmatrix} 0{,}0042 & 0{,}0182 & 0{,}0199 & 0{,}1202 & 0{,}0392 \\ 0{,}0245 & 0{,}0547 & 0{,}1137 & 0{,}1445 & 0{,}0638 \\ 0{,}0316 & 0{,}1059 & 0{,}0619 & 0{,}1121 & 0{,}1505 \\ 0{,}0316 & 0{,}0623 & 0{,}1137 & 0{,}1121 & 0{,}0163 \\ 0{,}0188 & 0{,}0710 & 0{,}0885 & 0{,}0292 & 0{,}0163 \\ 0{,}0188 & 0{,}1059 & 0{,}0199 & 0{,}0741 & 0{,}1312 \\ 0{,}0547 & 0{,}1293 & 0{,}1095 & 0{,}0874 & 0{,}1166 \\ 0{,}0668 & 0{,}2244 & 0{,}2268 & 0{,}2161 & 0{,}2690 \\ 0{,}0788 & 0{,}2282 & 0{,}2461 & 0{,}1042 & 0{,}1971 \end{pmatrix} x \begin{pmatrix} 0{,}3865 \\ 0{,}3176 \\ 0{,}1400 \\ 0{,}0938 \\ 0{,}0621 \end{pmatrix} = \begin{pmatrix} 0{,}0239 \\ 0{,}0603 \\ 0{,}0744 \\ 0{,}0594 \\ 0{,}0460 \\ 0{,}0588 \\ 0{,}0930 \\ 0{,}1658 \\ 0{,}1594 \end{pmatrix}$$

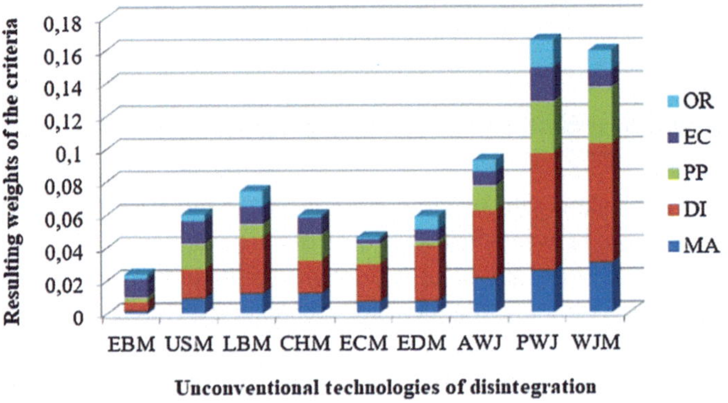

Fig. 8 The resulting graph ANP method

$$w_p.w_c = W_{\text{ANP}} \tag{4}$$

Finally, to select the most unconventional cutting technology for cutting the PMMA bone cement, a graphical dependence has been created (Fig. 8), where the result of the ANP calculation methods can be seen.

5 Conclusions

Selection of the most appropriate technology for unconventional disintegration of PMMA bone cement was based on ANP methods and calculations and the water jet cutting technology was selected, specifically the cutting by pulsating water jet. This technology has many advantages compared to other technologies, such as the use of low pressure (8–20 MPa), less water consumption, and associated lower power requirement. Despite the pressure pulsations a continuous stream of water exits the nozzle, which, however, due to the pulsation of the variable has axial velocity component. Therefore, there is a continuous water jet disintegration to partial clusters of water and thus it acts as a pulsating current. Sufficient water pressure, which is able to divide a bone cement, is based on the experimental work of Prof. Hloch et al. dealing not only with this issue. Pulsating water jet is one of the gentlest processes of cutting technology using water flows in such a case of cutting [9, 10].

Based on all these data, as well as decision by ANP method, the most appropriate and least intrusive way of cutting the bone cement is pulsating water jet technology.

Acknowledgement The present contribution has been prepared with direct support of the Ministry of Education, Science, Research and Sport of Slovak Republic through the projects KEGA 007TUKE-4/2018 and VEGA 1/0795/19.

References

1. Krehel, R., et al. (2009). Mathematical model of technological processes with prediction of operating determining value. *Acta Technica Corviniensis: Bulletin of Engineering, 2*(4), 39–42.
2. Tanasic, I., et al. (2012). Optical aspect of deformities analysis in the bone-denture complex. *Collegium Antropologicum, 36*(1), 173–178.
3. Bone Cement Homepage. http://www.vertebroplastika.sk/kostny-cement. Accessed 19 Aug 2018.
4. Ungureanu, M., et al. (2016). Innovation and technology transfer for business development. *Procedia Engineering, 149*, 495–500.
5. Mitrovic, N., et al. (2011). Application and mode of operation of non-contact stereometric measuring system of biomaterials. *FME Transactions, 39*(2), 55–60.
6. Krausova, L. (2013). *Application of ANP on the choice of technology for the disintegration of composite materials, Presov*, Diploma thesis, TU Kosice.
7. Kadarova, J., et al. (2013). Proposal of performance assessment by integration of two management tools. *Quality Innovation Prosperity, 17*(1), 88–103.
8. Baron, P., et al. (2016). Proposal of the knowledge application environment of calculating operational parameters for conventional machining technology. *Key Engineering Materials, 669*, 95–102.
9. Zezula, M. (2010). *Analysis of the flow properties of bone cement during curing*. Master's thesis. Faculty of Chemistry, Brno University of Technology, Brno.
10. Hloch, S., et al. (2008). Experimental study of surface topography created by abrasive water jet cutting. *Strojarstvo, 49*(4), 303–309.

Index

© Springer Nature Switzerland AG 2020

D. Cagáňová, N. Horňáková (eds.), *Mobility Internet of Things 2018*,
EAI/Springer Innovations in Communication and Computing,
https://doi.org/10.1007/978-3-030-30911-4

J

K

L

M

N

O

P

Q

R

S